Hypertension in the Dog and Cat

Jonathan Elliott • Harriet M. Syme •
Rosanne E. Jepson
Editors

Hypertension in the Dog and Cat

Editors
Jonathan Elliott
Department of Comparative
Biomedical Sciences
Royal Veterinary College
University of London
London, UK

Harriet M. Syme
Department of Clinical Sciences and Services
Royal Veterinary College
University of London
Hertfordshire, UK

Rosanne E. Jepson
Department of Clinical
Sciences and Services
Royal Veterinary College
University of London
Hertfordshire, UK

ISBN 978-3-030-33019-4   ISBN 978-3-030-33020-0 (eBook)
https://doi.org/10.1007/978-3-030-33020-0

© Springer Nature Switzerland AG 2020
This work is subject to copyright. All rights are reserved by the Publisher, whether the whole or part of the material is concerned, specifically the rights of translation, reprinting, reuse of illustrations, recitation, broadcasting, reproduction on microfilms or in any other physical way, and transmission or information storage and retrieval, electronic adaptation, computer software, or by similar or dissimilar methodology now known or hereafter developed.
The use of general descriptive names, registered names, trademarks, service marks, etc. in this publication does not imply, even in the absence of a specific statement, that such names are exempt from the relevant protective laws and regulations and therefore free for general use.
The publisher, the authors, and the editors are safe to assume that the advice and information in this book are believed to be true and accurate at the date of publication. Neither the publisher nor the authors or the editors give a warranty, expressed or implied, with respect to the material contained herein or for any errors or omissions that may have been made. The publisher remains neutral with regard to jurisdictional claims in published maps and institutional affiliations.

This Springer imprint is published by the registered company Springer Nature Switzerland AG.
The registered company address is: Gewerbestrasse 11, 6330 Cham, Switzerland

Preface

It is more than 30 years since clinical researchers at the University of Pennsylvania documented and reported hypertension in 5 dogs which stimulated the study of naturally occurring hypertension in dogs and cats. As intern at the University of Pennsylvania (1985–1986), I was inspired by the work of Drs Littman and Bovee on hypertension. The 1980s also saw significant advances in our understanding of vascular biology, with the discovery of nitric oxide as endothelium-derived relaxing factor. At the time, disruption of the L-arginine-nitric oxide pathway seemed likely to contribute to the pathophysiology of hypertension, creating excitement for any aspiring clinical pharmacologist commencing an academic research career. Joining the Royal Veterinary College in 1990, I was privileged to work alongside Bob Michell, a clinical physiologist with strong interest in measurement of kidney function and blood pressure in dogs. I was inspired by these people to establish a line of research into chronic kidney disease and hypertension in cats, receiving my first grant from BSAVA PetSavers in 1992.

When I was invited by Springer to write a book on hypertension in the dog and cat, it seemed like the time was right to look back over the last 30 years and review progress in the field. It has been an opportunity to review the literature in depth, critically evaluate and compare the published studies and, in some instances, present data that were not included in the original publications or re-analyse data that were. I am indebted to my co-editors for agreeing to indulge me and take on writing and editing this volume and helping me to identify specialists willing to work with us on this book. I also acknowledge the interest and support of Ceva Animal Health and Boehringer Ingelheim for allowing us to use unpublished data to provide a more complete picture of the drugs currently authorised for the management of hypertension in Europe.

As someone who has worked on feline hypertension and CKD for the last 30 years, it is reassuring to see that the field is thriving with an increasing number of early career clinician scientists making significant contributions.

London, UK   Jonathan Elliott

Contents

Part I Physiology, Pathophysiology and Epidemiology of Hypertension

1 Physiology of Blood Pressure Regulation and Pathophysiology of Hypertension.. 3
   Jonathan Elliott

2 Measurement of Blood Pressure in Conscious Cats and Dogs..... 31
   Rosanne E. Jepson

3 Epidemiology of Hypertension............................. 67
   Harriet M. Syme

4 Hypertension and Adrenal Gland Disease.................... 101
   Rosanne E. Jepson

5 Thyroid Gland Disease................................... 131
   Harriet M. Syme

6 Genetics of Hypertension: The Human and Veterinary Perspectives.. 145
   Rosanne E. Jepson

Part II Clinical and Pathological Consequences of Hypertension

7 Hypertension and the Kidney.............................. 171
   Jonathan Elliott and Cathy Brown

8 Hypertension and the Heart and Vasculature................ 187
   Amanda E. Coleman and Scott A. Brown

9 Hypertension and the Eye................................ 217
   Elaine Holt

10 Hypertension and the Central Nervous System............... 241
   Kaspar Matiasek, Lara Alexa Matiasek, and Marco Rosati

Part III Pharmacology and Therapeutic Use of Antihypertensive Drugs

11 Pharmacology of Antihypertensive Drugs 267
Jonathan Elliott and Ludovic Pelligand

12 Management of Hypertension in Cats 315
Sarah M. A. Caney

13 Management of Hypertension in Dogs 331
Sarah Spencer

Part IV Future Perspectives

14 Future Perspectives 371
Harriet M. Syme, Rosanne E. Jepson, and Jonathan Elliott

List of Contributors

Cathy Brown University of Georgia College of Veterinary Medicine, Athens, GA, USA

Scott A. Brown Department of Physiology and Pharmacology, University of Georgia College of Veterinary Medicine, Athens, GA, USA

Sarah M. A. Caney Midlothian Innovation Centre, Pentlandfield, Roslin, Edinburgh, UK

Amanda E. Coleman Department of Small Animal Medicine and Surgery, University of Georgia College of Veterinary Medicine, Athens, GA, USA

Jonathan Elliott Department of Comparative Biomedical Sciences, Royal Veterinary College, University of London, London, UK

Elaine Holt University of Pennsylvania School of Veterinary Medicine, Philadelphia, PA, USA

Rosanne E. Jepson Department of Clinical Sciences and Services, Royal Veterinary College, University of London, Hertfordshire, UK

Kaspar Matiasek Section of Clinical and Comparative Neuropathology, Centre for Clinical Veterinary Medicine, Ludwig-Maximilians-Universität München, Munich, Germany

Lara Alexa Matiasek Anicura Kleintierklinik, Babenhausen, Germany

Ludovic Pelligand Department of Comparative Biomedical Sciences, Royal Veterinary College, University of London, London, UK

Marco Rosati Section of Clinical and Comparative Neuropathology, Centre for Clinical Veterinary Medicine, Ludwig-Maximilians-Universität München, Munich, Germany

Sarah Spencer Department of Comparative Biomedical Sciences, Royal Veterinary College, London, UK

Harriet M. Syme Department of Clinical Sciences and Services, Royal Veterinary College, University of London, Hertfordshire, UK

List of Abbreviations

11-Beta-HSD	11 beta-hydroxysteroid dehydrogenase
AAMI	Association for the Advancement of Medical Instrumentation
ABPM	Ambulatory blood pressure monitoring
ACC	American College of Cardiology
ACE	Angiotensin-converting enzyme
ACE-1	Angiotensin-converting enzyme type 1
ACE-2	Angiotensin converting enzyme type 2
ACEi	Angiotensin-converting enzyme inhibitor
ACTH	Adrenocorticotrophic hormone
ACVIM	American College of Veterinary Internal Medicine
ADC	Apparent diffusion coefficient
ADMA	Asymmetric dimethylarginine
AHA	American Heart Association
AKI	Acute kidney injury
ALD	Aldosterone
AMP	Adenosine monophosphate
ANG II	Angiotensin II
Ang_{1-7}	$Angiotensin_{1-7}$ peptide
ANP	Atrial natriuretic peptide
AoR	Aortic root
APA	Adenoma of the aldosterone producing adrenal cortex
ARB	Angiotensin receptor blocker
AT_1 receptor	Angiotensin receptor type 1
AT_2 receptor	Angiotensin receptor type 2
AUC	Area under the curve
BBBB	Blood–brain barrier breakdown
BCS	Body condition score
BHS	British Hypertension Society
B_{max}	Maximum binding capacity of a protein (e.g. enzyme or receptor) for a drug
BMI	Body mass index
BNP	Brain natriuretic peptide
BP	Blood pressure

BRB	Blood retinal barrier
CASVD	Chronic arterial small-vessel disease
CC	Correlation coefficient
CCB	Calcium channel blocker
CHF	Congestive heart failure
CKD	Chronic kidney disease
C_{max}	Maximum plasma concentration (of a drug)
CNS	Central nervous system
CO	Cardiac output
COX-1	Cyclo-oxygenase 1
COX-2	Cyclo-oxygenase 2
CSF	Cerebrospinal fluid
CV	Coefficient of variation
DBP	Diastolic blood pressure
DDAH	Dimethylarginine dimethylaminohydrolase
DWI	Diffusion-weighted imaging
DOCP	Desoxycorticosterone pivalate
EBF	Encephalic blood flow
ECG	Electrocardiography
ED_{50}	Dose of drug giving rise to 50% of the maximal response
EDHF	Endothelium-dependent hyperpolarising factor
EDRF	Endothelium-dependent relaxing factor
EMA	European Medicines Agency
E_{max}	Maximum response (of a drug's pharmacodynamic effect)
ENaC	Epithelial sodium channel
eNOS	Endothelial nitric oxide synthase
EPP	Encephalic perfusion pressure
eQTL	Expression Quantitative Trait Loci
ESH	European Society of Hypertension
ET-1	Endothelin-1
ECFV	Extracellular fluid volume
EPO	Erythropoietin
EPR	Electronic patient record
ERG	Electroretinography
FDA	Food and Drug Administration
FGF23	Fibroblast growth factor 23
FL	Forelimb
FLAIR	Fluid-attenuated inversion recovery
GFR	Glomerular filtration rate
GI tract	Gastrointestinal tract
GM	Grey matter
GWAS	Genome-wide association studies
HAC	Hyperadrenocorticism
HBPM	Home blood pressure monitoring
HCM	Hypertrophic cardiomyopathy

HDO	High definition oscillometry
HE	Hypertensive encephalopathy
HL	Hind limb
HPBH	Hypertensive parenchymal brain haemorrhage
HT	Hypertension
HTH	Hyperthyroidism
ICP	Intracranial pressure
ISACHC	International Small Animal Cardiac Health Council classification
IPG	Impedance-plethysmography
IQR	Interquartile range
IRIS	International Renal Interest Society
ISO	International Organization for Standardization
IVS	Interventricular septum
IVS_d	End-diastolic interventricular septum wall thickness
JNC	Joint National Committee (on Prevention, Detection, Evaluation, and Treatment of High Blood Pressure in the USA)
Kd	Dissociation constant for a drug from its target binding site (a measure of the drugs affinity for the target)
Km	Substrate concentration at which the enzyme is half saturated (Michaelis–Menten kinetics)
LA	Left atrium
LLOQ	Lower limit of quantification of the assay
LOA	Limits of agreement
LV	Left ventricle
LVH	Left ventricular hypertrophy
MAP	Mean arterial pressure
MBP	Mean blood pressure
MEN	Multiple Endocrine Neoplasia
MR	Mineralocorticoid receptor
MRA	Mineralocorticoid receptor antagonist
MRI	Magnetic resonance imaging
NCC	Na^+/Cl^- ion cotransporter
NGS	Next generation sequencing
NKCC2	$Na^+/K^+/2Cl^-$ cotransporter 2
NO	Nitric oxide
NSAID	Non-steroidal anti-inflammatory drug
NT-pro-ANP	N-terminal pro-atrial natriuretic peptide
NT-pro-BNP	N-terminal pro-brain natriuretic peptide
OMIM®	Online Mendelian Inheritance in Man®
$PaCO_2$	Partial pressure of arterial blood carbon dioxide
PIIINP	Pro-collagen type III amino-terminal peptide
PCV	Packed cell volume
PK/PD	Pharmacokinetic/pharmacodynamics
PO	*per os (by mouth)*
PPARγ	Peroxisome proliferator-activated receptor-γ

PPG	Photoplethysmography
PRA	Plasma renin activity
PRES	Posterior reversible encephalopathy syndrome
PTT	Pulse transit time
PW_d	End-diastolic left ventricular posterior wall thickness
RAAS	Renin-angiotensin-aldosterone system
RALES study	Randomized Aldactone Evaluation Study
ROC	Receiver operating curve
ROMK	Renal outer medullary potassium channels
RPE	Retinal pigment epithelium
SABP	Systemic arterial blood pressure
SARD(S)	Sudden-acquired retinal degeneration (syndrome)
SBP	Systolic blood pressure
SBPOT	Time-averaged systolic blood pressure
SH	Systemic hypertension
SHR	Spontaneously hypertensive rat
SNGFR	Single nephron glomerular filtration rate
SNP	Single-nucleotide polymorphism
SPRINT	Systolic Blood Pressure Intervention Trial
SUN	Serum urea nitrogen
SVR	Systemic vascular resistance
T1W	T1 weighted
T2W	T2 weighted
T_3	Triiodothyronine
T_4	Thyroxine
T_{max}	Time to maximum plasma concentration (of a drug) following dosing
TPPG	Transmitted photoplethysmography
TPR	Total peripheral resistance
TOD	Target organ damage
TR	Thyroid hormone receptor
TRE	Thyroid hormone response element
TSH	Thyroid-stimulating hormone
UAC	Urine albumin to creatinine ratio
UPC	Urine protein to creatinine ratio
VEGF	Vascular endothelium growth factor
Vmax	Maximal enzyme velocity (Michaelis–Menten kinetics)

Part I

Physiology, Pathophysiology and Epidemiology of Hypertension

Physiology of Blood Pressure Regulation and Pathophysiology of Hypertension

Jonathan Elliott

1.1 Introduction

Systemic arterial blood pressure (SABP) is the product of cardiac output (CO) and total peripheral resistance (TPR), which resides mainly in the systemic arterioles of the vasculature. Cardiac output is dependent on venous return and heart rate. Since the heart is a demand pump, the volume of blood returning to the heart (venous return) multiplied by the heart rate determines CO. The circulating blood volume relative to the capacitance of the circulation (predominantly venous capacitance vessels) determines venous return. These facts are fundamental to circulatory physiology. What is far more complex is how SABP is regulated such that minute by minute it can vary to meet the tissue perfusion demands of the body but that over a 24 h period maintain the average pressure at a remarkably stable level to ensure all tissues in the body receive adequate blood perfusion.

The integrated interactions between the kidney, the heart and the vasculature and their regulation by the sympathetic nervous system, multiple endocrine mediators produced by adrenal and other glands and paracrine mediators produced by these organs themselves all play a role in controlling the complex feedback systems. These ultimately maintain stable SABP at a mean level of around 100 mmHg in all mammalian species, with the exception of the giraffe. Understanding of all the control points, interactions and feedback systems is not comprehensive, yet it is fundamental to begin to unravel why things go wrong and systemic hypertension occurs. Figure 1.1 is an attempt to summarise all these systems and their interactions (Bijsmans et al. 2012). It deliberately depicts the kidney at the centre of the integrated system.

The major control points for blood pressure regulation are:

J. Elliott (✉)
Department of Comparative Biomedical Sciences, Royal Veterinary College, University of London, London, UK
e-mail: jelliott@rvc.ac.uk

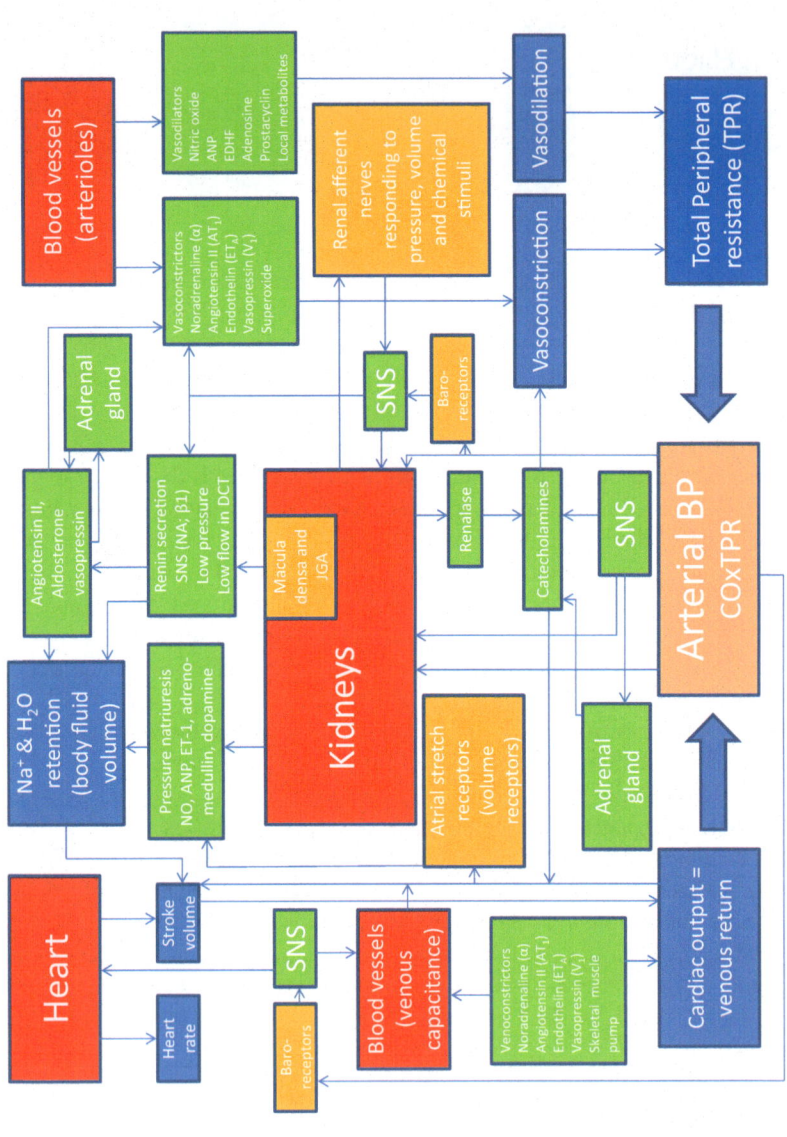

Fig. 1.1 Model of the arterial blood pressure integrated control system. The red boxes contain the effector organs, the blue boxes the physiological parameters affected by changes to these organ functions, the green boxes represent the regulators affecting the organ functions and the orange boxes contain the sensory

systems which detect pressure, volume and chemicals within the systems and feed that information to the central nervous system. For clarity, the central nervous system is not shown. This is key in integrating information received from the baroreceptors, atrial stretch receptors and renal afferent nerves. These signals are integrated in the pons and medulla of the brain stem and regulate the effector systems influencing sympathetic tone and vasopressin release. These brain centres have neurones that release angiotensin II and nitric oxide which stimulate and inhibit the activation of the sympathetic nervous system, respectively. No attempt has been made to indicate which direction the regulators change the physiological parameters they are influencing—see text for details. *ANP* atrial natriuretic peptide, *CO* cardiac output, *DCT* distal convoluted tubule, *EDHF* endothelium derived hyperpolarising factor, *ET-1* endothelin-1, *JGA* juxtaglomerular apparatus, *NA* noradrenaline, *NO* nitric oxide, *SNS* sympathetic nervous system, *TPR* total peripheral resistance

- Circulating fluid volume (a component of extracellular fluid volume determined by sodium homeostasis)
- Vascular resistance (total peripheral resistance)
- Vascular capacitance (volume of the circulation relative to the circulating fluid volume)
- Heart rate and force of contraction

These control points are influenced by sensory systems detecting changes in pressure or volume and initiating feedback control mechanisms. The main recognised sensory mechanisms illustrated in Fig. 1.1 are:

- Arterial stretch receptors (baroreceptors) which determine minute by minute variations in blood pressure but adapt rapidly
- Atrial stretch receptors which determine filling of the circulation and its relationship to vascular capacitance
- Renal afferent arteriolar stretch receptors determining the perfusion pressure of the kidney and chemoreceptors detecting ischaemia and osmolality
- Macula densa at the start of the distal convoluted tubule which senses the rate of chloride ion delivery, detects renal perfusion and adjusts renal haemodynamics to maintain stable tubular flow

Guyton views the kidney as the long-term regulator of SABP (Guyton and Hall 2015), and as will be clear later in this book, kidney disease or dysfunction underlies the pathophysiology of hypertension in many dogs and cats presenting with organ damage secondary to high blood pressure. This chapter will focus on each of the major control points for SABP, the current state of knowledge of their normal physiological regulation and how these may be impacted by kidney disease. No attempt is made to give a comprehensive overview of the pathophysiology of all forms of hypertension. With such a complex integrated feedback system, the reductionist approach to identifying drivers that increase blood pressure is a gross oversimplification as the system as a whole should respond to normalise pressure again. It is more likely that multiple factors, including genetics, environment and disease, combine to provide the circumstances in which hypertension leading to target organ damage occurs. It is also clear that events in early life (foetal development, maturation of the cardiorenal systems through to puberty) are likely to 'programme' the system for it to adapt to ageing. Thus, although monogenetic forms of hypertension do occur (see Chap. 6), most hypertension we see as veterinarians in dogs and cats will have a complex multifactorial pathophysiology, at least part of which may be rooted in early life events.

Life-time longitudinal studies of dogs and cats are starting to be undertaken, and it will be important that blood pressure is measured in these studies. There are data to suggest that, as in westernised human subjects, cats show an increase in blood pressure with age (Bijsmans et al. 2015a; Payne et al. 2017). Whether this is a product of domestication of cats and underlies why as a species they are susceptible to hypertension remains to be determined.

1.2 The Kidney and Body Fluid Volume Axis

Sodium is the major extracellular fluid cation (c140 mmol/L). With its attendant anions [chloride (c110 mmol/l) and bicarbonate (c25 mmol/l)], it makes up the osmolality of extracellular fluid (c280 mOsm/L). The concentration of sodium in the extracellular fluid is maintained at a stable level through osmoregulation, which ensures water intake each day is balanced by renal and extrarenal losses. Antidiuretic hormone (vasopressin) and the osmoreceptors in the hypothalamus provide the physiological feedback mechanisms to ensure water balance is achieved. Under these circumstances, the amount of sodium in the body at any given time is the determinant of extracellular fluid volume. SABP and kidney perfusion pressure are intrinsically linked to renal sodium ion excretion, creating a feedback loop whereby renal sodium excretion increases if SABP goes up and decreases if it goes down. This can occur in the absence of any neurohormonal influences on the kidney as the phenomenon of pressure natriuresis can be demonstrated in isolated perfused kidneys.

The steepness of the relationship between renal perfusion pressure and sodium ion excretion in urine is increased in the presence of functional neurohormonal systems. Most notably, the renin-angiotensin-aldosterone system (RAAS) is involved in steepening this renal function curve for sodium. Activation of the RAAS system starts with stimuli for renin secretion from the modified smooth muscle cells of the afferent arteriole, which make contact with the specialised cells in the distal tubule which form the macula densa in the kidney. Together these two groups of cells form what is termed the juxtaglomerular apparatus. Renin secretion is regulated by:

- Sympathetic nerve activity (stimulates renin secretion via beta1-adrenoceptor activation on the juxtaglomerular apparatus)
- Stretch of the smooth muscle cells of the afferent arteriole (which inhibits renin secretion)
- Signals from the macula densa of two forms
  - Nitric oxide and prostaglandins (PGE_2 and PGI_2) which are produced when low flow in the macula densa is sensed—these stimulate renin secretion and dilate the afferent arteriole.
  - Adenosine which is produced when there is high flow in the macula densa—this inhibits renin secretion and constricts the afferent arteriole.

Both angiotensin II (acting on the proximal tubule) and aldosterone (acting on the late distal tubule) have effects on the kidney which lead to sodium retention, and the RAAS system is a sodium-conserving system. If the animal is consuming sodium well in excess of its daily needs, inhibition of the RAAS system leads to increased sodium loss in the urine. Conversely, aldosterone-producing tumours lead to excess sodium retention (and potassium loss) and give rise to systemic hypertension (see Chap. 7). The study of the RAAS system is complicated because what is measured in the plasma (plasma renin activity, angiotensin II and plasma aldosterone concentrations) may not reflect what prevails within the kidney itself, as renin is produced locally within the kidney from the juxtaglomerular apparatus. In the systemic circulation, angiotensinogen is abundant, and plasma renin activity is limited, whereas the local levels of renin in the kidney are 100-fold higher than in the plasma. Tubular cells are also capable of synthesising angiotensinogen (Navar et al. 2002), and urinary angiotensinogen is thought to be derived from tubular secretion rather than filtration of systemic circulating angiotensinogen. Precise understanding of the regulation of the intrarenal RAAS is currently lacking. Whether aldosterone is produced locally within the kidney is also not known.

The sympathetic nervous system (renal sympathetic nerves) not only stimulates renin secretion and activates the RAAS system but also has direct effects on the vasculature and proximal tubule of the kidney to cause vasoconstriction and increase sodium and water reabsorption (Sata et al. 2018). These effects are mediated by noradrenaline, the neurotransmitter in the sympathetic postganglionic neurones, acting on alpha1-adrenoceptors predominantly although some of the effect may be through alpha2-adrenoceptors as well. In the collecting ducts, however, alpha2-adrenoceptors are linked to inhibition of the action of antidiuretic hormone leading to diuresis. Renal sympathetic nerves also release dopamine, and the renal sympathetic nerve endings, vasculature and tubular epithelium possess dopamine receptors which are linked to inhibition of noradrenaline release, vasodilation and sodium and water excretion, thus opposing or balancing the effects of noradrenaline released by the renal sympathetic nerves. As discussed below, the RAAS and sympathetic systems interact here having similar effects, and with angiotensin II-enhancing release of noradrenaline from sympathetic nerve endings, there is cross-talk between the two systems. There is clear evidence that renal sympathetic nerve activity is modulated by sodium intake, with reduction in activity occurring as sodium intake increases and an increase in activity occurring with salt deprivation.

The salt-conserving RAAS system (working in concert with the sympathetic nervous system) is the best characterised physiologically important sodium-regulating system, whereas the natriuretic system is less well characterised, and its physiological importance is not completely understood. Natriuretic peptides are circulating hormones released from the heart in response to stretch of the vasculature and increased filling pressure in the atria (overfilling). Atrial natriuretic (ANP; left atrium) and brain natriuretic peptides (BNP; left ventricle) circulate. They are stored as preprohormones in granules and require enzymatic cleavage prior to release into the circulation. The proteolytic cleavage of pro-ANP to ANP is dependent on a membrane-bound protease enzyme called corin, genetic variants of which have been associated with hypertension (Li et al. 2017). Natriuretic peptides inhibit aldosterone secretion from the adrenal gland, reduce RAAS and sympathetic nervous system activation, increase GFR through their ability to dilate the afferent arteriole and have direct effects on the tubular cells of the kidney to inhibit sodium uptake (Sarzani et al. 2017). Most of the physiological effects are mediated by ANP acting via NPRA and NPRB receptors. Concentrations of BNP tend only to increase to levels that would have effects when there is pathological ventricular dilatation as occurs in heart failure. It has been reported that corin co-localises with NPRA receptors in renal tubular cells suggesting local formation of ANP may occur close to its receptor enabling it to act in an autocrine fashion (Li et al. 2017).

In general, the physiological importance of ANP in regulating sodium balance is thought to be one of the fine-tuning effects. Nevertheless, in the early stages of the development of hypertension in human subjects, there is some evidence that deficiency in the ANP system may contribute (Macheret et al. 2012). Whether this is due to impaired release or increased clearance is not known.

Of the other mediators giving rise to natriuresis, the endothelin system is perhaps best characterised and studied. Endothelin-1 (ET-1) is a paracrine mediator best known for its ability to raise blood pressure through its systemic vascular actions where it is released abluminally from the endothelial cells of blood vessels and acts on receptors on vascular smooth muscle cells. Counterintuitively, ET-1 in the kidney appears to be a natriuretic factor (Kohan et al. 2011). The kidney produces large amounts of ET-1, much more than other organs, and many renal tubular cells express receptors for this peptide mediator. Although ET-1 reduces blood flow to the glomerulus, this peptide seems to dilate and increase blood flow through the vasa recta in the medulla. In addition, ET-1 working via the ET-B receptor inhibits sodium reabsorption in the thick ascending limb of the loop of Henle, an effect which is probably mediated, at least in part, by local production of nitric oxide. The overall effects of the endothelin system on blood pressure control are complex for three main reasons. First, there are different types of endothelin receptors which have opposing actions; second the systemic effects of administered endothelin peptides tend to raise blood pressure, whereas the local effects of ET-1 produced within the kidney appear to be predominantly natriuretic and therefore lower blood pressure; and finally, there appear to be species differences which also complicate the interpretation of the published scientific data.

Other natriuretic factors are known to exist. The best characterised are also local paracrine mediators in the kidney. Contrary to their actions in the cortex, prostaglandins and nitric oxide when produced in the medulla of the kidney induce natriuresis. Cyclooxygenase-2 (COX-2) is upregulated in the medulla in animals fed with high-sodium diets (Yang and Liu 2017). PGE_2 appears to be the major prostanoid-mediating natriuretic effect, and COX-2 inhibition removes this protective mechanism in animals that are salt loaded. However, this is not a simple situation in that products of COX-2 appear to assist the effect of angiotensin II in medullary areas of the kidney so the effect of prostanoids in this part of the kidney depends on the prevailing physiological conditions and the degree of activation of RAAS (Yang 2015). The role of nitric oxide in pressure natriuresis described above also appears to emanate from the medulla (O'Connor and Cowley Jr. 2010). The vasa recta, which are the blood vessels that run alongside the loops of Henle, show relatively poor autoregulation. As blood pressure increases, the flow through these vessels increases. This means that the capillaries are exposed to higher perfusion pressure and shear forces, increasing the release of nitric oxide from these vessels. NO produced locally has effects on the thick ascending limb of the loop of Henle reducing sodium transport. The source of NO is debated. Nitric oxide synthase (NOS) is extremely active in this area of the kidney, and in addition to the endothelium of the vasculature, there is also evidence that epithelial cells of the outer medullary collecting duct also express endothelial NOS (eNOS). Indeed, ET-1 produced locally within the kidney seems to stimulate the release of NO from these cells. Other natriuretic factors that have been identified include adrenomedullin and endogenous digitalis-like factor. The evidence that these mediators have a substantive role to play in the renal regulation of body fluid balance is not conclusive, however.

In summary, the interplay between the renal intrinsic mechanisms that link SABP to sodium ion excretion and the endogenous hormones and paracrine agents that steepen this relationship are hugely important in the long-term regulation of blood pressure. However, they are also extremely complex with significant redundancy built into multiple feedback systems making the reductionist approach of studying each in isolation nonviable. We have only really begun to scratch the surface of understanding the specific species adaptations of these systems that have evolved to serve their particular environmental contexts and in understanding how domestication and our imposed lifestyle changes on our pets impact on the functioning of these physiological systems designed to maintain body fluid balance.

1.3 Pathophysiological Mechanisms in Regulation of Body Fluid Volume Associated with Hypertension in Dogs and Cats

Much of the above understanding of renal physiology and blood pressure control involves use of the dog as a species for experimental investigation. However, when it comes to clinical studies of naturally occurring hypertension, the dog has not been studied in any detail despite eloquent articles being written strongly suggesting this resource would be ideally suited for hypertension research in general (Michell 1994, 2000). Not surprisingly, the use of renal mass reduction in the dog to induce CKD leads to persistently higher blood pressure (compared to control dogs). This is associated with increased plasma renin activity, plasma angiotensin I and aldosterone concentrations (Mishina and Watanabe 2008), suggesting the adaptive response in this model to loss of functioning nephrons results in activation of the RAAS system. Treating these experimental dogs with benazepril reduced the blood pressure and led to an associated reduction in plasma aldosterone concentration.

There are more published studies investigating the pathophysiology of naturally occurring feline hypertension than canine hypertension. Not surprisingly, these have taken the reductionist approach and examined individual components of the renal body fluid control systems rather than taking an integrated approach. In many instances, relatively small numbers of cases have been involved in these studies, making hard conclusions difficult to draw.

The question as to whether cats or dogs with naturally occurring hypertension have expanded extracellular fluid volumes associated with their high blood pressure has not been addressed comprehensively. Finch et al. (2015) measured extracellular fluid volumes (ECFV) in normal and azotaemic CKD cats, but the number of cats

1 Physiology of Blood Pressure Regulation and Pathophysiology of Hypertension

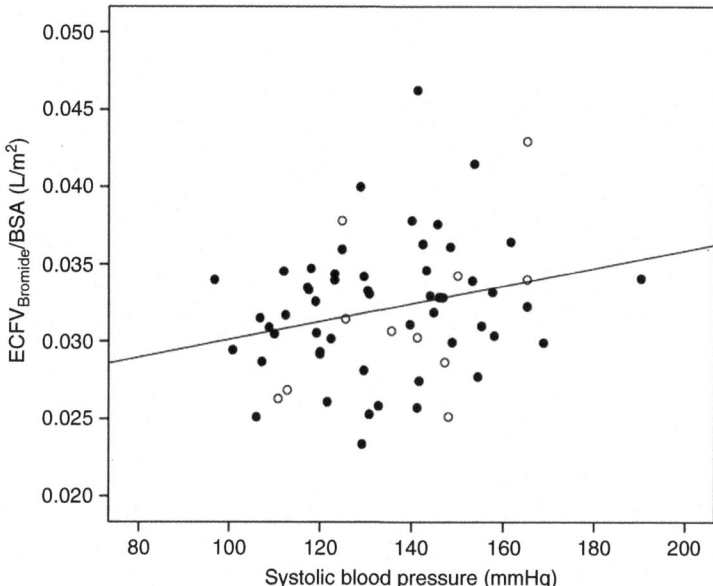

Fig. 1.2 Relationship between ECFV$_{Bromide}$/BSA and systolic blood pressure in 66 cats (55 non-azotaemic and 11 azotaemic). A weak positive relationship was identified ($R^2 = 0.063$, $P = 0.040$). Filled circles represent non-azotaemic cats and unfilled circles represent azotaemic cats. *BSA* body surface area. Reproduced with permission from Finch (2011)

involved in this study was probably too small to determine reliably any relationship between ECFV and blood pressure. Nevertheless, when these data were examined normalising ECFV to body surface area, a very weak positive relationship was found between ECFV and blood pressure (Fig. 1.2; Finch 2011), suggesting this is worthy of further investigation.

A number of clinical researchers have examined the state of activation of the RAAS system in naturally occurring feline hypertension. The issue with these studies is what to use as a comparator group and how to control for dietary sodium intake, and these are limitations of all of the studies. Jensen et al. (1997) reported data on plasma renin activity (PRA), plasma angiotensin I and plasma aldosterone concentration from 12 cats with untreated hypertension and compared these to 5 younger healthy cats. They found no difference in PRA or baseline angiotensin I concentrations, but plasma aldosterone concentrations were significantly higher in the hypertensive cats with 8 of the 12 cats having plasma aldosterone concentrations more than 2 SD above the mean value for the control group. Of these eight cats, three had low PRA, four had PRAs that were considered normal and one had a high PRA. These were the first published data to suggest that RAAS functioning is variable in cats with naturally occurring hypertension and that in some cats there may be a discordant change between PRA and plasma aldosterone concentration.

Jepson et al. (2014) reported data from a larger group of cats ($n = 193$) with naturally occurring hypertension ($n = 74$) divided into two groups based on whether

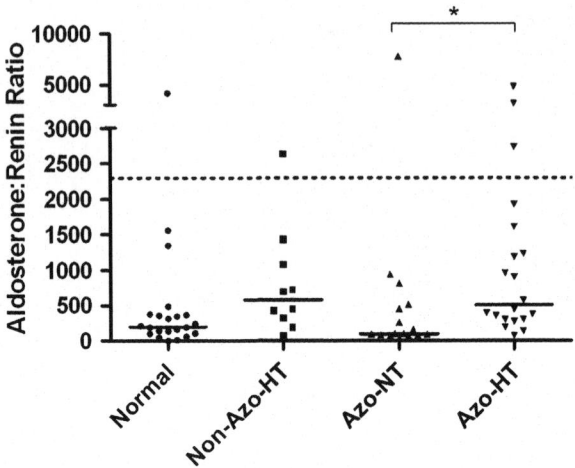

Fig. 1.3 Comparison of plasma aldosterone to plasma renin activity ratios (ARR) in cats grouped according to renal and blood pressure status. Horizontal bars represent the median ARR for each group. The dashed line represents the upper limit of the reference interval generated from the group of normal geriatric cats. Non-Azo-NT, non-azotaemic normotensive; non-Azo-HT, non-azotaemic cats with systemic hypertension; Azo-NT, azotaemic cats normotensive; Azo-HT, azotaemic cats with systemic hypertension. Asterisk indicates statistical significance ($p < 0.05$) was attained on post hoc analysis. Reproduced with permission from Jepson et al. (2014)

the cats had elevated plasma creatinine concentration or not. Two comparator groups were recruited, geriatric healthy cats (aged 10 years or older; $n = 68$) and normotensive azotaemic CKD patients ($n = 41$). This study confirmed the variability of these markers of the RAAS system and demonstrated that on average, non-azotaemic cats with naturally occurring hypertension have lower PRA but no difference in plasma aldosterone concentrations compared to normal healthy cats of similar age. Azotaemic hypertensive cats had higher plasma aldosterone concentration than the most relevant control group, namely, the azotaemic normotensive cats despite having similar PRA. Even though an attempt was made to have an age-matched healthy control group, the age of these cats was significantly lower than the other three groups. When calculating the aldosterone-to-renin ratio, this was significantly lower in the azotaemic hypertensive cats compared to the azotaemic normotensive cats, and although the same trend appeared to hold for the non-azotaemic hypertensive cats, this did not reach statistical significance (see Fig. 1.3). However, the marked heterogeneity demonstrated by this reasonably sized population suggests that generalisation of these findings to all cats with hypertension would be unwise. Lack of proper control for age, although understandable, is problematic in interpreting these data from the non-azotaemic hypertensive cats since they were, on average, 3 years older than their controls, and Javadi et al. (Javadi et al. 2004) have shown PRA decreases with age in the cat. Further data presented in the Jepson et al. (2014) paper also suggested a disconnect between PRA and plasma aldosterone concentration. A group of 20 hypertensive cats were treated with amlodipine and the dose adjusted to reduce their blood pressure to below 160 mmHg. In these cats, measurement of PRA and aldosterone was repeated when blood pressure was controlled. This treatment led to a significant increase in PRA but no change in plasma aldosterone concentration. The PRA increase is expected as a physiological response to a decrease in blood pressure, but one might expect plasma aldosterone concentration to also increase if angiotensin II produced by the increase in PRA is regulating aldosterone secretion. However, no measurements of angiotensin II were made in this study to confirm the change in PRA elicited with the expected increase in plasma angiotensin II.

The finding that plasma aldosterone concentration tends to be higher than the normotensive population, particularly in the azotaemic cat population, fits with the observation from an earlier study published by the same group that low plasma potassium concentration is a risk factor for hypertension in cats with CKD (Syme et al. 2002). However, the azotaemic cats in the Jepson et al. (2014) study did not have lower plasma potassium concentration compared to the azotaemic normotensive control group. The data Jepson and colleagues generated from hypertensive cats shows some similarities to the situation in black human patients who have low renin hypertension and elevated plasma aldosterone concentrations. These patients are described as being 'salt sensitive' and, similar to the findings in cats, respond poorly to ACE inhibitors and very well to calcium channel blockers as antihypertensive agents. Nevertheless, there is currently no evidence that cats as a species have high sensitivity to dietary salt intake. Indeed, cats aged between 5 and 14 years of age fed a diet containing three times the standard sodium chloride content for 2 years showed no harmful health effects and maintained a stable blood pressure and renal function throughout the study. They showed the expected reduced plasma aldosterone concentration and increased urinary excretion of sodium across the 2-year study period (Reynolds et al. 2013).

It has been proposed that feline hypertension is associated with adrenal gland pathology. Javadi et al. (2005) published a series of 11 cases describing non-tumorous pathology in the adrenal glands associated with hypokalaemia, severe hypertension and progressive kidney disease. However, this paper did not examine the adrenal gland histopathology of cats with CKD but without evidence of hypertension and hypokalaemia. A larger study of adrenal gland pathology from cats with CKD found no such association between adrenal gland pathology and hypertension (Keele et al. 2009). This study involved 67 cats, 37 of which had been treated for hypertension, 15 of which had high normal blood pressure (but did not receive antihypertensive treatment) and 15 of which had systolic arterial blood pressure below 140 mmHg and were described as low normotensive (see Fig. 1.4). The hyperplasia score and mean number of nodules did not differ between each of these three groups (see Figs. 1.5a, b).

In summary, the data on RAAS activity in cats with hypertension demonstrates heterogeneity of the phenotype with a tendency for many cats to have low PRA and normal or marginally high plasma aldosterone concentrations. The studies to date, although interesting, are limited in terms of the numbers of cats examined and are observational rather than interventional nature and reductionist, focusing on one

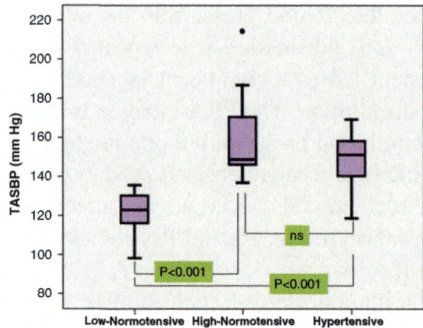

Fig. 1.4 Time-averaged systolic blood pressure (TASBP) from 67 cats where postmortem examinations had been carried out. Cats were divided into normotensive and hypertensive cats based on the clinical decision as to whether their systolic arterial blood pressure required treatment or not. All hypertensive cats ($n = 37$) were treated with oral amlodipine to control their blood pressure (posttreatment target 160 mmHg). The normotensive cats were divided into two groups—those with a TASBP below 140 mmHg referred to as low normotensive group ($n = 15$) and those with TASBP above 140 mmHg referred to as the high normotensive group ($n = 15$). The data for each group are presented as a box and whisker plot. TASBP was compared between each group using the Kruskal-Wallis test ($P < 0.01$) followed by post hoc comparisons using the Mann-Whitney U test. *ns* not significant. Data from Keele et al. (2009)

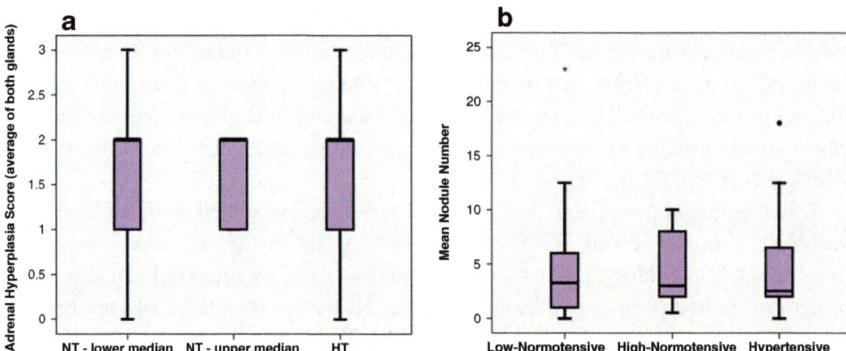

Fig. 1.5 (**a**) Adrenal gland pathology from cats of know blood pressure status undergoing routine postmortem examination. The groups of cats are defined in the legend to Fig. 1.4. A histopathology score for adrenal gland hyperplasia (scale 0 to 3) was applied by an expert veterinary pathologist who was unaware of the blood pressure status of the cats. Where both adrenal glands were available from a cat, an average hyperplasia score has been used. Data are presented as box and whisker plots. The groups were compared using the Kruskal-Wallis test, and no significant difference was found between groups. (**b**) The pathologist counted the number of nodules present in a gland, and where pathology was available from both glands of a cat, an average number of nodules per gland were used. The data are presented as box and whisker plots. No statistically significant difference was found between the three groups using a Kruskal-Wallis test

system. For example, the list of known factors influencing aldosterone secretion includes ACTH, natriuretic peptides, ET-1, dopamine and endogenous digitalis-like factor and adrenomedullin. Measurement of all these factors influencing the system in studies where an intervention is made to change dietary sodium intake, for example, would enable one to understand their integration.

Relatively little further information is available on the various factors affecting the body fluid homeostatic mechanisms in hypertension in the dog and cat. There have been some studies examining natriuretic peptides in cats with hypertension (Lalor et al. 2009; Bijsmans et al. 2017). These studies have demonstrated that at diagnosis of hypertension, NT-proBNP is higher in hypertensive cats than normotensive cats. However, the same was not true of NT-proANP (only measured by Lalor et al. 2009). Both of these markers are secreted at the same time as the active peptide is cleaved from the prohormone and so do not represent direct measurement of the biologically active hormone. Both markers seem to accumulate in the plasma in severely azotaemic cats which is to be expected as they are renally cleared. It seems likely that release of NT-proBNP in hypertensive cats occurred at least in part because of increased wall stress in the ventricular myocardium because when amlodipine was used to lower blood pressure and presumably reduce ventricular wall stress, NT-proBNP in the plasma decreased (Bijsmans et al. 2017).

Pelligand et al. (2015) demonstrated that both COX-1 and COX2 are located in the macula densa of the feline kidney and that both enzymes needed to be inhibited before the increase in renin secretion in response to furosemide treatment was reduced. Suemanotham et al. (2010) showed that CKD led to upregulation of both COX-1 and COX2 in kidneys of cats (harvested at postmortem following euthanasia). The degree of upregulation was not influenced by the presence of hypertension, proteinuria or the progressive nature (or not) of the CKD (see Figs. 1.6 and 1.7).

1.4 Cardiovascular Influences on Blood Pressure: Regulation of the Heart and Vasculature

The three remaining control points for blood pressure regulation, vascular resistance, vascular capacitance and cardiac rate and force of contraction are intrinsically linked within the cardiovascular system and so will be considered together. Many of the effector systems influence all three directly or indirectly. The function of the circulatory system is to provide each organ with sufficient blood, carrying nutrients and oxygen at a rate that meets the tissues' metabolic demands. Many of the same effector systems (nervous, endocrine and paracrine factors) discussed for the renal body fluid system are also involved in regulating the cardiovascular system and so influence blood pressure control from body fluid volume and volume of the vasculature perspectives.

When tissue metabolic requirements increase (e.g. muscle during exercise), the vasculature supplying that tissue dilates, and, provided blood pressure is maintained or increased, perfusion rate goes up. This will require mobilisation of blood from within vascular stores (constriction of the capacitance vessels; i.e. the veins and

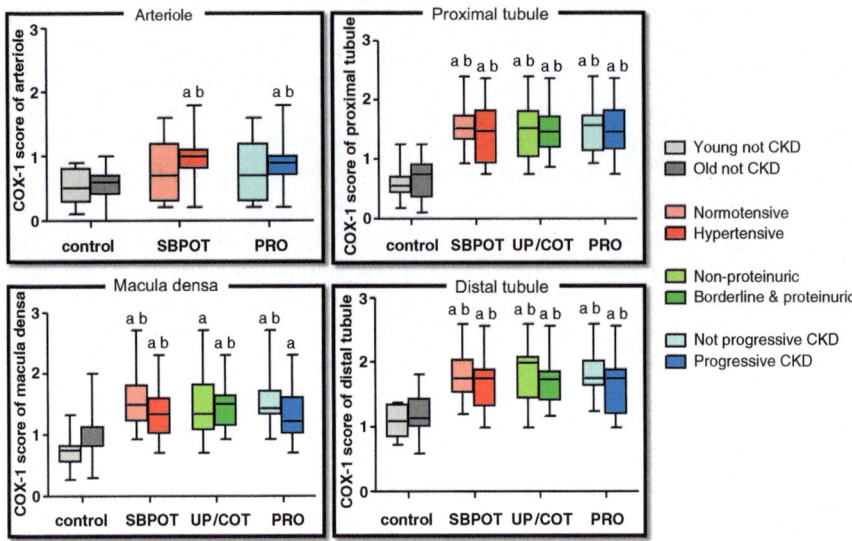

Fig. 1.6 Immunohistochemistry of kidneys from cats stained for cyclooxygenase 1. There were 22 cats with azotaemic chronic kidney disease, 10 cats that were < 5 years old that died of diseases other than CKD and were non-azotaemic at the time they were euthanased (young not CKD controls) and 11 cats >9 years old that died of diseases other than CKD and were non-azotaemia at the time of euthanasia (Old not CKD controls). The azotaemic CKD cats were grouped in three ways: (1) based on their time-averaged systolic blood pressure (SBPOT) as either normotensive ($n = 11$; not requiring antihypertensive treatment) or hypertensive ($n = 11$; on antihypertensive treatment at death), (2) non-proteinuric ($n = 9$; urine protein to creatinine ratio averaged over time (UP/COT) ≤ 0.2) or borderline and proteinuric ($n = 12$; UP/COT >0.2), (3) or not progressive ($n = 11$; change in plasma creatinine concentration over time was <25% from diagnosis to death) and progressive ($n = 11$; plasma creatinine increased by more than 25% between diagnosis and death). Data are presented as box and whisker plots. Each CKD group has been compared to the young and the old not CKD group using the Kruskal-Wallis test followed by the Mann-Whitney U tests as post hoc analysis. (a) denotes $P < 0.013$ vs. the young control group and (b) denotes $P < 0.013$ vs. the old control group

splenic contraction), increasing venous return to the heart and raising cardiac filling pressure. In the case of exercising skeletal muscle, many limb veins run deep between the muscle fascicles and are squeezed as the muscle contracts, massaging (mobilising) more of the blood within the veins back to the heart. Through Starling's law of the heart, increased end-diastolic volume of the heart leads to an increased force of contraction of the cardiac muscle (which is a demand pump—i.e. cardiac output = venous return). Even in the absence of any cardiac innervation (e.g. in transplant patients), cardiac output can increase threefold through these intrinsic mechanisms, primarily through increased stroke volume (Guyton and Hall 2015). Stretch of the atria also has an intrinsic effect on heart rate (Bainbridge reflex), increasing the number of beats per minute and contributing to the increase in cardiac output (which is the product of heart rate and stroke volume).

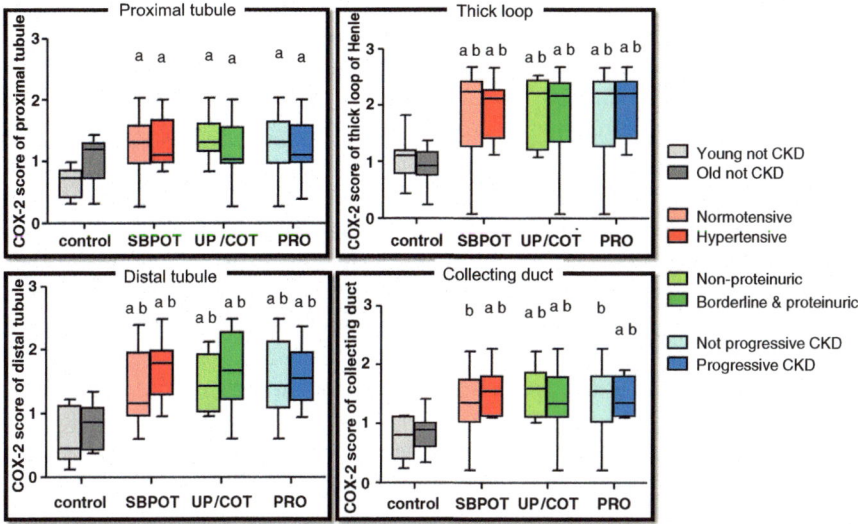

Fig. 1.7 Immunohistochemistry of kidneys from cats stained for Cyclo-oxygenase 2. The remaining details are the same as Fig. 1.6

If blood flow is needed preferentially by one particular tissue, the vascular tone to other tissues can increase, maintaining total peripheral resistance and reducing flow to less active tissues in preference to the metabolically very active tissue. Tissue metabolites (adenosine, AMP, lactate and so on) cause local dilatation of vascular beds opposing the sympathetic tonic vasoconstriction that exists. In addition, as flow through the vascular bed increases, as a result of downstream vasodilation of small resistance vessels (small arteries and arterioles), the shear force at the surface of the endothelial cells lining arteries gives rise to increased release of nitric oxide, leading to the phenomenon of flow-mediated vasodilatation (Zhou et al. 2014). Thus, as described for the kidney, mechanisms intrinsic to the heart and vasculature can increase tissue blood flow without the neuroendocrine control systems. However, these control systems make the responses faster and more coordinated and to some extent enable the body to anticipate and prepare for certain situations whilst maximally protecting the brain and the heart to ensure fluctuations in blood pressure of a significant magnitude do not occur that compromise brain and cardiac tissue perfusion.

1.5 The Nervous System and Control of Cardiovascular Function

The central nervous system is involved in integrating information received from the sensory elements, detecting blood pressure within the cardiovascular system (pressure and volume; see above) and making necessary adjustments, primarily via the autonomic nervous system to maintain a stable arterial blood pressure that means that tissues (particularly the brain and heart) are adequately perfused. Information is received from the stretch receptors in the aortic arch and carotid sinuses about the degree of stretch following each phase of systole of the heart when blood is ejected into the arterial circulation. The vasomotor centre in the pons controls the autonomic nervous system influence on both the vasculature and the heart. Low pressure receptors are located in the atria and pulmonary vasculature and provide information about the degree of filling of the circulation. When the atria and pulmonary circulatory system experience excessive filling, the integrating centres in the brain adjust the traffic of nerve impulses in the renal sympathetic nerves leading to afferent arteriolar vasodilation, increased glomerular filtration rate and reduced activation of the RAAS, leading to increased sodium and water excretion.

Renal afferent nerves are also thought to be important regulators of renal sympathetic nerve activity through reflexes at the level of the spinal cord, which are modulated by higher inputs (Zheng and Patel 2017). In addition, information from these nerves seems to influence the higher centres controlling sympathetic nerve activity in general. There is also evidence for reflex stimulation of vasopressin release from the posterior pituitary gland as a result of activation of renal afferent nerve pathways. The afferent nerves are activated by mechanoreceptors (sensing pressure and volume within the kidney) and chemoreceptors (with two types of receptor-sensing ischaemia and hypertonicity). They also seem to be activated by efferent sympathetic nerve activity and normally act as a negative feedback system, reducing efferent sympathetic nerve activity to the kidney. In CKD (and heart failure), it is thought that this feedback system becomes dysfunctional, and renal afferent nerve activity drives excessive sympathetic discharge. Hence, there is interest in renal denervation as a treatment for drug-resistant hypertension and heart failure, where overactivity of the sympathetic nervous system elicited by renal afferent nerve dysfunction is thought to occur (Thorp and Schlaich 2015).

1.5.1 The Concept of Sympathetic Vascular Tone

Tonic contraction of vascular smooth muscle of arterioles provides resistance within the circulatory system against which the heart pumps blood. Total peripheral resistance (TPR) is one major factor determining SABP, the other being cardiac output (CO) with SABP being the product of these two. Vascular tone is produced by tonic stimulation of sympathetic nerves supplying the vasculature and is balanced by continuous production of vasodilator mediators. Sympathetic nerves are distributed along all the major arteries of the circulatory system from the chain of paravertebral ganglion where the postganglionic neurones originate (Sheng and Zhu 2018). The release of noradrenaline from the varicosities innervating the vascular smooth muscle stimulates alpha-1 adrenoceptors in the majority of vascular beds to give rise to an increase in vascular tone. At low frequencies of stimulation, the cotransmitter, ATP, released by the sympathetic nerves, plays an increasingly important role in maintaining vascular tone.

If all sympathetic nerve stimulation is blocked (e.g. by spinal anaesthesia), SABP will drop from mean pressure of 100 mmHg to about 50 mmHg as a result primarily of a fall in TPR. Baroreceptors discussed above in the aortic arch or carotid sinuses are stretch receptors that detect pressure in these large arteries. They respond most to changes in pressure and at stable pressures adapt quite rapidly. They are sometimes referred to as 'buffer nerves' as they dampen down large fluctuations in blood pressure over a 24 h period, adjusting vasomotor tone in response to changes in blood pressure. If a sustained increase in blood pressure is required, however, these receptors adapt rapidly. So, for example, in extreme exercise, elevated blood pressure may be required to ensure adequate perfusion of exercising muscle, and the baroreceptors reset their set point. It used to be thought that because of this property, this sensory system only regulated very short-term changes in blood pressure. However, it is now recognised that these receptors do not completely adapt, and chronic stimulation above the set point of around 100 mmHg leads to adjustment of sympathetic activity in the renal nerves, and this is part of the way in which the renal body fluid system adjusts fluid volumes to regulate blood pressure in the longer term.

1.5.2 Vagal Tone

Although in many systems the parasympathetic and sympathetic systems have opposing actions, parasympathetic vascular innervation is minor, and the main role the parasympathetic system plays is in regulating heart rate through vagal innervation of the sinoatrial and atrioventricular nodes, slowing the intrinsic rhythm of the heart. Vagal tone predominates at rest, and just by inhibiting vagal tone, heart rate increases with activation of the cardiac sympathetic nerves to the heart increasing heart rate and force of contraction further. As heart rate increases, it is important for the effective functioning of the heart that sufficient time in the cardiac cycle is available for the heart to fill effectively. This means that speed of relaxation needs to increase to maintain diastole as long as possible at higher heart rates. Sympathetic stimulation of the heart not only increases the intrinsic rate and the force of contraction (positive chronotropic and inotropic effects respectively) but also increases the speed of relaxation (positive lusiotropic effect).

1.5.3 Vasodilator Tone: The Influence of the Endothelium

Whilst sympathetic vasomotor tone has been known about since the discovery of the sympathetic nervous system, the understanding that the vascular endothelium tonically produces vasodilator mediators (nitric oxide, hyperpolarising factor and prostacyclin) is a more recent concept. It began with the discovery of endothelium-derived relaxing factor (EDRF) in the early 1980s (Furchgott and Zawadzki 1980) and its identification as nitric oxide (NO) in the late 1980s (Palmer et al. 1988). The L-arginine-NO pathway was completely novel to biology, but it is now clear that the endothelium of vessels generates NO continuously from the metabolism of

L-arginine. Inhibition of this pathway [usually by inhibiting the enzyme and endothelial nitric oxide synthase (eNOS)] leads to a rise in total peripheral resistance and a consequent increase in blood pressure (Rees et al. 1989). The greater the flow of blood across the surface of the endothelium, the more NO is produced, so production is greater on the arterial than the venous side of the circulation. Not only are NO and the other endothelium-derived vasodilator mediators produced tonically, but increased production occurs to modulate the effects of vasoconstrictor mediators, many of which have receptors on the endothelium, mediating NO (and EDHF or prostacyclin) release, modulating the vasoconstriction produced by activation of the smooth muscle receptors and thus modulating against excessive vasoconstriction.

The endothelium, under certain circumstances, also produces mediators that cause vasoconstriction, most notably ET-1 and superoxide. These mediators act locally on vascular smooth muscle cells. Endothelin-1 in particular has a dual action, stimulating receptors (ET_B receptors) on the endothelial cell surface which enhance NO release as well as smooth muscle receptors (usually ET_A receptors) that mediate vasoconstriction. The balance between these two opposing effects may become disturbed in disease states where the endothelium becomes dysfunctional, and rather than ensuring the vasculature is maintained in a partially dilated state, vasoconstriction and excessive peripheral resistance predominate (see below).

1.5.4 Cross-Talk Between the Sympathetic Nervous System, the RAAS and the NO System

As discussed above, increased activity in the sympathetic nerve supplying the kidney is one of the signals for release of renin from the macula densa. Angiotensin II, in addition to its actions on sodium reabsorption by the proximal convoluted tubule of the kidney and its ability to stimulate aldosterone secretion, is also a potent vasoconstrictor. It acts on AT_1 receptors on both arteriolar and venous smooth muscle, raising TPR and increasing venous return, both factors that increase SABP. These direct actions are accompanied by effects on the sympathetic nervous system, which increase the amount of noradrenaline released per nerve impulse. AT_1 receptors are also located on the nerve terminals of postganglionic sympathetic nerves, and their activation enhances the release of noradrenaline. In this way, the RAAS system works in concert with the SNS to protect against a fall in blood pressure, both by actions on the cardiovascular and renal-body fluid systems.

Interactions between SNS, RAAS and NO systems also occur in the CNS. Both angiotensin II and NO are transmitters. Angiotensin II increases activation of the pathways in the CNS leading to increased sympathetic nerve activity, one of the effects of which would be to raise vascular tone. By contrast, NO activates pathways that lead to the reduction of sympathetic nerve activity, thus reducing vascular sympathetic tone.

1.5.5 Regulation to Prevent Excessive Vasoconstriction

There are a number of mechanisms acting on the effector systems discussed above that ensure excessive vasoconstriction or excessive stimulation of the heart does not occur. For example, presynaptic alpha$_2$-adrenoceptors activated by noradrenaline feedback and inhibit further noradrenaline release preventing very high concentrations being attained at the effector cells (vascular smooth muscle or cardiac muscle). As discussed above, many of the vasoconstrictors (noradrenaline, angiotensin II and ET-1) have receptors on the endothelium that are linked to NO production (alpha$_2$ adrenoceptors, AT$_2$ receptors and ET$_B$ receptors), and if the endothelium is functioning normally, this will lead to a lowering of vascular tone. As venous return increases through constriction of the capacitance vessels, then atrial stretch leads to increased release of atrial natriuretic peptide, which in addition to its natriuretic properties also has vasodilator actions, lowering venous and arterial tone and thus counterbalancing the vasoconstrictor and sodium-retaining systems.

In summary, from the above account it is clear that multiple systems work to counterbalance each other and ensure that active tissues receive an adequate supply of nutrients for their level of activity. SABP remains, on average, within relatively narrow limits that ensure the brain and heart are adequately perfused without leading to pathological damage to tissues such as the brain, eye, kidney and heart (see Chap. 7–10). It is generally accepted that regulation of the cardiovascular system adjusts the blood pressure regulatory system on a minute by minute basis, whereas the kidney and body fluid system regulate blood pressure over a longer time frame (hours to days) although this concept is an oversimplification. As discussed above, for the kidney and body fluid system, small defects in the functioning of multiple components of these control and effector systems regulating cardiovascular function could contribute to the development of hypertension. Although this separation between the cardiovascular system and the renal body fluid system in the regulation of blood pressure is convenient, it is evident that the two are part of an integrated whole and many of the regulating neuroendocrine and paracrine factors work on both.

The following sections focus on the sympathetic nervous system, endothelial cell dysfunction and cardiovascular mineralisation and their potential pathophysiological role in hypertension particularly in the context of the data available from veterinary species.

1.6 Sympathetic Nervous System Overactivity and Pathophysiology of Hypertension Associated with CKD

The dog has been used as a model organism to study regulation of blood pressure and to produce CKD models of hypertension where increased sympathetic nerve activity has been documented to play a role in the hypertension. Indeed, the dog is being used to trial novel treatments of renal denervation and baroreceptor stimulation in the management of hypertension. However, published data documenting overactivity of the sympathetic nervous system and its contribution to the pathophysiology of hypertension in dog (and cats) have not been found.

The evidence from human medicine is substantial (Kaur et al. 2017), initially demonstrating higher concentrations of noradrenaline in the plasma from hypertensive patients. These observations were confirmed by recording from sympathetic nerves supplying muscle, initially in haemodialysis patients but then in patients from earlier stages of CKD with muscle sympathetic nerve activity progressively increasing as GFR decreased (Grassi et al. 2011). The novel finding that the kidney is the major site of production of an enzyme, renalase, which metabolises catecholamines may also be important in this respect. Renalase activity in plasma is markedly reduced in human patients with CKD, possibly contributing to the high circulating concentrations of noradrenaline in these patients (Xu et al. 2005). This novel circulating amine oxidase enzyme may well be important in regulating overflow of noradrenaline from sympathetic nerve endings when sympathetic nerve activity is high, exceeding the normal neuronal and extraneuronal uptake processes that regulate noradrenaline concentration. The reduction in production and release of this enzyme from the kidney is another example of cross-talk between the cardiovascular and renal body fluid blood pressure regulatory systems and may be important in the context of the pathophysiology of hypertension associated with CKD.

This general finding of overactivity of the sympathetic nervous system in human CKD patients is most likely a contributing factor in the high prevalence of hypertension in CKD patients in human medicine. The prevalence in canine and feline CKD patients may not be as high (see Chap. 3); thus the study of normotensive veterinary patients with CKD would be of comparative interest.

Chronic stimulation of the cardiovascular system by sympathetic nerves leads to vascular smooth muscle hypertrophy and hyperplasia and endothelial cell dysfunction, making it harder for counterregulatory mechanisms to dilate the blood vessels and counterbalance the vasoconstriction. In the heart, ventricular muscle hypertrophy and fibrosis result, which is detrimental to cardiac relaxation, influencing the diastolic functioning of the heart. In the kidney, sympathetic nerve overactivity leads to podocyte injury, glomerulosclerosis and albuminuria.

The mechanism by which sympathetic overactivity in the CKD patient occurs is not well understood. The RAAS system interacts with the sympathetic nervous system in multiple ways as described above. Activation of RAAS will lead to increased sympathetic nerve activity, both through peripheral and central effects of angiotensin II. In addition, aldosterone has been linked to dysfunction of the baroreceptor system preventing this feedback reduction in sympathetic nerve activity from working effectively (Wang et al. 1992). Inhibition of RAAS with ACE inhibitors and angiotensin receptor blockers reduces but does not normalise muscle sympathetic nerve activity measured in human patients (Klein et al. 2003), suggesting that RAAS activation contributes to this phenomenon but possibly does not completely explain it.

Activity in renal afferent nerves increases in experimental renal reduction models (rats). These nerves are also stimulated by renal ischaemia possibly via release of adenosine. Selective removal of renal afferent nerves in these models inhibits the development of hypertension (Campese and Kogosov 1995). Removal of the diseased kidneys in patients undergoing transplants resulted in reduced muscle sympathetic nerve activity supporting the concept that the diseased kidneys were the source of the general increase in sympathetic nerve activity (Hausberg et al. 2002).

Another central mechanism that could lead to increased sympathetic nerve activity is the NO system in the brain. Data from experimental cats initially demonstrated that injection of inhibitors of the NO pathway into the rostral ventrolateral medulla, areas of the brain regulating sympathetic nerve activity, led to an increase in sympathetic renal nerve discharge and an increase in SABP (Shapoval et al. 1991). The possible mechanisms by which the L-arginine-NO pathway is inhibited in CKD are discussed below in connection with the phenomenon of endothelial cell dysfunction. These mechanisms might also lead to inhibition of NO pathways in the brain as the body of scientific evidence suggests these pathways are important in regulating sympathetic nerve outflow and therefore tonic cardiovascular stimulation.

In summary, the role of the sympathetic nervous system in the pathophysiology of hypertension associated with CKD has not been studied in veterinary patients. The evidence supporting its contribution to hypertension in human CKD patients and in animal models is strong. Activation of the RAAS system, stimulation of renal afferent nerves in diseased kidneys and inhibition of the CNS NO-arginine pathway are all likely to contribute to this increase in sympathetic nerve activity. Sympathetic nerve hyperactivity will contribute both to the occurrence of hypertension and to progressive renal injury (directly and as a consequence of the hypertension).

1.7 Endothelial Cell Dysfunction and Hypertension Associated with CKD

Normal functioning of the endothelium is essential as a balance to tonic constriction of the vasculature induced by sympathetic tone. Many of the risk factors for hypertension in human patients, including CKD, have a negative impact on endothelial cell function and induce at state of endothelial cell dysfunction. Other risk factors include smoking, hypercholesterolaemia, obesity and insulin-resistant diabetes. This is characterised by reduced ability of the endothelium to release vasodilator substances, including NO, and increased production of vasoconstrictor mediators such as ET-1. Chronic overstimulation of the vasculature with angiotensin II, aldosterone, ET-1 and sympathetic nerve activity induces a proinflammatory state and increased production of oxygen-free radicals which contribute to the state of endothelial cell dysfunction.

Methods which directly measure endothelial cell function in human patients, such as flow-mediated dilation, are difficult to apply to veterinary patients. This method has been used to demonstrate human CKD patients show greater fluctuation in blood pressure on exercise, the peak increase in blood pressure being inversely related to their flow-mediated dilation response (Downey et al. 2017). This study shows how important endothelial cell function is within the integrated blood pressure control system to counterbalance the vasoconstrictor systems that are activated to raise systemic blood pressure as a response to increased skeletal muscle activity.

A direct mechanism by which the L-arginine-NO pathway is inhibited in CKD is through accumulation of the inhibitor of this asymmetric dimethylarginine (ADMA) pathway. ADMA is a product of hydrolysis of methylated proteins by protein arginine methyltransferase and is produced in all cells, including those in the CNS. It is an inhibitor of nitric oxide synthase (NOS) and a competitive substrate of the cationic amino acid transporter, which transports arginine into cells to act as the NOS substrate (Teerlink et al. 2009). Plasma concentrations of ADMA increase in human patients with severity of CKD and are predictive of outcome in these patients (Zoccali et al. 2001). However, the prevailing plasma concentrations are not thought to be high enough to compete with L-arginine to inhibit its uptake or metabolism to NO. For this reason, it is proposed that locally produced ADMA may be important. Although ADMA is renally cleared, which is why plasma concentrations increase in CKD, its accumulation within cells is thought to occur because of reduced metabolism by the enzyme dimethylarginine dimethylaminohydrolase (DDAH) (Teerlink et al. 2009).

Little work has been published assessing endothelial cell function in dogs and cats with CKD and hypertension despite its well-documented occurrence in human CKD patients. Jepson et al. (2008) measured ADMA in a group of 69 cats with differing severities of kidney dysfunction, some of which were hypertensive and others of which were normotensive. Samples were collected at diagnosis and, for the

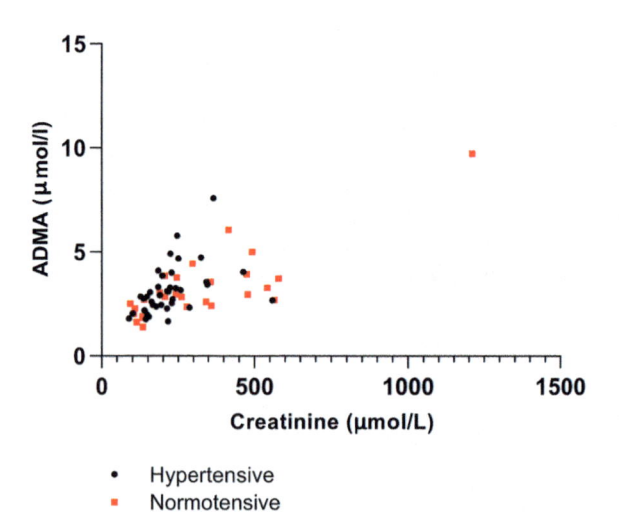

Fig. 1.8 Correlation between log ADMA and plasma creatinine concentration. Data are reproduced with permission from Jepson et al. (2008). In this study, ADMA was measured in plasma samples collected from 69 cats with varying degrees of azotaemia and differing systolic arterial blood pressure. No correlation was found between plasma ADMA concentration and blood pressure, but a moderate correlation was found between plasma ADMA and plasma creatinine concentrations ($r = 0.608$, $P < 0.001$)

hypertensive cats, following treatment with amlodipine. The data showed plasma ADMA concentrations were related to plasma creatinine (and so influenced by renal function) but did not differ between normotensive and hypertensive groups (see Fig. 1.8). Thus, these data did not support the hypothesis that accumulation of ADMA in feline CKD patients contributes to the pathophysiology of hypertension. Interestingly, plasma ADMA concentrations were considerably higher in the cats than reported in human subjects when measured using the same method.

Jepson et al. (2008) also measured plasma nitrite/nitrate concentrations in the same cats using the Griess reaction, a colorimetric assay. Contrary to the expected reduction in plasma nitrate/nitrite concentrations predicted in hypertensive cats with CKD, plasma nitrite/nitrate increased with severity of kidney disease and did not differ between normotensive and hypertensive groups. However, plasma nitrite/nitrate concentrations are an indirect measure of the L-arginine-NO pathway and could be influenced by diet, the presence of systemic inflammatory disease and other factors that were difficult to control in this exploratory study. In order to try to reduce some of these factors that influence plasma nitrite/nitrate concentrations, Bijsmans et al. (2015b) used a more sensitive and specific chemiluminescent assay for nitrite. In this case, no effect of either severity of kidney disease or presence of hypertension was found to influence the plasma nitrite concentration measured. Thus, to date, there is no evidence that supports the suggestion that reduction in NO by the endothelium contributes to the pathophysiology of feline hypertension associated with CKD.

Direct measurement of endothelial cell function in dogs and cats would be preferable to answer the question 'Does endothelial cell dysfunction occur in dogs and cats with CKD and does this contribute to hypertension?' Flow-mediated vasodilation has been used to investigate endothelial cell function in dogs with mitral valve disease (Moesgaard et al. 2012) but, to date, has not been applied in dogs with naturally occurring CKD and hypertension. Application of this technique to cats would be technically and practically challenging given the size of the femoral and brachial arteries that would need to be accurately measured by ultrasound.

1.8 Reduced Vascular Compliance, Cardiovascular Mineralisation and Hypertension Associated with CKD

One of the complications of CKD is the retention of phosphate, which results in bone and mineral disturbances of increasing severity with increasing stage of CKD. One result of this is deposition of calcium and phosphate within the subendothelial layer of blood vessels and heart valves. It is well accepted in human medicine that vascular calcification is linked to poor outcomes in renal dialysis patients and increases cardiovascular risk in hypertensive patients in general. Inhibitors of vascular calcification include magnesium ions, albumin and fetuin A, whereas risk factors that promote vascular calcification include the phosphaturic hormone FGF23 and its co-receptor alpha-Klotho (Paloian and Giachelli 2014).

The deposition of calcium and phosphate within the subendothelium leads to a change in the phenotype of the vascular smooth muscle cells such that they secrete

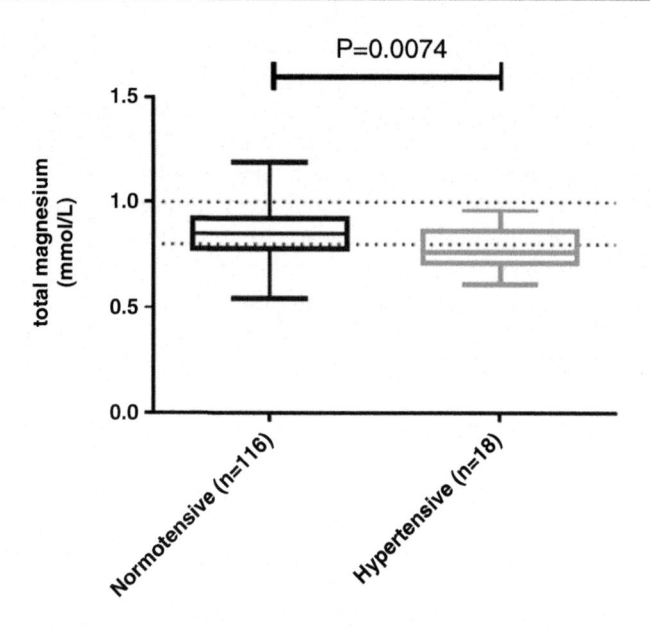

Fig. 1.9 Total plasma magnesium concentrations in cats with CKD comparing cats diagnosed with hypertension and those considered to be normotensive at diagnosis of CKD. Data are taken from van den Broek et al. (2018)

connective tissue collagen, taking on a more fibroblast phenotype. This leads to increased stiffness of the vasculature and a reduced ability to relax. Such a complication of CKD may contribute to the pathophysiology of hypertension with increasing severity of CKD.

Recent data derived from cats with CKD supports this contention. Magnesium deficiency is one of the most important factors associated with vascular calcification in human medicine (Kostov and Halacheva 2018) and is a significant risk factor for hypertension. van den Broek et al. (2018) recently showed that hypomagnesaemia was associated with increased risk of death and higher plasma FGF23 concentrations in cats with CKD. Furthermore, within this cohort of cats with CKD, hypertensive cats have lower plasma magnesium concentrations (see Fig. 1.9) than normotensive cats perhaps supporting the hypothesis that vascular calcification and the associated vascular dysfunction it causes contribute to the pathophysiology of hypertension in cats associated with CKD. Further studies are warranted to assess vascular calcification directly to determine the significance of this contribution.

1.9 Conclusions

The systems regulating systemic arterial blood pressure in the body, the renal body fluid system and the cardiovascular system work in an integrated way to maintain adequate blood flow to the organs of the body. The kidney is central to this system

Pathophysiology of hypertension in CKD patients

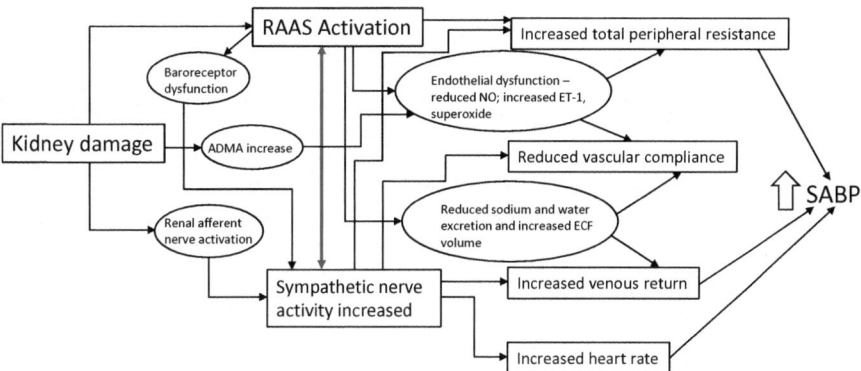

Fig. 1.10 Summary of the proposed pathophysiological mechanisms driving hypertension in chronic kidney disease. This figure shows the interactions between all the mechanisms thought to contribute to hypertension associated with CKD. These are increased ECF volume in relation to the volume of the circulation and increased TPR and reduced vascular compliance

acting as a sensor which influences sympathetic nerve activity via the CNS, a generator of the major regulating hormone angiotensin II and an effector organ regulating body fluid homeostasis through intrinsic mechanisms. It is not surprising, therefore, that in CKD blood pressure regulation tends to go awry and hypertension occurs. The multiple pathophysiological mechanisms by which this occurs are complex and interact in many ways, and their study requires a systems-based holistic rather than a reductionist approach (see Fig. 1.10). Knowledge of these pathophysiological mechanisms is in its infancy in veterinary medicine.

References

Bijsmans E, Jepson RE, Syme H, Elliott J (2012) The association between hypertension and chronic kidney disease in cats and the available management options. Companion Anim 1:4

Bijsmans ES, Jepson RE, Chang YM, Syme HM, Elliott J (2015a) Changes in systolic blood pressure over time in healthy cats and cats with chronic kidney disease. J Vet Intern Med 29(3):855–861

Bijsmans ES, Jepson RE, Syme HM, Elliott J (2015b) Nitric oxide in feline chronic kidney disease and hypertension. Proceedings of the British Small Animal Veterinary Association Congress, Birmingham April 2015

Bijsmans ES, Jepson RE, Wheeler C, Syme HM, Elliott J (2017) Plasma N-terminal Probrain natriuretic peptide, vascular endothelial growth factor, and cardiac troponin I as novel biomarkers of hypertensive disease and target organ damage in cats. J Vet Intern Med 31(3):650–660

Campese VM, Kogosov E (1995) Renal afferent denervation prevents hypertension in rats with chronic renal failure. Hypertension 25(4 Pt 2):878–882

Downey RM, Liao P, Millson EC, Quyyumi AA, Sher S, Park J (2017) Endothelial dysfunction correlates with exaggerated exercise pressor response during whole body maximal exercise in chronic kidney disease. Am J Physiol Renal Physiol 312(5):F917–F924

Finch NC (2011) Predicting the development of azotaemia in geriatric cats. PhD thesis. University of London

Finch NC, Heiene R, Elliott J, Syme HM, Peters AM (2015) Determination of extracellular fluid volume in healthy and azotemic cats. J Vet Intern Med 29(1):35–42

Furchgott RF, Zawadzki JV (1980) The obligatory role of endothelial cells in the relaxation of arterial smooth muscle by acetylcholine. Nature 288(5789):373–376

Grassi G, Quarti-Trevano F, Seravalle G, Arenare F, Volpe M, Furiani S, Dell'Oro R, Mancia G (2011) Early sympathetic activation in the initial clinical stages of chronic renal failure. Hypertension 57(4):846–851

Guyton AC, Hall JE (2015) Textbook of medical physiology, 13th edn. Elsevier, Philadelphia, PA

Hausberg M, Kosch M, Harmelink P, Barenbrock M, Hohage H, Kisters K, Dietl KH, Rahn KH (2002) Sympathetic nerve activity in end-stage renal disease. Circulation 106(15):1974–1979

Javadi S, Slingerland LI, van de Beek MG, Boer P, Boer WH, Mol JA, Rijnberk A, Kooistra HS (2004) Plasma renin activity and plasma concentrations of aldosterone, cortisol, adrenocorticotropic hormone, and alpha-melanocyte-stimulating hormone in healthy cats. J Vet Intern Med 18(5):625–631

Javadi S, Djajadiningrat-Laanen SC, Kooistra HS, van Dongen AM, Voorhout G, van Sluijs FJ, van den Ingh TS, Boer WH, Rijnberk A (2005) Primary hyperaldosteronism, a mediator of progressive renal disease in cats. Domest Anim Endocrinol 28(1):85–104

Jensen J, Henik RA, Brownfield M, Armstrong J (1997) Plasma renin activity and angiotensin I and aldosterone concentrations in cats with hypertension associated with chronic renal disease. Am J Vet Res 58(5):535–540

Jepson RE, Syme HM, Vallance C, Elliott J (2008) Plasma asymmetric dimethylarginine, symmetric dimethylarginine, l-arginine, and nitrite/nitrate concentrations in cats with chronic kidney disease and hypertension. J Vet Intern Med 22(2):317–324

Jepson RE, Syme HM, Elliott J (2014) Plasma renin activity and aldosterone concentrations in hypertensive cats with and without azotemia and in response to treatment with amlodipine besylate. J Vet Intern Med 28(1):144–153

Kaur J, Young BE, Fadel PJ (2017) Sympathetic Overactivity in chronic kidney disease: consequences and mechanisms. Int J Mol Sci 18(8):E1682

Keele SJ, Smith KC, Elliott J, Syme HM (2009) Adrenocortical morphology in cats with chronic kidney disease (CKD) and systemic hypertension. J Vet Intern Med 23(6):1328

Klein IH, Ligtenberg G, Oey PL, Koomans HA, Blankestijn PJ (2003) Enalapril and losartan reduce sympathetic hyperactivity in patients with chronic renal failure. J Am Soc Nephrol 14(2):425–430

Kohan DE, Rossi NF, Inscho EW, Pollock DM (2011) Regulation of blood pressure and salt homeostasis by endothelin. Physiol Rev 91(1):1–77

Kostov K, Halacheva L (2018) Role of magnesium deficiency in promoting atherosclerosis, endothelial dysfunction, and arterial stiffening as risk factors for hypertension. Int J Mol Sci 19(6):E1724

Lalor SM, Connolly DJ, Elliott J, Syme HM (2009) Plasma concentrations of natriuretic peptides in normal cats and normotensive and hypertensive cats with chronic kidney disease. J Vet Cardiol 1(11 Suppl):S71–S79

Li H, Zhang Y, Wu Q (2017) Role of corin in the regulation of blood pressure. Curr Opin Nephrol Hypertens 26(2):67–73

Macheret F, Heublein D, Costello-Boerrigter LC, Boerrigter G, McKie P, Bellavia D, Mangiafico S, Ikeda Y, Bailey K, Scott CG, Sandberg S, Chen HH, Malatino L, Redfield MM, Rodeheffer R, Burnett J Jr, Cataliotti A (2012) Human hypertension is characterized by a lack of activation of the antihypertensive cardiac hormones ANP and BNP. J Am Coll Cardiol 60(16):1558–1565

Michell AR (1994) Salt, hypertension and renal disease: comparative medicine, models and real diseases. Postgrad Med J 70(828):686–694

Michell AR (2000) Hypertension in dogs: the value of comparative medicine. J R Soc Med 93 (9):451–452

Mishina M, Watanabe T (2008) Development of hypertension and effects of benazepril hydrochloride in a canine remnant kidney model of chronic renal failure. J Vet Med Sci 70(5):455–460

Moesgaard SG, Klostergaard C, Zois NE, Teerlink T, Molin M, Falk T, Rasmussen CE, Luis Fuentes V, Jones ID, Olsen LH (2012) Flow-mediated vasodilation measurements in cavalier King Charles spaniels with increasing severity of myxomatous mitral valve disease. J Vet Intern Med 26(1):61–68

Navar LG, Harrison-Bernard LM, Nishiyama A, Kobori H (2002) Regulation of intrarenal angiotensin II in hypertension. Hypertension 39(2 Pt 2):316–322

O'Connor PM, Cowley AW Jr (2010) Modulation of pressure-natriuresis by renal medullary reactive oxygen species and nitric oxide. Curr Hypertens Rep 12(2):86–92

Palmer RM, Ashton DS, Moncada S (1988) Vascular endothelial cells synthesize nitric oxide from L-arginine. Nature 333(6174):664–666

Paloian NJ, Giachelli CM (2014) A current understanding of vascular calcification in CKD. Am J Physiol Renal Physiol 307(8):F891–F900

Payne JR, Brodbelt DC, Luis Fuentes V (2017) Blood pressure measurements in 780 apparently healthy cats. J Vet Intern Med 31(1):15–21

Pelligand L, Suemanotham N, King JN, Seewald W, Syme H, Smith K, Lees P, Elliott J (2015) Effect of cyclooxygenase(COX)-1 and COX-2 inhibition on furosemide-induced renal responses and isoform immunolocalization in the healthy cat kidney. BMC Vet Res 3(11):296

Rees DD, Palmer RM, Moncada S (1989) Role of endothelium-derived nitric oxide in the regulation of blood pressure. Proc Natl Acad Sci U S A 86(9):3375–3378

Reynolds BS, Chetboul V, Nguyen P, Testault I, Concordet DV, Carlos Sampedrano C, Elliott J, Trehiou-Sechi E, Abadie J, Biourge V, Lefebvre HP (2013) Effects of dietary salt intake on renal function: a 2-year study in healthy aged cats. J Vet Intern Med 27(3):507–515

Sarzani R, Spannella F, Giulietti F, Balietti P, Cocci G, Bordicchia M (2017) Cardiac natriuretic peptides, hypertension and cardiovascular risk. High Blood Press Cardiovasc Prev 24 (2):115–126

Sata Y, Head GA, Denton K, May CN, Schlaich MP (2018) Role of the sympathetic nervous system and its modulation in renal hypertension. Front Med (Lausanne) 5:82

Shapoval LN, Sagach VF, Pobegailo LS (1991) Nitric oxide influences ventrolateral medullary mechanisms of vasomotor control in the cat. Neurosci Lett 132(1):47–50

Sheng Y, Zhu L (2018) The crosstalk between autonomic nervous system and blood vessels. Int J Physiol Pathophysiol Pharmacol 10(1):17–28

Suemanotham N, Syme H, Berhane Y, Smith, K, Elliott J (2010) Immunohistochemical expression of cyclooxygenase (COX)-1 and -2 enzymes in feline chronic kidney disease. J Vet Intern Med 24(3):678–679 (abstract 28)

Syme HM, Barber PJ, Markwell PJ, Elliott J (2002) Prevalence of systolic hypertension in cats with chronic renal failure at initial evaluation. J Am Vet Med Assoc 220(12):1799–1804

Teerlink T, Luo Z, Palm F, Wilcox CS (2009) Cellular ADMA: regulation and action. Pharmacol Res 60(6):448–460

Thorp AA, Schlaich MP (2015) Device-based approaches for renal nerve ablation for hypertension and beyond. Front Physiol 6:193

van den Broek DHN, Chang YM, Elliott J, Jepson RE (2018) Prognostic importance of plasma total magnesium in a cohort of cats with azotemic chronic kidney disease. J Vet Intern Med 32 (4):1359–1371

Wang W, McClain JM, Zucker IH (1992) Aldosterone reduces baroreceptor discharge in the dog. Hypertension 19(3):270–277

Xu J, Li G, Wang P, Velazquez H, Yao X, Li Y, Wu Y, Peixoto A, Crowley S, Desir GV (2005) Renalase is a novel, soluble monoamine oxidase that regulates cardiac function and blood pressure. J Clin Invest 115(5):1275–1280

Yang T (2015) Crosstalk between (pro)renin receptor and COX-2 in the renal medulla during angiotensin II-induced hypertension. Curr Opin Pharmacol 21:89–94

Yang T, Liu M (2017) Regulation and function of renal medullary cyclooxygenase-2 during high salt loading. Front Biosci (Landmark Ed) 22:128–136

Zheng H, Patel KP (2017) Integration of renal sensory afferents at the level of the paraventricular nucleus dictating sympathetic outflow. Auton Neurosci 204:57–64

Zhou J, Li YS, Chien S (2014) Shear stress-initiated signaling and its regulation of endothelial function. Arterioscler Thromb Vasc Biol 34(10):2191–2198

Zoccali C, Bode-Böger S, Mallamaci F, Benedetto F, Tripepi G, Malatino L, Cataliotti A, Bellanuova I, Fermo I, Frölich J, Böger R (2001) Plasma concentration of asymmetrical dimethylarginine and mortality in patients with end-stage renal disease: a prospective study. Lancet 358(9299):2113–2117

Measurement of Blood Pressure in Conscious Cats and Dogs

Rosanne E. Jepson

2.1 Importance of Accurate Blood Pressure Measurement

Accurate, repeatable and reliable measurement of blood pressure (BP) is fundamental to establishing normotension and defining hypertension in dogs and cats. Erroneous or inaccurate measurement may have high clinical impact, preventing the appropriate diagnosis and management of those individuals with systemic hypertension, delaying administration of antihypertensive medication where indicated and also failing to identify complications associated with antihypertensive therapy, for example, if hypotension is not detected.

Direct arterial catheterisation is considered the gold standard for assessment of BP but is impractical for day-to-day use in the conscious patient, although continues to be used for research purposes, in patients within an intensive care setting and in patients undergoing general anaesthesia where evaluation of arterial BP is beneficial. Instead, for dogs and cats in either a consultation or general hospital ward environment, indirect BP measuring devices including oscillometry, high-definition oscillometry and Doppler sphygmomanometry are preferred.

However, indirect measurement of BP in the clinical setting provides the clinician with many challenges, and obtaining repeatable, reliable and accurate measurements is not always straightforward; machine, patient and operator factors all impact on the readings that are obtained. A good understanding of these factors may influence not only choice of machine and method of BP measurements but also the manner in which results are interpreted.

R. E. Jepson (✉)
Department of Clinical Sciences and Services, Royal Veterinary College, University of London, Hertfordshire, UK
e-mail: rjepson@rvc.ac.uk

2.2 Validation for Blood Pressure Monitoring Devices in Human and Veterinary Medicine

In human medicine, validation for BP monitoring devices has been ongoing since the 1980s. Independent groups including the Association for the Advancement of Medical Instrumentation (AAMI), British Hypertension Society (BHS) and European Society of Hypertension (ESH) and International Organization for Standardization (ISO) have all produced guidelines for BP monitoring devices (Brien et al. 2002; O'Brien et al. 2010). However, in 2018 AAMI/ESH/ISO published a collaborative and internationally accepted methodology for validation of BP measuring devices with consensus based on the evidence from previous validation studies published by these societies, with statistical input for establishing power and expert opinion (Stergiou et al. 2018). The key requirements are listed in Table 2.1 (O'Brien et al. 2010).

In veterinary medicine, such stringent requirements for BP machine validation do not exist. In 2007, the American College of Veterinary Internal Medicine (ACVIM) developed guidelines for the validation of indirect BP measuring devices which were loosely based on those that existed for human medicine at the time (Brown et al. 2012). Given that no indirect BP measuring device currently reaches the standards that are accepted by human medicine, the reference methodology identified by the ACVIM guidelines is direct arterial catheterisation until such time an indirect device has achieved appropriate validation, when that device could also be used as a reference standard. The guidelines recognise that the device is validated only for the species and conditions in which the validation test is performed, for example, a device that is validated for use in anaesthetised dogs is not automatically validated for conscious dogs and that validation may be attained for systolic BP (SBP), mean BP (MBP) or diastolic BP (DBP) or combination of the above. It is recommended that the AAMI criteria be followed in terms of patient selection, pressure ranges, number of observers, blinding of observations and reporting of the study. Specific details on the device recommendations are listed in Table 2.2.

The guidelines above were not updated in the 2018 revision of the ACVIM hypertension guidelines, and in 10 years since their publication, there has been little advancement in the validation of indirect devices for use in conscious dogs or cats despite numerous publications on this topic (Acierno et al. 2018). Nevertheless, when indirect machines are compared to direct arterial BP measurements, there are a number of key considerations which become important when assessing the available validation literature. Studies evaluating indirect BP measuring devices should consider the accuracy of the device when compared to direct arterial measurements. In the ACVIM 2007 consensus and in many early publications, this was assessed only by linear regression analysis or correlation (correlation coefficient; CC). However, there are clear limitations with these analyses because strong correlation does not preclude the possibility of either continuous or variable bias across the range of BP measurements. In addition, the 2007 ACVIM guidelines recommend providing the percentage of indirect measurements that fall within a given parameter (e.g. 10 mmHg and 20 mmHg). However, a further and beneficial step, which is

Table 2.1 Key requirements for a validation study for an indirect blood pressure monitoring device from AAMI/ESH/ISO (2018)

Aspect of interest	Requirement
Efficacy measure	≥85% of readings within ≤10 mmHg
Sample size	≥85 participants
General population versus special population	Machines should initially be validated for the general population (>12 years of age) before special populations are considered. Special populations may include the following: • Paediatric <3 years • Pregnancy • Arm >42 cm • Atrial fibrillation Special population studies may include ≥35 participants once general population validation has been achieved. Pregnancy population should include 45 participants (15 normotensive, 15 gestational hypertension, 15 preeclampsia). Validation for children (3–12 years) should include 35 children and 50 older participants
Cuff sizes	Minimum number of participants per cuff size
Patient positioning	Clear requirements provided for position of patient, i.e. seated, acclimatisation period and not crossing legs
Reference BP measurement	Reference BP measurements can be obtained with either mercury or non-mercury manometers providing they fulfil the ISO 81060-1 requirements with the accuracy of non-mercury manometers evaluated against a mercury manometer or calibrated and certified pressure device at the beginning of each validation study
Data collection	The same arm sequential BP measurement
Supervisors and observers	Reference methodology is auscultatory standard with dual Y-tube-connected stethoscopes permitting measurements to be obtained by two trained observers blinded to each other's readings and the test device measurements. Observers must be qualified to BHS protocol within 12 months of the validation study and undergo auditory testing within 3 years. A supervisor must be present in addition to the two observers
Pass criteria	Average BP difference and SD criteria 1 and 2 of ANSI/AAMI/ISO. Absolute BP differences ≤5, 10, 15 mmHg and scatterplots must be presented using Bland-Altman plots. Standardised reporting system provided

included in the validation of human BP measuring devices although not listed by the ACVIM guidelines, is evaluation of Bland-Altman plots (Giavarina 2015). This is a graphical methodology to compare two measurement techniques for a given parameter; the differences between the two techniques are plotted against the averages of the two techniques. This provides a simple and effective way to evaluate bias and to estimate an agreement interval within which 95% of the differences of the second method compared to the first one fall. The Bland-Altman plot only defines intervals of agreement, and biological and clinical principles must subsequently be applied to decide whether those intervals are acceptable for the clinical patient. BP machine

Table 2.2 ACVIM guidelines for the validation of indirect blood pressure monitoring devices

The mean difference of paired measurements for systolic and diastolic blood pressure treated separately is ±10 mmHg or less with a standard deviation of 15 mmHg or less
The correlation between paired measures for systolic and diastolic pressures treated separately is ≥0.9 across the range of measured values of BP
50% of all measurements for systolic and diastolic pressures treated separately lie within 10 mmHg of the reference method
80% of all measurements for systolic and diastolic pressures treated separately lie within 20 mmHg of the reference method
The study results have been accepted for publication in a refereed journal
The subject database contains no fewer than 8 animals for comparison with an intra-arterial method or 25 animals for comparison with a previously validated indirect device

validation studies may also wish to consider the precision of multiple sequential readings; this can be achieved by evaluation of coefficient of variance (CV). Similar analysis can also be used to evaluate both day-to-day and interoperator characteristics of the device.

2.3 Calibration of Blood Pressure Monitoring Devices

A topic which is frequently overlooked in practice is periodic calibration of BP monitoring devices. Ideally BP measuring devices would undergo at least 6 monthly calibration so that the static pressure output of the device can be compared against another device (e.g. mercury or aneroid manometer that has been approved ISO 81060-1) over a clinically relevant range of BP measurements (30–300 mmHg) (Binns et al. 1995). For some automated oscillometric devices, there may not be the facility for operators to perform such calibration, and in this situation, the device should be returned to the manufacturer periodically for calibration. For Doppler sphygmomanometers, calibration requires connection of the sphygmomanometer to Y-tubing and a reference manometer. The pressure should be inflated and a series of repeated readings made over the full range of BP measures that could be encountered with readings ideally falling within ~3 mmHg of the reference device. Calibration has been documented in some but not all studies that report in veterinary medicine on BP measuring devices. There are no data on how often veterinary surgeons perform routine calibration of their BP measuring devices although it can be anticipated to be rare in both general and specialist practice.

2.4 Direct Arterial Blood Pressure Monitoring

Direct arterial BP measurement in conscious dogs and cats is rarely used outside of an intensive care setting given the specialised skill and equipment required, the difficulties in maintaining peripheral arterial catheters in conscious patients that are moving freely and the potential for complications associated with this procedure,

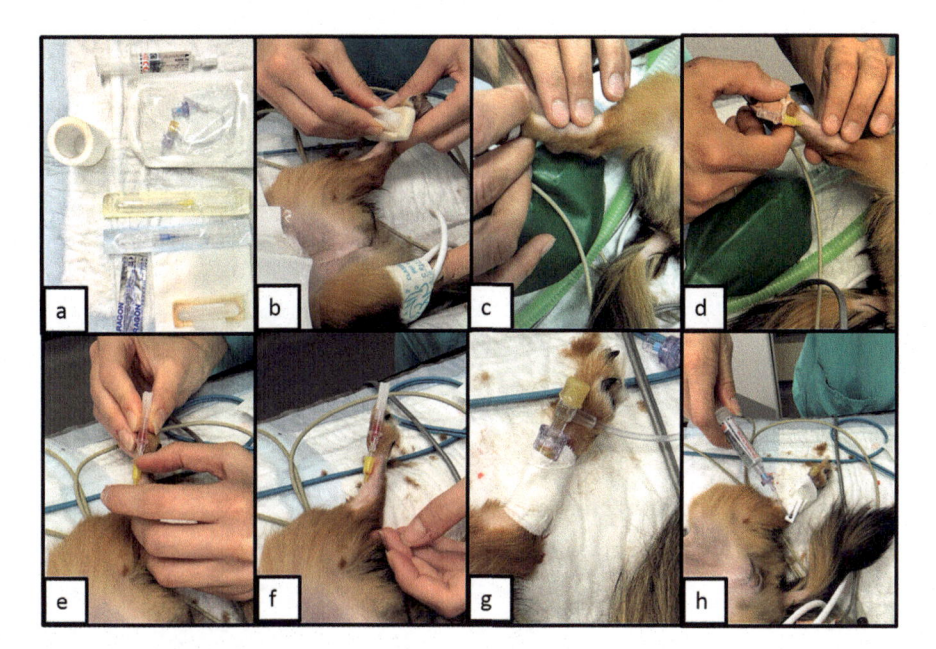

Fig. 2.1 Placement of a dorsal pedal arterial catheter. (**a**) Preparation of kit (heparinised saline flush, t-connector, arterial catheters, scalpel blade, aseptic preparation, tape). (**b**) Surgical preparation of area. (**c**) Feeling for the dorsal pedal arterial pulse. (**d**) Careful placement of arterial catheter. (**e**) Flashback of blood into arterial catheter hub. (**f**) Careful removal of stylet. (**g**) Placement of pre-flushed t-connector and taping of catheter. (**h**) Careful flush with heparinised saline to ensure and maintain patency

which include technical challenges and discomfort on initial placement of the arterial catheter, infection, embolus formation and haemorrhage. Direct arterial BP measurement most often involves placement of an arterial catheter (22 or 24 Ga) in the dorsal pedal artery although historically other arteries (e.g. femoral) have been used, more often for research purposes (Fig. 2.1) (Bosiack et al. 2010). The arterial catheter can then be connected via saline-filled low-compliance tubing to a pressure transducer (Fig. 2.2). The position of the transducer should be fixed at the level of the patient's heart with zero calibrations performed in accordance with the manufacturer's instructions. Appropriate positioning of the transducer is easiest in the recumbent patient. The transducer is then typically connected to a pressure monitor which provides a continuous arterial BP trace. When assessing arterial BP traces, it is important to ensure that there is no artefactual dampening of the trace by careful assessment of the anticipated waveform. Connection to a pressurised heparinised syringe allows for periodic flushing of the arterial catheter as required. Despite the difficulties associated with direct arterial catheter BP monitoring, this can be a very useful method for careful continuous monitoring of BP in critical patients, particularly when considering use of potent intravenous antihypertensive agents in the situation of hypertensive encephalopathy, e.g. phentolamine, hydralazine constant rate infusion or sodium nitroprusside. Direct arterial measurement allows continuous

Fig. 2.2 Arterial blood pressure transducer

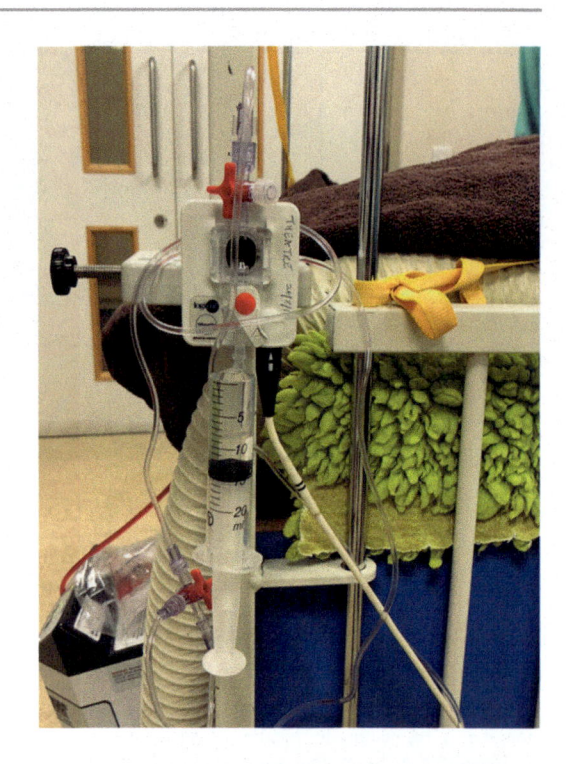

Fig. 2.3 Photograph of radiotelemetric Data Science International (TA11 PA-C40). Blood pressure monitoring device prior to placement. Photographs courtesy of Dr C. Schmeidt

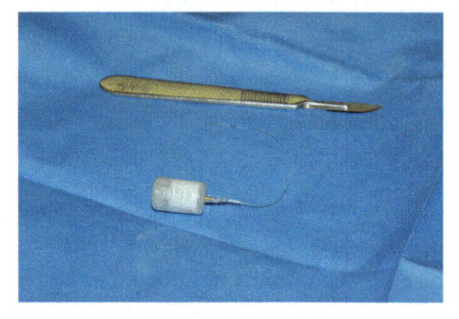

assessment of SBP, MAP, DBP and pulse pressures. Evidence suggests that for critical patients where hypotension is a concern, indirect devices including oscillometric and Doppler methodologies have limitations and do not always provide accurate representation of a patient's BP (Bosiack et al. 2010). In this situation, direct arterial measurements can prove very useful.

It is even rarer for direct arterial pressure monitoring to be performed in cats, even in the veterinary criticalist sphere due to the difficulties and challenges associated with placement and maintenance of a pedal arterial catheter. For research purposes in both dogs and cats, direct arterial BP measurements are often performed using radiotelemetric devices (Figs. 2.3 and 2.4). Studies that have performed comparisons in conscious cats between direct arterial measurements and indirect devices have

Fig. 2.4 Surgical placement of radiotelemetric blood pressure monitoring device placed in the left carotid artery of a cat. Photographs courtesy of Dr C. Schmeidt

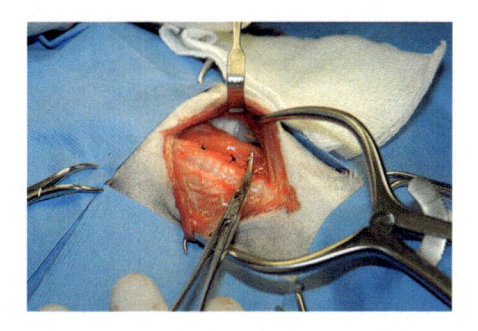

usually involved placement of radiotelemetric devices (Miller et al. 2000; Martel et al. 2013). In these procedures, a sensor is placed via the femoral artery up to the aorta, and a subcutaneous radiotelemetric device is placed subcutaneously to allow continuous transmission of recorded data. Placement of this type of arterial BP monitoring requires general anaesthesia but subsequently provides the ability to monitor direct arterial BP long term, and the technique has been widely used in research studies where SBP monitoring in conscious cats is required, facilitating the comparison between direct arterial and indirect BP measuring devices (Buranakarl et al. 2004; Belew et al. 1999; Haberman et al. 2004; Mathur et al. 2002, 2004).

2.5 Indirect Blood Pressure Monitoring

A number of indirect BP monitoring devices are available for use in conscious dogs and cats of which none to date have fulfilled the ACVIM guideline validation criteria. Available indirect machines measure BP either by oscillometry or Doppler technique. These machines provide a practical approach to the measurement of BP in the dog and the cat and can be used in a consulting or patient bedside manner. An ideal machine would be inexpensive to purchase; provide repeatable, reliable and accurate readings of SBP, DBP and MAP; be well tolerated by the patient; and be quick and easy to use. All of the machines available in veterinary medicine have limitations but, nevertheless, with experience provide the best approximation to BP that can be achieved for clinical patients.

2.6 Oscillometry

Oscillometric BP measurements are determined by analysis of a series of pressure pulses in the cuff whilst it is being deflated from a suprasystolic pressure to a subdiastolic pressure. The oscillometric readings are obtained by analysing the amplitude of the air oscillations generated from the peripheral pulse and sensed within the cuff. The pressure within the cuff is inflated to a target pressure; the cuff pressure is then regulated by a valve and deflated in a controlled manner in either a linear or stepwise manner. When the cuff is inflated on the limb and the artery is

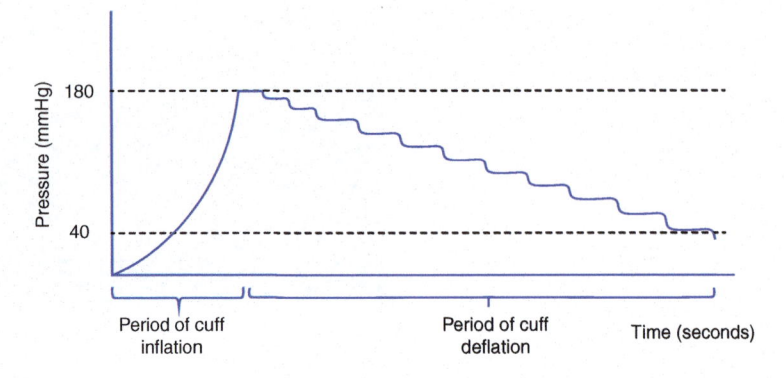

Fig. 2.5 Pressure trace showing change in pressure within the cuff during inflation and deflation using the oscillometric technique

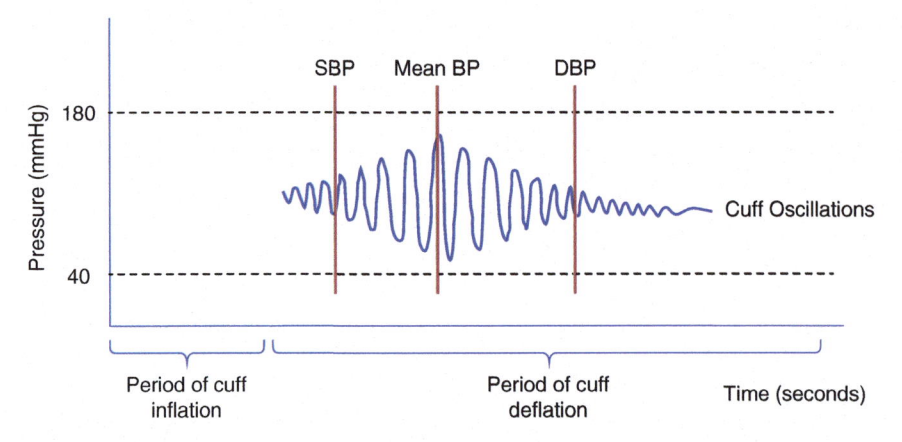

Fig. 2.6 Pressure trace showing change in amplitude of oscillometric pulses over the period of inflation and deflation of the cuff using an oscillometric device

either partially or fully occluded, pressure oscillations are detected in the cuff with the expansion of the arterial wall at each heartbeat (Figs. 2.5 and 2.6). The pressure oscillations are very small at cuff pressures greater than systolic BP but will get gradually larger when reaching MAP and then decrease again towards DBP. If pressure in the cuff is reduced in a stepwise manner, pressure is held static for a minimum of usually two heartbeats at each step allowing the device to record oscillometric pulses. In the continuous linear method, oscillometric pulses are recorded with each heartbeat. Once the recording of oscillometric impulses is complete, a curve can be fitted with the maximum point relating to the MAP. Algorithms specific to each manufacturer and machine are then used to calculate the SBP and DBP. These are based either on percentage of maximum amplitude or on the slopes or a combination of both. However, it should be recognised that SBP and DBP values are calculated extrapolations rather than true measurements of these values. The validity of measurements is therefore dependent on the reliability of the algorithms that are used in these machines and their validation for the population of interest (Babbs 2012; Alpert et al. 2014). For this reason, oscillometric machines that are designed for use in humans may not provide accurate or reliable assessment of BP in veterinary patients, particularly when the variability and higher heart rates are considered, e.g. in the cat. Ideally, where an oscillometric machine is used, it should be one that is specifically designed for the veterinary market. The initial point for pressure inflation of the cuff is usually set for oscillometric machines. For individuals that have particularly high SBP, the machine may fail to obtain a reading in the first instance if the cuff does not automatically inflate to a suprasystolic pressure. In some machines, this triggers an automatic reinflation procedure to a higher starting cuff pressure, whilst in other machines, it is possible to set the initial cuff inflation pressure if it is anticipated that a patient will have systolic pressures higher than anticipated.

There can be considerable impact of movement on BP measurement when using an oscillometric machine, which can make their use in conscious veterinary patients challenging. Although some of the modern devices (e.g. high-definition oscillometry) may allow for faster cuff inflation and deflation, it can take several minutes to obtain a single reading with older devices, and slow measurement can be a considerable contributor to measurement failure (Jepson et al. 2005; Wernick et al. 2012). Accurate analysis of small oscillations within the cuff requires the patient to remain very still as any movement, including muscle contraction, limb movement and trembling, can impair the ability of the oscillometric machine to obtain a reliable measurement (Jepson et al. 2005). Low-signal strength may also be a problem for small patients, those with poor peripheral pulses, e.g. hypotension and shock or arrhythmias. In addition, anatomic variability affecting arterial occlusion (e.g. chondrodystrophic breeds) and patients with peripheral oedema (e.g. nephrotic syndrome) present challenges for measurement of oscillometric pulse variation, which can lead to unreliable measurements or difficulties obtaining measurements with these devices.

2.6.1 Oscillometric Measurement in Dogs and Cats

There have been relatively few studies that have specifically evaluated performance of oscillometric machines comparing with direct arterial measurement in conscious dogs and cats (Table 2.3) (Haberman et al. 2004; Vachon et al. 2014; Bodey et al. 1996; Stepien et al. 2003; Chalifoux et al. 1985; Stepien and Rapoport 1999). However, collectively the published studies indicate limitations in terms of accuracy associated with direct arterial measurements and, irrespective of cuff location, a tendency towards underestimation in healthy dogs which becomes greater towards higher systolic pressures and with higher heart rates, thus raising potential concern for the identification of hypertension (Bodey et al. 1996). Further studies have evaluated oscillometric techniques in anaesthetised dogs, and compared to direct arterial BP measurements, again results are conflicting in terms of accuracy (Vachon

Table 2.3 Published studies comparing indirect blood pressure devices in *conscious* cats and dogs with direct arterial measurements

Study	Blood pressure evaluated	Population	Reference method	Indirect method (cuff position)	Blood pressure range explored	Correlation coefficient [r or r (O'Brien et al. 2010)]	Bias and limits of agreement (LOA)	% within 5 mmHg	% within 10 mmHg	% within 15 mmHg	% within 20 mmHg
Cats											
Haberman et al. (2004)		13 cats (8 healthy and 5 with reduced renal mass) simultaneous readings	Radiotelemetry	**Oscillometric Dinamap 8300 Critikon**	Atenolol and hydralazine given to reduce BP						
	Systolic	127 readings		Coccygeal	107–186	0.324	Difference direct–indirect 17.1 ± SD 29.8	11	26.8	ND	ND
	Mean	127 readings		Coccygeal	80–162	0.349	Difference direct–indirect 25.6 ± SD 29.8	6.3	12.6	ND	ND
	Diastolic	127 readings		Coccygeal	71–147	0.346	Difference direct–indirect 25.6 ± SD 22.5	7.1	12.6	ND	ND -
	Systolic	123 readings		Median	104–209	0.258	Difference direct–indirect 15.2 ± SD 28.2	14.6	27.6	ND	ND
	Mean	123 readings		Median	80–184	0.315	Difference direct–indirect 16.2 ± SD 28.4	8.1	16.3	ND	ND
	Diastolic	123 readings		Median	59–159	0.295	Difference direct–indirect 11.5 ± SD 28.3	9.8	25.2	ND	ND
				Doppler (811) Parks Medical Electronics	Atenolol and hydralazine given to reduce BP						
	Systolic	130 readings		Median		0.824	Difference direct–indirect 23.1 ± SD 19.4	7.7	19.2	ND	ND

Martel et al. (2013)		6 young cats	Radiotelemetry	HDO S + B MedVET	Amlodipine and phenylephrine administered				
	Systolic	25 readings		Coccygeal	~90–190	0.92	Bias: −2.2 ± 1.1, LOA: 21.6 ± 2.6	ND	88	92	ND
	Diastolic	25 readings		Coccygeal	~45–85	0.81	Bias: 22.3 ± 1.6 LOA: 22.6 ± 1.9	ND	13	38	ND
Dogs										
Chalifoux et al. (1985)		12 young dogs	Arterial catheterisation (femoral and carotid)	Infrasonde Oscillometric D4000					
	Systolic			Tibial artery	Hypotensive 128/88/101	0.76	ND	ND	ND	ND
	Diastolic			Tibial artery	Hypotensive 128/88/101	0.53	ND	ND	ND	ND
	Systolic			Tibial artery	Normotensive 154/96/115	0.86	ND	ND	ND	ND
	Diastolic			Tibial artery	Normotensive 154/96/115	0.51	ND	ND	ND	ND
	Systolic			Tibial artery	Hypertensive 197/124/149	0.32	ND	ND	ND	ND
	Diastolic			Tibial artery	Hypertensive 197/124/149	0.68	ND	ND	ND	ND
				Doppler (811) Parks Medical Electronics			ND			
	Systolic			Left ulnar	Hypotensive 128/88/101	0.52	ND	ND	ND	ND
	Systolic			Left ulnar	Normotensive 154/96/115	0.64	ND	ND	ND	ND
	Systolic			Left ulnar	Hypertensive 197/124/149	0.76	ND	ND	ND	ND

(continued)

Table 2.3 (continued)

Study	Population	Blood pressure evaluated		Reference method	Indirect method (cuff position)	Blood pressure range explored	Correlation coefficient [r or r (O'Brien et al. 2010)]	Bias and limits of agreement (LOA)	% within 5 mmHg	% within 10 mmHg	% within 15 mmHg	% within 20 mmHg
Bodey et al. (1996)	7 dogs			Radiotelemetry	Oscillometric Dinamap 1846SX Critikon (5 cuff positions evaluated; tail, distal HL, proximal HL, distal FL, proximal FL—for each parameter the best is presented)		Best option presented	Best option presented				
		Systolic	Standing			Approx 130–213	Prox FL: 0.727	Prox FL bias: 21.0	ND	ND	ND	ND
		Mean	Standing			Approx 80–125	Prox FL: 0.649	Tail: Bias: 8.1	ND	ND	ND	ND
		Diastolic	Standing			Approx 60–110	Tail: 0.576	Prox HL bias: 5.9	ND	ND	ND	ND
		Systolic	Lateral			Approx 140–245	Prox FL: 0.909	Dist HL bias: 30.8	ND	ND	ND	ND
		Mean	Lateral			Approx 90–139	Tail: 0.699	Tail bias: 1.5	ND	ND	ND	ND
		Diastolic	Lateral			Approx 70–98	Tail: 0.570	Dist HL bias: 1.2	ND	ND	ND	ND
Stepien and Rapoport (1999)	28 dogs			Direct arterial (femoral)	Oscillometric machine (device not identified)							
		Systolic	Lateral		Metatarsal	112–196	0.08	ND	ND	37	ND	59
		Mean	Lateral		Metatarsal	85–128	0.16	ND	ND	59	ND	89
		Diastolic	Lateral		Metatarsal	66–98	0.08	ND	ND	41	ND	70
		Systolic	Lateral		Doppler (811) Parks Medical Electronics			ND	ND		ND	
		Systolic			Metatarsal	112–196	0.12	ND	ND	48	ND	70
Haberman et al. (2006)	6 healthy beagles and 6 beagles post-renal mass reduction			Radiotelemetry	Dinamap 8300, Critikon	Hydralazine, atenolol and phenylephrine administered						
	Standing (Pavlovian sling)	Systolic			Tibial artery	106–222	0.251	ND	ND		ND	ND
	90 readings											

	Mean	Standing (Pavlovian sling) 90 readings		Tibial artery	76–170	0.342	ND	ND	ND	ND
	Diastolic	Standing (Pavlovian sling) 90 readings		Tibial artery	56–152	0.422	ND	ND	ND	ND
	Systolic	Standing (Pavlovian sling) 174 readings		Coccygeal	108–268	0.786	Bias: 20.0 ± 18.2	14.4	29.9	ND	ND
	Mean	Standing (Pavlovian sling) 174 readings		Coccygeal	65–190	0.740	Bias: 10.8 ± 15.6	24.7	48.9	ND	ND
	Diastolic	Standing (Pavlovian sling) 174 readings		Coccygeal	52–163	0.720	Bias: 5.3 ± 15.3	25.9	56.9	ND	ND
				Doppler (811) Parks Medical Electronics							
	Systolic	Standing (Pavlovian sling) 270 readings		Metatarsal	87–199	0.753	Bias: 11.6 ± 16.7	21	43	ND	ND
Meyer et al. (2010)		6 male dogs (686 readings)	Radiotelemetry		Torcetrapib administered						
	Systolic	Standing		Coccygeal HDO S + B MedVET			Bias: 10.4 LOA: −21.4-43.8	25	46	66	ND
	Mean	Standing		Coccygeal			Bias: 1.9 LOA: −27.5-33.0	25	46	66	ND
	Diastolic	Standing		Coccygeal			Bias: 5.7 LOA: −26.8-39.6	25	46	66	ND
Mitchell et al. (2010)		6 dogs	Radiotelemetry	HDO S + B MedVET	Hexamethonium administered						
	Systolic	Prone		Coccygeal		0.60	ND	ND	ND	ND	ND
	Mean	Prone		Coccygeal		0.75	ND	ND	ND	ND	ND

(continued)

Table 2.3 (continued)

Study	Blood pressure evaluated	Population	Reference method	Indirect method (cuff position)	Blood pressure range explored	Correlation coefficient [r or r (O'Brien et al. 2010)]	Bias and limits of agreement (LOA)	% within 5 mmHg	% within 10 mmHg	% within 15 mmHg	% within 20 mmHg
	Diastolic	Prone		Coccygeal		0.71	ND	ND	ND	ND	ND
				Cardell, Veterinary Monitor 9401							
	Systolic	Prone		Coccygeal		0.73	ND	ND	ND	ND	ND
	Mean	Prone		Coccygeal		0.71	ND	ND	ND	ND	ND
	Diastolic	Prone		Coccygeal		0.56	ND	ND	ND	ND	ND
Bosiak et al. (2010)		11 dogs	Direct arterial (pedal artery)	**Datascope passport, Datascope corp.**							
	Systolic	Lateral (n = 26 readings) (n = 33 readings)		Hindlimb	Hypotension 104 mmHg (SD 21.8)	0.5	Bias: −15.3	ND	ND	ND	ND
					Normotension 149.9 mmHg (SD 31.8)	0.17	Bias: 14.6				
					Hypertension 159.3 mmHg (SD 26.1)	−0.34	Bias: −3.9				
	Mean	Lateral (n = 26 readings) (n = 33 readings)		Hindlimb	Hypotension 72.5 mmHg (SD 8.6)	0.75	Bias: −12.1	ND	ND	ND	ND
					Normotension 89.8 mmHg (SD 7.6)	0.3	Bias: −5.2				
					Hypertension 108.2 mmHg (7.0)	0.81	Bias: −17.2				
	Diastolic	Lateral (n = 26 readings) (n = 33 readings)		Hindlimb	Hypotension 56.0 mmHg (SD 4.7)	0.84	Bias: −6.4	ND	ND	ND	ND
						0.21	Bias: 1.6				

				Normotension 69.3 mmHg (SD 5.9)	−0.28	Bias: −3.5		
				Hypertension 86.4 mmHg (SD 6.8)				
		Cardell, Veterinary Monitor 9401						
Systolic	Lateral (n = 31 readings) (n = 81 readings) (n = 33 readings)	Hindlimb		Hypotension (99 mmHg SD 16.7)	0.97	Bias: −22.5	ND	ND	ND
				Normotension 148.5 mmHg (SD 31.2)	0.47	Bias: 15.9		
				Hypertension 158.4 mmHg (SD 28.9)	0.22	Bias: 3.0		
Mean	Lateral (n = 31 readings) (n = 81 readings) (n = 33 readings)	Hindlimb		Hypotension 71.5 mmHg (SD 6.6)	0.96	Bias: −12.6	ND	ND	ND
				Normotension 89.7 mmHg (SD 7.0)	0.60	Bias: −1.3		
				Hypertension 106.7 mmHg (SD 5.8)	0.73	Bias: −5.8		
Diastolic	Lateral (n = 31 readings) (n = 81 readings) (n = 33 readings)	Hindlimb		Hypotension 55.8 mmHg (SD 2.6)	0.86	Bias: −15	ND	ND	ND
				Normotension 68.6 mmHg (SD 7.4)	0.57	Bias: −4.6		
				Hypertension 84.3 mmHg (SD 5.5)	0.40	Bias: −12.1		

(continued)

Table 2.3 (continued)

Study	Blood pressure evaluated	Population	Reference method	Indirect method (cuff position)	Blood pressure range explored	Correlation coefficient [r or r (O'Brien et al. 2010)]	Bias and limits of agreement (LOA)	% within 5 mmHg	% within 10 mmHg	% within 15 mmHg	% within 20 mmHg
				Doppler 811-B (Parks Medical Electronics)				ND	ND	ND	ND
	Systolic	Lateral		Hindlimb	Hypotension 104.3 mmHg (SD 18.0)	0.63	Bias: −16.6				
					Normotension 146.3 mmHg (SD 33.2)	−0.16	Bias: 16.1				
					Hypertension 164.4 mmHg (SD 26.4)	−0.12	Bias 8.2				
Vachon et al. (2014)		10 dogs (female) 50 paired readings/dog/device	Arterial catheterisation (dorsal pedal)	**Oscillometric petMAP, RAMSEY Medical Inc.**							
	Systolic	Lateral		Forelimb	112–189	ND	Bias: −5.8 LOA: −45.7–34.1	ND	50	ND	70
	Mean	Lateral		Forelimb	71–119	ND	Bias: 3.8 LOA: −12–19.9	ND	ND	ND	ND
	Diastolic	Lateral		Forelimb	55–94	ND	Bias: 4.6 LOA: −9.2–18.4	ND	70	ND	100
				Doppler 811-B (Parks Medical Electronics)							
	Systolic	Lateral		Forelimb	102–183	ND	Bias: −15.1 LOA: −64.7–34.5	ND	20	ND	40

The bold is indicating the machine that was used to measure SBP—all of the readings underneath each bold machine have been taken using the same machine
BP Blood pressure, *LOA* limits of agreement, *FL* forelimb, *HL* hindlimb, *prox* proximal, *dist* distal, *SD* standard deviation, *difference* difference between direct arterial and indirect measurement, *ND* not determined

et al. 2014; Sawyer et al. 1991; Seliškar et al. 2013; Garofalo et al. 2012; Bodey et al. 1994; Rysnik et al. 2013). Irrespective of device, studies which compare indirect and direct measurements in anaesthetised patients should be interpreted with caution in the context of diagnosis of hypertension as, ultimately, the diagnosis of hypertension is made in the conscious patient, and therefore, a device that performs well in the conscious patient is paramount. Some difference in measurements between any direct and indirect technique may not be unexpected; given that there are differences in the artery used, it is not possible to obtain absolutely simultaneous measurements, and there can be some second by second change in direct arterial measurements particularly associated with catheter placement (Bodey et al. 1996).

In general, in conscious patients, oscillometric diastolic readings are considered less reliable than systolic measurements; although given that there is currently no recognised clinical significance of diastolic hypertension in cats and dogs, this is of limited concern (Stepien and Rapoport 1999). Clinical studies in conscious dogs have also evaluated precision via coefficient of variation (CV) between series of readings using the oscillometric technique. In many instances, such studies suggest that the CVs may be larger than with other indirect techniques (e.g. Doppler or high-definition oscillometry) ranging between 9.0 and 30% for systolic measurements (Chalifoux et al. 1985; Hsiang et al. 2008; Rattez et al. 2010). It is possible that this reflects the extended period of time that it can take to obtain serial BP measurements using oscillometric devices.

Oscillometric devices have been used historically in cats although fewer studies have made comparisons between direct and indirect methodologies in conscious cats (Haberman et al. 2004). Results were similar to those in the dog suggesting weaker correlation than other indirect methodologies (Doppler technique) and a tendency towards underestimation of systolic BP (Haberman et al. 2004). Studies performed in anaesthetised cats with comparison to direct arterial measurement have shown better results in terms of accuracy and precision (Cerejo et al. 2017; Branson et al. 1997; Zwijnenberg et al. 2011; Binns et al. 1995), but overall there are concerns, particularly in the conscious cat, about the ability of standard oscillometric devices to identify cats with systemic hypertension due to the risk of underestimation of systolic BP particularly at higher ranges (Jepson et al. 2005). The time to obtain readings and 'failed' attempts may also be a cause of frustration when trying to use standard oscillometric devices in cats (Jepson et al. 2005).

2.6.2 High-Definition Oscillometry in Dogs and Cats

The high-definition oscillometry (HDO) is a veterinary-specific device manufactured by S + B medVET (Vet-HDO®-Monitor). Like other oscillometric devices, measurement is based on placement of a cuff which occludes a peripheral artery. Measurements of SBP, MAP and DBP are extrapolated from the analysis of waveforms of the recorded pressure oscillations within the cuff. The machine is considered superior based on real-time analysis of the pulse waveform (10,000 times per second) detected during cuff deflation giving improved validity of measurements. Deflation of the cuff is linear rather than stepwise which may also increase the sensitivity of the device to detect pressure oscillations, and the process of measurement is much faster (8–15 s), therefore typically better tolerated by dogs and cats. It has also been designed to cope with the greater range of heart rates that are appreciated in small veterinary patients. Although there are limited data relating to clinical application, the device offers the theoretical possibility to perform pulse wave analysis allowing assessment of systemic vascular resistance, stroke volume and stroke volume variance.

Clinical studies have explored the use of the HDO device in both the conscious dog and cat. In experimental studies involving conscious dogs, correlation of the HDO with telemetry measurements has been on average, for example, correlation coefficient 0.6 for systolic BP within normotensive ranges (Mitchell et al. 2010). In conscious dogs, Bland-Altman analysis of systolic pressures has demonstrated significant differences when compared to telemetric measurements (Table 2.3) (Mitchell et al. 2010; Meyer et al. 2010). Three studies have compared HDO and direct measurement in anaesthetised dogs (Seliškar et al. 2013; Rysnik et al. 2013; Wernick et al. 2010). In two of these studies, 36% and 70% versus 52.4 and 82.5% of systolic readings were within 10 and 20 mmHg, respectively, of the direct systolic measurement (Seliškar et al. 2013; Wernick et al. 2010). However, in all studies there were still concerns for bias from the Bland-Altman analysis particularly for systolic BP, the measurement of most interest for the diagnosis of systemic hypertension, and the HDO device, therefore, has not yet fulfilled the ACVIM validation criteria for either conscious or anaesthetised dogs, despite most likely performing better than some of the traditional oscillometric devices (Seliškar et al. 2013; Rysnik et al. 2013; Wernick et al. 2010).

In conscious cats, only one study has compared the HDO device with radiotelemetric BP monitoring (Table 2.3) (Martel et al. 2013). In this study, medical intervention (amlodipine or phenylephrine) was administered in order to generate a range of potential BP for measurement. Linear regression correlations for systolic BP obtained by the HDO and radiotelemetry were very good (correlation coefficient $r = 0.98$) with 88% and 96% of measurements being within 10 and 20 mmHg compared to the reference telemetry method. The HDO device had a bias towards overestimation of SBP in hypotensive conditions (-10.8 ± 11.3 mmHg) suggesting that it may make identification of hypotension difficult. But the bias for normal and high BP groups were $- 3.4 \pm 3.8$ and 2.9 ± 1.3 mmHg with limits of agreement across all systolic BP categories being 21.6 ± 2.2 mmHg (Martel et al. 2013). Studies in clinical patients suggest that the HDO machine is typically well tolerated in cats for use on both the coccygeal and radial artery but that tail measurements may be better tolerated (Cannon and Brett 2012). In anaesthetised cats, the HDO device was able to obtain readings on 90% of occasions; when compared to the Doppler technique (100%), the readings were precise (coefficient of variance for systolic BP 2.77 ± 2.96 mmHg $n = 543$ readings) (Petrič et al. 2010). However, in this study of anaesthetised cats, tail measurements were inconsistent and in some instances not attainable, particularly in cats with narrow tails, leading the authors to prefer use of a forelimb (Petrič et al. 2010). Furthermore, Bland-Altman evaluation raised concern with the HDO device for overestimation of low systolic BP and underestimation of high systolic BP when compared to the Doppler technique (Petrič et al. 2010). Nevertheless, the HDO device has been used for measurement of systolic BP in the European pivotal clinical trial supporting the product authorisation of amlodipine besylate (Huhtinen et al. 2015). Furthermore, when using the HDO device, ocular hypertensive injury was detected in cats where systolic BP was mostly >160 mmHg (Carter et al. 2014). However, in a clinical field study evaluating hypertensive ocular injury, it was not possible to obtain BP readings using the HDO device in 32/105 cats; for 17 cats, this was due to failure to read BP, whilst for 15 cats, it was because the readings lacked precision and were too widely spread for evaluation (Carter et al. 2014). The HDO device therefore offers some clear advantages to traditional oscillometric devices, however has yet to achieve the criteria that would be required for ACVIM validation of a BP monitoring device. Nevertheless, the HDO device is widely used in clinical practice for BP measurement in both dogs and cats.

2.7 Doppler Sphygmomanometry

Doppler sphygmomanometry has historically been the most widely used methodology both in clinical practice and in a research setting for evaluation of BP and diagnosis of systemic hypertension. Doppler sphygmomanometry involves placement of a BP cuff for arterial occlusion and is dependent on the detection of the Korotkoff sounds. High-frequency sounds emanate from the resonant motion of the arterial wall during stretching and relaxation of arterial pulsations. When a cuff is inflated to suprasystolic pressures, the arterial blood flow is completely occluded and the sounds disappear. Systolic BP equates to either just above or the point at which the Korotkoff sounds return, whilst disappearance of the arterial sounds equates to at or just above the level of diastolic pressure (Babbs 2015). Korotkoff described five phases of sounds associated with arterial sounds: K1, appearance; K2, softening; K3, sharpening; K4, muffling; and K5, disappearance. It is speculated that these arterial sounds relate to changes in net transmural pressure, but the exact biophysical mechanisms by which these audible changes occur remain poorly understood (Babbs 2015).

Detection of Korotkoff sounds or their loss can be performed in human subjects using an auscultatory method, i.e. stethoscope placed distal to the cuff on the antebrachium. However, dogs and cats are too small for this approach, and instead, a piezoelectric crystal is used in a Doppler probe in order to detect these audible changes. Whilst detecting the first appearance of Korotkoff sounds is relatively easy and therefore reliable in terms of identification of a SBP reading, detecting the muffling or change in sound with the onset of DBP is challenging and lacks both precision and repeatability (Gouni et al. 2015). Therefore, in veterinary medicine, DBP measurements are typically not attempted using the Doppler technique, particularly in conscious patients, and are not recommended. Currently, clinical associations between diastolic hypertension and target organ damage (TOD) have not been identified in veterinary medicine, and so the inability to evaluate DBP using the Doppler technique is not of great clinical concern. However, this does mean that the use of the Doppler technique is constrained to evaluation of SBP without simultaneous assessment of either DBP or MAP which is obtained by oscillometric and HDO devices.

Precision with the Doppler technique in patients that are anaesthetised has been reported to be very high (coefficient of variance ~1.5%) (Seliškar et al. 2013) in some studies but worse in others (CVs 17.3%) (Chalifoux et al. 1985). The Doppler technique is considered reliable in terms of being able to obtain systolic BP measurements in 100% of recordings when experienced operators are using the machine both in conscious dogs and cats (Petrič et al. 2010; Binns et al. 1995; Chetboul et al. 2010) although it potentially takes longer to gain experience with this machine than with some of the more automated oscillometric devices. This could influence repeatability of obtaining measurements in the inexperienced operator (Gouni et al. 2015). Nevertheless, the Doppler technique allows very rapid collection of sequential measurements with studies suggesting that it takes approximately 4–6 min to obtain a series of 5 readings (Jepson et al. 2005; Cannon and Brett 2012). Using the Doppler technique, coefficient of variation for within and between day measurements between observers in conscious dogs has been shown to be variable (3.1–15.4%) although overall acceptable when compared to other devices (Wernick et al. 2012; Chalifoux et al. 1985; Hsiang et al. 2008; Chetboul et al. 2010).

However, there are concerns relating to accuracy of the Doppler technique in both conscious and anaesthetised patients (Table 2.3). In anaesthetised dogs, when compared to direct arterial BP measurements, only 10% and 34% of systolic BP readings were reported to lie within 10 and 20 mmHg of the direct arterial measurements (Vachon et al. 2014; Seliškar et al. 2013). Doppler measurements have been reported to overestimate systolic readings in hypotensive dogs and underestimate for hypertensive dogs and to have poor agreement with systolic BP measurements >140 mmHg in anaesthetised dogs, particularly those of small body size (Bosiack et al. 2010; Garofalo et al. 2012; Moll et al. 2018). In anaesthetised cats, associations with direct arterial systolic BP measurements have also been highly variable. One study, with SBP ranging between 50 and 200 mmHg, reported correlation coefficients of 0.95 suggesting excellent correlation (Binns et al. 1995). However, other studies in anaesthetised cats suggest that, although there can be good correlation, there can simultaneously be significant negative bias and poor accuracy (Caulkett et al. 1998). Results of a more recent study where Bland-Altman analysis was performed comparing direct arterial measurements in anaesthetised cats indicated very poor agreement between direct SBP and Doppler measurements (Da Cunha et al. 2014). The Doppler technique underestimated true systolic BP measurements with a significant bias of -8.8 mmHg and limits of agreement between -39 and 21 mmHg. The negative bias also became larger with increasing SBP indicating that it would become harder to identify hypertension had the Doppler device been the sole method of BP assessment (Da Cunha et al. 2014). Some studies have suggested that correction factors could be applied to Doppler measurements for improved association with true SBP values. However, the application of such a formula relies on any bias being consistent across a range of BP values, and this does not seem to always be the case (Da Cunha et al. 2014; Grandy et al. 1992).

The Doppler technique therefore does not fulfil the criteria for the ACVIM BP machine validation guidelines. However, despite these obvious limitations, the Doppler methodology has been widely used for clinically applicable research both in the dog and the cat, and clinical associations have been drawn between the presence of hypertensive target organ damage and systolic measurements performed with the Doppler technique. It therefore remains one of the most commonly used methods for assessment of SBP in small animal practice despite the limitations discussed.

2.8 Cuffless Blood Pressure Monitoring

Historically and also more recently in human medicine, there has been interest in 'cuffless' BP monitoring devices. Ahlstrom and colleagues proposed the use of pulse transit time (PTT) to evaluate SBP (Ahlstrom et al. 2005). PTT is defined as the time from the end of left ventricular systole to reach the position of a specific peripheral artery. This methodology therefore requires at least two sensors: one on the chest and one on either the arm or leg. Some studies have used a combination of electrocardiography (ECG) and photoplethysmography (PPG), whilst others have used phonocardiography replacing the ECG. Photoplethysmography is a noninvasive optical measurement detecting changes in blood volume in the peripheral circulation, similar to technology used for evaluation of fingertip oxygen saturation monitors. When PTT is used together with PPG to estimate BP, PTT is calculated from the time delay between the R-wave of the ECG and the arrival of the pulse wave in the periphery assessed by PPG. The usual conformation is that the ECG is placed on the chest, whilst PPG is measured on the index finger. There is a negative relationship between PTT and BP such that BP can be calculated from PTT measurements.

Photoplethysmography can also be used alone to estimate BP. Here the PPG sensor is used to detect changes in blood volume in an arterial segment (typically the fingertip). A photoplethysmography unit contains both a light source and a photodetector unit. The light source transmits infrared LED or green LED which transmits through or reflects from the skin and blood cells and with either transmitted (transmitted photoplethysmography; TPPG) or reflected (reflected photoplethysmography; RPPG) light being detected by a photodiode. Red blood cells absorb more light intensity than body tissue; therefore, the red blood cells can be differentiated easily by observing the degree of reflected or transmitted light. A photoplethysmography signal is derived as a voltage signal generated by a photodetector for light with changes in the flow of blood. The alternating current component of the recording corresponds to pulsatile blood volume, whilst the direct current component corresponds to static blood flow. Systolic and diastolic measurements are then extrapolated from the wave forms that are generated (Taha et al. 2017).

Impedance plethysmography (IPG) is a more recently developed technique for cuffless BP measurement that measures the ionic conduction of a specific body segment contrasting it to electrical conduction characteristics. The electrical conductivity depends on the conditions of arteriovenous blood volume within a body segment. Thus, it can be used to detect changes in haemodynamic parameters such as blood flow and stroke volume which can then be correlated with BP measurements (Liu et al. 2017). To overcome the difficulties that are faced in human medicine between intermittent and hospital-acquired measurements of BP versus continuous BP monitoring, implantable continuous BP monitoring devices based on impedance plethysmography are also being developed (Theodor et al. 2014).

There has been little exploration of the use of PTT, PPG or IPG for use in veterinary patients although some of the experimental literature developing these devices has been performed in dogs and cats (Binns et al. 1995; Caulkett et al. 1998; Jeong et al. 2010; Heishima et al. 2016). These methodologies are not currently used in clinical practice.

2.9 Practical Aspects of Blood Pressure Measurement in the Clinic

In order to make an accurate assessment of BP, practical aspects of BP measurement should be considered. Notwithstanding the limitations in the indirect BP machines that are currently available, the manner in which a BP measurement is obtained can have substantial impact on the reliability of the measurements that are obtained. Every attempt should therefore be made to obtain BP measurements in a standardised manner with due consideration for the protocols that are being used. Factors associated with the measurement of BP which may impact are considered.

2.9.1 Location for Performing Blood Pressure Measurement and Acclimatisation

Given the potential impact of environment on BP, it is important to ensure that all BP measurements, irrespective of device, are performed in a quiet environment away from noise and disturbance. When possible, performing BP measurement with the owner present is preferable in order to reduce anxiety. Therefore, obtaining BP readings in a consultation environment prior to admitting patients to the hospital can be beneficial or, for example, making use of nurse-led clinics designed specifically for this purpose. If a BP measurement is to be made within a consultation however, it is important to ensure that it is performed before any more invasive procedures, e.g. physical examination or blood sampling.

It has been widely accepted that a period of acclimatisation is important prior to measurement of BP. This acclimatisation period allows the dog or cat to relax helping to reduce artefactual increases in BP and situational hypertension. Early studies in healthy cats with radiotelemetric devices that were taken on simulated visits to the veterinary clinic identified the importance of this acclimatisation period, as well as the marked variation in response between different cats to the clinic environment (Belew et al. 1999). The mean increase in systolic BP identified was 17.6 mmHg ±1.6 mmHg, but the range was between 75 and −27 mmHg for individual cats. This variability was also larger for cats with underlying renal disease when compared to the healthy normal cats (Belew et al. 1999). Provision of a 10 min acclimatisation period for cats has been shown to result in a significant reduction in systolic BP measurements, and this period of time is therefore often considered 'gold standard' for acclimatisation both in dogs and cats (Sparkes et al. 1999). Studies also support that situational hypertension may occur in the dog, and again that individual responses are very variable (Bodey and Michell 1997). From historical data, there is limited evidence that dogs and cats become habituated to BP measurement in the clinic (Belew et al. 1999; Bodey and Michell 1997) although it could be speculated that some of the placebo effect identified in recent antihypertensive studies could relate to repeated visits for BP assessment (Huhtinen et al. 2015). Due to the heterogeneity in response between individuals, serial assessment and repeated measurements are important as part of the diagnosis of systemic hypertension to ensure that readings are a true reflection of that individual's BP. Unfortunately, the clinical appearance of anxiety in an individual dog or cat often does not translate into the degree to which they may experience situational hypertension, and parameters such as heart rate do not perform well as indicators (Belew et al. 1999; Bodey and Michell 1997; Marino et al. 2011).

In human medicine, ambulatory BP measurement and home BP measurement are often used as part of the diagnostic route for ascertaining whether an individual has genuine versus situational hypertension. The option to perform home BP measurements should certainly be considered for dogs and cats where there is a clinical question about the possibility of situational hypertension as studies comparing home and clinic BP measurement have shown that systolic BP is typically lower in the home environment (Marino et al. 2011; Bragg et al. 2015; Kallet et al. 1997). However, the time and cost associated with such measurements are often prohibitive.

2.9.2 Obtaining Multiple Measurements

The ACVIM hypertension consensus recommends that the first reading should be discarded and that a series of 5–7 BP readings should then be obtained and the arithmetic mean calculated in order to obtain an average systolic BP. It is advocated that all measurements used for this average BP should be within 20 mmHg. However, there is a caveat placed on this process, recognising that in some patients BP measurements may begin to fall as the patient acclimatises to the process of BP measurement and that conversely anxiety may increase along with BP measurements in others. Therefore, in some patients, it can be beneficial to continue measuring BP for >5 readings in total until the measurements reach a plateau, and five consecutive readings should then be obtained which can be considered clinically the most valid to be averaged and recorded for that patient. Using the Doppler technique, the first reading has been shown to be a true indication of the subsequent average of five readings in cats although this was not the case for systolic BP obtained using an oscillometric technique in that study (Jepson et al. 2005). In dogs, repeated BP measurements improved precision and accuracy when comparing an oscillometric device to direct arterial measurements (Bodey et al. 1996). Therefore, the recommendation remains that between 5 and 7 readings should ideally be obtained and averaged irrespective of device.

2.9.3 Timing of Assessment of Blood Pressure

In human subjects, there is a marked diurnal variation in BP, with BP naturally falling during periods of sleep and a sharp increase around the time of wakening. This early morning surge in BP has been associated with an increased risk of cardiovascular events. Studies typically performed in experimental settings have explored diurnal variation in BP in both healthy dogs and cats. Studies in cats and dogs have given support for diurnal variation when radiotelemetric devices have been used for continuous BP evaluation (Mishina et al. 1999, 2006; Brown et al. 1997). However, in some studies, increasing BP has been associated with activity, whilst in dogs temporal association with feeding has also been noted (Mishina et al. 2006; Mochel et al. 2014). However, the variation is typically small, and therefore, in comparison to the inaccuracies associated with indirect BP measurement, timing is unlikely to play a major role in BP measurements for the diagnosis of systemic hypertension. A caveat to this however is logical consideration of timing of BP measurement for hospitalised patients to ensure that the ward environment is quiet. Busy ward environments may result in situational hypertension in some individuals, and measuring BP at such times should be avoided where possible.

2.9.4 Position of the Patient for Blood Pressure Measurement

It has been recognised for many years that the position of the patient and more specifically the location of the cuff in relation to the heart and the distance from the heart to the site of BP measurement will have effects on the BP readings obtained. In general, if the cuff is placed in a position above the heart then values that are lower than true BP will be obtained, whilst if the cuff is placed below the level of the heart higher values will be obtained. This has important implications for veterinary medicine particularly when considering whether patients are positioned in sternal or lateral recumbency or remain standing for BP measurement and also in relation to position of cuff placement discussed further.

Early work by Bodey and colleagues compared direct arterial BP measurements in dogs that were either recumbent (lateral) or standing with indirect oscillometric measurements (Bodey et al. 1996). They recognised that all indirect measurements were more accurate in recumbent dogs than when standing (Bodey et al. 1996). More recent work has indicated that in conscious dogs SBP is significantly higher in dogs that are in a sitting position compared to lateral and that results are less variable when dogs are positioned in lateral recumbency (Rondeau et al. 2013). Based on these findings, many studies in dogs have opted for lateral recumbency for BP assessment. However, the study by Bodey also discussed the potential difficulties enforcing lateral recumbency on less cooperative dogs and that this process may in itself lead to situational hypertension and inaccurate BP readings (Bodey et al. 1996). Body position for BP measurement in conscious cats has not been explored, probably because of difficulties trying to ensure a cat adopts a given position. Most cats will adopt either a sternal, sitting or standing position for indirect BP measurement. Ultimately, for both dogs and cats, the optimal position is the one in which they are most comfortable and relaxed and permits BP measurement to be performed without undue anxiety. It is however very important that position is recorded as part of the clinical records to ensure that the conditions of measurement can be replicated in the future for consistency of readings.

2.9.5 Position of the Cuff for Blood Pressure Assessment

In dogs and cats, when indirect BP measurements with a cuff-based technique are performed, different sites can be used. For oscillometric machines, the following sites have been reported:

- Proximal forelimb with cuff above carpus and artery arrow medial
- Distal forelimb with cuff applied around metacarpus with the artery arrow palmar
- Proximal hindlimb with cuff above hock and artery arrow craniomedial
- Distal hindlimb with cuff around mid-metatarsus and artery arrow craniomedial
- Tail cuff placement with the artery arrow placed ventrally

For Doppler technique measurements, it is most common to use either distal forelimb/hindlimb or tail base measurements with the piezoelectric crystal placed on the palmar/plantar aspects of the foot, over the dorsal pedal artery of the hindlimb or the ventral aspect of the tail.

When the BP literature is reviewed, many studies use different cuff positions for indirect BP measurement, and some will use multiple cuff positions even within the same study protocol. Differences have been identified in studies evaluating the accuracy of indirect BP versus direct BP measurements; however, such studies have typically used alternate sites for direct versus indirect measurements making the assumption, for example, that measurement of direct BP from the hindlimb should be identical to that of the forelimb. True differences in BP measured from different body locations could have impact on studies that try to determine the accuracy of indirect BP measuring devices. Acierno and colleagues compared direct arterial BP measurements in anaesthetised dogs obtained from four different arterial locations (superficial palmar arch and the contralateral dorsal pedal artery or the superficial palmar arch and median sacral artery) with dogs in two positions (lateral or dorsal) (Acierno et al. 2015). This study identified that hindlimb direct arterial systolic measurements were a mean of 16 mmHg higher when the dogs were in dorsal recumbency and 14 mmHg higher when in lateral recumbency than the simultaneous systolic measurements obtained from the forelimb (Acierno et al. 2015). Interestingly, the mean differences between coccygeal versus carpal systolic measurements were less marked (3 mmHg higher in dorsal recumbency and 4.67 mmHg higher in lateral recumbency), and measurements of diastolic and mean BP were much similar between locations. However, collectively, these results indicate that we must be cautious when making comparisons between direct and indirect BP measurements where readings are obtained from different body locations and such differences could be contributing to the lack of accuracy seen in some studies between direct and indirect methodologies. Such data has not yet been explored in cat.

Overall, the results from studies that have compared cuff position using indirect devices have been very variable. An early study using a standard oscillometric device reported that for conscious standing dogs the proximal forelimb position gave the most accurate representation of systolic BP when compared to direct arterial pressures although for diastolic readings the tail was superior (Bodey et al. 1996). In the same study, the results for conscious dogs in lateral recumbency were similar between the tail and the limbs when using the oscillometric device and comparing to direct measurements (Bodey et al. 1996). However, other studies evaluating conscious dogs with direct arterial measurements have shown greater concordance with the hindlimb and tail for oscillometric devices and favoured the hindlimb for Doppler measurement (Bodey et al. 1994; Haberman et al. 2006). Whilst in anaesthetised dogs comparing direct BP with both Doppler and standard oscillometric techniques, hindlimb cuff placement has been favoured (Garofalo et al. 2012). In conscious dogs, where direct BP measurements are not available, marked variation has been reported between forelimb and hindlimb measurements obtained by the Doppler leading to concern that trends in the same animal should always be performed using the same location (Scansen et al. 2014).

Data relating to cuff position in cats is also highly variable. In anaesthetised cats, where comparisons were made with direct arterial BP measurement, the highest accuracy was obtained with the Doppler device on the tail although differences from the hindlimb were small (Binns et al. 1995). In the same study, the highest bias was obtained with the oscillometric machine irrespective of cuff location (limb or tail) (Binns et al. 1995). Conversely in a study of conscious cats with direct radiotelemetric implants for continuous direct BP evaluation, the median artery showed the highest correlation for the Doppler technique, whilst coccygeal and median artery were equivalent for the oscillometric measurement (Haberman et al. 2004). In a randomised controlled study of 27 conscious cats (Syme unpublished data), where SBP was measured using the Doppler machine, values were on average 15 mmHg higher using the coccygeal artery (139 ± 30 mmHg) than when taken from the forelimb (126 ± 19 mmHg).

In addition to accuracy of readings from particular sites in comparison to direct arterial measurements, functionality of a particular site for indirect BP measurement is also important in the clinical patient. In cats, a study specifically evaluating the HDO device compared clinical utility of the radial artery to the coccygeal artery but without direct arterial comparisons available (Cannon and Brett 2012). In this study, approximately 16% of attempts to measure BP failed when using the coccygeal artery compared to 48% with the radial artery. Mean time to obtain five readings with the coccygeal artery was 4.35 min (ranging from 2 to 13 min), whilst for the radial artery, it was 8.1 min (ranging from 2 to 28 min). The conclusion was that for the HDO device the coccygeal artery may be better tolerated and more reliable in terms of obtaining a reading (Cannon and Brett 2012). Contrary to this however, in a study evaluating anaesthetised cats using the HDO device, the tail was excluded as an option for measurement of BP because results were too discordant and difficult to obtain, and instead, the forelimb was used (Petrič et al. 2010).

Ultimately, in the clinical situation, the best location for cuff placement is the one that is best tolerated by that individual animal and where consistent readings can be obtained. What is important however is that the site of cuff placement is recorded as part of the clinical records so that, given that there may be variability between sites, the same location can be used when BP measurements must be repeated. In cats there is also debate relating to whether or not clipping influences BP measurements. A study evaluating anaesthetised cats identified no significant difference in measurements whether or not clipping was performed (Branson et al. 1997). Therefore, particularly in conscious cats, it is not advised to clip given that the noise associated may contribute to situational hypertension.

2.9.6 Size of Cuff for Blood Pressure Assessment

Cuffs for both oscillometric and Doppler BP measurements should be between 30 and 40% of the limb circumference at the level at which the cuff is placed (Chalifoux et al. 1985; Valtonen and Eriksson 1970). This should be assessed by using a dressmaker tape measure to accurately calculate the ideal cuff width. Although in cats it is common for just one of two sizes of cuff to be required, the width of cuffs used in dogs can be very variable and cannot necessarily be extrapolated directly from body size. If a machine is manufactured with specific cuffs, then it is important that these are used and not interchanged with cuffs from other devices. It is recognised that using a cuff that is too small will result in measurements that are falsely increased, whilst a cuff that is too large will lead to false reduction in BP measurement. Cuff size used should be routinely recorded for every BP measurement so that the same cuff can be used every time the patient's BP is evaluated.

2.9.7 Accurate Recording of Blood Pressure Measurement in Clinical Records

As has already been alluded to, recording of BP measurements is very important as part of clinical record keeping. Standardised forms can assist in ensuring that all relevant information relating to obtaining a BP measurement is collected. Typically, this should include date and time of measurement, attitude of the patient, medications the patient is receiving, position of the patient, location of measurement, method of BP measurement/device used, position of cuff placement, size of cuff, serial recording of a minimum of six readings within indication that the first was discarded, calculated mean of the remaining five, reason for exclusion of any other readings obtained and box for veterinary surgeon clinical interpretation. These records help to ensure that serial assessments of BP measurement can be performed by different operators but in an identical manner allowing trending of measurements to be performed. Whether or not headphones are used for measurement with the Doppler technique should also be reported, particularly if it is considered that an individual patient is particularly noise phobic. However, most dogs and cats tolerate Doppler measurement well without the use of headphones, and at least one study in cats suggests that their use has no impact on BP measurement (Williams et al. 2010).

2.10 Patient-Associated Factors Influencing Accuracy of Blood Pressure Measurement

A number of different patient-related factors may impact on ease and accuracy of obtaining an indirect BP measurement. Cardiac arrhythmias may result in beat to beat variation in blood pressure due to variation in cardiac filling. This may make it more challenging to obtain readings, particularly when using the oscillometric devices where variable heart rate may impact on detection of cuff oscillations. Motion (e.g. fidgeting and/or kneading in cats) will also have a large impact on an operator's ability to obtain a Doppler measurement and also on the ability of an oscillometric machine to make a recording. In particular, for the oscillometric machine, muscle trembling and movement may cause the machine to fail to obtain a reading and prolong the process of trying to obtain a series of five readings. Careful restraint of the patient by the owner or veterinary nurse can greatly reduce movement and help in successful BP measurement. It has also been recognised anecdotally that certain physical attributes may have an impact on cuff-based BP measurement. This includes any structural lesion that impairs cuff placement (e.g. mass lesion) and prevents artery occlusion or detection of arterial oscillations (e.g. peripheral oedema). It should be recognised that in these situations, accurate BP measurements may not be achieved and, in severe instances, readings may not be possible. Careful selection of an alternative location for BP measurement, e.g. coccygeal artery, may need to be considered.

Recently, there has also been interest in the effect of body condition and muscle condition on the measurement of BP in both dogs and cats (Lin et al. 2006; Whittemore et al. 2017; Mooney et al. 2017). Theoretically, in both cats and dogs, body and muscle condition could affect Doppler measurements due to interference from surrounding tissue. Studies that have explored the relationship between obesity and BP have typically not identified associations, despite the relationship that exists in humans (see Chap. 3). In cats, one study identified that systolic Doppler BP measurements were significantly and positively correlated with age and negatively with muscle condition, whilst coccygeal measurements were not associated with age, muscle or body condition (Whittemore et al. 2017). However, extremes of body condition were not included in the study, particularly obese categories, where the greatest impact might have been anticipated. In a similar study using dogs, no association was identified between body weight, age and muscle or body condition with Doppler BP measurements (Mooney et al. 2017).

2.11 Clinical Protocols for Blood Pressure Measurement for the Diagnosis of Hypertension

There are a number of key situations when BP measurement should be performed. As systemic hypertension is considered a disease of the ageing population, routine screening of BP, particularly in cats, should be preserved for those >9 years of age (see Chap. 3). Routine assessment of BP in young cats without clinical indication (i.e. predisposing disease condition or clinical manifestations of hypertensive target organ damage) is more likely to result in identification of situational hypertension when high SBP measurements are obtained. In dogs, given the wider range of breed-related life expectancies, it is more challenging to stipulate an age at which screening BP could be of benefit although it is probable that BP increases with age in dogs as in other species. However, in both dogs and cats, SBP should be evaluated when there is evidence of a disease condition that could be associated with hypertension (see Chaps. 3–5) or when a drug that may be linked to secondary hypertension is administered (see Chap. 3). Similarly, measurement of systolic BP should always be performed in a dog or cat that presents with clinical manifestations that could be compatible with hypertensive TOD (see Chaps. 7–10).

Where evidence of hypertensive TOD exists, measurement of SBP on a single occasion may be sufficient to confirm the diagnosis of systemic hypertension. In patients where ocular or central nervous system TOD is not evident but a predisposing condition is present, repeated measurement of SBP on at least two occasions (i.e. 7–14 days apart depending on severity of clinical presentation) is required to confirm persistence of hypertension. However, for those patients where a diagnosis of idiopathic hypertension is probable (i.e. no secondary cause can be identified), measurement of SBP on ≥3 occasions may be required before a diagnosis can be made and situational hypertension excluded. The ACVIM hypertension guidelines and the International Renal Interest Society (IRIS) provide a staging system for the diagnosis of hypertension based on risk of developing TOD (Table 2.4). It should be noted that the evidence base that was used to define the

Table 2.4 ACVIM and IRIS staging of blood pressure

Blood pressure category	Systolic blood pressure	Risk of target organ damage
Normotensive	<140 mmHg	Minimal
Prehypertensive	140–159 mmHg	Mild
Hypertensive	160–179 mmHg	Moderate
Severely hypertensive	≥180 mmHg	Severe

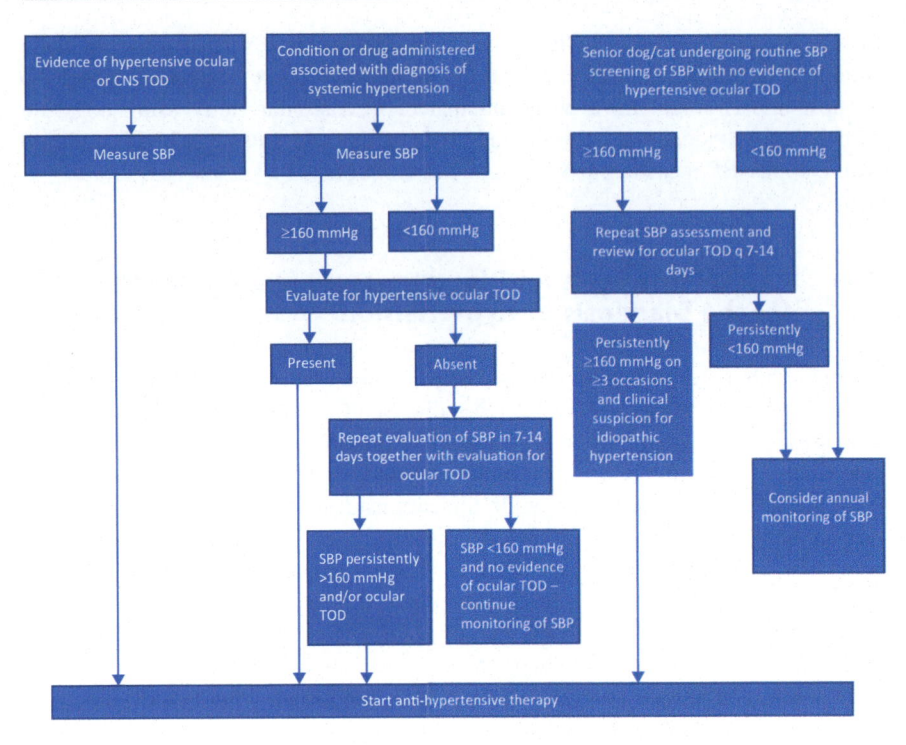

Fig. 2.7 Flow diagram for the diagnosis of systemic hypertension in cat or dog. *SBP* systolic blood pressure, *TOD* target organ damage, *CNS* central nervous system

BP stages has largely been obtained from SBP measurements performed using forelimb cuff placement. This should be considered when the criteria are extrapolated to use with cuffs in other positions, e.g. tail. Based on this information, antihypertensive therapy should be considered for patients where hypertensive TOD is noted and in those patients where SBP is persistently >160 mmHg and where there is clinical concordance that the measured pressures are likely to be a true reflection of systolic BP in that particular patient (Fig. 2.7).

Serial monitoring of BP is important for patients that may have conditions associated with hypertension. This is based on the premise that hypertension may not be present at initial diagnosis but may develop in the future, a phenomenon that has been clearly demonstrated in cats with chronic kidney disease and with treatment of hyperthyroidism (Bijsmans et al. 2015; Morrow et al. 2009). Serial monitoring is also fundamental to careful evaluation of the response to antihypertensive treatment (Chaps. 12 and 13).

For all of the situations where BP measurement should be performed, it is useful to have a clinic protocol established. This protocol should consider not only the practicalities of measurement and the machine used but also the environment in which the measurements are performed and structured recording of the data collected (Table 2.5). Such protocols help to ensure that BP measurement becomes part of standard veterinary clinical practice.

Table 2.5 Protocol for measurement of blood pressure in dogs and cats (Adapted from ACVIM hypertension guidelines 2018)

Calibration of the BP measuring device should be performed twice yearly either by the user where self-test modes are included or by the manufacturer
The environment where BP measurement is performed should be quiet and away from other animals and where possible the owner should be present. The patient should not be sedated and a period of 5–10 min should be allowed for acclimatisation
The animal should be allowed to adopt a comfortable position with gentle restraint, ideally in sternal or lateral recumbency in order to minimise the distance from cuff to heart base (if more than 10 cm a correction factor of +0.8 mmHg/cm should be applied to the measurement below or above the heart base)
The limb circumference should be measured and recorded and a cuff with a width of 30–40% of the circumference chosen
The site of cuff placement may be forelimb, hindlimb or tail based on operator and patient preference/conformation but should be recorded
An experienced operator should perform all BP measurements. Training is essential for operator confidence
The measurements should only be performed when the patient is quiet and relaxed
The first measurement should be discarded. A series of 5–7 consistent and consecutive readings should then be obtained and averaged. In some patients, measured BP may trend downwards during the measurement procedure. In this situation, measurement should continue until a plateau has been reached when the 5–7 readings should be taken
BP measurements should be repeated as necessary to ensure consistent and reliable values are obtained. This may include repeat measurement on separate occasions with 5–7 readings obtained on each occasion
Written records should be maintained on a standardised form and include operator, cuff size and site, values obtained, rationale for exclusion of values, final (mean) BP, attitude of the patient, whether headphones were used (Doppler) and interpretation of the results by a veterinary surgeon

References

Acierno MJ, Domingues ME, Ramos SJ, Shelby AM, Da Cunha AF (2015) Comparison of directly measured arterial blood pressure at various anatomic locations in anesthetized dogs. Am J Vet Res 76(3):266–271

Acierno MJ, Brown S, Coleman AE, Jepson RE, Papich M, Stepien RL et al (2018) ACVIM consensus statement: guidelines for the identification, evaluation, and management of systemic hypertension in dogs and cats. J Vet Intern Med 32(Aug):1803–1822

Ahlstrom C, Johansson A, Uhlin F, Länne T, Ask P (2005) Noninvasive investigation of blood pressure changes using the pulse wave transit time: a novel approach in the monitoring of hemodialysis patients. J Artif Organs 8(3):192–197

Alpert BS, Quinn D, Gallick D (2014) Oscillometric blood pressure: a review for clinicians. J Am Soc Hypertens 8(12):930–938

Babbs CF (2012) Oscillometric measurement of systolic and diastolic blood pressures validated in a physiologic mathematical model. Biomed Eng Online 11:1–22

Babbs CF (2015) The origin of Korotkoff sounds and the accuracy of auscultatory blood pressure measurements. J Am Soc Hypertens 9(12):935–950

Belew AM, Barlett T, Brown SA (1999) Evaluation of white-coat effects in cats. J Vet Intern Med 13:134–142

Bijsmans ES, Jepson RE, Chang YM, Syme HM, Elliott J (2015) Changes in systolic blood pressure over time in healthy cats and cats with chronic kidney disease. J Vet Intern Med 29(3):855–861

Binns SH, Sisson DD, Buoscio DA, Schaeffer DJ (1995) Doppler ultrasonographic, oscillometric sphygmomanometric, and photoplethysmographic techniques for non-invasive blood pressure measurement in anesthetized cats. J Vet Intern Med 9(6):405–414

Bodey AR, Michell AR (1997) Longitudinal studies of reproducibility and variability of indirect (oscillometric) blood pressure measurements in dogs: evidence for tracking. Res Vet Sci 63(1):15–21

Bodey AR, Young LE, Bartram DH, Diamond MJ, Michell AR (1994) A comparison of direct and indirect (oscillometric) measurements of arterial blood pressure in anaesthetised dogs, using tail and limb cuffs. Res Vet Sci 57(3):265–269

Bodey AR, Michell AR, Bovee KC, Buranakurl C, Garg T (1996) Comparison of direct and indirect (oscillometric) measurements of arterial blood pressure in conscious dogs. Res Vet Sci 61(1):17–21

Bosiack AP, Mann FA, Dodam JR, Wagner-Mann CC, Branson KR (2010) Comparison of ultrasonic Doppler flow monitor, oscillometric, and direct arterial blood pressure measurements in ill dogs. J Vet Emerg Crit Care 20(2):207–215

Bragg RF, Bennett JS, Cummings A, Quimby JM (2015) Evaluation of the effects of hospital visit stress on physiologic variables in dogs. J Am Vet Med Assoc 246(2):212–215

Branson KR, Wagner-Mann CC, Mann FA (1997) Evaluation of an oscillometric blood pressure monitor on anaesthetized cats and the effect of cuff placement and fur on accuracy. Vet Anesth 26:347–353

Brien EO, Pickering T, Asmar R, Myers M, Parati G, Staessen J (2002) Working group on blood pressure monitoring of the European Society of Hypertension International Protocol for validation of blood pressure measuring devices in adults Paolo Palatini j and with the statistical assistance of Neil Atkins a and William Gerin. Blood Press Monit 7:3–17

Brown SA, Langford K, Tarver S (1997) Effects of certain vasoactive agents on the long-term pattern of blood pressure, heart rate, and motor activity in cats. Am J Vet Res 58(6):647–652

Brown S, Atkins C, Bagley R, Carr A, Cowgill L, Davidson M et al (2012) Guidelines for the identification, evaluation, and Management of Systemic Hypertension in dogs and cats. Vet Clin North Am Small Anim Pract 42(4):542–558

Buranakarl C, Mathur S, Brown SA (2004) Effects of dietary sodium chloride intake on renal function and blood pressure in cats with normal and reduced renal function. Am J Vet Res 65(5):620–627

Cannon MJ, Brett J (2012) Comparison of how well conscious cats tolerate blood pressure measurement from the radial and coccygeal arteries. J Feline Med Surg 14(12):906–909

Carter JM, Irving AC, Bridges JP, Jones BR (2014) The prevalence of ocular lesions associated with hypertension in a population of geriatric cats in Auckland, New Zealand. N Z Vet J 62(1):21–29

Caulkett NA, Cantwell SL, Houston DM (1998) A comparison of indirect blood pressure monitoring techniques in the anaesthetized cat. Vet Anesth. 27:370–377

Cerejo SA, Teixeira-Neto FJ, Garofalo NA, Rodrigues JC, Celeita-Rodríguez N, Lagos-Carvajal AP (2017) Comparison of two species-specific oscillometric blood pressure monitors with direct blood pressure measurement in anesthetized cats. J Vet Emerg Crit Care 27(4):409–418

Chalifoux A, Dallaire A, Blais D, Larivière N, Pelletier N (1985) Evaluation of the arterial blood pressure of dogs by two noninvasive methods. Can J Comp Med 49(4):419–423

Chetboul V, Tissier R, Gouni V, De Almeida V, Lefebvre HP, Concordet D et al (2010) Comparison of Doppler ultrasonography in healthy awake dogs. Am J Vet Res 71(7):14–18

Da Cunha AF, Saile K, Beaufre H, Wolfson W, Seaton D, Acierno MJ (2014) Measuring level of agreement between values obtained by directly measured blood pressure and ultrasonic Doppler flow detector in cats. J Vet Emerg Crit Care 24(3):272–278

Garofalo NA, Neto FJT, Alvaides RK, de Oliveira FA, Pignaton W, Pinheiro RT (2012) Agreement between direct, oscillometric and Doppler ultrasound blood pressures using three different cuff positions in anesthetized dogs. Vet Anaesth Analg 39(4):324–334

Giavarina D (2015) Understanding bland Altman analysis. Biochem Med 25(2):141–151

Gouni V, Tissier R, Misbach C, Balouka D, Bueno H, Pouchelon JL et al (2015) Influence of the observer's level of experience on systolic and diastolic arterial blood pressure measurements using Doppler ultrasonography in healthy conscious cats. J Feline Med Surg 17(2):94–100

Grandy JL, Dunlop CI, Hodgson DS, Curtis CR, Chapman PL (1992) Evaluation of the Doppler ultrasonic method of measuring systolic arterial blood pressure in cats. Am J Vet Res 53(7):1166–1169

Haberman CE, Morgan JD, Kang CW, Brown S (2004) a. Evaluation of Doppler ultrasonic and Oscillometric methods of indirect blood pressure measurement in cats. Intern J Appl Res Vet Med 2(706):279–289

Haberman CE, Kang CW, Morgan JD, Brown SA (2006) Evaluation of oscillometric and Doppler ultrasonic methods of indirect blood pressure estimation in conscious dogs. Can J Vet Res 70:211–217

Heishima Y, Hori Y, Chikazawa S, Kanai K, Hoshi F, Itoh N (2016) Indirect arterial blood pressure measurement in healthy anesthetized cats using a device that combines oscillometry with photoplethysmography. J Vet Med Sci 78(7):2–5

Hsiang T-Y, Lien Y-H, Huang H-P (2008) Indirect measurement of systemic blood pressure in conscious dogs in a clinical setting. J Vet Med Sci 70(5):449–453

Huhtinen M, Derré G, Renoldi HJ, Rinkinen M, Adler K, Aspegrén J et al (2015) Randomized placebo-controlled clinical trial of a chewable formulation of amlodipine for the treatment of hypertension in client-owned cats. J Vet Intern Med 29(3):786–793

Jeong I, Jun S, Um D, Oh J, Yoon H (2010) Non-invasive estimation of systolic blood pressure and diastolic blood pressure using photoplethysmograph components. Yonsei Med J 51(3):345–353

Jepson RE, Hartley V, Mendl M, Caney SME, Gould DJ (2005) A comparison of CAT Doppler and oscillometric Memoprint machines for non-invasive blood pressure measurement in conscious cats. J Feline Med Surg 7(3):147–152

Kallet AJ, Cowgill LD, Kass PH (1997) Comparison of blood pressure measurements obtained in dogs by use of indirect oscillometry in a veterinary clinic versus at home. J Am Vet Med Assoc 210(5):651–654

Lin C-H, Yan C-J, Lien Y-H, Huang H-P (2006) Systolic blood pressure of clinically Normal and conscious cats determined by an indirect Doppler method in a clinical setting. J Vet Med Sci 68(8):827–832

Liu S-H, Cheng D-C, Su C-H (2017) A Cuffless blood pressure measurement based on the impedance Plethysmography technique. Sensors 17(6):1176

Marino CL, Cober RE, Iazbik MC, Couto CG (2011) White-coat effect on systemic blood pressure in retired racing greyhounds. J Vet Intern Med 25(4):861–865

Martel E, Egner B, Brown SA, King JN, Laveissiere A, Champeroux P et al (2013) Comparison of high-definition oscillometry—a non-invasive technology for arterial blood pressure measurement—with a direct invasive method using radio-telemetry in awake healthy cats. J Feline Med Surg 15(12):1104–1113

Mathur S, Syme HM, Brown CA, Elliott J, Moore PA, Newell MA et al (2002) Effects of the calcium channel antagonist amlodipine in cats with surgically induced hypertensive renal insufficiency. Am J Vet Res 63(6):833–839

Mathur S, Brown CA, Dietrich UM, Munday JS, Newell MA, Sheldon SE et al (2004) Evaluation of a technique of inducing hypertensive renal insufficiency in cats. Am J Vet Res 65(7):1006–1013

Meyer O, Jenni R, Greiter-Wilke A, Breidenbach A, Holzgrefe HH (2010) Comparison of telemetry and high-definition oscillometry for blood pressure measurements in conscious dogs: effects of torcetrapib. J Am Assoc Lab Anim Sci 49(4):464–471

Miller RH, Smeak DD, Lehmkuhl LB, Brown SB, DiBartola SP (2000) Radiotelemetry catheter implantation: surgical technique and results in cats. Contemp Top Lab Anim Sci 39(2):34–39

Mishina M, Watanabe T, Matsuoka S, Shibata K, Fujii K, Maeda H et al (1999) Diurnal variations of blood pressure in dogs. J Vet Med Sci 61(6):643–647

Mishina M, Watanabe N, Watanabe T (2006) Diurnal variations of blood pressure in cats. J Vet Med Sci 68(3):243–248

Mitchell AZ, McMahon C, Beck TW, Sarazan RD (2010) Sensitivity of two noninvasive blood pressure measurement techniques compared to telemetry in cynomolgus monkeys and beagle dogs. J Pharmacol Toxicol Methods 62(1):54–63

Mochel JP, Fink M, Bon C, Peyrou M, Bieth B, Desevaux C et al (2014 Jun 1) Influence of feeding schedules on the chronobiology of renin activity, urinary electrolytes and blood pressure in dogs. Chronobiol Int 31(5):715–730

Moll X, Aguilar A, García F, Ferrer R, Andaluz A (2018) Validity and reliability of Doppler ultrasonography and direct arterial blood pressure measurements in anaesthetized dogs weighing less than 5 kg. Vet Anaesth Analg 45(2):135–144

Mooney AP, Mawby DI, Price JM, Whittemore JC (2017) Effects of various factors on Doppler flow ultrasonic radial and coccygeal artery systolic blood pressure measurements in privately-owned, conscious dogs. PeerJ 5:e3101

Morrow LD, Adams VJ, Elliott J, Syme HM (2009) Hypertension in hyperthyroid cats: prevalence, incidence and predictors of its development. J Vet Intern Med 23:699

O'Brien E, Atkins N, Stergiou G, Karpettas N, Parati G, Asmar R et al (2010) European Society of Hypertension International Protocol revision 2010 for the validation of blood pressure measuring devices in adults. Blood Press Monit 15(1):23–38

Petrič AD, Petra Z, Jerneja S, Alenka S (2010) Comparison of high definition oscillometric and Doppler ultrasonic devices for measuring blood pressure in anaesthetised cats. J Feline Med Surg 12(10):731–737

Rattez EP, Reynolds BS, Concordet D, Layssol-Lamour CJ, Segalen MM, Chetboul V et al (2010) Within-day and between-day variability of blood pressure measurement in healthy conscious beagle dogs using a new oscillometric device. J Vet Cardiol 12(1):35–40

Rondeau DA, Mackalonis ME, Hess R (2013) Effect of body position on indirect measurement of systolic arterial blood pressure in dogs. J Am Vet Med Assoc 242(11):1523–1527

Rysnik MK, Cripps P, Iff I (2013) A clinical comparison between a non-invasive blood pressure monitor using high definition oscillometry (Memodiagnostic MD 15/90 pro) and invasive arterial blood pressure measurement in anaesthetized dogs. Vet Anaesth Analg 40(5):503–511

Sawyer DC, Brown M, Striler EL, Durham RA, Langham MA, Rech RH (1991) Comparison of direct and indirect blood pressure measurement in anaesthetized dogs. Lab Anim Sci 41(2):134–138

Scansen BA, Vitt J, Chew DJ, Schober KE, Bonagura JD (2014) Comparison of forelimb and hindlimb systolic blood pressures and proteinuria in healthy shetland sheepdogs. J Vet Intern Med 28(2):277–283

Seliškar A, Zrimšek P, Sredenšek J, Petrič AD (2013) Comparison of high definition oscillometric and Doppler ultrasound devices with invasive blood pressure in anaesthetized dogs. Vet Anaesth Analg 40(1):21–27

Sparkes AH, Caney SM, King MC, Gruffydd-Jones TJ (1999) Inter- and intraindividual variation in Doppler ultrasonic indirect blood pressure measurements in healthy cats. J Vet Intern Med 13(4):314–318

Stepien RL, Rapoport GS (1999) Clinical comparison of three methods to measure blood pressure in nonsedated dogs. J Am Vet Med Assoc 215(11):1623–1628

Stepien RL, Rapoport GS, Henik RA, Wenholz L, Thomas CB (2003) Comparative diagnostic test characteristics of Oscillometric and Doppler Ultrasonographic methods in the detection of systolic hypertension in dogs. J Vet Intern Med 17(1):65–72

Stergiou GS, Alpert B, Mieke S, Asmar R, Atkins N, Eckert S et al (2018) A universal standard for the validation of blood pressure measuring devices: Association for the Advancement of

Medical Instrumentation/European Society of Hypertension/International Organization for Standardization (AAMI/ESH/ISO) Collaboration Statement. Hypertension 71(3):368–374

Taha Z, Shirley L, Azraai M, Razman M (2017) A review on non-invasive hypertension monitoring system by using Photoplethysmography method. Mov Heal Exerc 6(1):47–57

Theodor M, Ruh D, Ocker M, Spether D, Förster K, Heilmann C et al (2014) Implantable impedance plethysmography. Sensors (Switzerland) 14(8):14858–14872

Vachon C, Belanger MC, Burns PM (2014) Evaluation of oscillometric and doppler ultrasonic devices for blood pressure measurements in anesthetized and conscious dogs. Res Vet Sci 97 (1):111–117

Valtonen MH, Eriksson LM (1970) The effect of cuff width on accuracy of indirect measurement of blood pressure in dogs. Res Vet Sci 11:358–361

Wernick M, Doherr M, Howard J, Francey T (2010) Evaluation of high-definition and conventional oscillometric blood pressure measurement in anaesthetised dogs using ACVIM guidelines. J Small Anim Pract 51(6):318–324

Wernick MB, Francey T, Howard J (2012) Comparison of arterial blood pressure measurements and hypertension scores obtained by using three indirect measurement devices in hospitalized dogs. J Am Vet Med Assoc 240(8):9–14

Whittemore JC, Nystrom MR, Mawby DI (2017) Effects of various factors on Doppler ultrasonographic measurements of radial and coccygeal arterial blood pressure in privately owned, conscious cats. J Am Vet Med Assoc 250(7):763–769

Williams T, Elliott J, Syme HM (2010) Measurement of systolic blood pressure (SBP) in cats by indirect Doppler technique is not altered by the use of headphones (Poster abstract), ECVIM, Conference Proceedings, Toulouse

Zwijnenberg RJ, del Rio CL, Cobb RM, Ueyama Y, Muir WW (2011) Evaluation of oscillometric and vascular access port arterial blood pressure measurement techniques versus implanted telemetry in anesthetized cats. Am J Vet Res 72(8):1015–1021

Epidemiology of Hypertension

Harriet M. Syme

The epidemiological study of hypertension in dogs and cats is in its infancy, whereas in human medicine there has been huge investment in enormous population-based studies that have characterised the risks associated with increased blood pressure (Mahmood et al. 2014; Kotchen 2011). For this review of the epidemiology of hypertension in veterinary patients, each section starts with a reflection on what is known from the study of humans so that analogies can be drawn.

3.1 Dichotomising Blood Pressure into Hypertension and Normotension Categories

Measures of blood pressure within any population, be that systolic, diastolic or pulse pressure (the difference between systolic and diastolic), range across a spectrum of values from very low to very high, with most individuals sitting somewhere in the middle of this continuum. Definitions of hypertension (and normotension) are an artificial construct, serving to dichotomise values into normal and abnormal. There is an implication in doing this that there is a clear segregation of values into two distinct populations, while in actual fact blood pressure within populations is usually normally distributed with a slight right skew to the data, and there is no natural segregation into distinct groups. The right skew to the data may be more marked in pathological states.

H. M. Syme (✉)
Department of Clinical Sciences and Services, Royal Veterinary College, University of London, Hertfordshire, UK
e-mail: hsyme@rvc.ac.uk

3.1.1 Humans

It has been said of blood pressure epidemiology in humans that the focus on hypertension is attractive to the population as a whole because it labels a deviant minority (although not that minor since hypertension is very common) as being abnormal and targets them for treatment in order to try to eliminate the high tail of the population distribution. This view is politically and sociologically attractive because it reassures the majority that they are 'normal' (Rose and Day 1990). It also suits the pharmaceutical industry because by focusing on the extremes of the population, the emphasis is placed on drug therapy. However, in truth, more would be accomplished for the population as a whole by achieving lifestyle changes (to reduce obesity, improve diet and increase physical activity as discussed below) if that shifted the blood pressure distribution even 1–2 mmHg to the left than by focusing treatment on the population extremes (Olsen et al. 2016). Of course, although lifestyle interventions are laudable goals, they are very difficult to achieve.

There are relatively few recent publications in human medicine (at least from developed countries) that report absolute blood pressure measurements, since many of the people studied are already receiving antihypertensive medications; instead, it is usually the proportion of the population that are 'hypertensive' that is reported (Beaney et al. 2018). Further, complexity is added to this situation by the fact that blood pressure cut-offs that constitute hypertension have progressively decreased over time. In 1948, a popular cardiology textbook defined hypertension as >180/110 mmHg (Evans 1948). In the 1960s, in spite of emerging evidence from the Framingham studies (which defined hypertension as >160/95 mmHg) about the associated risk of cardiovascular endpoints (Mahmood et al. 2014), hypertension was still considered part of the normal ageing process with systolic blood pressure of 'age + 100' commonly regarded as normal (Moser 2006). The first Joint National Committee (JNC; on Prevention, Detection, Evaluation, and Treatment of High Blood Pressure in the USA, published in 1977) recommended treatment if the diastolic blood pressure was >105 mmHg (Anonymous 1977). The emphasis subsequently shifted to principally focusing on systolic blood pressure targets due to evidence that this was more strongly associated with cardiovascular outcomes. The cut-offs taken to signify hypertension progressively reduced through each of the subsequent JNC reports to <140/90 mmHg in JNC 7 which was published in 2003 (Chobanian et al. 2003), before reaching a period of relative stability. The term prehypertension was also introduced at that time (systolic blood pressure 120–139 mmHg and/or diastolic 80–89 mmHg).

Historically, the European cut-offs, and those of other countries, have been broadly aligned with these recommendations from the USA, although with some differences, particularly for those aged over 60 years or those with renal disease or diabetes. However, since 2017, this has changed, due to the American College of Cardiology/American Heart Association (which now produces the guidelines, in place of JNC) abandoning the category of prehypertension and reducing the cut-off for what constitutes hypertension to 130/80 mmHg (Whelton et al. 2018), but most other countries are not following suit (Williams et al. 2018). A key driver to the change in the US guidelines was the publication of the results of the Systolic Blood Pressure Intervention Trial (SPRINT) which showed an improvement in outcome in patients without diabetes when the treatment target was a systolic pressure of <120 mmHg, rather than the conventional value of <140 mmHg (The SPRINT Research Group 2015). These results contrast with those of a similar study, this time of patients with diabetes, in which no benefit was evident (Group et al. 2010). The discordant results of these trials emphasise that the risk of target-organ damage due to hypertension, and of undesirable side effects from treatment, is not uniform but depends on many other factors, particularly concurrent disease. Notably, the SPRINT predominantly enrolled older patients at high risk of cardiovascular disease and used an unattended automated measurement technique, a method that often yields blood pressure values 10 to 20 mmHg lower than traditional attended office blood pressure readings, leading many authors to question whether this study should be used as the basis for a change in more general treatment recommendations (Garies et al. 2019; Kovesdy 2017).

The consequences of such changes are not trivial. Reducing hypertension cut-offs from 140/90 to 130/80 mmHg is estimated to have increased the number of Americans considered to be hypertensive by 31 million, with 4.2 million of these now recommended to receive drug therapy (Muntner et al. 2018). In addition, changes in blood pressure treatment targets meant that more than half of all hypertensive patients on therapy were then classified as inadequately controlled.

In this chapter, references to hypertension in humans will use the current European (pre-2017 ACC/AHA) cut-offs of 140/90 mmHg to signify hypertension, unless otherwise specified. The differences in the US and European systems for blood pressure classification in humans are outlined in Tables 3.1 and 3.2.

Table 3.1 2017 American College of Cardiology/American Heart Association (Humans)

Category	SBP (mmHg)		DBP (mmHg)
Normal	<120	and	<80
Elevated	120–129	and	<80
Hypertension stage 1	130–139	or	80–89
Hypertension stage 2	≥140	or	≥90

Table 3.2 2018 European Society of Cardiology/Hypertension Guidelines (Humans)

Category	SBP (mmHg)		DBP (mmHg)
Optimal	<120	and	<80
Normal	120–129	and/or	80–84
High normal	130–139	and/or	85–89
Grade 1 hypertension	140–159	and/or	90–99
Grade 2 hypertension	160–179	and/or	100–109
Grade 3 hypertension	≥180	and/or	≥110
Isolated systolic hypertension	≥140	and	<90

3.1.2 Dogs and Cats

Reports of blood pressure measurements in dogs and cats have often described the data in terms of mean and standard deviation (see Table 3.3) and have not always referenced tests for the normality of the data (Hoglund et al. 2012; Taffin et al. 2016; Paige et al. 2009; Mishina et al. 1997, 1998), or have contained too few individuals for any deviations from Gaussian distribution to be appreciated. However, with some exceptions (Rondeau et al. 2013; Hori et al. 2019), published studies of blood pressure containing more than 50 animals in which the distribution of the data is reported (or can be ascertained) have demonstrated a slight right skew to the blood pressure distribution in both dogs (Bodey and Michell 1996) and cats (Bodey and Sansom 1998; Payne et al. 2017; Sparkes et al. 1999). The degree to which the distribution is skewed may depend on its composition, for example, if blood pressure increases with age in the population studied, then inclusion of older animals may skew the distribution to the right. Skewedness of the blood pressure distribution may also be more marked in disease states. Figure 3.1 illustrates this with a display of systolic blood pressure measurements from two different populations of cats examined using the Doppler method, showing that the blood pressure of the cats with CKD is both generally higher and has a greater right skew. It should be noted, however, that although the studies used the same methodology and both originated from the Royal Veterinary College, they were conducted at different times, in separate clinics, with measurements made by different clinicians, so they may not be directly comparable (Payne et al. 2017; Syme et al. 2002).

In a similar manner to human medicine, cut-offs have been proposed for what constitutes hypertension in dogs and cats. The latest version of these, published in 2018 (Acierno et al. 2018), is given in Table 3.4 and can also be accessed via the IRIS website www.iris-kidney.com. The proposed values for normotension are 10 mmHg lower (now <140 mmHg) than stipulated in the original 2007 guidelines (Brown et al. 2007), in part as a response to the publication of the epidemiological study of 780 apparently healthy cats that had a median systolic blood pressure of 120.6 (interquartile range 110.4–132.4) mmHg (Payne et al. 2017). In addition, in the new guidelines, systolic blood pressure of 140–159 mmHg is now considered to be prehypertension in both dogs and cats. In one study, cats that developed hypertension over time had initial blood pressure values (median and interquartile ranges) that were higher than cats that remained normotensive: 145.6 (139.5, 154) mmHg in healthy, older cats and 147.2 (140.4, 156.1) mmHg in cats with CKD, indicating that cats with blood pressure in the prehypertension range may be more likely to go on to develop hypertension (Bijsmans et al. 2015). It is not intended that dogs and cats with prehypertension are treated, but these patients should be targeted for more frequent follow-up blood pressure measurements.

3 Epidemiology of Hypertension

Table 3.3 Arterial blood pressure measurements obtained from healthy, conscious dogs and cats

	n	Special characteristics	SBP (mmHg)	MBP	DBP	References
Dogs						
Telemetry	66		143 ± 5	105 ± 13	85 ± 11	Soloviev et al. (2006)
Telemetry	41 17	(Male) (Female)	147 ± 15 136 ± 12	106 ± 13 100 ± 10	82 ± 10 81 ± 11	Miyazaki et al. (2002)
Telemetry	7		125 ± 10	92 ± 8	75 ± 6	Mishina et al. (1999)
Arterial puncture	28		144 ± 16	104 ± 13	81 ± 9	Anderson and Fisher (1968)
Arterial catheter	12	Carotid Femoral	141 ± 23 154 ± 30	113 ± 16 115 ± 16	99 ± 15 96 ± 12	Chalifoux et al. (1985)
Arterial puncture	27		154 ± 20	107 ± 20	84 ± 9	Stepien and Rapoport (1999)
Oscillometric	158	Irish wolfhounds	116 (SE 1.4)	88 (SE 1.2)	69 (SE 1.2)	Bright and Dentino (2002)
Oscillometric	1267		131 ± 20	97 ± 16	74 ± 15	Bodey and Michell (1996)
Oscillometric	51		144 ± 27	110 ± 21	91 ± 20	Coulter and Keith Jr. (1984)
Oscillometric	89		139 ± 16		71 ± 13	Hoglund et al. (2012)
Doppler	72	Shetland sheepdogs	132 ± 20 (FL) 118 ± 20 (HL)			Scansen et al. (2014)
Doppler	51		147 ± 25			Rondeau et al. (2013)
Cats						
Telemetry	20		118 ± 11	95 ± 10	78 ± 9	Mishina et al. (2006)
Telemetry	6		125 ± 11	105 ± 10	89 ± 9	Brown et al. (1997)
Telemetry	6		126 ± 9	106 ± 10	91 ± 11	Belew et al. (1999)
Telemetry	20		132 ± 9	115 ± 8	96 ± 8	Slingerland et al. (2008)
Telemetry	6		160 ± 12	138 ± 11	116 ± 8	Zwijnenberg et al. (2011)
Telemetry	6		111 ± 4		75 ± 2	Jenkins et al. (2015)
Oscillometric	137		147 [134, 158]	117 [105, 127]	98 [82, 108]	Hori et al. (2019)

(continued)

Table 3.3 (continued)

	n	Special characteristics	SBP (mmHg)	MBP	DBP	References
Oscillometric	104		139 ± 27	99 ± 27	77 ± 25	Bodey and Sansom (1998)
Oscillometric	60		115 ± 10	96 ± 12	74 ± 11	Mishina et al. (1998)
Doppler	50		162 ± 19			Sparkes et al. (1999)
Doppler	53		134 ± 16			Lin et al. (2006)
Doppler	113[a]		133 ± 20			Taffin et al. (2017)
Doppler	87		131 ± 18			Paige et al. (2009)
Doppler	780		121 [110, 132]			Payne et al. (2017)

FL forelimb, *HL* hindlimb, *SE* standard error
For study of measurement using noninvasive methods, only studies with a minimum of 50 participants are included in the table. Data have been rounded to whole numbers where required. Data are presented in the format mean ± standard deviation, median [25th, 75th percentile] or mean (standard error)
[a]Includes three cats with renal azotaemia

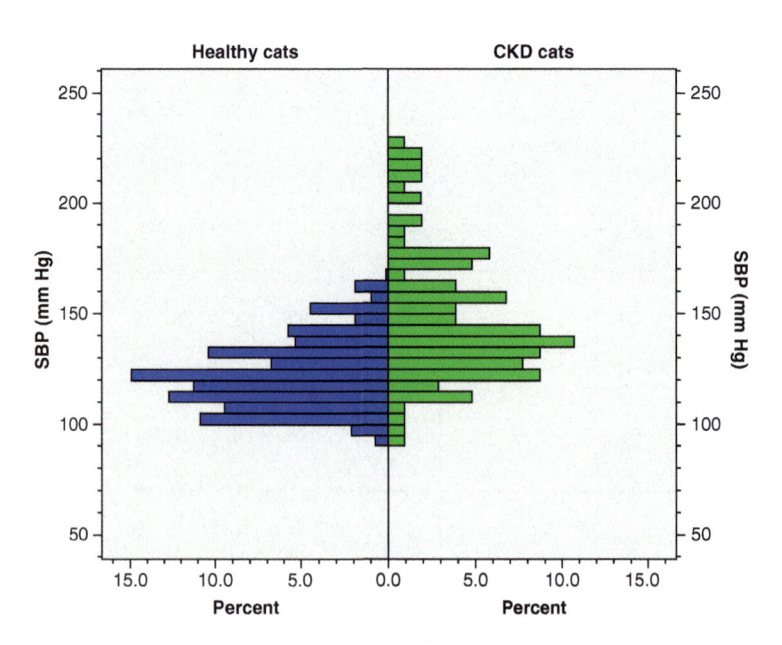

Fig. 3.1 Frequency histograms of systolic blood pressure measurements made by a Doppler method, in two populations of cats, the first healthy cats at rehoming centres [data from Payne et al. (2017)], although not included in the original publication in this format) and the second cats with CKD (data from Syme et al. (2002)]. Note that the distribution of blood pressure measurements in the cats with CKD is skewed upwards

Table 3.4 2018 International Renal Interest Society (IRIS)/ACVIM hypertension consensus statement (dogs and cats)

Category	Risk of TOD	SBP (mmHg)
Normotensive	Minimal	<140
Prehypertensive	Low	140–159
Hypertensive	Moderate	160–179
Severely hypertensive	High	≥180

3.2 Epidemiological Associations with Adverse Outcomes/Target Organ Damage

3.2.1 Humans

Hypertension currently affects more than 1 billion people worldwide and is projected to affect a third of the world's population by 2025, due to escalating obesity and population ageing. High blood pressure is responsible for approximately two-thirds of all strokes and half of all ischaemic heart disease. Only half of this burden occurs in people with hypertension as classically defined (i.e. blood pressure ≥ 140/90 mmHg) with the remainder due to the increased risk with supraoptimal blood pressure in the normotensive or prehypertensive range. Indeed, epidemiological data show a continuous positive relationship between the risk of death due to coronary artery disease and stroke down to values as low as 115 mmHg (Lewington et al. 2002).

There is strong evidence that when measures of blood pressure are considered individually, systolic pressure is the best predictor of cardiovascular risk (outperforming diastolic and pulse pressure) (Rahimi et al. 2015). There is also some evidence that the relative strength of systolic and diastolic blood pressure as a prognostic factor of cardiovascular disease varies by age (Stamler et al. 1993). In older individuals, pulse pressure is thought to be more, and diastolic pressure less, important, probably reflecting the risk associated with arterial stiffening (Steppan et al. 2011).

Abundant evidence from randomized controlled trials has shown benefit of antihypertensive drug treatment in reducing important adverse health outcomes (especially stroke, myocardial infarction and heart failure) in people with hypertension (Beckett et al. 2008; Xie et al. 2016). Treatment benefits appear to be broadly consistent among a range of subpopulations and with different antihypertensive agents (Neal et al. 2000). The main area of controversy is how low is optimal when considering blood pressure treatment targets (The SPRINT Research Group 2015; Xie et al. 2016; Cushman et al. 2010). Interestingly, however, while meta-analyses have confirmed the reduced risk of cardiovascular endpoints with effective antihypertensive therapy, intensive treatment has not been shown to reduce the risk of developing end-stage renal disease resulting in dialysis, transplantation or death (Ettehad et al. 2016). However, in the Modification of Diet in Kidney Disease (MDRD) Study, assignment to the lower blood pressure goal slowed GFR decline among the subset of patients that were most proteinuric, although not in the population as a whole (Klahr et al. 1994).

In some studies, mortality increases with very low diastolic pressures, resulting in a so-called 'J-shaped' curve for the relationship with cardiovascular disease, particularly myocardial infarction. There are various proposed reasons that this might occur: (1) reduced myocardial perfusion since coronary blood flow occurs in diastole; (2) decreased diastolic pressure that could, in fact, be a surrogate for arterial hardening in patients with isolated systolic hypertension; or (3) early deaths that might occur due to inverse causation (i.e. worsening heart function reducing blood pressure due to ill-health) (Flack et al. 1995; Messerli and Panjrath 2009). However, suboptimal control of blood pressure in patients treated for hypertension is exceedingly common and remains a much greater issue on a population basis than that of overzealous treatment.

Hypertension is an important cause of retinopathy and choroidopathy in humans (Klein et al. 1993). In addition to direct injury caused by high blood pressure, hypertension is also a risk factor for other forms of vascular eye disease (see Chap. 9) (Wong and Mitchell 2007). Signs of hypertensive retinopathy are also predictive of incident stroke (see Chap. 10), congestive heart failure and cardiovascular mortality (see Chap. 8), even after accounting for traditional risk factors such as blood pressure and smoking (Wong and McIntosh 2005).

3.2.2 Dogs and Cats

In dogs and cats, large-scale population studies akin to the Framingham study in humans are lacking. There are ongoing prospective longitudinal studies such as the Dog Aging Project (Kaeberlein et al. 2016), the Golden Retriever Lifetime Study (Guy et al. 2015) and the Bristol Cats Study (Murray et al. 2017) that are large in scale and ambitious in scope, but these have been set up relatively recently, and it will be some time before these yield results. In any case, measurement of blood pressure is not proposed in any of these studies at present, although it might be performed at the discretion of the veterinarian managing the case.

In contrast to the situation in humans, in dogs and cats no population-wide association has been established between hypertension and adverse cardiac outcomes, such as myocardial infarction and stroke, since these conditions are relatively uncommon in these species. In patients with mitral-valve disease, blood pressure decreases with increasing heart failure class (Petit et al. 2013; Borgarelli et al. 2008) and was not associated with survival time in a study of dogs with heterogeneous severity of disease (Borgarelli et al. 2008). However, in the EPIC study, a large, placebo-controlled study of pimobendan treatment in dogs with preclinical mitral valve disease, survival time was inversely related to the blood pressure at entry to the study (Boswood et al. 2016). Thus, in dogs with mitral-valve disease, adverse outcome is associated with hypotension, not hypertension (Chap. 8). Left ventricular hypertrophy is reportedly common in dogs with hypertension associated with glomerular disease due to leishmaniosis (Cortadellas et al. 2006).

Cats with hypertension have been documented to have mild left ventricular hypertrophy in several small observational and case-control studies (Chap. 8) (Snyder et al. 2001; Chetboul et al. 2003; Carlos Sampedrano et al. 2006; Nelson et al. 2002). In one of these studies, survival was reported and was not different in the hypertensive and normotensive groups, in spite of the fact the hypertensive cats were older (Chetboul et al. 2003). Conversely, in a large study of cats with hypertrophic cardiomyopathy, systolic blood pressure was higher in affected cats than cats with normal hearts (128.8 vs. 120.0 mmHg) in the univariate analysis, although this association was lost in the multivariate analysis. This was probably because male sex and increasing age were confounding and were associated with both hypertrophic cardiomyopathy and blood pressure (Payne et al. 2015). Cats with overt hypertension (systolic blood pressure $\geq$ 180 mmHg) were excluded from the study which aimed to include patients with primary rather than secondary myocardial disease.

Ocular lesions are present in many hypertensive cats (Chap. 9). It can be difficult to determine the proportion of cats with hypertension that develop ocular lesions because in the large ophthalmological case series, many of the cats were presented due to hyphema or blindness (Maggio et al. 2000); patients with hypertension that is asymptomatic may never be referred for evaluation by an ophthalmologist which tends to increase the apparent prevalence of ocular lesions in specialist centres. Conversely, when the ophthalmic examination is performed by non-specialists, subtle lesions may be overlooked. The other factor that is likely to affect the prevalence of retinal lesions is the blood pressure cut-off that is used to signify hypertension; if cats with relatively mild hypertension are included, the prevalence will decrease. One study found that most cats with hypertensive retinopathy had systolic blood pressure > 168 mmHg (Sansom et al. 2004). In another study, the blood pressure of cats with retinal lesions was significantly higher (262 ± 34 mmHg) than that of the hypertensive cats without retinal lesions (221 ± 34 mmHg) (Chetboul et al. 2003). Hypertensive retinal lesions and haemorrhages were only documented to worsen or progress to retinal detachment in 6 of 141 cats after initiation of antihypertensive medication when the target blood pressure on treatment was <160 mmHg, so this appears to be adequate to prevent ocular lesions from developing (Jepson et al. 2007). Results of another study suggest that vision may be restored with antihypertensive treatment, even if the blood pressure control is suboptimal since improvement was noted in most treated cats in spite of the fact that systolic blood pressure was only reduced to less than 180 mmHg in just over half of the cats treated (Young et al. 2019).

Hypertensive ocular lesions have also been reported to occur in dogs (Chap. 9); however, anecdotally they appear to be less common than in cats. In one study of dogs with chronic kidney disease, 3/14 dogs with hypertension had signs of retinopathy or choroidopathy (Jacob et al. 2003). In dogs with hypertension associated with glomerular disease due to leishmaniosis, the prevalence of hypertensive ocular lesions was only 6% (Cortadellas et al. 2006). It is possible that the apparently low incidence of ocular lesions in both studies was because the hypertension in many of the dogs was relatively mild, the dogs with fundic lesions having blood pressure that was generally higher than those without lesions, above 190 mmHg in all cases (Cortadellas et al. 2006; Jacob et al. 2003). In a larger case series of dogs with hypertension, 62% (26/42) had at least one fundic lesion. However, the prevalence of ocular lesions may have been increased because in many cases that was why the dog was referred (LeBlanc et al. 2011). In a case series of 83 dogs with punctate retinal haemorrhage, only 11 (13%) had hypertension as the underlying cause (Violette and Ledbetter 2018).

Hypertensive encephalopathy has been reported in both dogs and cats (Chap. 10) (O'Neill et al. 2013a; Church et al. 2019; Littman 1994). It appears to be relatively less common than hypertensive ocular disease, although it is possible that in its milder forms it may not be recognised. Estimates of the prevalence of neurological signs in cats in published studies range from 15 to 46% (Chetboul et al. 2003; Maggio et al. 2000; Littman 1994; Elliott et al. 2001). It is also proposed that neurological lesions are more likely to develop when the increase in blood pressure is acute, rather than chronic, since it has developed at relatively high frequency in cats following renal transplantation (Kyles et al. 1999) and experimental subtotal nephrectomy (Brown et al. 2005).

As discussed below renal disease is a potential cause of secondary systemic hypertension. The kidney is also a target organ that can be damaged. In cats (Syme et al. 2006), and dogs (Jacob et al. 2003), with CKD, it has been shown that hypertensive patients are more likely to be proteinuric and that proteinuria, in turn, is associated with shortened survival times and/or progression of renal disease (Jepson et al. 2007; Syme et al. 2006; Jacob et al. 2005; Chakrabarti et al. 2012). What is not clear is whether blood pressure, independent of proteinuria, has an effect on the progression of renal disease (see Chap. 7). In a study of hypertensive cats treated with amlodipine, many but not all of which had azotaemic CKD, the degree of blood pressure control did not affect survival if severity of proteinuria was also in the model (Jepson et al. 2007). Interestingly, the cats that had poorest blood pressure control were also those that were most proteinuric before initiating treatment. Similarly, in dogs, hypertension is only associated with survival in CKD if proteinuria is excluded from the model (Jacob et al. 2003). It is possible that antihypertensive therapy has not improved survival in dogs or cats with renal disease because the targets of blood pressure reduction have been too conservative and that more intensive therapy is required; but it should also be noted that survival benefit in humans is most evident in patients with proteinuria (Klahr et al. 1994). Dogs and cats may not show improvement in survival with antihypertensive therapy unless they have primary glomerular disease.

The target organ damage caused to the heart, the brain, the eye and the kidney by hypertension is discussed in detail in Chaps. 7–10.

3.3 Effect of Patient Characteristics (Age, Breed and Sex)

3.3.1 Humans

Blood pressure increases with age in humans, and the changes are not uniform but vary between the sexes and with different temporal patterns observed in systolic and diastolic pressure as shown in Fig. 3.2a. In general, diastolic blood pressure increases until middle age, after which it stabilises and in old age may actually decrease (Biino et al. 2013). Systolic blood pressure, on the other hand, continues to increase into old age and only declines on a population basis after the age of 70 years, due to the mortality excess of those with high blood pressure. These differences mean that there is widening of the pulse pressure with age leading to an increase in the prevalence of isolated systolic hypertension, in addition to the general increase in blood pressure (Bavishi et al. 2016). This is due to stiffening of the aorta and a more rapid return of the reflected pulse waves from the periphery. In young to middle-aged adults, the blood pressure of males exceeds that of females, but after menopause, there is less difference between the sexes with females' blood pressure slightly exceeding that of males over the age of 65 years. The Framingham Heart Study showed that even people that were normotensive at age 55 to 65 had a greater than 90% risk of eventually developing hypertension if they lived long enough (Vasan et al. 2002).

These age-related changes in blood pressure are thought to occur as a result of industrialisation of society. In nonindustrial, rural societies in which obesity does not occur, physical activity is frequent, and salt intake and consumption of processed food are very low, blood pressure has a very narrow distribution averaging around 115/75 mmHg and changes very little with age (Page et al. 1974).

Overall, the prevalence of hypertension is higher in people of African origin than in those of European origin (Lane et al. 2002), although this relation is confounded by socioeconomic status, which in turn is explained, at least in part, by differences in body mass index (Cappuccio et al. 2008).

3.3.2 Dogs and Cats

Blood pressure has been shown to increase with age in some (Bodey and Michell 1996; Bright and Dentino 2002), but not all (Mishina et al. 1997; Remillard et al. 1991; Meurs et al. 2000), population-based observational studies in dogs. Similarly, several studies have found that blood pressure is higher in male dogs (Mishina et al. 1997; Bodey and Michell 1996; Bright and Dentino 2002; Schellenberg et al. 2007), whereas in other studies no difference has been evident (Surman et al. 2012). In cats, again, the association of age with blood pressure has been inconsistent with some studies showing an increase (Mishina et al. 1998; Bodey and Sansom 1998; Payne et al. 2017; Sansom et al. 2004) and others showing no difference (Hori et al. 2019; Sparkes et al. 1999; Kobayashi et al. 1990). The majority of feline studies have not documented any difference in blood pressure with sex (Mishina et al. 1998; Sparkes et al. 1999; Sansom et al. 2004), although one study found that blood pressure was slightly higher in males (Payne et al. 2017). If variation in blood pressure with age and sex in dogs and cats is similar to that observed in humans, it may be difficult to detect in relatively small veterinary studies that often include fewer than 100 individuals, whereas population studies in humans often have many thousands of subjects. In general, it has been the largest studies of blood pressure in dogs and cats that have shown variation in association with age and, in some cases, sex (Bodey and Michell 1996; Bodey and Sansom 1998; Payne et al. 2017). This may suggest the other studies were underpowered to detect a difference. In addition, some studies have found that the increase in blood pressure with age is nonlinear which may also make detection of association in small data sets more difficult (Bodey and Sansom 1998). An alternative explanation is that since blood tests were not performed in the apparently healthy dogs and cats included in these larger veterinary studies, it is possible that some of the older animals had undetected disease resulting in a secondary increase in blood pressure (Bodey and Michell 1996; Bodey and Sansom 1998; Payne et al. 2017). In Fig. 3.2b–d, the changes in blood pressure with age in the larger canine and feline studies are illustrated for comparison with the graph of human data in Fig. 3.2a. Examining these graphs, there appear to be similar trends in blood pressure with age and sex in humans, dogs and cats. A recent study that reported repeated blood pressure measurements in 133 healthy older cats, and 160 cats with CKD that were initially normotensive, found that blood pressure increased progressively over time (with ageing) with 7% of the healthy and 17% of the cats with CKD eventually being diagnosed with hypertension (systolic blood pressure > 170 mmHg repeatedly or in association with fundic lesions) over the period of follow-up (Bijsmans et al. 2015).

Cats and dogs that are presented with signs of target organ damage due to severe systemic hypertension are usually old. The average age at diagnosis of hypertension in cats with hypertensive retinopathy is reportedly around 15 years (Maggio et al. 2000; Jepson et al. 2007; Littman 1994; Carter et al. 2014). Hypertensive retinopathy is rare in cats that are under the age of 9 years (Carlos Sampedrano et al. 2006; Maggio et al. 2000; Elliott et al. 2001). Studies in which apparently healthy senior and geriatric dogs have been screened for hypertension have yielded conflicting results: in one study none of the dogs were hypertensive (Meurs et al. 2000) and, in a second study, although many dogs were, hypertensive none had retinal lesions, and in many, the blood pressure decreased substantially with repeated measurement (Willems et al. 2017). The dogs were suspected to have white-coat hypertension.

Several studies have found that hypertensive ocular lesions are more common in females than males (Maggio et al. 2000; Young et al. 2019; Carter et al. 2014); however, reviewing the data from a series of 141 hypertensive cats reported by Jepson and colleagues (Jepson et al. 2007), there were approximately equal numbers of male and female cats, both in the study overall and in the subgroup (58 cats) with retinal lesions. Information regarding the patient sex was not reported in the original paper.

In most epidemiological studies of client-owned animals, the majority (especially cats) have been neutered making evaluation of the impact of this on blood pressure

3 Epidemiology of Hypertension

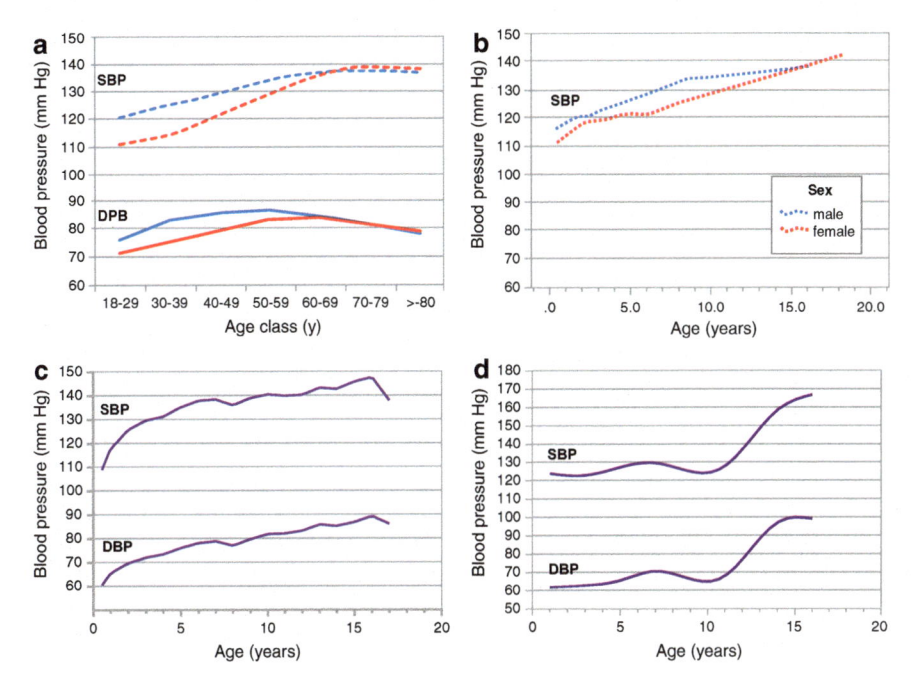

Fig. 3.2 Graphs to show variation in blood pressure with age. (**a**) Variation of systolic (SBP) and diastolic (DBP) in human males (in blue) and females (in red). Graph reproduced from Biino et al. (2013). (**b**) Variation of SBP (measured by Doppler method) by age and sex in cats. Data were kindly supplied by the authors of Payne et al. (2017), although not included in the original publication. (**c**) Variation of SBP and DBP by age in dogs with measurements using an oscillometric method. Data for males and females are combined. Lowess curve of data reported in Bodey and Michell (1996). (**d**) Variation of SBP and DBP by age in cats with measurements using an oscillometric method. Data for males and females are combined. Note the extended scale of the Y-axis compared with the other graphs. Spline curve of data reported in Bodey and Sansom (1998)

difficult to evaluate. One study of dogs did find that blood pressure of the entire females was lower, and entire males was higher, than in neutered dogs of the same sex (Bodey and Michell 1996). However, absolute differences were small (<5 mmHg). Systolic blood pressure of neutered cats is also higher than that of unneutered cats (Payne et al. 2017).

Breed reportedly affects blood pressure in dogs. The main reported difference is that most sighthounds, including greyhounds and deerhounds, have higher blood pressure than other breeds of dog (Bodey and Michell 1996). Approximately two-thirds of greyhounds have blood pressures that would be categorised as hypertensive (>160 mmHg) in a mongrel (Bright and Dentino 2002). The higher blood pressures in greyhounds have been documented in studies where direct blood pressure measurements have been obtained in the laboratory setting (Cox et al. 1976), as well as with indirect methods in client-owned animals. However, this increase in blood pressure is not true of all sighthounds because the blood pressure of Irish wolfhounds is reportedly quite low (Bright and Dentino 2002). No difference in blood pressure in different breeds of cats has been observed (Bodey and Sansom 1998; Payne et al. 2017; Lin et al. 2006).

3.4 Obesity and Dietary Factors

3.4.1 Humans

A health-related factor that accounts for a large portion of hypertension in developed countries is excess weight. In an analysis of the Framingham cohort, the population's attributable risk of new-onset hypertension due to excessive weight (overweight and obese categories combined) was 26% for men and 28% for women (Wilson et al. 2002). The global prevalence of obesity has doubled since the 1980s, and together with ageing of the population, this is thought to account for the current hypertension epidemic (Hall et al. 2019). The relationship between blood pressure and body mass index is linear and is not limited to those with obesity; it is also evident in patients that are lean, or only mildly overweight (Jones et al. 1994). Visceral adipose tissue appears to convey a greater risk for hypertension than fat deposited subcutaneously; high body mass index (BMI) and large waist circumference generally predict increased risk of hypertension in population studies.

Obstructive sleep apnoea is associated with an increased risk of hypertension and stroke; causation is unproven however, and confounders such as age and obesity weaken the association (Peppard et al. 2000).

Large, multinational observational studies have demonstrated a positive correlation between sodium intake and blood pressure in humans (Intersalt cooperative research group 1988). Population studies have also shown an inverse relation between potassium intake and blood pressure, the prevalence of hypertension, and the risk of stroke (Adrogué and Madias 2007). In clinical trials, reduction in sodium intake reduces blood pressure, although only by a small amount (less than 5 mmHg) (Aburto et al. 2013). Similarly, multiple randomized controlled trials have demonstrated that increased potassium intake reduces blood pressure in hypertensive people (Aburto et al. 2013). The sodium and potassium contents of foods tend to be inversely related: foods with high-sodium content tend to have low potassium and vice versa. This makes it difficult to discern the extent to which these dietary factors are distinct or overlapping. However, on balance, it seems that the ratio of sodium/potassium excretion is more closely related to blood pressure than either cation considered in isolation (Intersalt cooperative research group 1988; Chmielewski and Carmody 2017).

3.4.2 Dogs and Cats

In cats, an increase in blood pressure has been seen in association with increasing body condition score (BCS), although the differences noticed were small (Payne et al. 2017). In part, that may have been because relatively few overweight cats were included in the study. However, an association with body weight was not evident in another study (Bodey and Sansom 1998). In dogs, increases in blood pressure with increasing obesity have been reported in several studies (Bodey and Michell 1996; Mehlman et al. 2013; Jose Lahm Cardoso et al. 2016; Perez-Sanchez et al. 2015), although in other studies no relationship was evident (Willems et al. 2017; Mooney et al. 2017). In some dogs, the association between blood pressure and obesity may be indirect, due to underlying disease (Perez-Sanchez et al. 2015). Experimentally induced obesity also increases blood pressure in dogs (Granger et al. 1994; Hall et al. 1999).

Parallels have been drawn between brachycephalic dogs and humans with obstructive sleep apnoea. Accordingly, in a very small study, blood pressure of brachycephalic dogs was found to be higher than that of a control group of dogs with normal conformation (Hoareau et al. 2012).

Epidemiological studies evaluating the effect of sodium and/or potassium intake on blood pressure in dogs and cats eating their usual diet have not been reported. Intervention studies, where dogs and cats are fed low- or high-sodium diets, have not resulted in any convincing evidence that this alters blood pressure, either in healthy cats (Luckschander et al. 2004; Chetboul et al. 2014) or in dogs or cats subjected to subtotal nephrectomy (Buranakarl et al. 2004; Greco et al. 1994).

3.5 Primary and Secondary Hypertension

Secondary hypertension has an identifiable underlying cause. If this is something that can be effectively treated, then it is possible that the hypertension will resolve. In contrast, in primary, or idiopathic, hypertension there is no identifiable underlying disease. Guytonian physiology decrees that the kidney is key to the maintenance of hypertension since if pressure-induced natriuresis functions effectively, this should correct the increase in blood pressure (Coffman and Crowley 2008). As a result, the boundaries between renal hypertension and idiopathic hypertension are sometimes indistinct.

3.5.1 Humans

Secondary hypertension makes up only a small proportion (5–10%) of hypertension in humans (Charles et al. 2017). Screening for secondary hypertension is not generally recommended unless clinical signs suggest an underlying cause: hypertension is early (<30 years of age) or very abrupt in onset, hypertension is particularly severe or refractory to treatment, there is disproportionate target organ damage, or there is excessive or unprovoked hypokalaemia (Whelton et al. 2018). There are many different potential causes of secondary hypertension in adults as outlined in Table 3.5; renal parenchymal and vascular disease, obstructive sleep apnoea and primary hyperaldosteronism are the most common. In children and adolescents, hypertension is usually secondary, and renal parenchymal disease and coarctation of the aorta are the most common underlying causes.

Table 3.5 Causes of secondary hypertension in humans and their estimated prevalence

Cause	Prevalence
Renal parenchymal disease	1–2%
Renovascular disease	5%
Primary hyperaldosteronism	8%
Obstructive sleep apnoea*	25–50%
Drug or alcohol induced	2–4%
Pheochromocytoma	0.1–0.6%
Cushing's syndrome	<0.1%
Hypothyroidism	<1%
Hyperthyroidism	<1%
Aortic coarctation	<0.1%
Primary hyperparathyroidism	Rare
Congenital adrenal hyperplasia	Rare
Syndrome of apparent mineralocorticoid excess	Rare
Acromegaly	Rare

Percentage prevalence values are taken from the 2017 American College of Cardiology/American Heart Association Practice Guidelines and refer to the prevalence of these conditions in unselected adult patients presenting with hypertension (Whelton et al. 2018)

*Although obstructive sleep apnoea is considered a cause of hypertension this is controversial since trials of positive pressure ventilation do not always result in reduced a blood pressure

3.5.2 Dogs and Cats

Secondary hypertension is more common than primary in both cats and dogs. However, because systematic, population-wide measurement of blood pressure is rarely performed, it is difficult to estimate the relative importance of different diseases in causing hypertension; the data that are easier to obtain is the prevalence of hypertension in each individual condition. These data are presented in Table 3.6. Estimates of the relative importance of different underlying diseases can be made by considering their representation in case series of dogs and cats presenting with signs of target organ damage, most often ocular signs (Maggio et al. 2000; LeBlanc et al. 2011), or through geriatric screening programmes (Jepson et al. 2007). In dogs, the most common causes of hypertension are chronic kidney disease, hyperadrenocorticism, diabetes mellitus and pheochromocytoma. In cats, the most common causes of hypertension are chronic kidney disease, hyperthyroidism and primary hyperaldosteronism.

Hypertension is commonly documented in association with renal disease in both dog and cat and has been reported in both acute kidney injury (AKI) and with chronic kidney disease (CKD; Chap. 7). The prevalence of hypertension has been very variable between different studies, probably reflecting different aetiologies of the renal disease and also different criteria for diagnosing hypertension. It is suggested that hypertension is more common in dogs with glomerular disease, for

3 Epidemiology of Hypertension

Table 3.6 Prevalence of hypertension with different diseases in dogs and cats

Disease	Hypertension criteria* (mmHg)	Prevalence	References
Dogs			
Chronic kidney disease			
GFR <2 ml/kg/min	>175	6/66 (9%)	Michell et al. (1997)
Leishmaniosis with UPC >0.5, with or without azotaemia	>180 or >150 with TOD	25/52 (48%) 32/52 (62%)	Cortadellas et al. (2006)
Renal azotaemia	MBP > 120	6/31 (19%)	Buranakarl et al. (2007)
Leishmaniosis	>160	19/66 (29%)	Braga et al. (2015)
Chronic nephritis	> 160/90	13/14 (93%)	Anderson and Fisher (1968)
Chronic kidney disease (EPR study)	Variable	14/228 (6%)[a]	O'Neill et al. (2013b)
Chronic kidney disease	>160	14/45 (31%)	Jacob et al. (2003)
Protein losing nephropathy	>180	11/14 (79%)	Rapoport and Stepien (2001)
Acute kidney injury			
AKI grade V (requiring dialysis)	>150/95	47/54 (87%)	Francey and Cowgill (2004)
AKI (40/52 due to leptospirosis)	>160	19/52 (37%) at admission 42/52 (81%) during hospitalisation	Geigy et al. (2011)
Hyperadrenocorticism			
Hyperadrenocorticism (EPR study)	Variable	9/1519 (0.59%)[a]	Hoffman et al. (2018)
Hyperadrenocorticism	>160/100	21/26 (81%) PDH 10/10 (100%) ADH 17/21 (81%) poorly controlled 8/20 (40%) well controlled	Ortega et al. (1996)
Hyperadrenocorticism	>180	8/10 (80%)	Rapoport and Stepien (2001)
Pituitary-dependent hyperadrenocorticism	>160	7/20 (35%)	Chen et al. (2016)
Pituitary-dependent hyperadrenocorticism	>160	7/19 (37%) 6/19 (32%) after 6 months 5/13 (38%) after 12 months	Smets et al. (2012)

(continued)

Table 3.6 (continued)

Disease	Hypertension criteria* (mmHg)	Prevalence	References
Hyperadrenocorticism	>160	8/40 (20%) PDH 14/30 (47%) ADH	Lien et al. (2010)
Pituitary-dependent hyperadrenocorticism	>160/90	12/20 (60%) elevated SBP 16/20 (80%) elevated DBP	Vidal et al. (2018)
Pituitary-dependent hyperadrenocorticism	≥ 160	56/73 (77%)	Fracassi et al. (2015)
Hyperadrenocorticism	>200	9/13 (69.2%)	Goy-Thollot et al. (2002)
Hyperadrenocorticism	>150	8/12 (66.7%)	Novellas et al. (2008)
Pheochromocytoma			
	>160	6/7 (85.7%)	Gilson et al. (1994)
	>160	10/23 (43.5%)	Barthez et al. (1997)
Diabetes mellitus			
	>150	6/12 (50%)	Marynissen et al. (2016)
	>160	12/50 (24%)	Struble et al. (1998)
	>150/95	6/17 (35%)	Herring et al. (2014)
Cats			
Chronic kidney disease			
	>140/108	17/28 (61%)	Kobayashi et al. (1990)
	>160/100	15/23 (65%)	Stiles et al. (1994)
	>175 with OL or twice	20/103 (19.4%)[b]	Syme et al. (2002)
	>160 with OL or > 170 twice	105/265 (39.6%)	Bijsmans et al. (2015)
	≥180	25/77 (32.5%)	Hori et al. (2018)
Acute kidney injury			
	>180	8/43 (18.6%) at presentation 13/46 (28.2%) during hospitalisation	Cole et al. (2017)

(continued)

3 Epidemiology of Hypertension

Table 3.6 (continued)

Disease	Hypertension criteria* (mmHg)	Prevalence	References
Hyperthyroidism			
	>140/108	34/39 (87%)	Kobayashi et al. (1990)
	>160/100	3/13 (23%)	Stiles et al. (1994)
	≥ 160	16/84 (19%)	Stepien et al. (2003)
	>170	1/15 (6.7%)	Sansom et al. (2004)
	>180	9/40 (22.5%)	Sartor et al. (2004)
	≥ 160	10/21 (47.6%)	van Hoek et al. (2009)
	>160 with OL or > 170 twice	Prior 13/300 (4.3%) Incident 20/300 (15%)	Williams et al. (2010)
	>175 with OL or twice	Prior 3 (3%)[c] Incident 9 (9%)	Syme and Elliott (2003)
	≥ 160 twice	50/140 (35.7%)	Watson et al. (2018)
Hyperaldosteronism			
	>180	8/10 (80%)	Lo et al. (2014)
	>170	11/12 (91.6%)	Ash et al. (2005)
	>199	10/10 (100%)	Javadi et al. (2005)
Diabetes mellitus			
	>180	0/14 (0%)	Sennello et al. (2003)
	≥160	10/66 (15%)	Al-Ghazlat et al. (2011)
	>150	10/17 (59%)	Priyanka et al. (2018)
	>160	0/10 (0%)	Norris et al. (1999)

EPR electronic patient record, *OL* ocular lesions
[a]BP may not have been measured and/or reported in many of the dogs included in the study
[b]Many of the cats included in this study would also have been included in the Bijsmans (2015) study as well, but the criteria for defining hypertension were different
[c]Many of the cats included in this study would also have been included in the Williams (2010) study, but the criteria for defining hypertension were different
*SBP unless otherwise stated

example, due to leishmaniosis (Cortadellas et al. 2006), than in patients with tubulointerstitial disease; however, there is substantial overlap in prevalence between groups. Renal disease may be less commonly associated with hypertension-induced ocular injury in dog than cat, because only 11/42 (26%) dogs (LeBlanc et al. 2011), but 54/69 (78%) of cats (Maggio et al. 2000), diagnosed with hypertensive retinopathy had evidence of abnormal renal function. Whether this is a reflection of CKD being more common in cat than dog (Conroy et al. 2019), higher blood pressure in cats with CKD or a difference in the vulnerability of the eye of the cat and the dog to hypertension-induced injury is uncertain.

As noted above, azotaemia is documented in about three-quarters of hypertensive cats, at the time of diagnosis. Conversely, the prevalence of hypertension in cats with CKD has been estimated at between 19% and 65% (Syme et al. 2002; Kobayashi et al. 1990; Stiles et al. 1994). Notably, in the study with the prevalence of 19%, the systolic blood pressure had to be >175 mmHg either repeatedly or in association with retinal lesions for the cat to be classed as hypertensive. Using the ACVIM/IRIS, blood pressure categories that are in use today would increase the prevalence to around 30%. The prevalence and severity of hypertension do not appear to be related to the severity of azotaemia, at least in cats with CKD, with most cats being mildly azotaemic (IRIS CKD stage 2) at time of diagnosis (Syme et al. 2002; Bijsmans et al. 2015; Jepson et al. 2007). During longitudinal follow-up, cats with CKD have been shown to develop hypertension more frequently than old, healthy cats (Bijsmans et al. 2015), although this appears to be because their baseline blood pressure is already higher, rather than the blood pressure increasing at a faster rate. When the clinical presentation and routine biochemical data of hypertensive and normotensive cats with CKD are compared, there are few differences between them. Interestingly, one biochemical parameter that has been shown to be different in some studies is the plasma/serum potassium concentration, leading to the suggestion that hyperaldosteronism may play a role in aetiopathogenesis (Syme et al. 2002).

Endocrine causes of hypertension are also recognised in both dog and cat. Hypertension is often present in dogs with hyperadrenocorticism (either pituitary- or adrenal-dependent forms) and does not always resolve with treatment (Chap. 4) (Ortega et al. 1996; Smets et al. 2012). Increases in blood pressure have also occurred in dogs receiving exogenous glucocorticoid, although the changes are relatively minor and did not reach the threshold where dogs would be considered hypertensive (Schellenberg et al. 2008). Pheochromocytoma and primary hyperaldosteronism are uncommon disorders, but when they do occur, they are frequently associated with hypertension (Chap. 4) (Ash et al. 2005; Bouayad et al. 1987; Gilson et al. 1994). These should be considered in patients with adrenal masses with former being more likely in dogs and the later more likely in cats. Diabetes mellitus appears to be more commonly associated with hypertension in dogs than cats (Sennello et al. 2003; Struble et al. 1998), although it is rarely associated with target organ damage in either species.

Hypertension has been documented in 10–23% of cats with hyperthyroidism at the time of diagnosis (Chap. 5) (Stiles et al. 1994; Williams et al. 2010; Syme and Elliott 2003). Interestingly, although hyperthyroidism is extremely common,

Table 3.7 Therapeutic agents and toxins that have been documented to cause hypertension in dogs and/or cats

Therapeutic agents
Erythropoiesis-stimulating agents
Mineralocorticoids
Glucocorticoids
Phenylpropanolamine
Ephedrine, pseudoephedrine
Phenylephrine
Toceranib
Toxins
Cocaine
5-hydroxytryptophan
Amphetamines

relatively few cats with hyperthyroidism present with hypertensive retinopathy/choroidopathy (Maggio et al. 2000; Carter et al. 2014), and when this occurs, many of the cats have concurrent renal disease. Additionally, some cats that are initially normotensive develop hypertension after instigating treatment for hyperthyroidism (Morrow et al. 2009). This may be as many as a quarter of all cats treated and not only cats that develop azotaemia (Morrow et al. 2009).

It is reported that 13–20% of hypertensive cats have idiopathic (or primary) hypertension where no underlying cause can be identified (Maggio et al. 2000; Jepson et al. 2007; Elliott et al. 2001). Further work is required to determine whether these patients have non-azotaemic chronic kidney disease (CKD) and, if so, what role that might play in causing the hypertension. There are also isolated case reports and small case series of dogs with idiopathic hypertension (Tippett et al. 1987). In a series of 42 dogs presenting for hypertensive ocular disease, no cause was identified in 10 (24%) of the dogs (LeBlanc et al. 2011).

Secondary causes of hypertension are discussed further in Chaps. 4 and 5. Secondary hypertension may also occur due to treatment with therapeutic agents or ingestion of toxins; some of these are listed in Table 3.7.

3.6 White-Coat, Situational and Masked Hypertension

3.6.1 Humans

Hypertension in humans is usually symptomless and is diagnosed on the basis of blood pressure measurement. Readings made in the clinical environment have poor reproducibility due to a combination of measurement errors (as discussed in Chap. 2) and true variance. Variance in turn is due to the fact that blood pressure is continuously changing due to varying physiological states such as exercise and excitement or with anxiety. In particular, blood pressure may increase due to the so-called white-coat effect in the clinical environment; when this results in misdiagnosis, it is referred to as 'white-coat hypertension'. The white-coat effect is very common and is particularly marked in patients with true hypertension (Banegas et al. 2018), the elderly (Conen et al. 2014), patients with hypertension who are seemingly refractory to treatment (Armario et al. 2017) and when the blood pressure measurements are made by a doctor rather than a nurse (Clark et al. 2014). It is worth emphasising, however, that the white-coat effect (i.e. a transient rise in blood pressure associated with the stress of the measurement procedure itself) occurs in most individuals, including those classified as normotensive.

To overcome the misclassification of patients that can occur due to the white-coat effect, there is increased use of ambulatory (ABPM) and home (HBPM) blood pressure monitoring in humans. ABPM is performed using devices that are preprogrammed to measure blood pressure at set intervals (usually 20–60 min) during the day and night. The advantage of these devices is that in addition to identifying white-coat and masked hypertension, they can also further characterise the decrease in blood pressure that is expected to occur at night (see diurnal variation below). ABPM is a stronger predictor of all cause and cardiovascular mortality than clinic blood pressure (Banegas et al. 2018). The disadvantages of ABPM are that it causes sleep disruption and is more expensive than HBPM, both in terms of instrumentation and also the time taken to interpret the results, a cost that may not be reimbursed by health-care providers. HBPM by comparison is relatively simple and cheap to perform, requiring measurement of blood pressure in the home environment daily for a minimum of 3 days. Although in some studies this has involved medical personnel visiting the home, more often in routine clinical practice the patient measures their own blood pressure using an automated oscillometric device (Parati et al. 2010). The results are relatively simple to interpret, and in addition to identifying white-coat and masked hypertension, the involvement of the patient in collecting the readings has the advantage of increasing subsequent compliance with antihypertensive therapy (Cappuccio et al. 2004). The disadvantage of HBPM is that diurnal variation cannot be detected. Both European (Parati et al. 2010), and more recently US (Siu 2015), treatment guidelines have recommended that elevated clinic blood pressure measurements are confirmed using out-of-clinic measurement devices prior to initiating or intensifying drug therapy.

The results of existing literature are mixed with regard to long-term cardiovascular risk associated with white-coat hypertension. Some studies have found that patients with white-coat hypertension have the same risk as normotensive individuals (Hanninen et al. 2012), whereas in other studies their risk is intermediate between normotensive and sustained hypertension groups (Tientcheu et al. 2015). This is likely to occur because in most studies the home blood pressure measurements of white-coat hypertensives are intermediate between these two groups, and it is known that increased cardiovascular risk does not start at the boundary of hypertension, but in fact there is a graded increase in risk that extends down to systolic blood pressures as low as 115 mmHg (Lewington et al. 2002).

In contrast to white-coat hypertension, masked hypertension is characterised by normal office readings but elevated ABPM or HBPMs. In many cases, this does not mean that it is actually higher in the home environment, rather that blood pressure does not show the expected decrease in the nonclinical setting; the cut-offs for what constitutes hypertension in each of these settings are slightly different: typically 140/90 for clinic, 135/85 for day time or HBPM and 130/80 mmHg for 24-h ABPM.

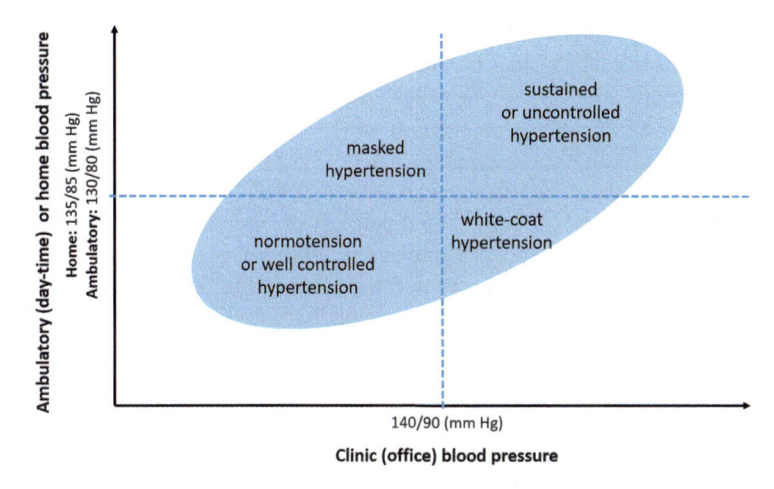

Fig. 3.3 Figure to show the relationship between clinic and ambulatory or home blood pressure measurements, showing the population of individuals with white-coat and masked hypertension

These different blood pressure categories are depicted graphically in Fig. 3.3. The proportion of patients with masked hypertension varies between studies but can be as high as 30% (Tientcheu et al. 2015; Diaz et al. 2015). Patients with untreated masked hypertension are at increased risk of adverse cardiovascular events compared to normotensive individuals, and patients on treatment with masked inadequate blood pressure control are at increased risk compared with those with controlled hypertension on therapy (Stergiou et al. 2014). Notably, masked hypertension has been associated with an increased risk of development and progression of kidney disease, possibly because the elevation in blood pressure tends to be longstanding (Drawz et al. 2016).

3.6.2 Dogs and Cats

Several small studies have compared blood pressure measurements obtained in the hospital and the clinical environment in dogs (Remillard et al. 1991; Bragg et al. 2015; Kallet et al. 1997) and cats (Quimby et al. 2011), using standard, clinically applicable noninvasive methods. In dogs, the results of the studies have been discordant with some studies finding no difference in blood pressure measurements in the two environments (Remillard et al. 1991; Kallet et al. 1997), while another found that the blood pressure increased by an average of 16% (95% confidence intervals 8.8–24%) in the clinical environment with one dog having an increase in systolic blood pressure of 68 mmHg (Bragg et al. 2015). A study of greyhounds found that their blood pressure was lower at home regardless of whether it was measured by the study investigator (a relative stranger to them) or by the owner (Marino et al. 2011). The authors concluded that the reason that greyhounds have high blood pressure measurements could be a profound white-coat effect in this breed, although the authors did not present any evidence to show that the same increase does not occur in other breeds. A training effect has also been reported in laboratory beagles in response to repeated blood pressure measurement with a decrease in values being documented over 4 or 5 measurement sessions, even without a change in their environment (Schellenberg et al. 2007).

Although statistically significant increases in systolic blood pressure have been documented in cats in the clinic compared with the home environment, the differences reported were small (6 mmHg) (Quimby et al. 2011). The authors of the study observed that measurement in the home environment was not as straightforward as they had anticipated; cats were actually more likely to object to manipulation by struggling and vocalising than when they were in the clinic.

Studies in instrumented cats having blood pressure measured using telemetric methods have demonstrated that there can be marked increases in blood pressure in a simulated hospital environment (Belew et al. 1999). Overall systolic blood pressure during a mock hospital visit exceeded the 24-h average by nearly 18 mmHg, but marked variability in response was evident, with one cat having an increase of 75 mmHg, while in a few cats the blood pressure actually decreased. The magnitude and duration of the white-coat effect was greater in cats with induced (subtotal nephrectomy) renal disease than in control animals with intact kidneys (Belew et al. 1999).

The recently published ACVIM hypertension consensus statement advocates use of the term 'situational hypertension' rather than 'white-coat' hypertension to reflect the fact that blood pressure can also increase due to excitement in nonclinical situations (Acierno et al. 2018). However, for consistency with terminology in humans, the term 'white coat' has been used here.

In summary, there are parallels with blood pressure measurements in humans in that blood pressure may increase or decrease in the hospital environment. Unfortunately, it does not seem feasible to reduce the stress of the measurement process by excluding the veterinarian and/or nurse as has been done with out-of-clinic blood pressure measurement in humans. Even if the owner could be trained to perform blood pressure measurements, one study somewhat paradoxically found that the blood pressures of cats being petted by unfamiliar people were actually lower than when the same thing was done by someone familiar to them (Slingerland et al. 2008). However, some comfort should be taken from the fact that the inherent limitations of blood pressure measurement in the clinical environment are also profound in humans and in spite of that still have value for predicting adverse outcomes.

3.7 Diurnal Variation

3.7.1 Humans

In humans, there is an established diurnal pattern of blood pressure variation with the highest values typically occurring in the morning. Blood pressure usually falls by 10–20% when subjects are asleep, a pattern referred to as 'dipping'

(O'Brien et al. 1988). A non-dipping pattern where the blood pressure does not decrease, or reverse dipping where the blood pressure increases, is associated with increased cardiovascular risk (Brotman et al. 2008; Ohkubo et al. 2002). Whether this is due to intrinsic circadian rhythms or is due to environmental and behavioural effects is debatable (Chen and Yang 2015).

3.7.2 Dogs and Cats

Patterns of blood pressure variability over the day have not been studied in client-owned animals. However, diurnal variations in blood pressure in instrumented, laboratory-housed dogs and cats have been reported (Mishina et al. 1999; Miyazaki et al. 2002; Brown et al. 1997). Blood pressure elevation is associated with activity and feeding in both species, with no consistent differences in day and night (light and dark) blood pressure if these times are excluded from the analysis. Increases in blood pressure were approximately 10 mmHg when averaged over a 2-h feeding period; transiently, the increases were much greater than this (Mishina et al. 1999).

References

Aburto NJ, Ziolkovska A, Hooper L et al (2013) Effect of lower sodium intake on health: systematic review and meta-analyses. BMJ 346:f1326
Acierno MJ, Brown S, Coleman AE et al (2018) ACVIM consensus statement: guidelines for the identification, evaluation, and management of systemic hypertension in dogs and cats. J Vet Intern Med 32:1803–1822
Adrogué HJ, Madias NE (2007) Sodium and potassium in the pathogenesis of hypertension. N Engl J Med 356:1966–1978
Al-Ghazlat SA, Langston CE, Greco DS et al (2011) The prevalence of microalbuminuria and proteinuria in cats with diabetes mellitus. Top Companion Anim Med 26:154–157
Anderson LJ, Fisher EW (1968) The blood pressure in canine interstitial nephritis. Res Vet Sci 9:304–313
Anonymous (1977) Report of the joint National Committee on detection, evaluation, and treatment of high blood pressure. A cooperative study. JAMA 237:255–261
Armario P, Calhoun DA, Oliveras A et al (2017) Prevalence and clinical characteristics of refractory hypertension. J Am Heart Assoc 6:e007365
Ash RA, Harvey AM, Tasker S (2005) Primary hyperaldosteronism in the cat: a series of 13 cases. J Feline Med Surg 7:173–182
Banegas JR, Ruilope LM, de la Sierra A et al (2018) Relationship between clinic and ambulatory blood-pressure measurements and mortality. N Engl J Med 378:1509–1520
Barthez PY, Marks SL, Woo J et al (1997) Pheochromocytoma in dogs: 61 cases (1984-1995). J Vet Intern Med 11:272–278
Bavishi C, Goel S, Messerli FH (2016) Isolated systolic hypertension: an update after SPRINT. Am J Med 129:1251–1258
Beaney T, Schutte AE, Tomaszewski M et al (2018) May measurement month 2017: an analysis of blood pressure screening results worldwide. Lancet Glob Health 6:e736–e743
Beckett NS, Peters R, Fletcher AE et al (2008) Treatment of hypertension in patients 80 years of age or older. N Engl J Med 358:1887–1898

Belew AM, Barlett T, Brown SA (1999) Evaluation of the white-coat effect in cats. J Vet Intern Med 13:134–142

Biino G, Parati G, Concas MP et al (2013) Environmental and genetic contribution to hypertension prevalence: data from an epidemiological survey on Sardinian genetic isolates. PLoS One 8: e59612

Bijsmans ES, Jepson RE, Chang YM et al (2015) Changes in systolic blood pressure over time in healthy cats and cats with chronic kidney disease. J Vet Intern Med 29:855–861

Bodey AR, Michell AR (1996) Epidemiological study of blood pressure in domestic dogs. J Small Anim Pract 37:116–125

Bodey AR, Sansom J (1998) Epidemiological study of blood pressure in domestic cats. J Small Anim Pract 39:567–573

Borgarelli M, Savarino P, Crosara S et al (2008) Survival characteristics and prognostic variables of dogs with mitral regurgitation attributable to Myxomatous valve disease. J Vet Intern Med 22:120–128

Boswood A, Häggström J, Gordon SG et al (2016) Effect of Pimobendan in dogs with preclinical Myxomatous mitral valve disease and cardiomegaly: the EPIC study—a randomized clinical trial. J Vet Intern Med 30:1765–1779

Bouayad H, Feeney DA, Caywood DD et al (1987) Pheochromocytoma in dogs: 13 cases (1980-1985). J Am Vet Med Assoc 191:1610–1615

Braga ET, Leite JH, Rosa FA et al (2015) Hypertension and its correlation with renal lesions in dogs with leishmaniosis. Revista brasileira de parasitologia veterinaria = Brazilian journal of veterinary parasitology : Orgao Oficial do Colegio Brasileiro de Parasitologia Veterinaria 24:45–51

Bragg RF, Bennett JS, Cummings A et al (2015) Evaluation of the effects of hospital visit stress on physiologic variables in dogs. J Am Vet Med Assoc 246:212–215

Bright JM, Dentino M (2002) Indirect arterial blood pressure measurement in nonsedated Irish wolfhounds: reference values for the breed. J Am Anim Hosp Assoc 38:521–526

Brotman DJ, Davidson MB, Boumitri M et al (2008) Impaired diurnal blood pressure variation and all-cause mortality. Am J Hypertens 21:92–97

Brown SA, Langford K, Tarver S (1997) Effects of certain vasoactive agents on the long-term pattern of blood pressure, heart rate, and motor activity in cats. Am J Vet Res 58:647–652

Brown CA, Munday JS, Mathur S et al (2005) Hypertensive encephalopathy in cats with reduced renal function. Vet Pathol 42:642–649

Brown S, Atkins C, Bagley R et al (2007) Guidelines for the identification, evaluation, and management of systemic hypertension in dogs and cats. J Vet Intern Med 21:542–558

Buranakarl C, Mathur S, Brown SA (2004) Effects of dietary sodium chloride intake on renal function and blood pressure in cats with normal and reduced renal function. Am J Vet Res 65:620–627

Buranakarl C, Ankanaporn K, Thammacharoen S et al (2007) Relationships between degree of azotaemia and blood pressure, urinary protein:creatinine ratio and fractional excretion of electrolytes in dogs with renal azotaemia. Vet Res Commun 31:245–257

Cappuccio FP, Kerry SM, Forbes L et al (2004) Blood pressure control by home monitoring: meta-analysis of randomised trials. BMJ 329:145

Cappuccio FP, Kerry SM, Adeyemo A et al (2008) Body size and blood pressure: an analysis of Africans and the African diaspora. Epidemiology 19:38–46

Carlos Sampedrano C, Chetboul V, Gouni V et al (2006) Systolic and diastolic myocardial dysfunction in cats with hypertrophic cardiomyopathy or systemic hypertension. J Vet Intern Med 20:1106–1115

Carter JM, Irving AC, Bridges JP et al (2014) The prevalence of ocular lesions associated with hypertension in a population of geriatric cats in Auckland, New Zealand. N Z Vet J 62:21–29

Chakrabarti S, Syme HM, Elliott J (2012) Clinicopathological variables predicting progression of Azotemia in cats with chronic kidney disease. J Vet Intern Med 26:275–281

Chalifoux A, Dallaire A, Blais D et al (1985) Evaluation of the arterial blood pressure of dogs by two noninvasive methods. Can J Comp Med 49:419–423

Charles L, Triscott J, Dobbs B (2017) Secondary hypertension: discovering the underlying cause. Am Fam Physician 96:453–461

Chen L, Yang G (2015) Recent advances in circadian rhythms in cardiovascular system. Front Pharmacol 6:71

Chen HY, Lien YH, Huang HP (2016) Association of Renal Resistive Index, renal Pulsatility index, systemic hypertension, and albuminuria with survival in dogs with pituitary-dependent Hyperadrenocorticism. Int J Endocrinol 2016:3814034

Chetboul V, Lefebvre HP, Pinhas C et al (2003) Spontaneous feline hypertension: clinical and echocardiographic abnormalities, and survival rate. J Vet Intern Med 17:89–95

Chetboul V, Reynolds BS, Trehiou-Sechi E et al (2014) Cardiovascular effects of dietary salt intake in aged healthy cats: a 2-year prospective randomized, blinded, and controlled study. PLoS One 9:e97862

Chmielewski J, Carmody JB (2017) Dietary sodium, dietary potassium, and systolic blood pressure in US adolescents. J Clin Hypertens (Greenwich) 19:904–909

Chobanian AV, Bakris GL, Black HR et al (2003) Seventh report of the joint National Committee on prevention, detection, evaluation, and treatment of high blood pressure. Hypertension 42:1206–1252

Church ME, Turek BJ, Durham AC (2019) Neuropathology of spontaneous hypertensive encephalopathy in cats. Vet Pathol 56(5):778–782

Clark CE, Horvath IA, Taylor RS et al (2014) Doctors record higher blood pressures than nurses: systematic review and meta-analysis. Br J Gen Pract 64:e223–e232

Coffman TM, Crowley SD (2008) Kidney in hypertension: guyton redux. Hypertension 51:811–816

Cole L, Jepson R, Humm K (2017) Systemic hypertension in cats with acute kidney injury. J Small Anim Pract 58:577–581

Conen D, Aeschbacher S, Thijs L et al (2014) Age-specific differences between conventional and ambulatory daytime blood pressure values. Hypertension 64:1073–1079

Conroy M, Brodbelt DC, O'Neill D et al (2019) Chronic kidney disease in cats attending primary care practice in the UK: a VetCompass study. Vet Rec 184:526–526

Cortadellas O, del Palacio MJ, Bayon A et al (2006) Systemic hypertension in dogs with leishmaniasis: prevalence and clinical consequences. J Vet Intern Med 20:941–947

Coulter DB, Keith JC Jr (1984) Blood pressures obtained by indirect measurement in conscious dogs. J Am Vet Med Assoc 184:1375–1378

Cox RH, Peterson LH, Detweiler DK (1976) Comparison of arterial hemodynamics in the mongrel dog and the racing greyhound. Am J Phys 230:211–218

Cushman WC, Evans GW, Byington RP et al (2010) Effects of intensive blood-pressure control in type 2 diabetes mellitus. N Engl J Med 362:1575–1585

Diaz KM, Veerabhadrappa P, Brown MD et al (2015) Prevalence, determinants, and clinical significance of masked hypertension in a population-based sample of African Americans: the Jackson heart study. Am J Hypertens 28:900–908

Drawz PE, Alper AB, Anderson AH et al (2016) Masked hypertension and elevated Nighttime blood pressure in CKD: prevalence and association with target organ damage. Clin J Am Soc Nephrol 11:642–652

Elliott J, Barber PJ, Syme HM et al (2001) Feline hypertension: clinical findings and response to antihypertensive treatment in 30 cases. J Small Anim Pract 42:122–129

Ettehad D, Emdin CA, Kiran A et al (2016) Blood pressure lowering for prevention of cardiovascular disease and death: a systematic review and meta-analysis. Lancet 387:957–967

Evans W (1948) Hypertension. In: Cardiology. Paul B. Hoeber, Inc, London, England, p 204

Flack JM, Neaton J, Grimm R Jr et al (1995) Blood pressure and mortality among men with prior myocardial infarction. Multiple risk factor intervention trial research group. Circulation 92:2437–2445

Fracassi F, Corradini S, Floriano D et al (2015) Prognostic factors for survival in dogs with pituitary-dependent hypercortisolism treated with trilostane. Vet Rec 176:49

Francey T, Cowgill LD (2004) Hypertension in dogs with severe acute renal failure. J Vet Intern Med 18:418

Garies S, Hao S, McBrien K et al (2019) Prevalence of hypertension, treatment, and blood pressure targets in Canada associated with the 2017 American College of Cardiology and American Heart Association Blood Pressure Guidelines Hypertension, treatment, and blood pressure targets and ACC/AHA guidelines in Canada Hypertension, treatment, and blood pressure targets and ACC/AHA guidelines in Canada. JAMA Netw Open 2:e190406

Geigy CA, Schweighauser A, Doherr M et al (2011) Occurrence of systemic hypertension in dogs with acute kidney injury and treatment with amlodipine besylate. J Small Anim Pract 52:340–346

Gilson SD, Withrow SJ, Wheeler SL et al (1994) Pheochromocytoma in 50 dogs. J Vet Intern Med 8:228–232

Goy-Thollot I, Pechereau D, Keroack S et al (2002) Investigation of the role of aldosterone in hypertension associated with spontaneous pituitary-dependent hyperadrenocorticism in dogs. J Small Anim Pract 43:489–492

Granger JP, West D, Scott J (1994) Abnormal pressure natriuresis in the dog model of obesity-induced hypertension. Hypertension 23:I8–11

Greco DS, Lees GE, Dzendzel G et al (1994) Effects of dietary sodium intake on blood pressure measurements in partially nephrectomized dogs. Am J Vet Res 55:160–165

Group AS, Cushman WC, Evans GW et al (2010) Effects of intensive blood-pressure control in type 2 diabetes mellitus. N Engl J Med 362:1575–1585

Guy MK, Page RL, Jensen WA et al (2015) The Golden retriever lifetime study: establishing an observational cohort study with translational relevance for human health. Philos Trans R Soc Lond Ser B Biol Sci 370

Hall JE, Brands MW, Henegar JR (1999) Mechanisms of hypertension and kidney disease in obesity. Ann N Y Acad Sci 892:91–107

Hall JE, do Carmo JM, da Silva AA et al (2019) Obesity, kidney dysfunction and hypertension: mechanistic links. Nat Rev Nephrol 15:367–385

Hanninen MR, Niiranen TJ, Puukka PJ et al (2012) Prognostic significance of masked and white-coat hypertension in the general population: the Finn-home study. J Hypertens 30:705–712

Herring IP, Panciera DL, Werre SR (2014) Longitudinal prevalence of hypertension, proteinuria, and retinopathy in dogs with spontaneous diabetes mellitus. J Vet Intern Med 28:488–495

Hoareau GL, Jourdan G, Mellema M et al (2012) Evaluation of arterial blood gases and arterial blood pressures in brachycephalic dogs. J Vet Intern Med 26:897–904

Hoffman JM, Lourenco BN, Promislow DEL et al (2018) Canine hyperadrenocorticism associations with signalment, selected comorbidities and mortality within north American veterinary teaching hospitals. J Small Anim Pract 59:681–690

Hoglund K, Hanas S, Carnabuci C et al (2012) Blood pressure, heart rate, and urinary catecholamines in healthy dogs subjected to different clinical settings. J Vet Intern Med 26:1300–1308

Hori Y, Heishima Y, Yamashita Y et al (2018) Relationship between indirect blood pressure and various stages of chronic kidney disease in cats. J Vet Med Sci 80:447–452

Hori Y, Heishima Y, Yamashita Y et al (2019) Epidemiological study of indirect blood pressure measured using oscillometry in clinically healthy cats at initial evaluation. J Vet Med Sci 81:513–516

Intersalt cooperative research group (1988) Intersalt: an international study of electrolyte excretion and blood pressure. Results for 24 hour urinary sodium and potassium excretion. Intersalt cooperative research group. BMJ 297:319–328

Jacob F, Polzin DJ, Osborne CA et al (2003) Association between initial systolic blood pressure and risk of developing a uremic crisis or of dying in dogs with chronic renal failure. J Am Vet Med Assoc 222:322–329

Jacob F, Polzin DJ, Osborne CA et al (2005) Evaluation of the association between initial proteinuria and morbidity rate or death in dogs with naturally occurring chronic renal failure. J Am Vet Med Assoc 226:393–400

Javadi S, Djajadiningrat-Laanen SC, Kooistra HS et al (2005) Primary hyperaldosteronism, a mediator of progressive renal disease in cats. Domest Anim Endocrinol 28:85–104

Jenkins TL, Coleman AE, Schmiedt CW et al (2015) Attenuation of the pressor response to exogenous angiotensin by angiotensin receptor blockers and benazepril hydrochloride in clinically normal cats. Am J Vet Res 76:807–813

Jepson RE, Elliott J, Brodbelt D et al (2007) Effect of control of systolic blood pressure on survival in cats with systemic hypertension. J Vet Intern Med 21:402–409

Jones DW, Kim JS, Andrew ME et al (1994) Body mass index and blood pressure in Korean men and women: the Korean National Blood Pressure Survey. J Hypertens 12:1433–1437

Jose Lahm Cardoso M, Fagnani R, Zaghi Cavalcante C et al (2016) Blood pressure, serum glucose, cholesterol, and triglycerides in dogs with different body scores. Vet Med Int 2016:8675283

Kaeberlein M, Creevy KE, Promislow DE (2016) The dog aging project: translational geroscience in companion animals. Mamm Genome 27:279–288

Kallet AJ, Cowgill LD, Kass PH (1997) Comparison of blood pressure measurements obtained in dogs by use of indirect oscillometry in a veterinary clinic versus at home. J Am Vet Med Assoc 210:651–654

Klahr S, Levey AS, Beck GJ et al (1994) The effects of dietary protein restriction and blood-pressure control on the progression of chronic renal disease. N Engl J Med 330:877–884

Klein R, Klein BE, Moss SE et al (1993) Blood pressure, hypertension and retinopathy in a population. Trans Am Ophthalmol Soc 91:207–226

Kobayashi DL, Peterson ME, Graves TK et al (1990) Hypertension in cats with chronic renal failure or hyperthyroidism. J Vet Intern Med 4:58–62

Kotchen TA (2011) Historical trends and milestones in hypertension research. Hypertension 58:522–538

Kovesdy CP (2017) Hypertension in chronic kidney disease after the systolic blood pressure intervention trial: targets, treatment and current uncertainties. Nephrol Dial Transplant 32:ii219–ii223

Kyles AE, Gregory CR, Wooldridge JD et al (1999) Management of hypertension controls postoperative neurologic disorders after renal transplantation in cats. Vet Surg 28:436–441

Lane D, Beevers DG, Lip GY (2002) Ethnic differences in blood pressure and the prevalence of hypertension in England. J Hum Hypertens 16:267–273

LeBlanc NL, Stepien RL, Bentley E (2011) Ocular lesions associated with systemic hypertension in dogs: 65 cases (2005–2007). J Am Vet Med Assoc 238:915–921

Lewington S, Clarke R, Qizilbash N et al (2002) Age-specific relevance of usual blood pressure to vascular mortality: a meta-analysis of individual data for one million adults in 61 prospective studies. Lancet 360:1903–1913

Lien YH, Hsiang TY, Huang HP (2010) Associations among systemic blood pressure, microalbuminuria and albuminuria in dogs affected with pituitary- and adrenal-dependent hyperadrenocorticism. Acta Vet Scand 52:61

Lin CH, Yan CJ, Lien YH et al (2006) Systolic blood pressure of clinically normal and conscious cats determined by an indirect Doppler method in a clinical setting. J Vet Med Sci 68:827–832

Littman MP (1994) Spontaneous systemic hypertension in 24 cats. J Vet Intern Med 8:79–86

Lo AJ, Holt DE, Brown DC et al (2014) Treatment of aldosterone-secreting adrenocortical tumors in cats by unilateral adrenalectomy: 10 cases (2002-2012). J Vet Intern Med 28:137–143

Luckschander N, Iben C, Hosgood G et al (2004) Dietary NaCl does not affect blood pressure in healthy cats. J Vet Intern Med 18:463–467

Maggio F, DeFrancesco TC, Atkins CE et al (2000) Ocular lesions associated with systemic hypertension in cats: 69 cases (1985-1998). J Am Vet Med Assoc 217:695–702

Mahmood SS, Levy D, Vasan RS et al (2014) The Framingham heart study and the epidemiology of cardiovascular disease: a historical perspective. Lancet 383:999–1008

Marino CL, Cober RE, Iazbik MC et al (2011) White-coat effect on systemic blood pressure in retired racing greyhounds. J Vet Intern Med 25:861–865

Marynissen SJ, Smets PM, Ghys LF et al (2016) Long-term follow-up of renal function assessing serum cystatin C in dogs with diabetes mellitus or hyperadrenocorticism. Vet Clin Pathol 45:320–329

Mehlman E, Bright JM, Jeckel K et al (2013) Echocardiographic evidence of left ventricular hypertrophy in obese dogs. J Vet Intern Med 27:62–68

Messerli FH, Panjrath GS (2009) The J-curve between blood pressure and coronary artery disease or essential hypertension: exactly how essential? J Am Coll Cardiol 54:1827–1834

Meurs KM, Miller MW, Slater MR et al (2000) Arterial blood pressure measurement in a population of healthy geriatric dogs. J Am Anim Hosp Assoc 36:497–500

Michell AR, Bodey AR, Gleadhill A (1997) Absence of hypertension in dogs with renal insufficiency. Ren Fail 19:61–68

Mishina M, Watanabe T, Fujii K et al (1997) A clinical evaluation of blood pressure through non-invasive measurement using the oscillometric procedure in conscious dogs. J Vet Med Sci 59:989–993

Mishina M, Watanabe T, Fujii K et al (1998) Non-invasive blood pressure measurements in cats: clinical significance of hypertension associated with chronic renal failure. J Vet Med Sci 60:805–808

Mishina M, Watanabe T, Matsuoka S et al (1999) Diurnal variations of blood pressure in dogs. J Vet Med Sci 61:643–647

Mishina M, Watanabe N, Watanabe T (2006) Diurnal variations of blood pressure in cats. J Vet Med Sci 68(3):243–248

Miyazaki H, Yoshida M, Samura K et al (2002) Ranges of diurnal variation and the pattern of body temperature, blood pressure and heart rate in laboratory beagle dogs. Exp Anim 51:95–98

Mooney AP, Mawby DI, Price JM et al (2017) Effects of various factors on Doppler flow ultrasonic radial and coccygeal artery systolic blood pressure measurements in privately-owned, conscious dogs. PeerJ 5:e3101

Morrow LD, Adams VJ, Elliott J et al (2009) Hypertension in hyperthyroid cats: prevalence, incidence and predictors of its development. J Vet Intern Med 23:700

Moser M (2006) Historical perspectives on the Management of Hypertension. The Journal of Clinical Hypertension 8:15–20

Muntner P, Carey RM, Gidding S et al (2018) Potential US population impact of the 2017 ACC/AHA high blood pressure guideline. Circulation 137:109–118

Murray JK, Casey RA, Gale E et al (2017) Cohort profile: the 'Bristol cats Study' (BCS)-a birth cohort of kittens owned by UK households. Int J Epidemiol 46:1749–1750e

Neal B, MacMahon S, Chapman N (2000) Effects of ACE inhibitors, calcium antagonists, and other blood-pressure-lowering drugs: results of prospectively designed overviews of randomised trials. Blood pressure lowering treatment Trialists' collaboration. Lancet 356:1955–1964

Nelson L, Reidesel E, Ware WA et al (2002) Echocardiographic and radiographic changes associated with systemic hypertension in cats. J Vet Intern Med 16:418–425

Norris CR, Nelson RW, Christopher MM (1999) Serum total and ionized magnesium concentrations and urinary fractional excretion of magnesium in cats with diabetes mellitus and diabetic ketoacidosis. J Am Vet Med Assoc 215:1455–1459

Novellas R, de Gopegui RR, Espada Y (2008) Determination of renal vascular resistance in dogs with diabetes mellitus and hyperadrenocorticism. Vet Rec 163:592–595

O'Neill J, Kent M, Glass EN et al (2013a) Clinicopathologic and MRI characteristics of presumptive hypertensive encephalopathy in two cats and two dogs. J Am Anim Hosp Assoc 49:412–420

O'Brien E, Sheridan J, O'Malley K (1988) Dippers and non-dippers. Lancet 2:397

Ohkubo T, Hozawa A, Yamaguchi J et al (2002) Prognostic significance of the nocturnal decline in blood pressure in individuals with and without high 24-h blood pressure: the Ohasama study. J Hypertens 20:2183–2189

Olsen MH, Angell SY, Asma S et al (2016) A call to action and a lifecourse strategy to address the global burden of raised blood pressure on current and future generations: the lancet commission on hypertension. Lancet 388:2665–2712

O'Neill DG, Elliott J, Church DB et al (2013b) Chronic kidney disease in dogs in UK veterinary practices: prevalence, risk factors, and survival. J Vet Intern Med 27:814–821

Ortega TM, Feldman EC, Nelson RW et al (1996) Systemic arterial blood pressure and urine protein/creatinine ratio in dogs with hyperadrenocorticism. J Am Vet Med Assoc 209:1724–1729

Page LB, Damon A, Moellering RC Jr (1974) Antecedents of cardiovascular disease in six Solomon Islands societies. Circulation 49:1132–1146

Paige CF, Abbott JA, Elvinger F et al (2009) Prevalence of cardiomyopathy in apparently healthy cats. J Am Vet Med Assoc 234:1398–1403

Parati G, Stergiou GS, Asmar R et al (2010) European Society of Hypertension practice guidelines for home blood pressure monitoring. J Hum Hypertens 24:779–785

Payne JR, Brodbelt DC, Luis FV (2015) Cardiomyopathy prevalence in 780 apparently healthy cats in rehoming centres (the CatScan study). J Vet Cardiol 17(Suppl 1):S244–S257

Payne JR, Brodbelt DC, Luis Fuentes V (2017) Blood pressure measurements in 780 apparently healthy cats. J Vet Intern Med 31:15–21

Peppard PE, Young T, Palta M et al (2000) Prospective study of the association between sleep-disordered breathing and hypertension. N Engl J Med 342:1378–1384

Perez-Sanchez AP, Del-Angel-Caraza J, Quijano-Hernandez IA et al (2015) Obesity-hypertension and its relation to other diseases in dogs. Vet Res Commun 39:45–51

Petit AM, Gouni V, Tissier R et al (2013) Systolic arterial blood pressure in small-breed dogs with degenerative mitral valve disease: a prospective study of 103 cases (2007–2012). Vet J 197:830–835

Priyanka M, Jeyaraja K, Thirunavakkarasu PS (2018) Abnormal renovascular resistance in dogs with diabetes mellitus: correlation with glycemic status and proteinuria. Iranian Journal of Veterinary Research 19:304–309

Quimby JM, Smith ML, Lunn KF (2011) Evaluation of the effects of hospital visit stress on physiologic parameters in the cat. J Feline Med Surg 13:733–737

Rahimi K, Emdin CA, MacMahon S (2015) The epidemiology of blood pressure and its worldwide management. Circ Res 116:925–936

Rapoport GS, Stepien R (2001) Direct arterial blood pressure measurement in 54 dogs presented for systemic hypertension screening 1998–2001. In: ESVIM Congress, Dublin, Ireland, 2001, p 62 (abstract)

Remillard RL, Ross JN, Eddy JB (1991) Variance of indirect blood pressure measurements and prevalence of hypertension in clinically normal dogs. Am J Vet Res 52:561–565

Rondeau DA, Mackalonis ME, Hess RS (2013) Effect of body position on indirect measurement of systolic arterial blood pressure in dogs. J Am Vet Med Assoc 242:1523–1527

Rose G, Day S (1990) The population mean predicts the number of deviant individuals. BMJ 301:1031–1034

Sansom J, Rogers K, Wood JL (2004) Blood pressure assessment in healthy cats and cats with hypertensive retinopathy. Am J Vet Res 65:245–252

Sartor LL, Trepanier LA, Kroll MM et al (2004) Efficacy and safety of transdermal Methimazole in the treatment of cats with hyperthyroidism. J Vet Intern Med 18:651–655

Scansen BA, Vitt J, Chew DJ et al (2014) Comparison of forelimb and hindlimb systolic blood pressures and proteinuria in healthy Shetland sheepdogs. J Vet Intern Med 28:277–283

Schellenberg S, Glaus TM, Reusch CE (2007) Effect of long-term adaptation on indirect measurements of systolic blood pressure in conscious untrained beagles. Vet Rec 161:418–421

Schellenberg S, Mettler M, Gentilini F et al (2008) The effects of hydrocortisone on systemic arterial blood pressure and urinary protein excretion in dogs. J Vet Intern Med 22:273–281

Sennello KA, Schulman RL, Prosek R et al (2003) Systolic blood pressure in cats with diabetes mellitus. J Am Vet Med Assoc 223:198–201

Siu AL (2015) Screening for high blood pressure in adults: U.S. preventive services task force recommendation statement. Ann Intern Med 163:778–786

Slingerland LI, Robben JH, Schaafsma I et al (2008) Response of cats to familiar and unfamiliar human contact using continuous direct arterial blood pressure measurement. Res Vet Sci 85:575–582

Smets PM, Lefebvre HP, Meij BP et al (2012) Long-term follow-up of renal function in dogs after treatment for ACTH-dependent hyperadrenocorticism. J Vet Intern Med 26:565–574

Snyder PS, Sadek D, Jones GL (2001) Effect of amlodipine on echocardiographic variables in cats with systemic hypertension. J Vet Intern Med 15:52–56

Soloviev MV, Hamlin RL, Shellhammer LJ et al (2006) Variations in hemodynamic parameters and ECG in healthy, conscious, freely moving telemetrized beagle dogs. Cardiovasc Toxicol 6:51–62

Sparkes AH, Caney SM, King MC et al (1999) Inter- and intraindividual variation in Doppler ultrasonic indirect blood pressure measurements in healthy cats. J Vet Intern Med 13:314–318

Stamler J, Stamler R, Neaton JD (1993) Blood pressure, systolic and diastolic, and cardiovascular risks. US population data. Arch Intern Med 153:598–615

Stepien RL, Rapoport GS (1999) Clinical comparison of three methods to measure blood pressure in nonsedated dogs. J Am Vet Med Assoc 215:1623–1628

Stepien RL, Rapoport GS, Henik RA et al (2003) Effect of measurement method on blood pressure findings in cats before and after therapy for hyperthyroidism. J Vet Int Med 17:754. (abstract)

Steppan J, Barodka V, Berkowitz DE et al (2011) Vascular stiffness and increased pulse pressure in the aging cardiovascular system. Cardiol Res Pract 2011:263585

Stergiou GS, Asayama K, Thijs L et al (2014) Prognosis of white-coat and masked hypertension: international database of home blood pressure in relation to cardiovascular outcome. Hypertension 63:675–682

Stiles J, Polzin DJ, Bistner DI (1994) The prevalence of retinopathy in cats with systemic hypertension and chronic renal failure or hyperthyroidism. J Am Anim Hosp Assoc 30:654–572

Struble AL, Feldman EC, Nelson RW et al (1998) Systemic hypertension and proteinuria in dogs with diabetes mellitus. J Am Vet Med Assoc 213:822–825

Surman S, Couto CG, Dibartola SP et al (2012) Arterial blood pressure, proteinuria, and renal histopathology in clinically healthy retired racing greyhounds. J Vet Intern Med 26:1320–1329

Syme HM, Elliott J (2003) The prevalence of hypertension in hyperthyroid cats at diagnosis and following treatment. J Vet Intern Med 17:754–755

Syme HM, Barber PJ, Markwell PJ et al (2002) Prevalence of systolic hypertension in cats with chronic renal failure at initial evaluation. J Am Vet Med Assoc 220:1799–1804

Syme HM, Markwell PJ, Pfeiffer D et al (2006) Survival of cats with naturally occurring chronic renal failure is related to severity of proteinuria. J Vet Intern Med 20:528–535

Taffin ER, Paepe D, Ghys LF et al (2016) Systolic blood pressure, routine kidney variables and renal ultrasonographic findings in cats naturally infected with feline immunodeficiency virus. J Feline Med Surg 19(6):672–679

Taffin ER, Paepe D, Ghys LF et al (2017) Systolic blood pressure, routine kidney variables and renal ultrasonographic findings in cats naturally infected with feline immunodeficiency virus. J Feline Med Surg 19:672–679

The SPRINT Research Group (2015) A randomized trial of intensive versus standard blood-pressure control. N Engl J Med 373:2103–2116

Tientcheu D, Ayers C, Das SR et al (2015) Target organ complications and cardiovascular events associated with masked hypertension and white-coat hypertension: analysis from the Dallas heart study. J Am Coll Cardiol 66:2159–2169

Tippett FE, Padgett GA, Eyster G et al (1987) Primary hypertension in a colony of dogs. Hypertension 9:49–58

van Hoek I, Lefebvre HP, Peremans K et al (2009) Short- and long-term follow-up of glomerular and tubular renal markers of kidney function in hyperthyroid cats after treatment with radioiodine. Domest Anim Endocrinol 36:45–56

Vasan RS, Beiser A, Seshadri S et al (2002) Residual lifetime risk for developing hypertension in middle-aged women and men: the Framingham heart study. JAMA 287:1003–1010

Vidal PN, Miceli DD, Arias ES et al (2018) Decrease of nitric oxide and increase in diastolic blood pressure are two events that affect renal function in dogs with pituitary dependent hyperadrenocorticism. Open Vet J 8:86–95

Violette NP, Ledbetter EC (2018) Punctate retinal hemorrhage and its relation to ocular and systemic disease in dogs: 83 cases. Vet Ophthalmol 21:233–239

Watson N, Murray JK, Fonfara S et al (2018) Clinicopathological features and comorbidities of cats with mild, moderate or severe hyperthyroidism: a radioiodine referral population. J Feline Med Surg 20:1130–1137

Whelton PK, Carey RM, Aronow WS et al (2018) 2017 ACC/AHA/AAPA/ABC/ACPM/AGS/APhA/ASH/ASPC/NMA/PCNA guideline for the prevention, detection, evaluation, and Management of High Blood Pressure in adults: executive summary: a report of the American College of Cardiology/American Heart Association task force on clinical practice guidelines. Circulation 138:e426–e483

Willems A, Paepe D, Marynissen S et al (2017) Results of screening of apparently healthy senior and geriatric dogs. J Vet Intern Med 31:81–92

Williams TL, Peak KJ, Brodbelt D et al (2010) Survival and the development of Azotemia after treatment of hyperthyroid cats. J Vet Intern Med 24:863–869

Williams B, Mancia G, Spiering W et al (2018) 2018 practice guidelines for the management of arterial hypertension of the European Society of Cardiology and the European Society of Hypertension. Blood Press 27:314–340

Wilson PW, D'Agostino RB, Sullivan L et al (2002) Overweight and obesity as determinants of cardiovascular risk: the Framingham experience. Arch Intern Med 162:1867–1872

Wong TY, McIntosh R (2005) Hypertensive retinopathy signs as risk indicators of cardiovascular morbidity and mortality. Br Med Bull 73–74:57–70

Wong T, Mitchell P (2007) The eye in hypertension. Lancet 369:425–435

Xie X, Atkins E, Lv J et al (2016) Effects of intensive blood pressure lowering on cardiovascular and renal outcomes: updated systematic review and meta-analysis. Lancet 387:435–443

Young WM, Zheng C, Davidson MG et al (2019) Visual outcome in cats with hypertensive chorioretinopathy. Vet Ophthalmol 22:161–167

Zwijnenberg RJ, del Rio CL, Cobb RM et al (2011) Evaluation of oscillometric and vascular access port arterial blood pressure measurement techniques versus implanted telemetry in anesthetized cats. Am J Vet Res 72:1015–1021

Hypertension and Adrenal Gland Disease

Rosanne E. Jepson

The adrenal gland is composed of two regions, cortex and medulla with the cortex divided histopathologically and functionally into three layers: from outer to inner the zona glomerulosa, zona fasciculata and zona reticularis. The outermost layer of the cortex, the zona glomerulosa, is responsible for the production of the mineralocorticoid aldosterone due to the presence of aldosterone synthase in this region. Aldosterone has important actions within the kidney for sodium and volume status and therefore plays an important role in blood pressure (BP) regulation and the pathogenesis of hypertension. The zona fasciculata is the central layer of the cortex with cells in this region responsible for the production of glucocorticoids, e.g. cortisol. The innermost layer of the cortex, the zona reticularis, is primarily responsible for the production of androgens. Only cells within the zona reticularis and zona fasciculata can synthesize 17-alpha-hydroxypregnenolone and 17-alpha-hydroxyprogesterone due to the presence of 17-alpha-hydroxylase in these regions which represent the precursors for the production of cortisol and adrenal androgens, respectively. These two zones are regulated by adrenocorticotropic hormone (ACTH), produced in the pituitary gland, which stimulates the production of both cortisol and androgens.

The adrenal medulla is a neuroendocrine organ composed of chromaffin cells where the major secretory products are the catecholamines: epinephrine and non-epinephrine. Major biological effects of the catecholamines include increased myocardial inotropy and chronotropy and peripheral vasoconstriction with direct effects on BP regulation. Catecholamines mediate the flight and fight response and therefore its associated 'situational hypertension' (Kemppainen and Behrend 1997).

Key conditions affecting the adrenal gland in both dogs and cats manifest with adrenal gland enlargement or masses and have been associated with systemic hypertension. These conditions include hyperadrenocorticism, hyperaldosteronism

R. E. Jepson (✉)
Department of Clinical Sciences and Services, Royal Veterinary College, University of London, Hertfordshire, UK
e-mail: rjepson@rvc.ac.uk

© Springer Nature Switzerland AG 2020
J. Elliott et al. (eds.), *Hypertension in the Dog and Cat*,
https://doi.org/10.1007/978-3-030-33020-0_4

(Conn's syndrome) and pheochromocytoma. This chapter will review these conditions in relation to the pathophysiology, prevalence and considerations for management of the associated hypertension. For more specific information relating to the diagnosis and management of the individual conditions, readers are directed to comprehensive general medical texts (Behrend 2015, 2017; Reusch 2015; Feldman 2015; Perez-Alenza and Melian 2017).

4.1 Hyperadrenocorticism

Hyperadrenocorticism (HAC) or Cushing's disease has been described in both dogs and cats. The estimated prevalence of HAC in dogs from first-opinion practices is 0.28% (O'Neill et al. 2016). HAC is a condition typically affecting older dogs with a mean age at diagnosis of 11 years (range 6–20). In dogs with HAC, 80–85% demonstrate a pituitary-dependent form of disease with >90% of such dogs demonstrating an observable pituitary mass (Meij et al. 1998). Excessive production of ACTH from such pituitary mass lesions results in bilateral adrenal hyperplasia, adrenal gland enlargement and chronic hypersecretion of cortisol. Pituitary tumours may be classified as adenomas, invasive adenomas or carcinomas. In 15–20% of dogs, a primary adrenal tumour is identified with autonomous cortisol secretion. When a unilateral cortisol secreting mass is identified, the contralateral gland is typically atrophied together with atrophy of any non-neoplastic tissue in the affected adrenal gland (Perez-Alenza and Melian 2017). No clinical or biochemical parameters allow differentiation between adenoma and carcinoma to be made; however, for those dogs with a functional adrenal mass lesion >2 cm and with vascular invasion, carcinoma is more likely (Labelle et al. 2004). Although unilateral cortisol secreting masses are most commonly identified, bilateral cortisol secreting masses, bilateral masses of different origin (i.e. cortisol and pheochromocytoma) and the combination of pituitary and adrenal HAC have all been reported in dogs (Behrend 2015; Greco et al. 1999; von Dehn et al. 1995). Ectopic ACTH production from neoplastic tissue other than the adrenal or pituitary gland is also described in humans (e.g. small cell lung carcinoma, carcinoma of intestine, liver, pancreas, ovary, thymoma, thyroid carcinoma) (Hayes and Grossman 2018). This scenario has been reported in a single German Shepherd Dog with a pancreatic neuroendocrine tumour (Galac et al. 2005). Finally, the clinical manifestations of hypercortisolism can also occur in those dogs and cats that receive exogenous steroids for protracted periods of time.

4.1.1 Epidemiology of Hypertension Associated with Hyperadrenocorticism in Dogs and Cats

Retrospective studies suggest that between 31 and 86% of dogs with HAC will also demonstrate systemic hypertension (Lien et al. 2010; Nichols 1997; Smets et al. 2012a; Goy-Thollot et al. 2002). A study by Lien and colleagues suggests that the prevalence of hypertension may be higher in dogs with adrenal dependent HAC (46.7%) than pituitary (20%) and that overall dogs with adrenal dependent HAC may have significantly higher systolic BP than those with pituitary dependent HAC (Lien et al. 2010). However, accurate estimates of the prevalence of hypertension in studies evaluating HAC are hindered as standardised BP measurement is infrequently performed (Hoffman et al. 2018). In dogs treated with exogenous steroids (hydrocortisone), significant increase in systolic BP has been reported but measurements typically remain within the normotensive category (Schellenberg et al. 2008). Therefore, without concurrent disease and considering the typical duration of use, the risk of development of hypertension with exogenous steroid use in dogs is unlikely.

There is limited information in the veterinary literature about the role of hypertension in dogs with HAC and its effect on target organ damage (TOD). Given that both chronic kidney disease (CKD) and HAC affect the ageing population, it may be expected that some degree of decline in renal function could be identified concurrently with HAC. In human medicine, patients with Cushing's syndrome demonstrate decreased glomerular filtration rate (GFR), focal segmental glomerulosclerosis and increased prevalence of nephrolithiasis (Hsieh et al. 2007; Haentjens et al. 2005; Faggiano et al. 2003). Cortisol is essential for maintenance of normal renal blood flow and GFR, although there may be inter-species differences in terms of effect (Smets et al. 2010). The specific effects of HAC on renal function in dogs are poorly described, but glucocorticoids are expected to have an effect on renal function (Smets et al. 2010, 2012a, b). In aged experimental beagle dogs, where hydrocortisone was administered over a 16-week period, GFR significantly increased compared to controls but normalised on discontinuing this therapy by 24 weeks (Smets et al. 2012c). In naturally occurring pituitary-dependent HAC, there is some evidence that GFR measured by plasma clearance of exogenous creatinine decreases with trilostane treatment with concomitant increases in both urea and creatinine although both remaining within reference interval. This implies an increase in GFR in the HAC state in dogs and subsequent 'normalisation' of GFR with successful treatment (Smets et al. 2012a).

Renal proteinuria is usually the consequence of increased glomerular capillary pressures, glomerular filtration barrier injury or altered function, or a decrease in tubular reabsorption of filtered proteins. Hypertension may contribute to the increase in glomerular capillary pressures, glomerulosclerosis and therefore proteinuria. In experimental dogs where hydrocortisone was administered, significant but mild increases in proteinuria were identified and occurred concurrently with small but significant increases in systolic BP (Schellenberg et al. 2008). Proteinuria is reported to occur in approximately 44–75% of dogs with HAC with the magnitude of urine protein to creatinine ratio (UPC) values ranging between 0.1 and 5.0 (Smets et al. 2010; Hurley and Vaden 1998; Mazzi et al. 2008; Ortega et al. 1996). Collectively, studies suggest that approximately 45% of dogs with HAC will have a UPC >1.0 and approximately 70% > 0.5 (Smets et al. 2012a; Hurley and Vaden 1998; Mazzi et al. 2008; Ortega et al. 1996). Some dogs with HAC do demonstrate a reduction in proteinuria with treatment (Caragelasco et al. 2017) although Smets and colleagues suggest that up to 40% of dogs remain proteinuric after treatment (Smets et al. 2012a). Systemic hypertension has been proposed as a mechanism contributing to the development of proteinuria in dogs with HAC. However, given that some dogs with HAC are normotensive and yet exhibit proteinuria this cannot be the full explanation (Schellenberg et al. 2008; Smets et al. 2010).

Dogs with HAC have been included in studies where ophthalmic examination has been performed to assess the presence of ocular TOD associated with hypertension (Leblanc et al. 2011). However, they comprise only a small subset of patients, and therefore ascertaining the risk of hypertensive ocular TOD in association specifically with HAC is not possible from currently available veterinary data. It is worth noting that recent publications by Violette and Ledbetter reported punctate retinal haemorrhages and intracorneal stromal haemorrhage in dogs with HAC and other systemic conditions indicating that systemic hypertension is just one of many conditions that may be associated with the finding of retinal and intracorneal stromal haemorrhage, although hypertension remains an important differential (Violette and Ledbetter 2017, 2018). Similarly, a recent study has evaluated echocardiographic changes in dogs with HAC. This study identified that 68% of dogs with HAC demonstrated left ventricular hypertrophy but that only 27% of these dogs were hypertensive (Takano et al. 2015). As in humans, HAC may therefore be associated with the development of left ventricular hypertrophy, and the impact of this should be carefully considered if echocardiographic assessment is being used to assess for evidence of systemic hypertension and cardiac TOD (Muiesan et al. 2003). Hypertensive encephalopathy is rarely reported in dogs (O'Neill et al. 2013). However, it is important to remember that where central neurological manifestations are identified in dogs with HAC, this could be the result of progressive enlargement of a pituitary mass rather than being associated with systemic hypertension. In this scenario, measurement of BP and advanced intracranial imaging are recommended.

Hyperadrenocorticism is a rare condition in the cat (Feldman 2015). In general, cats are regarded to be less sensitive to the effects of glucocorticoid excess than dogs, which may partially account for the reduced prevalence of this condition. When naturally occurring HAC is identified, the proportion of cats with pituitary versus adrenal dependent disease is similar to that in dogs, with data suggesting that approximately 80% of cases will be pituitary in origin and 20% adrenal (adenoma versus carcinoma) (Feldman 2015). Most, but not all cats diagnosed with HAC have a concurrent diagnosis of diabetes mellitus (DM). Iatrogenic HAC has been reported in cats. In experimental studies, periods of 4–9 weeks of steroid administration do not often result in clinical manifestations of HAC, although a minority may begin to show typical clinical signs, e.g. polydipsia, polyuria, abdominal enlargement, thin skin, polyphagia and curling of the ear tips (Lowe et al. 2008). Some cats treated with long-term steroids may exhibit mild hyperglycaemia, and a few will develop overt hyperglycaemia and secondary DM. The median age at diagnosis of pituitary HAC in cats is 11 years (range 5–17 years) and for adrenal HAC 12 years (range 8–15 years) (Feldman 2015).

There is virtually no data available on the blood pressure distribution of cats that are diagnosed with iatrogenic, pituitary or adrenal HAC. However, given the associations that are identified in both dogs and humans with hypertension in HAC, routine BP assessment should be performed in all cats where HAC is either suspected or diagnosed. A single case report identified systemic hypertension in a cat with pituitary-dependent HAC which resolved after bilateral adrenalectomy but recurred after 19 months and with the onset of azotaemic CKD (Brown et al. 2012). Given the age of onset of HAC in cats, concurrent CKD could be hypothesised to be contributory to the pathogenesis of hypertension in some of the cases where it occurs. Evaluation of renal function is therefore recommended in all older cats where a diagnosis of HAC is being made along with BP monitoring.

4.1.2 Pathophysiology of Hypertension in Hyperadrenocorticism

The pathogenesis of hypertension associated with HAC or hypercortisolism is complex and incompletely understood. Hypertension is identified in approximately 70–85% of adult humans with endogenous hypercortisolism, whilst the prevalence is closer to 20% of those patients who are treated with long-term exogenous corticosteroids (Isidori et al. 2015). However, interestingly the presence of hypertension in those individuals with endogenous hypercortisolism can persist beyond the biochemical and clinical resolution of disease (Isidori et al. 2015). The pathophysiology of hypertension associated with hypercortisolism is believed to be multifactorial and is summarised in Table 4.1. Little information is available specifically relating to the pathogenesis of hypertension associated with HAC in the dog, and no data are available for the cat.

Renin-Angiotensin System
In humans with endogenous hypercortisolism, a number of changes in the renin-angiotensin system may contribute to the development of hypertension. These include increase in angiotensinogen, the substrate for the production of angiotensin II, and increases in expression of angiotensin II receptors (type 1A) leading to an enhanced pressor response, e.g. on infusion of angiotensin II (Isidori et al. 2015; Suzuki et al. 1986; Shibata et al. 1995; Yasuda et al. 1994). This apparent dysregulation, in addition to a good response to ACEi and ARB medications,

Table 4.1 Mechanisms proposed to be involved in the pathogenesis of hypertension associated with hypercortisolism in humans

• Activation of the renin-angiotensin system
• Mineralocorticoid activity of cortisol
• Enhanced pressor response to catecholamines, vasopressin and angiotensin II
• Increased beta-adrenergic sensitivity
• Inhibition of vasodilatory mechanisms, e.g. nitric oxide synthase, prostacyclin and kinin-kallikrein system
• Increased vascular resistance, atherosclerosis and metabolic syndrome
• Sleep apnoea

supports the involvement of the renin-angiotensin system in the pathogenesis of hypertension in humans with hypercortisolism. In humans, ACEi and ARB are therefore considered first-line anti-hypertensive agents for individuals with endogenous hypercortisolism (Isidori et al. 2015). To date there is no data exploring specifically angiotensinogen, angiotensin II, plasma renin activity or angiotensin II receptor density in dogs with hyperadrenocorticism.

Mineralocorticoids
Mineralocorticoid activity is also believed to play a role in the development of hypertension in hypercortisolism. The mineralocorticoid receptor (MR) is a nuclear receptor identified primarily within renal tissue. It can be stimulated by either aldosterone or cortisol. In health, the MR is protected from activation by cortisol due to the presence of the enzyme 11 beta-hydroxysteroid dehydrogenase (11beta-HSD), which catalyses the conversion of cortisol to cortisone which is not able to stimulate the MR. However, severe hypercortisolism overwhelms this protective mechanism and excessive concentrations of cortisol are able to bind to the MR inducing both sodium retention and potassium excretion (Isidori et al. 2015; Yasuda et al. 1994). However, in human medicine, whilst it is accepted that this mechanism may be contributory to the development of hypokalaemia, there continues to be debate with regard to the extent that it contributes to sodium retention and the development of hypertension (Isidori et al. 2015). It is hypothesised that this mechanism may be more relevant in the situation of acute glucocorticoid excess but less relevant for chronic hypercortisolism where sodium concentrations and excretion may be normal (Connell et al. 1987; Bailey et al. 2009). Another isoform of 11beta-HSD (type 1) is identified in other regions of the body including the liver, adipose tissue, heart, vascular endothelial and smooth muscle cells and catalyses the re-activation of cortisone to cortisol, therefore modulating local bioavailability of cortisol (Quinkler and Stewart 2003). In addition, glucocorticoid receptors may also play a role in renal handling of sodium and enhancing insertion and activity of the sodium channels (ENaC) (Bailey et al. 2009). Evidence for this comes from individuals from hypercortisolism that respond better to selective glucocorticoid receptor antagonists in terms of their hypertension (e.g. mifepristone) than to aldosterone antagonists (Nisula et al. 1985). Vascular MRs may also play an important role in determining vascular tone and cardiac remodelling that is seen with hypercortisolism (Isidori et al. 2015).

In dogs, aldosterone concentrations have been explored as an initiating or perpetuating factor in the development of hypertension associated with pituitary-dependent HAC (Goy-Thollot et al. 2002). Thirteen dogs were reviewed before and after starting mitotane therapy. In the untreated pituitary-dependent HAC dogs, aldosterone was significantly decreased, whilst cortisol, sodium and systolic BP were significantly increased compared to healthy control dogs (Goy-Thollot et al. 2002). Similar results in terms of aldosterone concentrations in pituitary-dependent HAC have been identified in the study by Javadi and colleagues although in just five dogs with adrenal-dependent HAC, aldosterone concentrations were significantly increased compared to the pituitary cases and healthy controls (Javadi et al. 2003).

However, BP assessment was not available in this study, and therefore any relationship between aldosterone and hypertension cannot be explored. In the study by Javadi and colleagues, there were no differences in plasma renin activity between control dogs and those with pituitary- and adrenal-dependent HAC (Javadi et al. 2003). In the study by Goy-Thollot, despite a significant reduction in cortisol after mitotane treatment, there was no significant reduction in systolic BP with dogs remaining hypertensive and aldosterone concentrations remained low (Goy-Thollot et al. 2002). This does not support the supposition that excessive aldosterone concentrations are involved in the pathogenesis of hypertension in dogs with pituitary-dependent HAC.

Sympathetic Nervous System
Historically, there was interest in the sympathetic nervous system being intrinsically involved in the pathogenesis of hypertension associated with hypercortisolism in humans. However, in humans with Cushing's disease studies suggest no alteration in concentrations of circulating catecholamines (e.g. epinephrine and nor-epinephrine) and no alteration in adrenergic receptor density (Isidori et al. 2015). Pressor responses to the administration of adrenergic agonists have been variable with some studies suggesting an increased response and others no difference from control patients (Heaney et al. 1999; McKnight et al. 1995). Based on the lack of evidence, beta-blockers are not considered first line anti-hypertensive agents in humans with hypercortisolism (Isidori et al. 2015).

A study by Martinez has explored the pressor response to infusion of norepinephrine in dogs with iatrogenic HAC. In this experimental study, dogs with iatrogenic HAC through administration of oral hydrocortisone had a more pronounced pressor response to norepinephrine infusion than the control dogs, and indeed the infusions had to be stopped in 7/8 dogs due to the development of severe hypertension (Martínez et al. 2005). It is possible that the role of the sympathetic nervous system in the pathogenesis of hypertension with HAC is different between species. However, the use of a short-term experimental model may also not extrapolate to findings in naturally occurring disease. The role of the sympathetic nervous system in hypertension in dogs associated with HAC therefore remains uncertain.

Regulation of Vascular Tone
Altered response to vasoactive mediators or density of vasoactive mediator receptors is also proposed as a mechanism involved in the pathogenesis of hypertension associated with HAC. There has been suggestion for increased circulating endothelin-1 concentrations, potential for down-regulation of the sodium-calcium transporter in aortic smooth muscle and a negative impact of glucocorticoids on the nitric oxide synthase pathway (Isidori et al. 2015). In particular, glucocorticoids may interact and impair the nitric oxide synthase pathway by down-regulating nitric oxide synthase, impairing the substrate (L-arginine) transport system and suppressing tetrahydrobiopterin synthesis which is an important co-factor for nitric oxide synthase (Isidori et al. 2015). In human patients with hypercortisolism, reduced urinary and plasma nitric oxide metabolites (e.g. nitrite/nitrate) have been

reported (Kelly et al. 1998). The role of other vasoactive mediators such as prostaglandins, prostacyclin, vascular endothelial growth factor and the kallikrein-kinin system remain to be fully elucidated (Isidori et al. 2015).

To date there is extremely limited information on the role of vasoactive mediators in dogs in relation to the development of hypertension in HAC. One study has explored cortisol, nitric oxide (nitrate/nitrite) and systolic BP measurements in 20 dogs with pituitary-dependent HAC and 12 control dogs (Vidal et al. 2018). In this study, significantly lower nitrate/nitrite concentrations were reported in the dogs with HAC compared to controls, and there was a significant negative correlation with systolic BP (Vidal et al. 2018). These data suggest that similar to humans, reduced NO availability might be part of the pathogenesis of hypertension in dogs with HAC but further work would be required to confirm this hypothesis.

In humans there is also concern regarding the relationship between metabolic syndrome, hypercortisolism and hypertension. Data support a link between the development of visceral adiposity, insulin resistance and premature atherosclerosis in patients with Cushing's disease. However, the relevance of these findings to dogs remains uncertain at this time.

4.1.3 Treatment of Hypertension in Dogs and Cats with Hyperadrenocorticism

It is recommended that all dogs diagnosed with HAC have regular BP assessment performed, not only at diagnosis but also as part of routine monitoring of their condition. In those dogs where severe hypertension is diagnosed (Chap. 3) instituting anti-hypertensive therapy together with treatment for HAC is recommended. Specific information on the management of hypertension in dogs can be found in Chap. 13. Given that many dogs with HAC may exhibit concurrent hypertension and proteinuria, first-line anti-hypertensive therapy with either an angiotensin receptor blocker (ARB), e.g. telmisartan, or angiotensin-converting enzyme inhibitor (ACEi) is recommended. Dogs should be carefully monitored in terms of reduction in both systolic BP and proteinuria having started this therapy. However, control of hypertension in dogs, including those with HAC, can be challenging, and it is therefore frequently necessary to consider use of a second anti-hypertensive agent such as the calcium channel blocker amlodipine besylate. Careful monitoring of systolic BP should be performed after every dose adjustment (typically after a period of 7–14 days) or in association with any change in clinical condition. Target systolic BP should be reduction to a minimum of the ACVIM pre-hypertensive category (<160 mmHg) and preferably to the normotensive category (<140 mmHg) (Acierno et al. 2018a). Titration of ARB or ACEi therapy may also be required for management of proteinuria in order to ensure a minimum of 50% reduction and ideally return to a non-proteinuric status (UPC < 0.2). Careful monitoring of renal function (urea, creatinine, symmetric dimethylarginine) and potassium concentrations should be performed where above-licensed doses of either ARB or ACEi are used and/or where there is concern relating to renal function.

For those dogs with HAC where there may be doubt in relation to the diagnosis of hypertension or the initial measurements fall in a pre-hypertensive category (e.g. circa 160 mmHg), provided there is no evidence of ocular TOD, initiating therapy for HAC with careful monitoring of systolic BP may be the preferred route given that some dogs with HAC may show a small reduction in SBP with HAC treatment. Should these dogs develop persistent hypertension then anti-hypertensive therapy can be introduced.

For those dogs that undergo surgical treatment (e.g. unilateral or bilateral adrenalectomy or hypophysectomy), there is currently no information available in the literature to indicate whether, if those dogs have been hypertensive due to HAC pre-treatment, long-term anti-hypertensive therapy continues to be required. In human medicine, after 'curative' surgery systolic and diastolic BP are reported to fall but approximately 30% of patients continue to have systolic hypertension (Cicala and Mantero 2010). In this situation, careful monitoring of systolic BP in the post-operative period is indicated. If there is a possibility that the aetiology of hypertension may not only be related to HAC (i.e. concurrent CKD), then long-term anti-hypertensive therapy may be required. For those patients where HAC is considered the only factor contributing to the development of hypertension then careful downwards titration of anti-hypertensive agents may be possible with subsequent withdrawal providing systolic BP remains stable.

In cats where systemic hypertension is diagnosed with HAC anti-hypertensive therapy should be initiated (Chap. 12). For the cat, two authorised products are available for the treatment of hypertension: the ARB telmisartan and the calcium channel blocker amlodipine besylate. Either of these drugs can be used as a first-line agent for cats with systemic hypertension with concurrent HAC. Anti-hypertensive therapy should be started whilst initiating therapy concurrently for HAC, diabetes mellitus when present and/or other underlying conditions which may be identified. Careful monitoring is required to ensure control of systolic BP is achieved together with monitoring of the underlying disease conditions. The target for anti-hypertensive therapy should be a minimum of reaching the ACVIM pre-hypertensive category (<160 mmHg) and ideally normotensive (<140 mmHg) (Acierno et al. 2018a). If a cat is diagnosed with systemic hypertension and HAC and undergoes surgery (either hypophysectomy or adrenalectomy), there is no information available regarding continued requirements for anti-hypertensive therapy post-surgery. In this situation, careful case-by-case evaluation and monitoring of systolic BP should be performed in the peri- and post-operative period, with adjustment of anti-hypertensive medication based on BP readings to ensure maintenance of normotension whilst avoiding hypotension (<100 mmHg).

4.2 Pheochromocytoma and Paraganglioma

Pheochromocytoma refers to tumours which arise from chromaffin cells in the sympathetic nervous tissue of the adrenal medulla and produce one or more of the catecholamines, epinephrine, norepinephrine or dopamine. The name 'pheochromocytoma' derives from the Greek words 'phaios', meaning 'dark', and 'chroma', meaning 'colour', and refers to the dark discolouration that chromaffin cells show when stained with chromium salts (Kantorovich and Pacak 2010). In human pheochromocytomas, the predominant catecholamine produced is norepinephrine and occasionally it may be the only catecholamine produced (Eisenhofer et al. 2011). Rarely epinephrine- or dopamine-only producing pheochromocytomas are reported in humans, and identification of such individuals can be important from a management perspective (Gallen et al. 1994; Bozin et al. 2017). Which catecholamine or combination of catecholamines that are produced and whether this is in a continuous or pulsatile manner is believed to relate to differences in the expression of genes that control biosynthesis and secretion processes (Eisenhofer et al. 2011).

Extra-adrenal chromaffin cell tumours, with potential to secrete catecholamines, can also arise from the paravertebral sympathetic ganglia of the thorax, abdomen or pelvis, where they are named 'paragangliomas'. However, it should be noted that 'paraganglioma' can also refer to tumours originating from the parasympathetic ganglia of the glossopharyngeal and vagal nerves in the neck, which therefore lack catecholamine-secreting ability (Lenders et al. 2014). In recent years there has been a lot of interest in the molecular and hereditary aspects of pheochromocytomas and paragangliomas, with paragangliomas having the highest degree of heritability of any endocrine neoplasm (Chap. 6) (Crona et al. 2017).

Pheochromocytoma and paraganglioma are highly uncommon in both people and dogs. Their incidence in people is estimated to be 0.04–0.21 cases/100,000 person-years (Berends et al. 2018), with approximately 30–40% of cases having a hereditary basis. These familial forms can occur alongside other malignancies, including other endocrine tumours, and are associated with several monogenic diseases such as multiple endocrine neoplasia (MEN) syndromes 1 and 2 and von Hippel-Lindau syndrome (Turchini et al. 2018). The incidence of pheochromocytoma in dogs has not been well established, but it is estimated to account for 32% of all neoplastic lesions of the canine adrenal gland (Labelle and Cock 2005). Pheochromocytoma is likely even rarer in cats. It is estimated to account for approximately 10% of neoplastic feline adrenal lesions, but only five cases have been comprehensively described in veterinary literature (Labelle and Cock 2005; Cervone 2017; Wimpole et al. 2010; Calsyn et al. 2010; Chun et al. 1997; Henry et al. 1993). Canine pheochromocytoma can occur at any age but is typically identified in older dogs with two large case series reporting a mean age at diagnosis of 10–12 years (Gilson et al. 1994; Barthez et al. 1997). In the five reported cases of feline pheochromocytoma, patient age ranged from 7 to 15 years. There is no apparent sex or breed predisposition in either species. Diagnosis is usually based on identification of an adrenal mass together with assessment of catecholamine metabolites, either plasma- or urine-free normetanephrine concentrations (Gostelow et al. 2013; Quante et al. 2010; Salesov et al. 2015). Readers are directed to other general medicine texts for more specific information on the diagnosis of pheochromocytoma (Reusch 2015; Galac and Korpershoek 2017).

Paragangliomas are exceptionally rare in both dogs and cats. They most frequently occur in the aortic and carotid bodies where they are referred to as 'chemodectomas' (Galac and Korpershoek 2017) but have been documented in various other locations, including the retroperitoneal space, right atrium and cauda equina (Robat et al. 2016; Wey and Moore 2012; Davis et al. 1997). Due to this chapter's focus on adrenal disease, the rest of this section refers to pheochromocytoma. However, paragangliomas should not be overlooked as an extremely rare extra-adrenal cause of hypertension in small animals (Wey and Moore 2012).

4.2.1 Epidemiology of Hypertension Associated with Pheochromocytoma and Paraganglioma

Catecholamine release from pheochromocytomas can be persistent or paroxysmal, and as such the clinical manifestations of patients can be chronic or episodic, or alternatively, patients may present with a sudden onset of catecholamine crisis. The severity of clinical signs can range from vague and mild through to dramatic and life threatening. The variability in the presence and severity of clinical signs has meant that in humans pheochromocytoma is often not considered as a diagnosis and indeed may be an incidental discovery either at laparotomy or post mortem examination (Stenström and Svärdsudd 1986; Sutton et al. 1981). In one study, of those individuals diagnosed with a pheochromocytoma after death, only 54% had demonstrated hypertension when alive (Sutton et al. 1981). The same is true in dogs in relation to the variable and episodic presentation with case series indicating that in 48–57% of cases the diagnosis of pheochromocytoma was an incidental finding (Gilson et al. 1994; Barthez et al. 1997). Table 4.2 reviews some of the most common clinical signs related to body system (Bravo and Tagle 2003). In dogs, the combination of vague and intermittent clinical signs should prompt the clinician to consider pheochromocytoma as a differential and to pursue BP assessment and abdominal imaging if deemed indicated.

In humans with pheochromocytoma, 80–90% will demonstrate hypertension. In approximately 45–48% of cases, hypertension is paroxysmal, whilst in between 29 and 50%, it is persistent and 5–13% normal blood pressure is documented

Table 4.2 Clinical manifestations reported in dogs with pheochromocytoma

Body system	Clinical signs associated
Cardiorespiratory	Systemic hypertension, epistaxis, hypertensive ocular target injury (e.g. retinal haemorrhage/hyphaema/acute onset blindness), hypertensive encephalopathy (e.g. depression, behavioural change, seizure), tachycardia, tachyarrhythmias, pale mucous membranes (cardiovascular shock), tachypnoea, panting
Neuromuscular	Weakness, anxiety, pacing, muscle tremors, seizures
Non-specific signs	Anorexia, weight loss, polyuria and polydipsia, gastrointestinal signs (vomiting and diarrhoea), abdominal discomfort, heat intolerance, anxiety
Related to intra-abdominal mass	Intra-abdominal or retroperitoneal haemorrhage, ascites, oedema (hind limb), abdominal distension

(Bravo and Tagle 2003; Orvieto and Gancar 2014). Sustained hypertension has most often been associated with pheochromocytoma secreting norepinephrine and paroxysmal hypertension with the combination of epinephrine and norepinephrine, and where tumours are large and cystic, release of catecholamines may be minimised due to intra-mass metabolism of catecholamines such that lower concentrations of catecholamine are released into the systemic circulation (Bravo and Tagle 2003).

In patients with *sustained hypertension*, there is reported to be a strong relationship between norepinephrine concentrations and daytime and night-time BP measurements (Zuber et al. 2011). Orthostatic or postural hypotension occurs more commonly in individuals with sustained hypertension. In this situation BP usually falls to normal levels but syncope can occur (Orvieto and Gancar 2014; Ueda et al. 2005; Shahbaz et al. 2018). Excretion of fluid volume by the kidney in response to sustained hypertension may contribute to apparent decreased blood volume. This in addition to the inability to venoconstrict on standing as blood pools in dependent veins will result in falling venous return and contributing to the postural hypotension observed (Zuber et al. 2011). It can be hypothesised that this orthostatic hypotension does not occur in dogs because the effect of standing is less marked in quadrupeds.

Episodes of *paroxysmal hypertension* can be triggered by many factors and results from the sudden release of excessive concentrations of catecholamines, most often epinephrine (Orvieto and Gancar 2014). Factors associated with paroxysmal hypertension in humans include physical activity, smoking, abdominal pressure, postural changes, anxiety, certain foods or drinks with high tyramine concentrations, drugs (e.g. tricyclic anti-depressants, histamine) and procedures both with and without general anaesthesia. However, often these paroxysmal episodes are unpredictable. Paroxysmal cardiovascular manifestations can be variable both in terms of duration (minutes to hours) and frequency ranging from multiple times per day or only sporadically, i.e. recurring only after days to weeks (Bravo and Tagle 2003). Dramatic increases in both systolic and diastolic BP are reported in these individuals (Zuber et al. 2011). Rarely paroxysmal hypertension may cycle with hypotension and syncope (Ionescu et al. 2008; Kobal et al. 2008). The pathophysiology underpinning these cycles of hypertension and hypotension are unclear; however, it is hypothesised that it may be due to a rapid rise in BP stimulating baroreceptors activating a negative feedback loop via both the sympathetic and parasympathetic nervous system resulting in rapid reduction in BP and hypotension (Zuber et al. 2011).

Normotension is most often reported in individuals with familial pheochromocytoma, very small mass lesions with low-level catecholamine production or dopamine-producing tumours which are commonly extra-adrenal. These are more likely to be identified as an incidental finding due to reduced or lack of apparent clinical signs, and any increase in BP may be minimal. Normotension is also more common with parasympathetic paragangliomas of the head and neck which do not secrete catecholamines (Zuber et al. 2011).

Target organ manifestations of hypertension are widely reported in humans (Zuber et al. 2011). Structural changes to the peripheral vasculature result in arterial stiffening, whilst in the heart, coronary vasoconstriction and inotropy result in myocardial hypoxia, ischaemia and infarction (Petrák et al. 2010; Zhang et al. 2017). Cardiomyopathy induced by pheochromocytoma can be either chronic with left ventricular hypertrophy or can be acute and ishaemic (Zhang et al. 2017). Hypertensive encephalopathy is also reported and manifests with headache, papilloedema, altered mentation and increased risk of stroke (Majic and Aiyagari 2008). During periods of sustained hypertension, there can be breakthrough of the cerebrovascular autoregulation which normally maintains cerebral perfusion. In paroxysmal hypertension, acute onset haemorrhage or stroke may occur whilst in postural hypotension there may be episodes of acute ischaemic stroke (Zuber et al. 2011). Renal injury has also been reported both as a consequence of renal artery stenosis and also secondary to reduced perfusion to muscles, rhabdomyolysis and acute tubular necrosis secondary to myoglobinuria (Zuber et al. 2011; Celik et al. 2014). Hypertensive ocular injury in humans can include retinal haemorrhage and retinal vessel aneurysms (Zuber et al. 2011).

In dogs with pheochromocytoma, approximately 30% have been reported to demonstrate cardiovascular manifestations of catecholamine release (e.g. tachycardia and arrhythmia) with bradyarrhythmias also documented (Galac and Korpershoek 2017; Brown et al. 2007). Blood pressure measurement is inconsistently performed in the available literature, but where performed 50–86% demonstrate systolic hypertension with systolic BP ranging from 160 to >300 mmHg (Gilson et al. 1994; Barthez et al. 1997; Galac and Korpershoek 2017; Herrera et al. 2008; Kook et al. 2010; Williams and Hackner 2001; Whittemore et al. 2001). Little information is available on whether dogs have sustained versus paroxysmal hypertension as it is rare for serial assessment of BP to be reported. Nevertheless, lack of hypertension should not be used to exclude the diagnosis of a pheochromocytoma.

Where identified, the diagnosis of hypertension in most cases has been based on BP measurement. Only relatively few cases have reported possible clinical manifestations of hypertensive TOD. Fundic examinations have rarely been part of the described investigation of cases although retinal lesions have been described (Gilson et al. 1994). Epistaxis was reported in 2/61 dogs in one case series although contributory metastatic lesions of the nasal cavity were also identified in these cases (Barthez et al. 1997). Similarly, neurological manifestations have been reported (e.g. seizures) and associated with intra-cranial haemorrhage or without apparent cause at post-mortem examination, and hypertensive encephalopathy may therefore have been contributory (Gilson et al. 1994; Barthez et al. 1997). However, in some dogs intra-cranial metastatic lesions were identified, and therefore, the relative contribution of BP versus metastatic disease in the development of seizures is uncertain (Barthez et al. 1997). Echocardiography has also been infrequently performed in the available case series. However, left ventricular hypertrophy and left ventricular dilation have been reported (Gilson et al. 1994). One study has explored the echocardiographic and histopathological cardiovascular findings in dogs with pheochromocytoma in greater detail. This study indicated that left ventricular hypertrophy was identified in ~44% and histopathological changes (multifocal cardiomyocyte necrosis, cardiomyocyte degeneration, myocardial haemorrhage, lymphohistiocytic myocarditis and interstitial fibrosis) in ~66% of cases (Edmondson et al. 2015).

Of the five case reports in cats with pheochromocytoma, BP data was not available in two cases, but all three cats from the other reports were hypertensive (range 180–360 mmHg) (Cervone 2017; Wimpole et al. 2010; Calsyn et al. 2010; Chun et al. 1997; Henry et al. 2003). One cat in the study by Wimpole and colleagues presented with transient hyphaema and retinal lesions related to systemic hypertension (Wimpole et al. 2010). One of the two cats where BP was not performed because the cat was too aggressive had a baseline systolic BP of 150 mmHg at the time of anaesthesia, and this cat together with one other was reported to exhibit severe hypotension during surgery (Calsyn et al. 2010; Henry et al. 2003). The cat in the study by Wimpole demonstrated persistent hypertension (~220 mmHg) after adrenalectomy, but metastatic disease was suspected on the basis of persistently elevated normetanephrine concentrations (Wimpole et al. 2010).

4.2.2 Pathophysiology of Hypertension in Pheochromocytoma and Paraganglioma

An understanding of the effect of catecholamines on the pathogenesis of hypertension requires an understanding of receptor interactions. Norepinephrine and epinephrine act through the G-protein-coupled adrenoreceptors $\alpha 1$, $\alpha 2$, $\beta 1$ and $\beta 2$. Hypertension in pheochromocytoma patients results predominantly from the action of excessive catecholamine secretion. In the peripheral vasculature, stimulation of $\alpha 1$ adrenoreceptors in vascular smooth muscle leads to peripheral vasoconstriction, increased total peripheral resistance and increased systemic BP together with reduced tissue and organ perfusion.

Activation of $\beta 1$ receptors on cardiomyocytes has a positive inotropic and chronotropic effect, producing an increase in stroke volume and cardiac output. The $\alpha 1$ adrenergic receptors can have the same effect on cardiomyocytes but to a lesser extent than $\beta 1$ activation. In addition, stimulation of $\beta 1$ receptors in the juxtaglomerular apparatus results in renin secretion, which promotes hypertension through increased angiotensin II production and subsequent release of aldosterone. These effects can be mediated both by norepinephrine and epinephrine (Zuber et al. 2011).

In health and with normal circulating concentrations of dopamine, the dopaminergic D1 and D2 receptors are stimulated. Activation of D1 receptors results in renal artery vasodilation whilst D2 inhibits release of norepinephrine from the sympathetic ganglia and a negative inotropic effect on cardiomyocytes. These actions of dopamine help to explain why individuals with dopamine-secreting pheochromocytomas and paragangliomas do not exhibit hypertension. However, in the event of excessive production and at high concentrations, dopamine can stimulate both α and β receptors and therefore could contribute to the development of systemic hypertension (Zuber et al. 2011).

4.2.3 Treatment of Hypertension Associated with Pheochromocytomas and Paragangliomas

Definitive treatment for pheochromocytoma is either surgical or laparoscopic adrenalectomy (Lenders et al. 2014). However, in order to control hypertension and also to try and improve outcome related to the surgical and post-operative period, medical management to combat the effect of excessive catecholamine release is recommended. In human medicine, the current recommendation in the pre-operative period is to begin α blockade. There is conflicting information whether use of a selective α1 blockade (e.g. prazosin or doxazosin) versus non-selective blockade (e.g. phenoxybenzamine; irreversible competitive inhibitor of α1 and α2) is superior (Lenders et al. 2014; Kocak et al. 2002). Some studies have supported that use of selective blockade may lower pre-operative diastolic BP, lower intraoperative heart rates, give better post-operative haemodynamic stability and give fewer complications relating to reactive tachycardia and hypotension but other studies have suggested no difference (Lenders et al. 2014; Kocak et al. 2002). This is based on the premise that selective blockade would leave pre-synaptic α2 adrenoceptors available to control release of norepinephrine from sympathetic nerve endings, hence reducing the risk of tachycardia. Irrespective of drug choice, it is recommended that α blockade is provided for 7–14 days pre-operatively in order to try and normalise blood pressure. Individuals should also consume a higher sodium diet and plenty of fluid in order to combat the catecholamine-driven volume contraction that can contribute to post-operative hypotension although evidence from randomised controlled trials is not available to support this (Lenders et al. 2014). As short and alternative protocols, intravenous administration of phenoxybenzamine for 5 h for 3 consecutive days has also been described as a pre-operative protocol in humans. For the intra-operative period, BP control with nitroprusside and nitroglycerin has been described although does not appear widely used (Bravo and Tagle 2003; Chew 2004; Jankovic et al. 2007). Intra-operative intravenous use of phentolamine, a reversible non-selective α-blocker, has also been described where cardiovascular complications are occurring either with or without prior medication with phenoxybenzamine (Yu et al. 2018; McMillian et al. 2011).

Recommended targets for BP control in the pre-operative period in humans are 130/80 mmHg. Additional anti-hypertensive therapy with a calcium channel blocker, e.g. amlodipine, has been used in order to achieve this goal. Monotherapy with calcium channel blockers is not recommended unless the patient has very mild pre-operative hypertension or suffers from marked orthostatic hypotension related to α-blockade (Lenders et al. 2014). For those patients where there is persistent tachycardia despite α blockade, either β1- or non-selective β-blockade has been recommended (e.g. propranolol or atenolol), provided α-blockade has been on board for a minimum of 3 days. Use of β-blockade without α-blockade on board is not recommended due to the potential for this to result in a hypertensive crisis due to unopposed activation of α-receptors. Finally, metyrosine has been used in humans as an inhibitor of tyrosine hydroxylase, therefore reducing the production of catecholamines. It has been used both pre-operatively and also in individuals with metastatic disease (Wachtel et al. 2015).

Whichever anti-hypertensive protocol is used, it is cautioned that the risk of intraoperative hypertension and tachycardia can never be completely removed. Careful monitoring of individuals in the pre-, peri- and post-operative period is always required, particularly for heart rate, blood pressure, potassium, blood glucose and volume status, a principle that is also true for dogs and cats.

In dogs it is widely recommended that the cardiovascular complications associated with pheochromocytomas should be managed medically prior to either surgical or laparoscopic adrenalectomy (Reusch 2015). This is most commonly achieved by administering phenoxybenzamine for between 7 and 14 days pre-operatively (0.25 mg/kg q8–12 h or 0.5 mg/kg q24 h), although during periods when phenoxybenzamine has been unavailable selective α1 blockers (e.g. prazosin 0.5–2 mg/kg q8–12 h) have been used (personal communication) (Acierno et al. 2018b). This pre-medication advice is based on the study by Herrera and colleagues who reported a significantly reduced mortality rate associated with phenoxybenzamine administration [median dose 0.6 mg/kg BID PO (range 0.1–2.5 mg/kg BID PO) for median 20 days (range 7–120 days)] (Herrera et al. 2008). However, the role of increasing surgical experience in improved outcome cannot be ascertained from this study, as a greater number of the phenoxybenzamine-treated dogs were recruited during the later years of the study. Nevertheless, this study and the evidence base for benefit from human medicine forms the foundation for the recommendation for a period of premedication with phenoxybenzamine. In acute situations intraoperatively, phentolamine has also been used in dogs (loading dose, 0.1 mg/kg IV; continuous rate infusion, 1 to 2 μg/kg/min) to manage catecholamine-induced hypertensive crisis (Kyles et al. 2003). The concurrent use of a second agent, e.g. amlodipine besylate, could be considered if pre-operative BP targets have not been achieved with phenoxybenzamine alone (Chap. 13).

There is little information in the literature about the chronic management of pheochromocytoma patients where adrenalectomy is not performed. In this situation, chronic administration of an α-blocker such as phenoxybenzamine is suggested with careful monitoring of clinical signs, BP, heart rate and response to therapy. Additional anti-hypertensive agents, as used in human medicine, e.g. amlodipine besylate, could be considered if hypertension is not controlled with α-blockade alone and use of β-blockade (e.g. atenolol) if tachycardia persists.

There is limited information about the pre-operative management of cats with pheochromocytoma. One cat with concurrent multiple myeloma and probable pheochromocytoma that did not undergo adrenalectomy was managed chronically with amlodipine besylate (Cervone 2017). The cat in the study by Wimpole and colleagues with severe hypertension (systolic BP 360 mmHg) was initially managed pre-operatively with phenoxybenzamine, atenolol and amlodipine besylate. Systolic BP post-operatively remained elevated (220 mmHg) due to suspected metastatic disease and the cat remained on the combination of phenoxybenzamine and amlodipine with reported good BP control (Wimpole et al. 2010). Based on the evidence from both dogs and humans, pre-medication for at least 7 days with an α-blocker such as phenoxybenzamine (2.5 mg/cat q8–12 h or 0.5 mg/kg q 24 h) would seem prudent with use of additional anti-hypertensive drugs as required in order to achieve pre-operative BP control (Chap. 12) (Acierno et al. 2018b).

4.3 Hyperaldosteronism (Conn's Syndrome)

Aldosterone is the principal mineralocorticoid secreted from the zona glomerulosa. Primary hyperaldosteronism is caused by autonomous excess secretion of aldosterone from the adrenal gland and has also been termed 'Conn's syndrome' after Jerome Conn who first described the condition in human patients. Primary hyperaldosteronism is an uncommon to rare condition in the cat (Behrend 2017; Reusch et al. 2010; Ash et al. 2005; Flood et al. 1999; Moore et al. 2000; Declue et al. 2005; Rijnberk et al. 2001a; MacKay et al. 1999; Rose et al. 2007; Lo et al. 2014) and a very rare condition in the dog being described only as individual case reports (Reusch et al. 2010; Breitschwerdt et al. 1985; Gójska-Zygner et al. 2012; Johnson et al. 2006).

In cats, primary hyperaldosteronism is usually the result of a unilateral cortical adenoma or carcinoma although bilateral adenomas have been reported and with malignant mass lesions being more common than benign (Behrend 2017). The median age at diagnosis of an aldosterone secreting mass is ~13 years with most reports indicating cats >10 years of age. Clinical signs at presentation most often relate to hypokalaemia (hypokalaemic polymyopathy resulting in weakness and cervical ventroflexion given the absence of a nuchal ligament in the cat), polyuria and polydipsia, and ocular target organ manifestations of systemic hypertension (Behrend 2017). Some cats with primary hyperaldosteronism and an adrenal mass may, at least initially, be able to maintain normokalaemia as observed in a case report where an adrenal mass was identified as an incidental finding but where primary hyperaldosteronism together with hypokalaemia was identified approximately 4 months later (Renschler and Dean 2009). Individual cases of hyperaldosteronism together with hyperprogesteronism and hyperaldosteronism with a contralateral pheochromocytoma have also been reported in the cat (Calsyn et al. 2010; Briscoe et al. 2009).

Readers are referred to general medical texts for information relating to the laboratory diagnostic investigation of cats with primary hyperaldosteronism (Behrend 2017); however, it should be noted that, together with identification of marked hypokalaemia and an adrenal mass via diagnostic imaging (e.g. ultrasound or computed tomography), plasma aldosterone concentrations are usually markedly elevated (>800 pmol/l) and, where it is possible to quantify, plasma renin activity appropriately suppressed (Behrend 2017; Reusch et al. 2010).

Hyperaldosteronism can also occur secondary to other underlying conditions such as CKD. In cats with azotaemic CKD, aldosterone concentrations are typically <400 pmol/l but can overlap with concentrations identified in cats diagnosed with primary hyperaldosteronism (Jepson et al. 2013). Data also supports that cats with both idiopathic hypertension and hypertension associated with CKD have significantly higher plasma aldosterone concentrations than normotensive age-matched cats with and without CKD (Jepson et al. 2013). Clinicians must therefore be cautious when interpreting aldosterone concentrations in hypertensive cats with azotaemic CKD and evidence of mild to moderate hypokalaemia and ensure that there is sufficient evidence for an adrenal lesion before considering a diagnosis of primary hyperaldosteronism.

In human medicine, it is reported that between 5 and 15% of hypertensive individuals fulfil the criteria for primary aldosteronism although this is dependent on how strictly criteria are applied (Funder 2017; Calhoun 2006). The prevalence increases to nearer 20% for those individuals who have resistant hypertension (Calhoun 2006). Aldosteronism in humans is most commonly (>60%) caused by bilateral micro- or macronodular adrenal gland hyperplasia (BAH) with a smaller proportion of individuals having an aldosterone-producing adenoma (APA) and rarely unilateral adrenal carcinoma, unilateral hyperplasia or familial forms of hyperaldosteronism (Chap. 6). Differentiation in humans is important given that surgical or laparoscopic adrenalectomy may be curative for those with APA whilst chronic medical management is necessary for those with BAH. Combinations of computed tomography, adrenal vein sampling, targeted functional imaging using a radiolabelled metomidate PET tracer and measurement of peripheral venous steroid profiles are used to diagnose and differentiate (Lenders et al. 2017). However, primary aldosteronism does not always result in hypertension and indeed in human patients potassium concentrations can be within reference interval in some individuals with BAH (Funder 2017; Dick et al. 2017).

BAH has been reported in the cat, and it has been suggested that this pathology could contribute to the development of idiopathic hypertension and progression of renal disease in older cats (Javadi et al. 2005). However, Keele and colleagues documented that 97% of cats >9 years at post-mortem examination have evidence of bilateral micronodular hyperplasia and that there was no significant difference in the prevalence of this histopathological finding between normotensive ($n = 30$) and hypertensive cats ($n = 37$), confounding this hypothesis (Keele et al. 2009). Further work is required to determine whether there could be functional differences between normotensive and hypertensive cats with adrenal gland hyperplasia. Based on the available data, aldosterone is hypothesised to be contributory to the pathogenesis of hypertension in cats with idiopathic hypertension and those with renal disease associated hypertension, but the specific role of BAH requires further investigation (Jepson et al. 2013).

Hypertension together with some of the biochemical manifestations of hyperaldosteronism can also be seen with adrenal mass lesions that are producing excess quantities of aldosterone precursors. Experimental administration of deoxycorticosterone to dogs has been reported to result in hypertension (Ferrario et al. 1988). An adrenal mass lesion secreting the aldosterone precursor deoxycorticosterone has been reported in the literature in a dog and associated with hypertension (Reine et al. 1999). In such patients, although clinical manifestations may be similar, aldosterone concentrations are within reference interval. Therefore, systolic BP should be evaluated in any dog with an adrenal mass and suspicion for production of either aldosterone or deoxycorticosterone (Reine et al. 1999).

4.3.1 Epidemiology of Hypertension Associated with Hyperaldosteronism in Dogs and Cats

The prevalence of hypertension in primary hyperaldosteronism in the cat is difficult to ascertain as only individual case reports and small case series exist but ranges between 50 and 100% depending on the study (Reusch et al. 2010). Ocular hypertensive target organ damage has been reported as a primary presenting clinical sign in cats ultimately diagnosed with primary hyperaldosteronism, and therefore, hyperaldosteronism should be a differential for secondary hypertension, particularly if identified together with moderate to marked hypokalaemia and muscle weakness (Ash et al. 2005). The true prevalence of hypertension in dogs with hyperaldosteronism or excessive deoxycorticosterone production cannot be ascertained from the available literature, but hypertension is reported in all dogs where BP was measured (Gójska-Zygner et al. 2012; Yamamoto et al. 1993). Fundic examinations are not described in the available canine case reports although blindness is not listed as a presenting clinical sign (Breitschwerdt et al. 1985; Gójska-Zygner et al. 2012; Johnson et al. 2006; Reine et al. 1999; Yamamoto et al. 1993; Rijnberk et al. 2001b).

4.3.2 Pathophysiology of Hypertension Associated with Hyperaldosteronism

Aldosterone is the main mineralocorticoid hormone which stimulates sodium reabsorption from the cortical connecting tubule (collecting duct and lesser activity in the distal convoluted tubule) and potassium excretion. The actions of aldosterone are mediated via the cytoplasmic MR which when bound moves to the nucleus. In the distal tubule, the apical membrane of cells contains the epithelial sodium channel (ENaC) which mediates sodium transport from the tubular filtrate to the cytosol (i.e. sodium reabsorption). From the cytosol, sodium is transported across the basolateral membrane driven by a $Na^+K^+ATPase$ where sodium is exchanged for potassium. Intracellular potassium movement is driven by a negative luminal electrochemical gradient from the cytosol to the tubular filtrate via renal outer medullary potassium channels (ROMK). The activity of ENaC and transcription and insertion of ENaC are under the control of the MR and aldosterone which also increases expression of Na/K ATPase in the basolateral membrane. The primary roles of aldosterone therefore are regulation of extracellular volume, BP and potassium homeostasis (Stowasser and Gordon 2016). Non-epithelial and non-genomic actions of aldosterone are also described. MR are identified in a number of non-epithelial tissues including endothelial cells, vascular smooth muscle cells, macrophages,

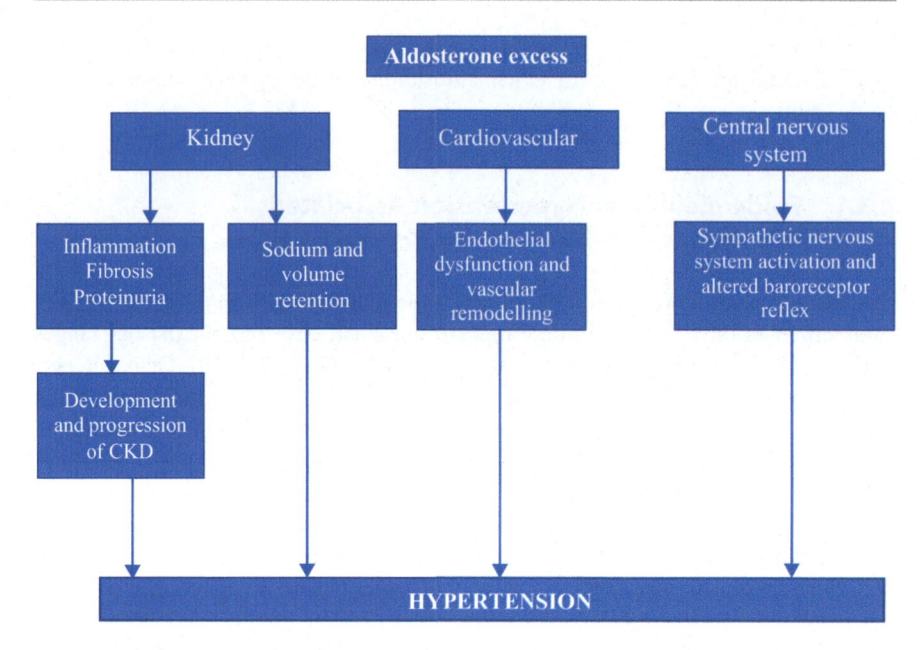

Fig. 4.1 Proposed pathophysiological mechanisms involved in hypertension associated with hyperaldosteronism

adipocytes, baroreceptors, cardiomyocytes and the hippocampus. Experimental studies suggest that activation of MR in the brain and the vasculature (endothelial cells and vascular smooth muscle) results in increases in BP through increased vascular reactivity and increased sympathetic nervous system output (see Chaps. 1 and 11) (Stowasser and Gordon 2016).

The most obvious association between excess aldosterone and hypertension is the concept of sodium retention leading to increased extracellular volume, increasing stroke volume and therefore cardiac output. However, the pathogenesis of hypertension associated with hyperaldosteronism is likely to be much more complex with many of the pathways associated with the development of hypertension involving MR activation in tissues other than the renal tubules and particularly the vasculature and brain (Fig. 4.1). Many of these pathways have been discussed in Chap. 1 in association with activation of RAAS in CKD and its role in the pathophysiology of hypertension associated with CKD.

Aldosterone is able to induce endothelial dysfunction and vascular remodelling, hence potentially modifying peripheral vascular resistance (Struthers et al. 2002; Briet and Schiffrin 2013), increase increasing sympathetic output (Huang et al. 2010) and also impairing the baroreceptor reflex (Monahan et al. 2006). Aldosterone is also recognised as a pro-inflammatory and pro-fibrotic molecule affecting multiple different tissues (Remuzzi et al. 2008; Struthers and MacDonald 2004; Nguyen Dinh Cat and Jaisser 2012; Gilbert and Brown 2010). Within the cardiovascular system excess aldosterone can result in vascular and cardiac inflammation, fibrosis and remodelling (Struthers et al. 2002; Briet and Schiffrin 2013). These effects are at least partly independent of the effects of hypertension on the cardiovascular system but may nonetheless be contributory to the development of hypertension. In human medicine, individuals with primary hyperaldosteronism have evidence of increased left ventricular measurements, carotid intima thickness, femoral pulse wave velocity and risk of cardiovascular events, e.g. arrhythmia, myocardial infarction and stroke, when compared to individuals with essential hypertension (Pimenta et al. 2011; Bernini et al. 2008; Catena et al. 2008). Identification of individuals with familial forms of hyperaldosteronism via genetic screening has also allowed evaluation of cardiac manifestations before the onset of systemic hypertension. Echocardiographic evaluation in these individuals documents left ventricular hypertrophy and reduced diastolic function compared to age- and sex-matched normotensive controls (Stowasser et al. 2005). It is also recognised that aldosterone has inflammatory and pro-fibrotic effects on the renal vasculature and interstitium, again at least partially independent of the specific effects of hypertension on the kidney (Remuzzi et al. 2008; Rossi et al. 2006; Fourkiotis et al. 2012). Alteration in renal function associated with chronic hyperaldosteronism may also therefore be contributory to the pathogenesis of hypertension and also in progression of renal disease. The potential role of aldosterone in the progression of renal disease in cats has previously been discussed (Jepson 2016; Javadi et al. 2005). To date there are no studies that have specifically explored the pathogenesis of hypertension associated with primary hyperaldosteronism in either the dog or the cat. However, it must be assumed that mechanisms described from experimental and human data are likely to be comparable in other mammalian species.

4.3.3 Treatment of Hypertension in Dogs and Cats Associated with Hyperaldosteronism

In dogs and cats diagnosed with hyperaldosteronism, initial medical management will focus on correction of electrolyte imbalances, i.e. hypokalaemia with either parenteral or enteral supplementation depending on severity, together with inhibition of excessive aldosterone concentration with an aldosterone antagonist, e.g. spironolactone. Readers are directed to medical texts for more specific information on management of hyperaldosteronism (Behrend 2017). However, for dogs and cats that are identified to be hypertensive, it is rare for this specific therapy to be sufficient to achieve adequate BP control. In most situations, concurrent anti-hypertensive therapy will be indicated. There is no evidence base to support which anti-hypertensive therapy is superior in the management of hypertension associated with hyperaldosteronism in dogs and cats, and therefore, therapy should follow the recommendations made in Chaps. 12 and 13.

In cats, concurrent use of the calcium channel blocker, amlodipine besylate, with spironolactone has been reportedly well tolerated (Ash et al. 2005). Data relating to the concurrent use of telmisartan is lacking due to the relatively recent authorisation of this product for use as an anti-hypertensive agent but is considered an equivalent alternative. Careful monitoring of BP should be performed and targets for anti-hypertensive therapy aligned to the ACVIM guidelines with reduction to a minimum of <160 mmHg (borderline hypertension) and preferably <140 mmHg (normotension) (Acierno et al. 2018b). Once stabilised, surgical or laparoscopic adrenalectomy is considered the definitive treatment of choice (Lo et al. 2014; Smith et al. 2012).

There is little data available in the literature to advise on the requirement for ongoing anti-hypertensive therapy after surgery. In the study by Lo and colleagues, six cats had systolic BP data available pre- and 6–15 days post-surgery (Lo et al. 2014). Systolic BP measurements were reported to be 'normal' post-operatively, and medical anti-hypertensive therapy was not continued. However, the range of SBP post-operatively in these cats was reported to be 100–180 mmHg; therefore, according to the ACVIM definition, at least some of these cats potentially remained hypertensive and may have benefited from further BP monitoring and/or ongoing anti-hypertensive therapy (Lo et al. 2014). Given that most cats with primary aldosteronism are >10 years of age at diagnosis, the potential for concurrent disease, particularly CKD, should always be considered and, when present, this may increase the chance that long-term anti-hypertensive therapy is required even if successful surgery is performed. Therefore, it is recommended that cats are monitored carefully in terms of their systolic BP in the peri- and post-operative period. If no other underlying aetiology for secondary hypertension can be identified and successful surgery is performed, it may be possible to taper and discontinue anti-hypertensive therapy provided the patient is carefully monitored. For those cats where surgery is not an option, long-term medical management with combined potassium supplementation, aldosterone antagonist (e.g. spironolactone) and anti-hypertensive therapy can be successfully used to manage both the clinical manifestations of hyperaldosteronism and systemic hypertension. In such patients, careful monitoring of BP should be performed as in any hypertensive cat together with monitoring of potassium.

When recognised, hypertension in the dog associated with primary aldosteronism or excessive deoxycorticosterone production should be treated and anti-hypertensive therapy started. A single case report describes the use of amlodipine besylate as an anti-hypertensive agent in a dog with suspected deoxycorticosterone excess (Gójska-Zygner et al. 2012). However, alternative or if necessary combined anti-hypertensive therapies, e.g. angiotensin receptor blockers or angiotensin-converting enzyme inhibitors, could be considered (Chap. 13). Targets for anti-hypertensive therapy should be as per the ACVIM guidelines (Acierno et al. 2018b). If successful adrenalectomy is performed, it is possible that long-term anti-hypertensive therapy can be discontinued provided there are no other underlying conditions that could be contributing to the pathogenesis of hypertension in the individual patient. Careful monitoring in the peri- and post-operative period is required to ascertain this and to ensure normotension is maintained.

References

Acierno M, Brown S, Coleman A, Jepson RE, Papich M, Stepien R et al (2018a) Guidelines for the identification, evaluation and management of systemic hypertension in dogs and cats. J Vet Intern Med 21(3):542–558

Acierno MJ, Brown S, Coleman AE, Jepson RE, Papich M, Stepien RL et al (2018b) ACVIM consensus statement: guidelines for the identification, evaluation, and management of systemic hypertension in dogs and cats. J Vet Intern Med 32(August):1803–1822

Ash AR, Harvey AM, Tasker S (2005) Primary hyperaldosteronism in the cat: a series of 13 cases. J Feline Med Surg 7(3):173–182

Bailey MA, Mullins JJ, Kenyon CJ (2009) Mineralocorticoid and glucocorticoid receptors stimulate epithelial sodium channel activity in a mouse model of Cushing syndrome. Hypertension 54 (4):890–896

Barthez PY, Marks SL, Woo J, Feldman EC, Matteucci M (1997) Pheochromocytoma in dogs: 61 cases (1984–1995). J Vet Intern Med 1(5):272–278

Behrend EN (2015) Canine hyperadrenocorticism. In: Feldman EC, Nelson RW, Reusch CE, Scott-Moncrieff C, Behrend E (eds) Canine and feline endocrinology, 4th edn. Elsevier, St. Louis, pp 377–451

Behrend EN (2017) Non-cortisol-secreting adrenocortical tumors and incidentalomas. In: Ettinger SJ, Feldman EC, Cote E (eds) Textbook of veterinary internal medicine, 8th edn. Elsevier, St. Louis, pp 1819–1825

Berends AMA, Buitenwerf E, de Krijger RR, Veeger NJGM, van der Horst-Schrivers ANA, Links TP et al (2018) Incidence of pheochromocytoma and sympathetic paraganglioma in the Netherlands: a nationwide study and systematic review. Eur J Intern Med 51:68–73

Bernini G, Galetta F, Franzoni F, Bardini M, Taurino C, Bernardini M et al (2008) Arterial stiffness, intima-media thickness and carotid artery fibrosis in patients with primary aldosteronism. J Hypertens 26(12):2399–2405

Bozin M, Lamb A, Putra LJ (2017) Pheochromocytoma with negative metanephrines: a rarity and the significance of dopamine secreting tumors. Urol Case Rep 12:51–53

Bravo EL, Tagle R (2003) Pheochromocytoma: state-of-the-art and future prospects. Endocr Rev 24 (4):539–553

Breitschwerdt EB, Meuten DJ, Greenfield CL, Anson LW, Cook CS, Fulghum RE (1985) Idiopathic hyperaldosteronism in a dog. J Am Vet Med Assoc 187(8):841–845

Briet M, Schiffrin EL (2013) Vascular actions of aldosterone. J Vasc Res 50(2):89–99

Briscoe K, Barrs VR, Foster DF, Beatty JA (2009) Hyperaldosteronism and hyperprogesteronism in a cat. J Feline Med Surg 11:758–762

Brown AJ, Alwood AJ, Cole SG (2007) Malignant pheochromocytoma presenting as a bradyarrhythmia in a dog. J Vet Emerg Crit Care 17(2):164–169

Brown AL, Beatty JA, Lindsay SA, Barrs VR (2012) Severe systemic hypertension in a cat with pituitary-dependent hyperadrenocorticism. J Small Anim Pr 53(2):132–135

Calhoun DA (2006) Aldosteronism and hypertension. Clin J Am Soc Nephrol 1(5):1039–1045

Calsyn JDR, Green RA, Davis GJ, Reilly CM (2010) Adrenal pheochromocytoma with contralateral adrenocortical adenoma in a cat. J Am Anim Hosp Assoc 46:36–42

Caragelasco DS, Kogika MM, Martorelli CR, Kanayama KK, Simões DMN (2017) Urine protein electrophoresis study in dogs with pituitary dependent hyperadrenocorticism during therapy with trilostane. Pesqui Vet Bras 37(7):734–740

Catena C, Colussi G, Nadalini E, Chiuch A, Baroselli S, Lapenna R et al (2008) Cardiovascular outcomes in patients with primary aldosteronism after treatment. Arch Intern Med 168(1):80–85

Celik H, Celik O, Guldiken S, Inal V, Puyan FO, Tugrul A (2014) Pheochromocytoma presenting with rhabdomyolysis and acute renal failure: a case report. Ren Fail 36(1):104–107

Cervone M (2017) Concomitant multiple myeloma probable phaeochromocytoma cat. J Feline Med Surg Open Rep 3(2):205511691771920
Chew SL (2004) Recent developments in the therapy of phaeochromocytoma AU. Expert Opin Investig Drugs 13(12):1579–1583
Chun R, Jakovljevic S, Morrison WB (1997) Apocrine gland adenocarcinoma and pheochromocytoma in a cat. J Am Anim Hosp Assoc 33:33–36
Cicala MV, Mantero F (2010) Hypertension in Cushing's syndrome: from pathogenesis to treatment. Neuroendocrinology 92(Suppl 1):44–49
Connell JM, Whitworth JA, Davies DL, Lever AF, Richards AM, Fraser R (1987) Effects of ACTH and cortisol administration on blood pressure, electrolyte metabolism, atrial natriuretic peptide and renal function in normal man. J Hypertens 5(4):425–433
Crona J, Taïeb D, Pacak K (2017) New perspectives on pheochromocytoma and paraganglioma: toward a molecular classification. Endocr Rev 38(6):489–515
Davis WP, Watson GL, Koehler KL, Brown CA (1997) Malignant cauda equina paraganglioma in a cat. Vet Pathol 34:243–246
Declue AE, Breshears LA, Pardo ID, Kerl ME, Perlis J, Cohn LA (2005) Hyperaldosteronism and hyperprotgesteronism in a cat with adrenocortical carcinoma. J Vet Intern Med 19:355–358
Dick S, Queiroz M, Bernardi B, Dall'Agnol A, Brondani L, Silveiro S (2017) Update in diagnosis and management of primary aldosteronism. Clin Chem Lab Med 56(3):360–372
Edmondson EF, Bright JM, Halsey CH, Ehrhart EJ (2015) Pathologic and cardiovascular characterization of pheochromocytoma-associated cardiomyopathy in dogs. Vet Pathol 52(2):338–343
Eisenhofer G, Pacak K, Huynh T-T, Qin N, Bratslavsky G, Linehan WM et al (2011) Catecholamine metabolomic and secretory phenotypes in phaeochromocytoma. Endocr Relat Cancer 18(1):97–111
Faggiano A, Pivonello R, Melis D, Filippella M, Di Somma C, Petretta M et al (2003) Nephrolithiasis in Cushing's disease: prevalence, etiopathogenesis, and modification after disease cure. J Clin Endocrinol Metab 88(5):2076–2080
Feldman EC (2015) Hyperadrenocorticism in cats. In: Feldman EC, Nelson RW, Reusch CE, Scott-Moncrieff JCR, Behrend EN (eds) Canine and feline endocrinology, 4th edn. Elsevier, St. Louis, pp 452–484
Ferrario CM, Mohara O, Ueno Y, Brosnihan KB (1988) Hemodynamic and neurohormonal changes in the development of DOC hypertension in the dog. Am J Med Sci 295(4):352–359
Flood SM, Randolph JF, Gelzer JF (1999) Primary hyperaldosteronism in two cats. J Am Anim Hosp Assoc 35:411–416
Fourkiotis VG, Hanslik G, Hanusch F, Lepenies J, Quinkler M (2012) Aldosterone and the kidney. Horm Metab Res 44(03):194–201
Funder JW (2017) Primary aldosteronism: the next five years. Horm Metab Res 49(12):977–983
Galac S, Korpershoek E (2017) Pheochromocytomas and paragangliomas in humans and dogs. Vet Comp Oncol 15(4):1158–1170
Galac S, Kooistra HS, Voorhout G, Van Den Ingh TSGAM, Mol JA, Van Den Berg G et al (2005) Hyperadrenocorticism in a dog due to ectopic secretion of adrenocorticotropic hormone. Domest Anim Endocrinol 28(3):338–348
Gallen IW, Taylor RS, Salzmann MB, Tooke JE (1994) Twenty-four hour ambulatory blood pressure and heart rate in a patient with a predominantly adrenaline secreting phaeochromocytoma. Postgrad Med J 70(826):589–591
Gilbert KC, Brown NJ (2010) Aldosterone and inflammation. Curr Opin Endocrinol Diabetes Obes 17(3):199–204
Gilson SD, Withrow SJ, Wheeler SL, Twedt DC (1994) Pheochromocytoma in 50 dogs. J Vet Intern Med 8(3):228–232
Gójska-Zygner O, Lechowski R, Zygner W (2012) Functioning unilateral adrenocortical carcinoma in a dog. Can Vet J 53:623–625

Gostelow R, Bridger N, Syme HM (2013) Plasma-free metanephrine and free normetanephrine measurement for the diagnosis of pheochromocytoma in dogs. J Vet Intern Med 27:83–90

Goy-Thollot I, Pechereau D, Keroack S, Dzempte J-V, Bonnet JM (2002) Investigation of the role of aldosterone in hypertension associated with spontaneous pituitary-dependent hyperadrenocorticism in dogs. J Small Anim Pract 43(11):489–492

Greco DC, Peterson ME, Davidson AP, Feldman EC, Komurek K (1999) Concurrent pituitary and adrenal tumors in dogs with hyperadrenocorticism: 17 cases (1978–1995). J Am Vet Med Assoc 214(9):1349–1353

Haentjens P, De Meirleir L, Abs R, Verhelst J, Poppe K, Velkeniers B (2005) Glomerular filtration rate in patients with Cushing's disease: a matched case-control study. Eur J Endocrinol 153(6):819–829

Hayes AR, Grossman AB (2018) The ectopic adrenocorticotropic hormone syndrome: rarely easy, always challenging. Endocrinol Metab Clin North Am 47(2):409–425

Heaney AP, Hunter SJ, Sheridan B, Atkinson AB (1999) Increased pressor response to noradrenaline in pituitary dependent Cushing's syndrome. Clin Endocrinol 51(3):293–299

Henry CJ, Brewer WG, Montgomery RD (1993) Adrenal pheochromocytoma in a cat. J Vet Intern Med 7:199–201

Henry CJ, Brewer WG, Montgomery RD, Groth AH, Cartee RG, Griffin KS (2003) Clinical vignette adrenal pheochromocytoma. J Vet Intern Med 7(3):119–201

Herrera MA, Mehl ML, Kass PH, Pascoe PJ, Feldman EC, Nelson RW (2008) Predictive factors and the effect of phenoxybenzamine on outcome in dogs undergoing adrenalectomy for pheochromocytoma. J Vet Intern Med 22:1333–1339

Hoffman JM, Lourenço BN, Promislow DEL, Creevy KE (2018) Canine hyperadrenocorticism associations with signalment, selected comorbidities and mortality within North American veterinary teaching hospitals. J Small Anim Pract 59(11):681–690

Hsieh CK, Hsieh YP, Wen YK, Chen ML (2007) Focal segmental glomerulosclerosis in association with Cushing's disease. Clin Nephrol 67(2):109–113

Huang BS, Ahmadi S, Ahmad M, White RA, Leenen FHH (2010) Central neuronal activation and pressor responses induced by circulating ANG II: role of the brain aldosterone-"ouabain" pathway. Am J Physiol Circ Physiol 299(2):H422–H430

Hurley KJ, Vaden SL (1998) Evaluation of urine protein content in dogs with pituitary-dependent hyperadrenocorticism. J Am Vet Med Assoc 212(3):644

Ionescu CN, Sakharova OV, Harwood MD, Caracciolo EA, Schoenfeld MH, Donohue TJ (2008) Cyclic rapid fluctuation of hypertension and hypotension in pheochromocytoma. J Clin Hypertens 10(12):936–940

Isidori AM, Graziadio C, Paragliola RM, Cozzolino A, Ambrogio AG, Colao A et al (2015) The hypertension of Cushing's syndrome: controversies in the pathophysiology and focus on cardiovascular complications. J Hypertens 33(1):44–60

Jankovic RJ, Konstantinovic SM, Milic DJ, Mihailovic DS, Stosic BS (2007) Can a patient be successfully prepared for pheochromocytoma surgery in three days? A case report. Minerva Anestesiol 73:245–248

Javadi S, Kooistra HS, Mol JA, Boer P, Boer WH, Rijnberk A (2003) Plasma aldosterone concentrations and plasma renin activity in healthy dogs and dogs with hyperadrenocorticism. Vet Rec 153(17):521–525

Javadi S, Djajadiningrat-Laanen SC, Kooistra HS, Van Dongen AM, Voorhout G, Van Sluijs FJ et al (2005) Primary hyperaldosteronism, a mediator of progressive renal disease in cats. Domest Anim Endocrinol 28(1):85–104

Jepson RE (2016) Current understanding of the pathogenesis of progressive chronic kidney disease in cats. Vet Clin North Am Small Anim Pract 46(6):1015–1048

Jepson RE, Syme HM, Elliott J (2013) Plasma renin activity and aldosterone concentrations in hypertensive cats with and without azotemia and in response to treatment with amlodipine besylate. J Vet Intern Med 28(1):144–153

Johnson KD, Henry CJ, McCaw DL, Turnquist SE, Stoll MR, Kiupel M et al (2006) Primary hyperaldosteronism in a dog with concurrent lymphoma. J Vet Med A Physiol Pathol Clin Med 53(9):467–470

Kantorovich V, Pacak K (2010) Chapter 15—Pheochromocytoma and paraganglioma. In: Martini LBT (ed) Progress in brain research. Neuroendocrinology. Elsevier, St. Louis, pp 343–373

Keele SJ, Smith KC, Elliott J, Syme HM (2009) Adrenocortical morphology in cats with chronic kidney disease (CKD) and systemic hypertension. J Vet Intern Med 23(6):1328

Kelly JJ, Tam SH, Williamson PM, Lawson J, Whitworth JA (1998) The nitric oxide system and cortisol-induced hypertension in humans. Clin Exp Pharmacol Physiol 25(11):945–946

Kemppainen RJ, Behrend EN (1997) Adrenal physiology. Vet Clin North Am Small Anim Pract 27 (2):173–186

Kobal SL, Paran E, Jamali A, Mizrahi S, Siegel RJ, Leor J (2008) Pheochromocytoma: cyclic attacks of hypertension alternating with hypotension. Nat Clin Pract Cardiovasc Med 5:53

Kocak S, Aydintug S, Canakci N (2002) Alpha blockade in preoperative preparation of patients with pheochromocytomas. Int Surg 87(3):191–194

Kook PH, Grest P, Quante S, Boretti FS, Reusch CE (2010) Urinary catecholamine and metadrenaline to creatinine ratios in dogs with a phaeochromocytoma. Vet Rec 166(6):169–174

Kyles AE, Feldman EC, De Cock HE, Kass PH, Mathews KG, Hardie EM et al (2003) Surgical management of adrenal gland tumors with and without associated tumor thrombi in dogs: 40 cases (1994–2001). J Am Vet Med Assoc 223(5):654–662

Labelle P, De Cock HEV (2005) Metastatic tumors to the adrenal glands in domestic animals. Vet Pathol 42:52–58

Labelle P, Kyles AE, Farver TB, De Cock HE (2004) Indicators of malignancy of canine adrenocortical tumors: histopathology and proliferation index. Vet Pathol 41:490–497

Leblanc NL, Stepien RL, Bentley E (2011) Ocular lesions associated with systemic hypertension in dogs: 65 cases (2005–2007). J Am Vet Med Assoc 238(7):915–921

Lenders JWM, Duh QY, Eisenhofer G, Gimenez-Roqueplo AP, Grebe SKG, Murad MH et al (2014) Pheochromocytoma and paraganglioma: an endocrine society clinical practice guideline. J Clin Endocrinol Metab 99(6):1915–1942

Lenders JWM, Eisenhofer G, Reincke M, Lenders J (2017) Subtyping of Patients with Primary Aldosteronism: an update authors anatomical imaging by CT and MRI. Horm Metab Res 49 (12):922–928

Lien YH, Hsiang TY, Huang HP (2010) Associations among systemic blood pressure, microalbuminuria and albuminuria in dogs affected with pituitary- and adrenal-dependent hyperadrenocorticism. Acta Vet Scand 52(1):61

Lo AJ, Holt DE, Brown DC, Schlicksup MD, Orsher RJ, Agnello KA (2014) Treatment of aldosterone-secreting adrenocortical tumors in cats by unilateral adrenalectomy: 10 cases (2002–2012). J Vet Intern Med 28(1):137–143

Lowe AD, Campbell KL, Barger A, Schaeffer DJ, Borst L (2008) Clinical, clinicopathological and histological changes observed in 14 cats treated with glucocorticoids. Vet Rec 162(24):777–783

MacKay AD, Holt PE, Sparkes AH (1999) Successful surgical treatment of a cat with primary aldosteronism. J Feline Med Surg 1:117–122

Majic T, Aiyagari V (2008) Cerebrovascular Manifestations of Pheochromocytoma and the Implications of a Missed Diagnosis. Neurocrit Care 9(3):378–381

Martínez NI, Panciera DL, Abbott JA, Ward DL (2005) Evaluation of pressor sensitivity to norepinephrine infusion in dogs with iatrogenic hyperadrenocorticism. Pressor sensitivity in dogs with hyperadrenocorticism. Res Vet Sci 78(1):25–31

Mazzi A, Fracassi F, Dondi F, Gentilini F, Famigli Bergamini P (2008) Ratio of urinary protein to creatinine and albumin to creatinine in dogs with diabetes mellitus and hyperadrenocorticism. Vet Res Commun 32(1):299–301

McKnight JA, Rooney DP, Whitehead H, Atkinson AB (1995) Blood pressure responses to phenylephrine infusions in subjects with Cushing's syndrome. J Hum Hypertens 9(10):855–858

McMillian WD, Trombley BJ, Charash WE, Christian RC (2011) Phentolamine continuous infusion in a patient with pheochromocytoma. Am J Heal Pharm 68(2):130–134

Meij BP, Voorhout G, Van Den Ingh TSGAM, Hazewinkel HAW, Teske E, Rijnberk A (1998) Results of transsphenoidal hypophysectomy in 52 dogs with pituitary-dependent hyperadrenocorticism. Vet Surg 27(3):246–261

Monahan KD, Leuenberger UA, Ray CA (2006) Aldosterone impairs baroreflex sensitivity in healthy adults. AJP Hear Circ Physiol 292(1):H190–H197

Moore LE, Biller DS, Smith TA (2000) Use of abdominal ultrasonography in the diagnosis of primary hyperaldosteronism in a cat. J Am Vet Med Assoc 217(2):213–215

Muiesan ML, Lupia M, Salvetti M, Grigoletto C, Sonino N, Boscaro M et al (2003) Left ventricular structural and functional characteristics in Cushing's syndrome. J Am Coll Cardiol 41(12):2275–2279

Nguyen Dinh Cat A, Jaisser F (2012) Extrarenal effects of aldosterone. Curr Opin Nephrol Hypertens 21(2):147–156

Nichols R (1997) Complications and concurrent disease associated with canine hyperadrenocorticism. Vet Clin North Am Small Anim Pract 27(2):309–320

Nisula BC, Bardin CW, Kellner C, Loriaux DL, Chrousos GP, Merriam GR et al (1985) Successful treatment of Cushing's syndrome with the glucocorticoid antagonist RU 486∗. J Clin Endocrinol Metab 61(3):536–540

O'Neill J, Kent M, Glass EN, Platt SR (2013) Clinicopathologic and MRI characteristics of presumptive hypertensive encephalopathy in two cats and two dogs. J Am Anim Hosp Assoc 49(6):412–420

O'Neill DG, Scudder C, Faire JM, Church DB, McGreevy PD, Thomson PC et al (2016) Epidemiology of hyperadrenocorticism among 210,824 dogs attending primary-care veterinary practices in the UK from 2009 to 2014. J Small Anim Pract 57(7):365–373

Ortega TM, Feldman EC, Nelson RW, Willits N, Cowgill LD (1996) Systemic arterial blood pressure and urine protein/creatinine ratio in dogs with hyperadrenocorticism. J Am Vet Med Assoc 209(10):1724–1729

Orvieto C, Gancar J (2014) Pheochromocytoma. Osteopath Fam Physician 6(3):33–41

Perez-Alenza D, Melian C (2017) Hyperadrenocorticism in dogs. In: Ettinger SJ, Feldman EC, Cote E (eds) Textbook of veterinary internal medicine, 8th edn. Elsevier, St. Louis, pp 1795–1811

Petrák O, Trauch B, Zelinka T, Rosa J, Holaj R, Vránková A et al (2010) Factors influencing arterial stiffness in pheochromocytoma and effect of adrenalectomy. Hypertens Res 33(5):454–459

Pimenta E, Gordon RD, Ahmed AH, Cowley D, Leano R, Marwick TH et al (2011) Cardiac dimensions are largely determined by dietary salt in patients with primary aldosteronism: results of a case-control study. J Clin Endocrinol Metab 96(9):2813–2820

Quante S, Boretti FS, Kook PH, Mueller C, Schellenberg S, Zini E et al (2010) Urinary catecholamine and metanephrine to creatinine ratios in dogs with hyperadrenocorticism or pheochromocytoma, and in healthy dogs. J Vet Intern Med 24(5):1093–1097

Quinkler M, Stewart PM (2003) Hypertension and the cortisol-cortisone shuttle. J Clin Endocrinol Metab 88(6):2384–2392

Reine NJ, Hohenhaus AE, Peterson ME, Patnaik AK (1999) Deoxycorticosterone-secreting adrenocortical carcinoma in a dog. J Vet Intern Med 1(3):386–390

Remuzzi G, Cattaneo D, Perico N (2008) The aggravating mechanisms of aldosterone on kidney fibrosis. J Am Soc Nephrol 19(8):1459–1462

Renschler JS, Dean GA (2009) What is your diagnosis? Abdominal mass aspirate in a cat with an increased Na:K ratio. Vet Clin Pathol 38:69–72

Reusch C (2015) Pheochromocytoma and multiple endocrine neoplasia. In: Feldman EC, Nelson RW, Reusch C, Scott-Moncrieff C, Behrend EN (eds) Canine and feline endocrinology, 4th edn. Elsevier, St. Louis, pp 521–554

Reusch CE, Schellenberg S, Wenger M (2010) Endocrine hypertension in small animals. Vet Clin North Am Small Anim Pract 40(2):335–352

Rijnberk A, Voorhout G, Kooistra HS, van der Waarden RJ, van Sluijs FJ, Jzer IJ et al (2001a) Hyperaldosteronism in a cat with metastasised adrenocortical tumour. Vet Q 23(1):38–43

Rijnberk A, Kooistra HS, Van Vonderen IK, Mol JA, Voorhout G, Van Sluijs FJ et al (2001b) Aldosteronoma in a dog with polyuria as the leading symptom. Domest Anim Endocrinol 20 (3):227–240

Robat C, Houseright R, Murphey J, Sample S, Pinkerton M (2016) Paraganglioma, pituitary adenoma, and osteosarcoma in a dog. Vet Clin Pathol 45(3):484–489

Rose SA, Kyles AE, Labelle P, Pypendop BH, Mattu JS, Foreman O et al (2007) Adrenalectomy and caval thrombectomy in a cat with primary hyperaldosteronism. J Am Anim Hosp Assoc 43 (4):209–214

Rossi GP, Bernini G, Desideri G, Fabris B, Ferri C, Giacchetti G et al (2006) Renal damage in primary aldosteronism. Hypertension 48(2):232–238

Salesov E, Boretti FS, Sieber-Ruckstuhl NS, Rentsch KM, Riond B, Hofmann-Lehmann R et al (2015) Urinary and plasma catecholamines and metanephrines in dogs with pheochromocytoma, hypercortisolism, nonadrenal disease and in healthy dogs. J Vet Intern Med 29 (2):597–602

Schellenberg S, Mettler M, Gentilini F, Portmann R, Glaus TM, Reusch CE (2008) The effects of hydrocortisone on systemic arterial blood pressure and urinary protein excretion in dogs. J Vet Intern Med 22(2):273–281

Shahbaz A, Aziz K, Fransawy Alkomos M, Nabi U, Zarghamravanbakhsh P, Sachmechi I (2018) Pheochromocytoma secreting large quantities of both epinephrine and norepinephrine presenting with episodes of hypotension and severe electrolyte imbalance. Cureus 10(7):1–7

Shibata H, Suzuki H, Maruyama T, Saruta T (1995) Gene expression of angiotensin II receptor in blood cells of Cushing's syndrome. Hypertension 26(6):1003–1010

Smets P, Meyer E, Maddens B, Daminet S (2010) Cushing's syndrome, glucocorticoids and the kidney. Gen Comp Endocrinol 169(1):1–10

Smets PMY, Lefebvre HP, Meij BP, Croubels S, Meyer E, Van de Maele I et al (2012a) Long-term follow-up of renal function in dogs after treatment for ACTH-dependent hyperadrenocorticism. J Vet Intern Med 26(3):565–574

Smets PMY, Lefebvre HP, Kooistra HS, Meyer E, Croubels S, Maddens BEJ et al (2012b) Hypercortisolism affects glomerular and tubular function in dogs. Vet J 192(3):532–534

Smets PMY, Lefebvre HP, Aresu L, Croubels S, Haers H, Piron K et al (2012c) Renal function and morphology in aged beagle dogs before and after hydrocortisone administration. PLoS One 7 (2):e31702

Smith RR, Mayhew PD, Berent AC (2012) Laparoscopic adrenalectomy for management of a functional adrenal tumor in a cat. J Am Vet Med Assoc 241(3):368–372

Stenström G, Svärdsudd K (1986) Pheochromocytoma in Sweden 1958–1981. An analysis of the National Cancer Registry Data. Acta Med Scand 220(3):225–232

Stowasser M, Gordon RD (2016) Primary aldosteronism: changing definitions and new concepts of physiology and pathophysiology both inside and outside the kidney. Physiol Rev 96 (4):1327–1384

Stowasser M, Sharman J, Leano R, Gordon RD, Ward G, Cowley D et al (2005) Evidence for abnormal left ventricular structure and function in normotensive individuals with familial hyperaldosteronism type I. J Clin Endocrinol Metab 90(9):5070–5076

Struthers AD, MacDonald TM (2004) Review of aldosterone- and angiotensin II-induced target organ damage and prevention. Cardiovasc Res 61(4):663–670

Struthers AD, Hospital N, Kingdom U, Struthers AD, Hospital N, Kingdom U (2002) Impact of aldosterone on vascular pathophysiology. Congest Heart Fail 8(1):18–22

Sutton MG, Sheps SG, Lie JT (1981) Prevalence of clinically unsuspected pheochromocytoma. Review of a 50-year autopsy series. Mayo Clin Proc 56(6):354–360

Suzuki H, Kondo K, Handa M, Senba S, Saruta T, Igarashi YU (1986) Multiple factors contribute to the pathogenesis of hypertension in Cushing's syndrome. J Clin Endocrinol Metab 62 (2):275–279

Takano H, Kokubu A, Sugimoto K, Sunahara H, Aoki T, Fijii Y (2015) Left ventricular structural and functional abnormalities in dogs with hyperadrenocorticism. J Vet Cardiol 17(3):173–181

Turchini J, Cheung VKY, Tischler AS, De Krijger RR, Gill AJ (2018) Pathology and genetics of phaeochromocytoma and paraganglioma. Histopathology 72(1):97–105

Ueda T, Oka N, Matsumoto A, Miyazaki H, Ohmura H, Kikuchi T et al (2005) Pheochromocytoma presenting as recurrent hypotension and syncope. Intern Med 44(3):222–227

Vidal PN, Miceli DD, Arias ES, Anna ED, García JD, Castillo VA (2018) Decrease of nitric oxide and increase in diastolic blood pressure are two events that affect renal function in dogs with pituitary dependent hyperadrenocorticism. Open Vet J 8(1):86–95

Violette NP, Ledbetter EC (2017) Intracorneal stromal hemorrhage in dogs and its associations with ocular and systemic disease: 39 cases. Vet Ophthalmol 20(1):27–33

Violette NP, Ledbetter EC (2018) Punctate retinal hemorrhage and its relation to ocular and systemic disease in dogs: 83 cases. Vet Ophthalmol 21(3):233–239

von Dehn BJ, Nelson RW, Feldman EC, Griffey SM (1995) Pheochromocytoma and hyperadrenocorticism in dogs: six cases (1982–1992). J Am Vet Med Assoc 207(3):322–324

Wachtel H, Kennedy EH, Zaheer S, Bartlett EK, Fishbein L, Roses RE et al (2015) Preoperative metyrosine improves cardiovascular outcomes for patients undergoing surgery for pheochromocytoma and paraganglioma. Ann Surg Oncol 22(3):646–654

Wey AC, Moore FM (2012) Right atrial chromaffin paraganglioma in a dog. J Vet Cardiol 14 (3):459–464

Whittemore JC, Preston CA, Kyles AE, Hardie EM, Feldman EC (2001) Nontraumatic rupture of an adrenal gland tumor causing intra-abdominal or retroperitoneal hemorrhage in four dogs. J Am Vet Med Assoc 219(3):329–333

Williams JE, Hackner SG (2001) Pheochromocytoma presenting as acute retroperitoneal hemorrhage in a dog. J Vet Emerg Crit Care 11(3):221–227

Wimpole JA, Adagra CFM, Billson MF, Pillai DN, Foster DF (2010) Plasma free metanephrines in healthy cats, cats with non-adrenal disease and a cat with suspected phaeochromocytoma. J Feline Med Surg 12:435–440

Yamamoto A, Naroda T, Kagawa S, Umaki Y, Shintani Y, Sano T et al (1993) Deoxycorticosterone-secreting adrenocortical carcinoma. Endocr Pathol 4(3):165–168

Yasuda G, Shionoiri H, Umemura S, Takisaki I, Ishii M (1994) Exaggerated blood pressure resonse to angiotensin II in patients with Cushing's syndrome due to adrenocortical adenoma. Eur J Endocrinol 131(6):585–588

Yu M, Han C, Zhou Q, Liu C, Ding Z (2018) Clinical effects of prophylactic use of phentolamine in patients undergoing pheochromocytoma surgery. J Clin Anesth 44:119

Zhang R, Gupta D, Albert SG (2017) Pheochromocytoma as a reversible cause of cardiomyopathy: analysis and review of the literature. Int J Cardiol 249:319–323

Zuber SM, Kantorovich V, Pacak K (2011) Hypertension in pheochromocytoma: characteristics and treatment. Endocrinol Metab Clin North Am 40(2):295–311

Thyroid Gland Disease

5

Harriet M. Syme

The relationship between thyroid function and blood pressure is complex. Both hyper- and hypothyroidism have been associated with hypertension in humans, with hyperthyroidism reportedly causing isolated systolic hypertension and hypothyroidism more closely associated with diastolic hypertension and atherosclerosis (Klein and Ojamaa 2001). However, although numerous reviews on the subject of cardiovascular manifestations of thyroid disorders state that hypertension is common (Klein and Danzi 2007; Mazza et al. 2011; Osuna et al. 2017; Udovcic et al. 2017), there is a dearth of good epidemiological data to support this claim. This may be because of the difficulty of proving an association between the conditions when hypertension is in any case very common in the general population, or due to prompt treatment of hypo- and hyperthyroidism with resolution and a consequent lack of recording of the transient hypertension, or because whilst changes in blood pressure do occur they are small in magnitude and do not often result in diagnosis of hypertension.

5.1 Physiological Effects of Thyroid Hormones on the Cardiovascular System

The thyroid gland produces two hormones, thyroxine (T_4) and triiodothyronine (T_3), with T_4 being produced in greater amounts and mainly acting as a pro-hormone. Conversion of T_4 to the biologically active form of the hormone, T_3, occurs by the action of deiodinases, with D1 enzymes expressed on cell membranes and D2 in the endoplasmic reticulum. D1 acts to maintain the circulating concentration of T_3 in the plasma (together with D2) and may be important in adaptation to iodine deficiency,

---

H. M. Syme (✉)
Department of Clinical Sciences and Services, Royal Veterinary College, University of London, Hertfordshire, UK
e-mail: hsyme@rvc.ac.uk

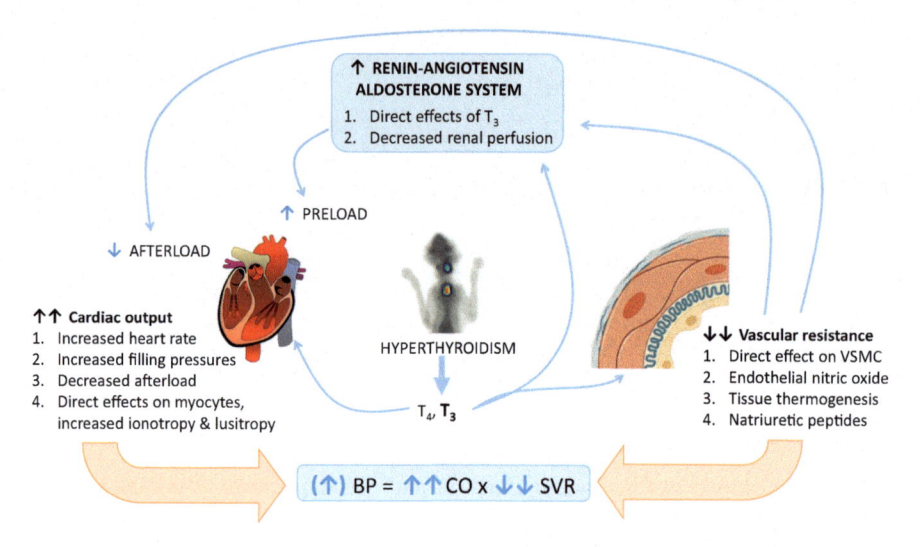

Fig. 5.1 Schematic diagram illustrating the effects of hyperthyroidism on cardiovascular factors influencing blood pressure regulation. T_4 thyroxine, T_3 triiodothyronine, *BP* blood pressure, *CO* cardiac output, *SVR* systemic vascular resistance, *VSMC* vascular smooth muscle cells, ↑ increased, ↑↑ marked increase, ↓ decreased, ↓↓ markedly decreased, (↑) mildly increased

whilst D2 acts intracellularly in target tissues (Bianco et al. 2002). A third deiodinase, D3, inactivates thyroid hormones.

The classical mechanism of action for T_3, once it is produced in the cell, is to interact with nuclear receptors that up- or down-regulate gene expression. There are two thyroid hormone receptors, TRα and TRβ, each of which in turn has several different isoforms. At the target gene promoter, the thyroid hormone receptor interacts with a distinct DNA sequence, the thyroid hormone response element (TRE) as a homodimer, or more often, as a heterodimer with retinoid X receptor (Kahaly and Dillmann 2005). Binding to the TRE can cause positive or negative regulation of gene transcription. Non-genomic actions of thyroid hormones have also been described, mainly acting at the plasma membrane modulating ion flux into the cell or activating various second-messenger systems (Bassett et al. 2003). These non-genomic mechanisms are thought to be responsible for the relatively rapidly occurring actions of thyroid hormones, such as changes in heart rate.

Thyroid hormones exert marked effects on the cardiovascular system that result from direct effects of the hormones on the cardiac myocyte as well as effects on the peripheral vasculature. Arterial blood pressure is a product of both cardiac output and systemic vascular resistance (SVR) as illustrated in Fig. 5.1 (see also Chap. 1). Cardiac output, in turn, is determined by heart rate and stroke volume. Broadly speaking in hyperthyroidism SVR is reduced and cardiac output increases, and in hypothyroidism the opposite is true. These changes have opposing effects on blood pressure, limiting the severity of the hypertension that develops with either endocrinopathy.

Table 5.1 Triiodothyronine (T_3)-regulated cardiac genes

Positively regulated	Negatively regulated
α-Myosin heavy chain (MHC)	β-Myosin heavy chain (MHC)
Voltage-gated K^+ channels	Na^+/Ca^{2+} exchanger (KCX1)
SERCA2	Phospholamban
β-Adrenergic receptor	Adenylyl cyclase types V, VI
Na^+/K^+ ATPase	Thyroid hormone receptor α-1
Adenine nucleotide translocase (ANT1)	Thyroid hormone transporters (MCT8, MCT10)

Adapted from Razvi et al. (2018)

The major effect of T_3 on the heart is to increase the force and speed of systolic contraction and also the speed of diastolic relaxation (Klein and Danzi 2016). Within the cardiac myocyte, thyroid hormones regulate numerous genes that are intimately related to contractile function (Table 5.1). One of the key events controlling systolic contraction and diastolic relaxation is the rate at which free calcium appears and disappears from the cytosol, so limiting its availability to bind to troponin C on the cardiac myofibrils. Thyroid hormones activate the gene expression of sarco/endoplasmic reticulum Ca^{2+}- ATPase (SERCA2a) and concomitantly repress expression of its inhibitor phospholamban, promoting myocyte Ca^{2+} cycling and as a result exerting positive ionotropic and lusitropic[1] effects (Jiang et al. 2000; Belakavadi et al. 2010). Thyroid hormones can also alter the expression of genes encoding structural proteins within the myocyte, such as myosin heavy chain, at least in rodent models of hyperthyroidism (Izumo et al. 1987). Although these, and numerous other cellular mechanisms, may result in intrinsic increases in contractility in hyperthyroidism, there is some evidence that the effects may be relatively trivial compared to those induced by changes in hemodynamic loading conditions. Experimental studies in rodents that have a heterotopically transplanted heart that is perfused by the abdominal aorta have been used to demonstrate that hyperthyroidism induces increases in heart rate and altered gene expression in the transplanted heart, but that cardiac hypertrophy only develops in the heart that is hemodynamically loaded (Klein and Hong 1986). T_3 also increases the rates of both depolarization and repolarization of the sinoatrial node, thus increasing heart rate.

Thyroid hormones act both through endothelium-dependent and endothelium-independent mechanisms to cause relaxation of vascular smooth muscle. Thyroid hormones stimulate the production of endothelial nitric oxide (Carrillo-Sepúlveda et al. 2009) and also have a direct (negative) vasomotor effect on resistance arteries (Park et al. 1997). In addition to its direct effects, T_3 also reduces expression of the angiotensin II type 1 receptor (AT_1 receptor) in smooth muscle cells to reduce the contractile response to angiotensin II (Fukuyama et al. 2003).

The reduction in SVR induced by the actions of thyroid hormones results in reduced renal perfusion pressure and activates the renin-angiotensin system leading

---

[1] Lusiotropy is the rate at which the muscle relaxes; thus, a mediator with a positive lusiotropic effect speeds relaxation.

to an expansion in blood volume. In addition, T_3 directly stimulates production of angiotensinogen from the liver and may also stimulate local production of angiotensin II in target tissues (Vargas et al. 2012). These haemodynamic changes result in atrial stretch and secretion of atrial natriuretic peptide causing yet more vasodilation (Wong et al. 1989).

5.2 Effect of Hyperthyroidism on Blood Pressure in Humans

Hyperthyroidism is characterised by increases in resting heart rate, blood volume, stroke volume, myocardial contractility, and ejection fraction and an improvement in diastolic relaxation (Osuna et al. 2017). The net effect of these changes is an increase in systolic blood pressure and a widening of pulse pressure (Prisant et al. 2006), although the effect size may be relatively small. Arterial stiffness may also be increased (Palmieri et al. 2004).

It is reported that the prevalence of hypertension in humans with thyrotoxicosis is 20–30%. However, hypertension is so common in the population as a whole it is not clear how much of the risk can be attributed to hyperthyroidism. One study from 1931, of 458 hyperthyroid patients, reported a prevalence of hypertension (using systolic blood pressure of ≥ 150 mmHg as the cut-off) of 26% (Hurxthal 1931). Another study compared the blood pressure of 446 untreated hyperthyroid and 549 euthyroid patients. The prevalence of hypertension was increased in young-to-middle aged patients (<50 years) compared to controls, but in the older patients there was no difference between the groups (Saito and Saruta 1994). Following treatment with thiamazole, systolic blood pressure and cardiac output decreased, and total peripheral resistance and diastolic blood pressure increased, in a group of 50 hyperthyroid patients (Marcisz et al. 2002). The changes in systolic blood pressure occurred faster than the changes in diastolic blood pressure. Notably, the changes in total peripheral resistance were much more marked than the changes in blood pressure.

Recently attention has shifted to the consideration of whether people with sub-clinical hyperthyroidism (low TSH, normal T_4) should be treated and whether this will alter blood pressure. In a large prospective cohort study, in which 2910 subjects were monitored for the development of hypertension over a 5-year period, the incidence of hypertension was greater in those with subclinical hyperthyroidism compared with those without (31.4 vs. 19.2%). However, when the analyses were adjusted for confounders (such as age, obesity and other common risk factors), the groups did not differ with respect to the risk of developing hypertension (Völzke et al. 2009). Similarly a meta-analysis of seven cross-sectional studies did not find that subclinical hyperthyroidism was associated with increases in systolic or diastolic pressure (Cai et al. 2011). Attempts have also been made to link within reference range increases in free T_4 to the occurrence of white-coat hypertension, but the data are unconvincing (Cai et al. 2019).

5.3 Effect of Hypothyroidism on Blood Pressure in Humans

Hypothyroidism is associated with decreased cardiac output due to impaired relaxation of vascular smooth muscle and decreased availability of endothelial nitric oxide. This increases arterial stiffness resulting in increased systemic vascular resistance. Thus in hypothyroidism diastolic blood pressure increases and pulse pressure is narrow (Udovcic et al. 2017). Hypothyroidism is also linked to hypercholesterolaemia and development of atherosclerosis, so worsening the cardiovascular risk profile of affected individuals (Cappola and Ladenson 2003).

Reports of blood pressure measurements obtained from humans with overt hypothyroidism are surprisingly infrequent and contain only small numbers of patients. A group of 12 normotensive patients that were hypothyroid due to thyroid ablation (thyroidectomy and radioactive iodine treatment) following treatment for thyroid carcinoma were studied following thyroid hormone withdrawal and again after re-implementation of thyroid hormone replacement (Fommei and Iervasi 2002). There was no difference in night-time measurements, but daytime measurements were increased when the patients were hypothyroid (125.5/84.6 vs. 120.4/76.4 mmHg). In 477 patients with chronic thyroiditis, of whom 169 were hypothyroid and 308 euthyroid, the prevalence of hypertension (blood pressure > 160/95 mmHg) was higher in the hypothyroid group (14.8% vs. 5.5%; $P < 0.01$), and diastolic blood pressure measurements were weakly, inversely, correlated with thyroid hormone measurements (Saito et al. 1983). When adjustments were made for age, the blood pressure was different in hypothyroid and euthyroid patients older than 50 years. In a small subgroup ($n = 14$), the blood pressure significantly decreased with thyroid replacement therapy.

When considering patients with subclinical hypothyroidism (increased TSH, normal T_4), the studies are substantially larger than those of patients with overt hypothyroidism. In a meta-analysis of over 50,000 patients with subclinical hypothyroidism, systolic blood pressure was marginally elevated (by 1.47 mmHg), but diastolic blood pressure was not different from a group of euthyroid controls (Ye et al. 2014). The observed differences in systolic blood pressure may have been due to differences in age between the groups. A study to evaluate the association between thyroid hormones and components of the metabolic syndrome, including blood pressure, in 1423 participants in Korea did not find any association between sub-clinical (or in a small number of patients with overt) hypothyroidism and the occurrence of hypertension, and the association with triglyceride levels was only weak (Jang et al. 2018). Thyroid hormone supplementation does not appear to lower blood pressure in patients with sub-clinical hypothyroidism. In a randomised, double-blind, placebo-controlled clinical trial involving 737 patients followed up for 1 year, there was no difference in blood pressure between the groups; thyroid supplementation was of no apparent benefit (Stott et al. 2017).

Interestingly, cross-sectional population-based studies of hypertensive humans, in which a search for secondary causes of hypertension was performed, have not indicated that hyper- and hypothyroidism are common causes of hypertension. In a recent study of nearly 4000 hypertensive patients in mainland China, of which 75%

had thyroid hormone measurement, only 0.8% were hypothyroid and 0.08% hyperthyroid (Wang et al. 2017). This is in spite of the fact that the prevalence of hyperthyroidism is relatively high, and hypothyroidism moderately common, in comparison with other countries (Taylor et al. 2018). In a recent review outlining the recommended diagnostic work-up for patients with suspected secondary hypertension, thyroid disorders were barely mentioned (Charles et al. 2017). It seems that thyroid hormone measurement is not usually performed in hypertensive humans unless clinical signs are suggestive for hyper- or hypothyroidism (Omura et al. 2004). The importance of hypothyroidism as a cause of hypertension in humans has recently been the subject of debate in the human medical literature (Hofstetter and Messerli 2018).

5.4 Hypertension in Dogs and Cats with Hyperthyroidism

Hyperthyroidism is second only to chronic kidney disease (CKD) as the disease most commonly associated with the diagnosis of hypertension in cats. In reported case series of cats diagnosed with systemic hypertension, the prevalence of hyperthyroidism is reported to be between 7.2 and 24.5%, as detailed in Table 5.2 (Littman 1994; Maggio et al. 2000; Chetboul et al. 2003; Carter et al. 2014; Young et al. 2018; Conroy et al. 2018). In the clinical trials that have been performed recently to support the licensing of amlodipine and telmisartan for treatment of hypertension, the prevalence of hyperthyroidism among the recruited cats has been similar, ranging from 10.9 to 24.7%, even though the cats needed to have been rendered euthyroid before they were eligible for inclusion (Glaus et al. 2018; Coleman et al. 2019a; Huhtinen et al. 2015). Conversely, when cats diagnosed with hyperthyroidism are screened for hypertension the documented prevalence has ranged from 6.7 to 87% (Table 5.3) (Kobayashi et al. 1990; Stiles et al. 1994; Stepien et al. 2003; Sansom et al. 2004; Sartor et al. 2004; van Hoek et al. 2009; Williams et al. 2010; Syme and Elliott 2003; Watson et al. 2018). However, the study with a prevalence of 87% diagnosed hypertension when the systolic blood pressure was >140 mmHg which is not reflective of the cut-offs that are currently employed, so if this study is excluded then the range of reported prevalence narrows from 6.7 to 47.6%. As might be expected, in general, the studies that classify cats as hypertensive at lower blood pressure values appear to have higher reported prevalence.

Although hypertension is reportedly common in hyperthyroid cats, relatively few of them seem to develop blindness or other signs associated with severe hypertensive retinopathy/choroidopathy (Peterson et al. 1983). In a series of 100 hyperthyroid cats examined by a veterinary ophthalmologist, abnormalities were frequent, but only 2 cats had findings (retinal haemorrhage, retinal detachment) that were considered to be suggestive of hypertension (Woerdt and Peterson 2000). Unfortunately blood pressure was not measured in that study. There are various reasons why hyperthyroid cats may exhibit fewer signs of hypertensive retinopathy than might be anticipated: the elevation in blood pressure might not be as high as occurs with other causes of hypertension such as CKD; or development of ocular lesions might be dependent not

5 Thyroid Gland Disease

Table 5.2 Prevalence of hyperthyroidism in cats diagnosed with systemic hypertension

	Publication	n	Hyperthyroid (%)	Criteria (mmHg)	Method	Ocular signs
Case series	Littman (1994)	24	3 (12.5)	>160/100	Doppler (17), direct (15)	Most
	Maggio et al. (2000)	69[a]	HTH 1 (1.4) HTH/CKD 4 (5.8)	>170	Doppler	Most
	Chetboul et al. (2003)	58[a]	3 (7.9)	>170	Doppler	Half
	Carter et al. (2014)	13[a]	2 (15.4)	>160/100	HDO	Uncertain
	Young et al. (2018)	88	16 (18)	>180	Doppler	Most
EPR	Conroy et al. (2018)	282	69 (24.5)	Variable	variable	Half
Drug trials	Huhtinen et al. (2015)	77	19 (24.7)	165–200[b, c]	HDO	No
	Glaus et al. (2018)	285[d]	HTH 21 (7.4) HTH/CKD 14 (3.5)	>160[b,e]	Doppler	No
	Coleman et al. (2019b)	288[d]	HTH 7 (2.4) HTH/CKD 33 (11.5)	160–200[b,c]	Doppler	No

EPR study utilising electronic patient records, *HTH* hyperthyroidism alone, *HTH/CKD* hyperthyroidism and chronic kidney disease diagnosed simultaneously, *n* total study population, *HDO* high-definition oscillometry

[a]Not all cats were tested for hyperthyroidism
[b]Measured on two or more occasions
[c]Due to the possibility of receiving a placebo cats with very severe blood pressure elevation were excluded
[d]Hyperthyroidism had to be controlled for the patient to be included in the study
[e]Patients with severe blood pressure elevation were included at the discretion of the supervising clinician

only on blood pressure elevation but also other changes, for example, to vascular permeability, that increase the risk of lesions developing (Heidbreder et al. 1987; Grosso et al. 2005); or it is possible that the blood pressure in hyperthyroid cats is only transiently elevated due to white coat (situational) hypertension. In one study where blood pressure was measured in hyperthyroid cats either in a secluded setting by a highly experienced veterinary technician, or by a variety of operators in the clinic environment, the values obtained were significantly different (Stepien et al. 2003). It seems reasonable to postulate that hyperthyroid cats could be especially vulnerable to transient, stress-induced, increases in blood pressure.

Hyperthyroidism is the most common feline endocrine disease with an overall prevalence of 2.4% in cats registered at veterinary practices in the UK, rising to 8.7% in cats over 10 years old (Stephens et al. 2014). The average age of cats that are diagnosed with systemic hypertension is around 15 years (Maggio et al. 2000;

Table 5.3 Prevalence of hypertension in cats diagnosed with hyperthyroidism

Publication	n	Hypertensive (%)	Criteria (mmHg)	Method	Ocular signs
Kobayashi et al. (1990)	39	34 (87%)	>140/108	Doppler	No
Stiles et al. (1994)	13	3 (23)	>160/100	Doppler	1/3
Stepien et al. (2003)	84	16 (19)	≥160	Doppler	Not reported
Sansom et al. (2004)	15	1 (6.7)	>170	Oscillometry	All
Sartor et al. (2004)	40	9 (22.5)	>180	Doppler	Not reported
van Hoek et al. (2009)	21	10 (47.6)	≥160	Doppler	No
Williams et al. (2010)	300	Prior 13 (4.3) Incident 20 (15)	>160 with FL or >170 twice	Doppler	Not reported
Syme and Elliott (2003)[a]	100	Prior 3 (3) Incident 9 (9)	>175 with FL or twice	Doppler	5/9
Watson et al. (2018)	140	50 (35.7)	≥160 twice	Doppler	Not reported

FL fundic lesions
[a]Many, if not all, of these cats would also have been included in the group above

Table 5.4 Changes in blood pressure with changes in thyroid function

Publication	n	Initial	Follow-up	P	Method
Syme and Elliott (2003)[a]	59	144.9 ± 16.8	139.8 ± 21.3	0.04	Doppler
Menaut et al. (2012)[a]	60	155 [138, 167]	146 [137, 159]	0.006	Doppler
Stock et al. (2017)	42	161.2 ± 29.8	162.7 ± 31.8	0.793	Doppler
Williams et al. (2014)	19	130.6 [122.0, 140.7][b]	135.4 [129.8, 140.7][b]	0.81	Doppler

Unless otherwise indicated the blood pressure at the initial visit is when the cat was presented with hyperthyroidism, and the follow-up visit is following restoration of euthyroidism
Values represent mean ± standard deviation, median [25th, 75th percentiles]
[a]There are some cats that are included in both of these studies
[b]Initial visit is when the cats were iatrogenically hypothyroid, and the follow-up visit is at restoration of euthyroidism

Jepson et al. 2007). It is therefore to be expected that a proportion of these cats will have hyperthyroidism and vice versa. Similar to the situation in humans, it is possible that the concurrent occurrence of hyperthyroidism and hypertension is largely coincidental and reflects only that the same group of patients are predisposed to development of both conditions. If the link between the conditions is coincidental, then blood pressure would not be expected to change very much when hyperthyroidism is treated. As can be seen from the data displayed in Table 5.4, the decrease in blood pressure that occurs with treatment of hyperthyroidism is small (Syme and

Elliott 2003; Menaut et al. 2012) or undetectable (van Hoek et al. 2009; Stock et al. 2017). A recent study also showed that the prevalence of hypertension was not different between cats with mild, moderate or severe hyperthyroidism (Watson et al. 2018). Although, taken together, these observations may suggest that blood pressure does not change very much with treatment of hyperthyroidism, it is also possible that once hypertension is established, chronic arteriolar remodelling occurs and that this change is not rapidly reversible.

Blood pressure may not change very much with treatment of hyperthyroidism in cats, but it is clear that marked cardiovascular changes are occurring, just as have been described in humans and experimental animal models. The echocardiographic features of hyperthyroidism in the cat have been well described (Moise and Dietze 1986; Bond et al. 1988). Tachycardia is a common clinical feature of feline hyperthyroidism. Concentrations of N-terminal pro brain natriuretic peptide (NT-proBNP) are significantly increased in hyperthyroid cats and decrease rapidly with treatment (Menaut et al. 2012; Sangster et al. 2013). There is also evidence that the renin-angiotensin system is activated in hyperthyroid cats, in as much as treating the hyperthyroidism causes both plasma renin activity (PRA) and, at least in normotensive cats, plasma aldosterone concentration to fall (Williams et al. 2013). These results indicate that hyperthyroid cats are volume expanded, and together with the high heart rate and increases in contractility, this will contribute to a high cardiac output. For blood pressure to remain relatively stable these changes must be offset by a reduction in systemic vascular resistance although this is difficult to measure directly.

In view of the small changes in blood pressure that can be anticipated due to treatment of hyperthyroidism, it is apparent that if a cat is significantly hypertensive when it is presented to the veterinarian, then anti-hypertensive drug therapy will be required alongside treatment for the hyperthyroidism. Beta-blockers are sometimes recommended for the treatment of hypertension in hyperthyroid human patients. Atenolol was used as monotherapy to treat 20 hypertensive, hyperthyroid cats, but although the blood pressure decreased, as well as the heart rate, the target blood pressure (<160 mmHg) was not reached in 70% of the cats (Henik et al. 2008). It is therefore recommended that treatments such as amlodipine, or telmisartan, are used in preference (see Chap. 12).

It has been noted in clinical patients that a small number of cats that are initially normotensive when diagnosed with hyperthyroidism will develop hypertension with long-term treatment. In a series of 215 cats that had longitudinal monitoring of their blood pressure, 49 (22.8%; 95% confidence interval 17.5–29.1) developed hypertension a median of 5.3 months (95% CI: 3.2–9.9) after treatment for hyperthyroidism began (Morrow et al. 2009). Some, but by no means all, of the cats that develop hypertension with treatment for hyperthyroidism are azotaemic. However, even those cats that remain non-azotaemic are likely to have a reduction in GFR following treatment; the cat's eventual renal status once euthyroid presumably depends on the extent of any pre-existing kidney disease. Increases in renal blood flow in hyperthyroid cats have been demonstrated both by using radionucleotide-based techniques

(Adams et al. 1997) and through the use of contrast-enhanced ultrasound and are rapidly reversed with treatment (Stock et al. 2017).

Although the association between hypertension and hyperthyroidism has been known about for over 20 years, it seems that more work needs to be done to encourage veterinarians working in general practice to measure blood pressure routinely. Two recent surveys, one of practitioners based in the UK and the other in Australia, indicate that only about half of all hyperthyroid cats are having their blood pressure measured at diagnosis, and only about a quarter after treatment is instituted (Higgs et al. 2014; Kopecny et al. 2017). This is disappointing because the veterinarians that respond to such surveys are likely to be the most diligent and informed.

Hyperthyroidism is a relatively rare condition in dogs, most often occurring in patients with functional carcinoma. In spite of this there have been individual case reports describing dogs with hypertension as a consequence of thyroid carcinoma (Simpson and McCown 2009; Looney and Wakshlag 2017).

> **Clinical Recommendations**
> All hyperthyroid cats should have their blood pressure measured both at time of diagnosis and during follow-up.
> The decrease in blood pressure that occurs with restoration of euthyroidism in hyperthyroid cats is small, and so treatment with amlodipine or telmisartan is usually recommended if cats are hypertensive.

5.5 Hypertension in Dogs and Cats with Hypothyroidism

There do not appear to have been any systematic studies to date of blood pressure measurements in hypothyroid dogs. There is a single case report of hypertension in a hypothyroid dog but it also had kidney disease so it is not clear what role the hypothyroidism played, or even whether the dog might have actually had sick euthyroidism (Gwin et al. 1978).

In a group of 19 cats with iatrogenic hypothyroidism due to overtreatment with carbimazole or methimazole, the blood pressure was relatively low at baseline and did not change significantly as they returned to euthyroidism (Table 5.4) (Williams et al. 2014).

5.6 Conclusions

Thyroid hormones have marked effects on the cardiovascular system, changing both SVR and cardiac output. The opposing effect of these changes means that systemic arterial blood pressure is impacted very little in hyperthyroid or hypothyroid states. Feline hyperthyroidism is an extremely common disease of the aged cat and hypertension is a problem that afflicts cats in old age, thus it is not surprising that these two conditions frequently co-exist. Hyperthyroid cats that are hypertensive will require anti-hypertensive treatment as management of their hyperthyroidism is unlikely to lead to resolution of their hypertension. In addition, monitoring of blood pressure in cats undergoing treatment for hyperthyroidism is strongly recommended as about 20% do show an increase in blood pressure once they are euthyroid to a level that warrants commencement of treatment to protect target organs against damage. The influence of clinical hypothyroidism and it management in dogs on blood pressure has not been studied to any extent and so represents a gap in our current knowledge.

References

Adams WH, Daniel GB, Legendre AM (1997) Investigation of the effects of hyperthyroidism on renal function in the cat. Can J Vet Res 61:53–56

Bassett JH, Harvey CB, Williams GR (2003) Mechanisms of thyroid hormone receptor-specific nuclear and extra nuclear actions. Mol Cell Endocrinol 213:1–11

Belakavadi M, Saunders J, Weisleder N et al (2010) Repression of cardiac phospholamban gene expression is mediated by thyroid hormone receptor-{alpha}1 and involves targeted covalent histone modifications. Endocrinology 151:2946–2956

Bianco AC, Salvatore D, Gereben BZ et al (2002) Biochemistry, cellular and molecular biology, and physiological roles of the iodothyronine selenodeiodinases. Endocr Rev 23:38–89

Bond BR, Fox PR, Peterson ME et al (1988) Echocardiographic findings in 103 cats with hyperthyroidism. J Am Vet Med Assoc 192:1546–1549

Cai Y, Ren Y, Shi J (2011) Blood pressure levels in patients with subclinical thyroid dysfunction: a meta-analysis of cross-sectional data. Hypertens Res 34:1098

Cai P, Peng Y, Chen Y et al (2019) Association of thyroid function with white coat hypertension and sustained hypertension. J Clin Hypertens (Greenwich) 21:674–683

Cappola AR, Ladenson PW (2003) Hypothyroidism and atherosclerosis. J Clin Endocrinol Metab 88:2438–2444

Carrillo-Sepúlveda MA, Ceravolo GS, Fortes ZB et al (2009) Thyroid hormone stimulates NO production via activation of the PI3K/Akt pathway in vascular myocytes. Cardiovasc Res 85:560–570

Carter JM, Irving AC, Bridges JP et al (2014) The prevalence of ocular lesions associated with hypertension in a population of geriatric cats in Auckland, New Zealand. N Z Vet J 62:21–29

Charles L, Triscott J, Dobbs B (2017) Secondary hypertension: discovering the underlying cause. Am Fam Physician 96:453–461

Chetboul V, Lefebvre HP, Pinhas C et al (2003) Spontaneous feline hypertension: clinical and echocardiographic abnormalities, and survival rate. J Vet Intern Med 17:89–95

Coleman AE, Brown SA, Traas AM et al (2019a) Safety and efficacy of orally administered telmisartan for the treatment of systemic hypertension in cats: Results of a double-blind, placebo-controlled, randomized clinical trial. J Vet Intern Med 33:478–488

Coleman AE, Brown SA, Stark M et al (2019b) Evaluation of orally administered telmisartan for the reduction of indirect systolic arterial blood pressure in awake, clinically normal cats. J Feline Med Surg 21:109–114

Conroy M, Chang Y-M, Brodbelt D et al (2018) Survival after diagnosis of hypertension in cats attending primary care practice in the United Kingdom. J Vet Intern Med 32:1846–1855

Fommei E, Iervasi G (2002) The role of thyroid hormone in blood pressure homeostasis: evidence from short-term hypothyroidism in humans. J Clin Endocrinol Metab 87:1996–2000

Fukuyama K, Ichiki T, Takeda K et al (2003) Downregulation of vascular angiotensin II type 1 receptor by thyroid hormone. Hypertension 41:598–603

Glaus TM, Elliott J, Herberich E et al (2018) Efficacy of long-term oral telmisartan treatment in cats with hypertension: results of a prospective European clinical trial. J Vet Intern Med 33(2):413–422

Grosso A, Veglio F, Porta M et al (2005) Hypertensive retinopathy revisited: some answers, more questions. Br J Ophthalmol 89:1646–1654

Gwin RM, Gelatt KN, Terrell TG et al (1978) Hypertensive retinopathy associated with hyperthyroidism, hypercholesterolemia, and renal failure in a dog. J Am Anim Hosp Assoc 14:200–209

Heidbreder E, Huller U, Schafer B et al (1987) Severe hypertensive retinopathy. Increased incidence in renoparenchymal hypertension. Am J Nephrol 7:394–400

Henik RA, Stepien RL, Wenholz LJ et al (2008) Efficacy of atenolol as a single antihypertensive agent in hyperthyroid cats. J Feline Med Surg 10:577–582

Higgs P, Murray JK, Hibbert A (2014) Medical management and monitoring of the hyperthyroid cat: a survey of UK general practitioners. J Feline Med Surg 16(10):788–795

Hofstetter L, Messerli FH (2018) Hypothyroidism and hypertension: fact or myth? Lancet 391:29–30

Huhtinen M, Derre G, Renoldi HJ et al (2015) Randomized placebo-controlled clinical trial of a chewable formulation of amlodipine for the treatment of hypertension in client-owned cats. J Vet Intern Med 29:786–793

Hurxthal LM (1931) Blood pressure before and after operation in hyperthyroidism. Arch Intern Med 47:167–181

Izumo S, Lompre AM, Matsuoka R et al (1987) Myosin heavy chain messenger RNA and protein isoform transitions during cardiac hypertrophy. Interaction between hemodynamic and thyroid hormone-induced signals. J Clin Invest 79:970–977

Jang J, Kim Y, Shin J et al (2018) Association between thyroid hormones and the components of metabolic syndrome. BMC Endocr Disord 18:29

Jepson RE, Elliott J, Brodbelt D et al (2007) Effect of control of systolic blood pressure on survival in cats with systemic hypertension. J Vet Intern Med 21:402–409

Jiang M, Xu A, Tokmakejian S et al (2000) Thyroid hormone-induced overexpression of functional ryanodine receptors in the rabbit heart. Am J Physiol Heart Circ Physiol 278:H1429–H1438

Kahaly GJ, Dillmann WH (2005) Thyroid hormone action in the heart. Endocr Rev 26:704–728

Klein I, Danzi S (2007) Thyroid disease and the heart. Circulation 116:1725–1735

Klein I, Danzi S (2016) Thyroid disease and the heart. Curr Probl Cardiol 41:65–92

Klein I, Hong C (1986) Effects of thyroid hormone on cardiac size and myosin content of the heterotopically transplanted rat heart. J Clin Invest 77:1694–1698

Klein I, Ojamaa K (2001) Thyroid hormone and the cardiovascular system. N Engl J Med 344:501–509

Kobayashi DL, Peterson ME, Graves TK et al (1990) Hypertension in cats with chronic renal failure or hyperthyroidism. J Vet Intern Med 4:58–62

Kopecny L, Higgs P, Hibbert A et al (2017) Management and monitoring of hyperthyroid cats: a survey of Australian veterinarians. J Feline Med Surg 19:559–567

Littman MP (1994) Spontaneous systemic hypertension in 24 cats. J Vet Intern Med 8:79–86

Looney A, Wakshlag J (2017) Dietary management of hyperthyroidism in a dog. J Am Anim Hosp Assoc 53:111–118

Maggio F, DeFrancesco TC, Atkins CE et al (2000) Ocular lesions associated with systemic hypertension in cats: 69 cases (1985–1998). J Am Vet Med Assoc 217:695–702

Marcisz C, Jonderko G, Kucharz E (2002) Changes of arterial pressure in patients with hyperthyroidism during therapy. Med Sci Monit 8:Cr502–Cr507

Mazza A, Beltramello G, Armigliato M et al (2011) Arterial hypertension and thyroid disorders: what is important to know in clinical practice? Ann Endocrinol 72:296–303

Menaut P, Connolly DJ, Volk A et al (2012) Circulating natriuretic peptide concentrations in hyperthyroid cats. J Small Anim Pract 53(12):673–678

Moise NS, Dietze AE (1986) Echocardiographic, electrocardiographic, and radiographic detection of cardiomegaly in hyperthyroid cats. Am J Vet Res 47:1487–1494

Morrow LD, Adams VJ, Elliott J et al (2009) Hypertension in hyperthyroid cats: prevalence, incidence and predictors of its development. J Vet Intern Med 23:700

Omura M, Saito J, Yamaguchi K et al (2004) Prospective study on the prevalence of secondary hypertension among hypertensive patients visiting a general outpatient clinic in Japan. Hypertens Res 27:193–202

Osuna PM, Udovcic M, Sharma MD (2017) Hyperthyroidism and the heart. Methodist Debakey Cardiovasc J 13:60–63

Palmieri EA, Fazio S, Palmieri V et al (2004) Myocardial contractility and total arterial stiffness in patients with overt hyperthyroidism: acute effects of beta1-adrenergic blockade. Eur J Endocrinol 150:757–762

Park KW, Dai HB, Ojamaa K et al (1997) The direct vasomotor effect of thyroid hormones on rat skeletal muscle resistance arteries. Anesth Analg 85:734–738

Peterson ME, Kintzer PP, Cavanagh PG et al (1983) Feline hyperthyroidism: pretreatment clinical and laboratory evaluation of 131 cases. J Am Vet Med Assoc 183:103–110

Prisant LM, Gujral JS, Mulloy AL (2006) Hyperthyroidism: a secondary cause of isolated systolic hypertension. J Clin Hypertens (Greenwich) 8:596–599

Razvi S, Jabbar A, Pingitore A et al (2018) Thyroid hormones and cardiovascular function and diseases. J Am Coll Cardiol 71:1781–1796

Saito I, Saruta T (1994) Hypertension in thyroid disorders. Endocrinol Metab Clin N Am 23:379–386

Saito I, Ito K, Saruta T (1983) Hypothyroidism as a cause of hypertension. Hypertension 5:112–115

Sangster JK, Panciera DL, Abbott JA et al (2013) Cardiac biomarkers in hyperthyroid cats. J Vet Intern Med 28(2):465–472

Sansom J, Rogers K, Wood JL (2004) Blood pressure assessment in healthy cats and cats with hypertensive retinopathy. Am J Vet Res 65:245–252

Sartor LL, Trepanier LA, Kroll MM et al (2004) Efficacy and safety of transdermal methimazole in the treatment of cats with hyperthyroidism. J Vet Intern Med 18:651–655

Simpson AC, McCown JL (2009) Systemic hypertension in a dog with a functional thyroid gland adenocarcinoma. J Am Vet Med Assoc 235:1474–1479

Stephens MJ, O'Neill DG, Church DB et al (2014) Feline hyperthyroidism reported in primary-care veterinary practices in England: prevalence, associated factors and spatial distribution. Vet Rec 175:458

Stepien RL, Rapoport GS, Henik RA et al (2003) Effect of measurement method on blood pressure findings in cats before and after therapy for hyperthyroidism. J Vet Int Med 17:754. (abstract)

Stiles J, Polzin DJ, Bistner DI (1994) The prevalence of retinopathy in cats with systemic hypertension and chronic renal failure or hyperthyroidism. J Am Anim Hosp Assoc 30:654–572

Stock E, Daminet S, Paepe D et al (2017) Evaluation of renal perfusion in hyperthyroid cats before and after radioiodine treatment. J Vet Intern Med 31(6):1658–1663

Stott DJ, Rodondi N, Kearney PM et al (2017) Thyroid hormone therapy for older adults with subclinical hypothyroidism. N Engl J Med 376:2534–2544

Syme HM, Elliott J (2003) The prevalence of hypertension in hyperthyroid cats at diagnosis and following treatment. J Vet Intern Med 17:754–755

Taylor PN, Albrecht D, Scholz A et al (2018) Global epidemiology of hyperthyroidism and hypothyroidism. Nat Rev Endocrinol 14:301–316

Udovcic M, Pena RH, Patham B et al (2017) Hypothyroidism and the heart. Methodist Debakey Cardiovasc J 13:55–59

van der Woerdt A, Peterson ME (2000) Prevalence of ocular abnormalities in cats with hyperthyroidism. J Vet Intern Med 14:202–203

van Hoek I, Lefebvre HP, Peremans K et al (2009) Short- and long-term follow-up of glomerular and tubular renal markers of kidney function in hyperthyroid cats after treatment with radioiodine. Domest Anim Endocrinol 36:45–56

Vargas F, Rodríguez-Gómez I, Vargas-Tendero P et al (2012) The renin–angiotensin system in thyroid disorders and its role in cardiovascular and renal manifestations. J Endocrinol 213:25–36

Völzke H, Ittermann T, Schmidt CO et al (2009) Subclinical hyperthyroidism and blood pressure in a population-based prospective cohort study. Eur J Endocrinol 161:615

Wang L, Li N, Yao X et al (2017) Detection of secondary causes and coexisting diseases in hypertensive patients: OSA and PA are the common causes associated with hypertension. Biomed Res Int 2017:8295010

Watson N, Murray JK, Fonfara S et al (2018) Clinicopathological features and comorbidities of cats with mild, moderate or severe hyperthyroidism: a radioiodine referral population. J Feline Med Surg 20:1130–1137

Williams TL, Peak KJ, Brodbelt D et al (2010) Survival and the development of azotemia after treatment of hyperthyroid cats. J Vet Intern Med 24:863–869

Williams TL, Elliott J, Syme HM (2013) Renin-angiotensin-aldosterone system activity in hyperthyroid cats with and without concurrent hypertension. J Vet Intern Med 27:522–529

Williams TL, Elliott J, Syme HM (2014) Effect on renal function of restoration of euthyroidism in hyperthyroid cats with iatrogenic hypothyroidism. J Vet Intern Med 28:1251–1255

Wong NL, Huang D, Guo NS et al (1989) Effects of thyroid status on atrial natriuretic peptide release from isolated rat atria. Am J Phys 256:E64–E67

Ye Y, Xie H, Zeng Y et al (2014) Association between subclinical hypothyroidism and blood pressure—a meta-analysis of observational studies. Endocr Pract 20:150–158

Young WM, Zheng C, Davidson MG et al (2018) Visual outcome in cats with hypertensive chorioretinopathy. Vet Ophthalmol 22(2):161–167

Genetics of Hypertension: The Human and Veterinary Perspectives

Rosanne E. Jepson

The health burden associated with hypertension in the world is very large. Current estimates suggest that there are over 1 billion individuals with hypertension worldwide with the World Health Organisation estimating that this will increase to 1.5 billion by 2025. Between 35% and 45% of individuals over 25 years demonstrate hypertension with prevalence varying with geographical location and socioeconomic status.[1] In humans, a continuous and incremental risk exists between blood pressure (BP) and cardiovascular disease, stroke and kidney disease. The significant health and economic impact of hypertension has, therefore, driven the requirement for a better understanding of factors, including genetic factors, that may contribute to the development of hypertension. An understanding of these genetic associations might not only improve individual and population risk prediction but, more importantly, may identify novel pathophysiological pathways that are involved in BP regulation and which may also act as pathway targets for new anti-hypertensive drug development assisting with personalisation of anti-hypertensive therapy.

As a physiological function, BP is the product of total peripheral resistance and cardiac output but the determinants of an individual's BP are much more complex and multifactorial with BP regulation requiring a complex interplay between extracellular fluid volume, cardiac contractility and vascular tone with input from renal, neural and endocrine systems (see Chap. 1). The mosaic theory of hypertension was first proposed by Page in 1960, suggesting that varied facets that compose BP control, including chemical, neuronal, elasticity, cardiac output, viscosity, vascular calibre, volume and reactivity, are all in equilibrium determining the final BP level

---

[1] http://www.who.int/cardiovascular_diseases/publications/global_brief_hypertension/en/

R. E. Jepson (✉)
Department of Clinical Sciences and Services, Royal Veterinary College, University of London, Hertfordshire, UK
e-mail: rjepson@rvc.ac.uk

(Page 1960). Today, the same construct can be used indicating that BP and hypertension are complex traits but with the inclusion of genetics as a contributing factor (Padmanabhan et al. 2015).

Early heritability studies in humans exploring familial associations in BP suggest that heritability of measurement of systolic BP (SBP) in the clinic is between 15% and 40%, diastolic BP 15–30% and 50–60% for long-term (i.e. ambulatory) systolic or diastolic BP measurements (Snieder et al. 2003; Kupper et al. 2005; Hottenga et al. 2005; Havlik et al. 1979; Fuentes et al. 2000; Bochud et al. 2005). These studies have been performed either using comparisons between singletons and twins (Snieder et al. 2003; Kupper et al. 2005; Hottenga et al. 2005) or by looking at familial association between parents and offspring (Havlik et al. 1979; Fuentes et al. 2000; Bochud et al. 2005). However, it should be remembered that heritability from such studies estimates the degree of phenotypic similarity between relatives and, therefore, depends not only on genetic similarities but also on other exposure factors, including the environment and its interaction with the genome as well as on the accuracy of measurement of BP (see Chap. 2) (Padmanabhan et al. 2015; Zuk et al. 2012). Environmental impacts can vary between populations; therefore, heritability estimates are valid only for the population studied. In addition, environmental changes over time may result in variation in environmental exposures between generations could and, therefore, influence apparent heritability of a given trait (Padmanabhan et al. 2015; Polderman et al. 2015).

In the early 1950s there was controversy known as the Platt-Pickering debate, exploring the concept of whether there was a unimodal or bimodal distribution of BP in the population resulting in hypertension (Fig. 6.1). Work by Platt measuring BP in normotensive and hypertensive individuals and their relatives favoured the bimodal distribution, leading him to propose that hypertension was a simple Mendelian trait (Oldham et al. 1960) whilst Pickering's work studying BP between the second and eighth decades in first-degree relatives supported a unimodal distribution, indicating that BP was a quantitative phenotype and, therefore, most likely to be polygenic and non-Mendelian (Pickering 1955). It has now been widely accepted that BP in the

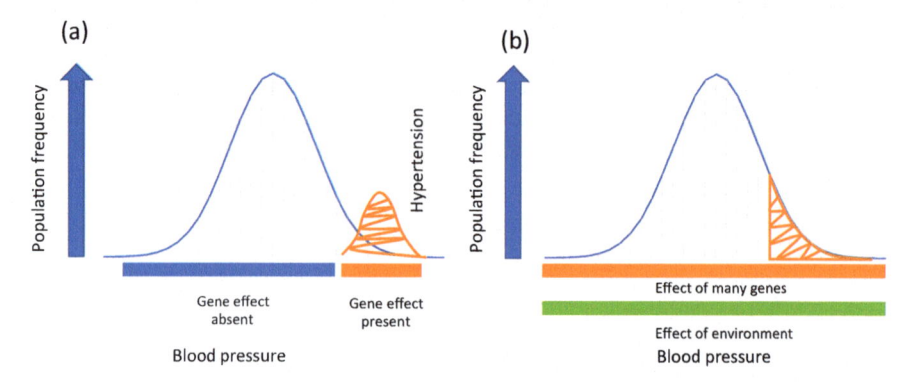

Fig. 6.1 The Platt-Pickering debate: hypertension as a discrete population (**a**) versus a range of blood pressures within a population (**b**)

general population has a complex, non-Mendelian mode of inheritance, meaning that no single-gene exerts an over-riding effect on BP but that many variants cumulatively will have a relatively small effect (common disease-common variant hypothesis). However, since the early 1930s individuals with rare conditions resulting in extreme hypertension and hypotension have been identified, supporting Platt's theory. These individuals typically present early in life and in many instances have now been linked to Mendelian modes of inheritance with rare genetic mutations that result in hypertension. Many of the genes implicated in these Mendelian monogenic hypertensive/hypotensive disorders are related to either renal or adrenal gland function, highlighting the importance of these two organs in BP regulation and the pathogenesis of hypertension. Interestingly, the role that rare variants may play in the development of hypertension in the general population is also making a re-emergence given the discrepancy in terms of the degree of heritability that can be explained by genetic discoveries to date and the new technologies that are enhancing our ability to explore the human genome, e.g. next-generation sequencing (NGS) and the common disease-rare variant hypothesis.

Animal models of hypertension have been used for decades to support research into the pathogenesis and treatment of hypertension (Lerman et al. 2005). These models have included surgically induced hypertension, e.g. Goldblatt's two-kidney one-clip canine model of hypertension with hypertension evoked by unilateral occlusion of the renal artery but have also included exploration of transgenic and gene-targeted models of hypertension (Lerman et al. 2005; Goldblatt et al. 1934; Pinto et al. 1998). The spontaneously hypertensive rat (SHR) is an example of a phenotypically driven genetic rodent model of hypertension based on the Wistar strain of rats and produced by selective breeding of a given phenotype (i.e. hypertensive) over several generations and then sib-mating until genetic homogeneity is achieved (Okamoto and Aoki 1963). Many of these phenotypically driven models of hypertension in rodents carry traits that are useful for the exploration of human hypertension, e.g. salt sensitivity in the salt-sensitive Dahl rat (Pinto et al. 1998) or obesity in the Sprague-Dawley (Dobrian et al. 2003) or Wistar fatty rats (Imai et al. 2003). However, other genetic models of hypertension are genotype driven. Mice have been used for both gene overexpression (transgenic) or deletion (knockout) studies to explore specific physiological pathways important in BP regulation and the pathogenesis of hypertension, e.g. transgenic and knockout mouse models for components of the renin-angiotensin system (Kimura et al. 1992; Krege et al. 1995). However, to date, although hypertension is recognised in companion animal medicine with many clinically observable similarities between hypertension in cats, dogs and humans, there has been very little exploration of phenotypic and genetic similarities between these species. This is an area of potential interest for the future given that companion animals share many of the same environmental influences as humans but also in light of species differences that we recognise in the way cats and dogs differ in expression of target organ damage (TOD) and response to anti-hypertensive therapies.

6.1 Mendelian Blood Pressure Syndromes

A Mendelian or monogenic trait is one whose presence is dependent on the genotype at a single locus (region of the genome). There are five basic forms of Mendelian inheritance including autosomal dominant, autosomal recessive, X-linked dominant, X-linked recessive and Y-linked with all Mendelian forms of hypertension that are discussed below being either autosomal dominant or recessive. These rare inherited mutations have a comparatively large effect on phenotype (6–10 mmHg systolic BP), sufficient to result in hyper- or hypotension, and in many instances occur at relatively young age. To date >25 rare mutations have been identified using linkage analysis and exome sequencing which result in syndromes of either hyper- or hypotension (Padmanabhan et al. 2012, 2015; Luft et al. 2003). The conditions resulting in hypertension typically involve pathways for sodium homeostasis and hence volume expansion and low renin hypertension secondary to sodium retention (Padmanabhan et al. 2012; Lifton et al. 2001). Further information regarding Mendelian-inherited conditions in man can be found at OMIM® (Online Mendelian Inheritance in Man®).[2] Understanding the exact molecular basis of these forms of hypertension, whether they impact genes from mineralocorticoid, glucocorticoid or sympathetic pathways, means that specific and targeted therapy can be provided (Table 6.1). To date, monogenic or Mendelian forms of hypertension have not been reported in the dog or the cat. However, this is not to say that such conditions do not or cannot exist, just that to date we have not identified these patients. Nevertheless, the concepts behind some of the monogenic causes of hypertension in humans are of interest, given their direct relation to pathways that are involved in the regulation of BP, and as discussed further below, they provide some interesting hypothesis generating ideas to better understand the pathogenesis of hypertension in our veterinary patients.

Many of the mongenic conditions that have been recognised in humans relate to altered activity of the mineralocorticoid pathway. These include familial hyperaldosteronism type 1 (glucocorticoid-remediable aldosteronism) where the presence of a chimeric gene formed from fusion of 11β-hydroxylase (*CYP11B1*) and aldosterone synthase (*CYP11B2*) results in ACTH rather than angiotensin II or potassium-stimulating aldosterone secretion (Lifton et al. 1992). Specific therapy for individuals with this condition includes low-dose glucocorticoid therapy in order to suppress ACTH production and, therefore, aldosterone secretion.

The syndrome of apparent mineralocorticoid excess represents another form of low-renin hypertension where patients exhibit both a metabolic alkalosis and hypokalaemia (Cerame and New 2000). This condition has historically sparked interest in feline medicine based on the observation that hypertensive cats have significantly lower potassium concentrations than their normotensive counterparts (Syme et al. 2002). In humans, it results from a mutation in 11β-hydroxysteroid dehydrogenase leading to reduced or absence of activity of this enzyme (Cerame and

---

[2]https://www.omim.org

Table 6.1 Monogenic syndromes of hypertension

Monogenic syndrome of inheritance	Nearest gene(s)	Genomic and phenotypic annotation	Specific therapeutics	References
Mineralocorticoid pathway				
Glucocorticoid-remediable aldosteronism/familial hyperaldosteronism type 1 (FH-1, OMIM 103900)	*CYP11B1, CYP11B2*	Autosomal dominant syndrome: HT caused by increased aldosterone secretion driven by ACTH. Chimeric gene containing 5′ regulatory sequences of *CYP11B1* (conferring ACTH responsiveness) fused with distal coding sequence of aldosterone synthase (*CYP11B2*) such that ACTH not angiotensin II or potassium is the main regulator of aldosterone synthesis	Low-dose glucocorticoids to suppress ACTH production or amiloride to inhibit ENaC or spironolactone (aldosterone inhibitor)	Lifton et al. (1992)
Corticosterone methyloxidase II deficiency (OMIM 610600)	*CYP11B1, CYP11B2*	Autosomal dominant		Padmanabhan et al. (2012)
Steroid 11β-hydroxylase deficiency (OMIM 202010)	*CYP11B1, CYP11B2*	Autosomal recessive		Padmanabhan et al. (2012)
Familial hyperaldosteronism type 2 (FH-II, OMIM 605635)	Not determined	Autosomal dominant syndrome caused by hyperplasia or adenoma of the aldosterone-producing adrenal cortex	Spironolactone, eplerenone	Lafferty et al. (2000)
Apparent mineralocorticoid excess (OMIM 218030)	*HSD11B2*	Low renin hypertension syndrome characterised by	Low-sodium diet and spironolactone blocking	Cerame and New (2000)

(continued)

Table 6.1 (continued)

Monogenic syndrome of inheritance	Nearest gene(s)	Genomic and phenotypic annotation	Specific therapeutics	References
		hypokalaemia and metabolic alkalosis. Absence or reduced activity of 11β-hydroxysteroid dehydrogenase. Normally both cortisol and aldosterone have MCR activity but cortisol is rapidly metabolised by *HSD11B2* preventing binding to the MCR	binding of both aldosterone and cortisol to MR	
Hypertension exacerbated in pregnancy (OMIM 605115)	*NR3C2*	Autosomal dominant; missense mutation (serine to leucine) in MR causing receptor to be constitutively active and altering specificity such that hormones that are normally antagonistic to the MR (e.g. progesterone and cortisone) act as agonists. Women with these mutations develop severe hypertension during pregnancy due to increase in progesterone concentrations	Spironolactone contraindicated	Geller et al. (2000)
Pseudohypoaldosteronism type II (Gordon's syndrome)	*WNK1, WNK 4*	Autosomal dominant form of hypertension associated with hyperkalaemia, non-anion	Low-salt diet or thiazide diuretics	Wilson et al. (2001) and Unseld et al. (1997)

6 Genetics of Hypertension: The Human and Veterinary Perspectives 151

		gap metabolic acidosis and increased salt reabsorption by the kidney. WNK (with no lysine [K]) plays central roles in regulating BP by initiating a signalling pathway that controls activity of the NCC (Na$^+$Cl$^-$ cotransporter) and NKCC2 (Na+/K+/2Cl- transporter). Mutations have been identified in both WNK1 and WNK4. However, this explains only a small proportion of individuals with pseudohypoaldosteronism type II.		
	KLHL3	Dominant and recessive mutations in Kelch-like 3 increase activity of the NCC via interactions with WNK		Schumacher et al. (2014), Louis-Dit-Picard et al. (2012), Ohta et al. (2013), and Boyden et al. (2012)
	CUL3	CUL3 acts as a scaffolding protein, binding to KLHL3 and may impact ubiquitination pathways for WNK4 and other WNK isoforms		Schumacher et al. (2014), Jensen et al. (1997), Ohta et al. (2013), and Boyden et al. (2012)
Liddle's syndrome (OMIM 177200)	SCNN1B, SCNN1G, NEDD4L	Autosomal dominant condition with hypertension	Amiloride or triamterene (block ENaC)	Shimkets et al. (1994) and Hansson et al. (1995)

(continued)

Table 6.1 (continued)

Monogenic syndrome of inheritance	Nearest gene(s)	Genomic and phenotypic annotation	Specific therapeutics	References
		and aldosterone excess but with low aldosterone and renin levels. Caused by mutations in the genes coding for the β or γ subunits of ENaC. These regions facilitate binding of Nedd4–2. The inability of β or γ subunits to bind Nedd4 results in constitutive expression of sodium channels and prolongation of half-life of ENaC in the distal tubule, increasing sodium reabsorption		
Primary aldosteronism				
Primary aldosteronism	*KCNJ5*	<40% of aldosterone-producing adenomas have a gain of function somatic mutation in KCNJ5, a potassium channel, which results in membrane depolarisation and enhanced aldosterone production. The mutations are believed to be located around the selectivity filter of the K^+ channel and may cause loss of selectivity,	Patients with resistance to medical therapy require bilateral adrenalectomy	Scholl and Lifton (2013)

			allowing sodium entry, membrane depolarisation and opening of voltage-gated Ca^{2+} channels		
		ATP1A1	In a further 7% of aldosterone-producing adenomas, mutations in ATP1A1 encoding for the α1-subunit of Na$^+$/K$^+$ ATPase, ATP2B3 encoding a plasma membrane Ca^2 $^+$ATPase 3 and CACNA1D, encoding an L-type Ca^{2+} channel (Ca$_v$1.3) may be present. These mutations may play important roles in membrane depolarisation and opening of voltage-dependent calcium channels which is the molecular target for L-type calcium channel blockers	Adrenalectomy	Scholl et al. (2013), Beuschlein et al. (2013), Fernandes-Rosa et al. (2015), and Azizan et al. (2013)
		ATP2B3			
		CACNA1D			
Glucocorticoid pathways					
Congenital adrenal hyperplasia (defects in enzymes of the cortisol biosynthesis pathways)	CYP11B1 (OMIM 202010), HSD3B2 (OMIM 613890), CYP17A1 (OMOM 609300), CYP11A1 (OMIM 118485)		In this group of syndromes, plasma ACTH concentrations increase in an attempt to produce cortisol, and some of the aberrant products of the affected cortisol pathway accumulate and may result in the		Padmanabhan et al. (2015) and New and Levine 1980

(continued)

Table 6.1 (continued)

Monogenic syndrome of inheritance	Nearest gene(s)	Genomic and phenotypic annotation	Specific therapeutics	References
		development of hypertension. Genes reported to be affected include CYP11B1 encoding for 11β-hydroxylase, HSD3B2 encoding for 3β-hydroxysteroid dehydrogenase, CYP17A1 encoding for 17α-hydroxylase and CYP11A1 encoding for cholesterol desmolase. The mutations are autosomal recessive		

Sympathetic pathway

Pheochromocytomas and paragangliomas	*RET* *VHL* *SDHA*, *SDHB*, *SDHC* and *SDHD* *KIF1Bbeta* *PHD2* *SDHAF2*	Approximately 30% of pheochromocytomas and paragangliomas may be the result of germline mutations. In some instances, an autosomal dominant inheritance of *RET* pro-to-oncogene mutations may predispose. Other susceptibility genes that have been observed include the tumour suppressor gene *VHL* (von Hippel-Lindau	Alpha-blockers followed by beta-blockers and adrenalectomy (see Chap. 4)	Iacobone et al. (2011), Opocher and Schiavi (2010), and Welander et al. (2011)

		syndrome), SDHA, SDHB, SDHC and SDHD which code for subunits of succinate dehydrogenase A-D, KIF1Bbeta, PHD2 and SDHAF2	
Other			
Hypertension associated with PPARγ mutations	PPARγ	Autosomal dominant leading to malfunction of the nuclear receptor PPARγ. Associated with hypertension, insulin resistance and type 2 diabetes mellitus	Barroso et al. (1999)
Hypertension and brachydactyly	PDE3A	Autosomal dominant mutations in PDE3A encoding for phosphodiesterase 3A, associated with hypertension and brachydactyly	Schuster et al. (1996) and Maass et al. (2015)

ENaC epithelial sodium channel, *MR* mineralocorticoid receptor, *ACTH* adrenocorticotrophic hormone

New 2000). In health, both cortisol and aldosterone can stimulate the intracellular mineralocorticoid receptor (MR). However, the former is rapidly degraded by 11-β-hydroxysteroid dehydrogenase, preventing binding to the MR. Deficiency in the action of 11β-hydroxysteroid dehydrogenase can, therefore, lead to increased mineralocorticoid activity, sodium retention, volume expansion and low-renin hypertension (Padmanabhan et al. 2012, 2015). These individuals respond well clinically to use of a low sodium diet together with spironolactone, a mineralocorticoid receptor antagonist. In cats, the role of the 11β-hydroxysteroid dehydrogenase pathway was explored via evaluation of urinary cortisol:cortisone ratios with no apparent difference identified between normotensive and hypertensive cats, suggesting that, at least in the population of cats studied, this pathophysiological mechanism does not appear to be involved in the development of feline hypertension (Walker et al. 2009).

Pseudohyperaldosteronism type 2 (Gordon's syndrome) is an autosomal dominant form of hypertension characterised by hyperkalaemia, non-anion gap mediated metabolic acidosis and increased renal sodium reabsorption (Padmanabhan et al. 2015). This condition has been linked with mutations in *WNK1* and *WNK4* and associated genes *KLHL3* and *CUL3*, encoding for With No Lysine [K^+] kinases, Kelch-like 3 and Cullin 3, respectively (Schumacher et al. 2014; Louis-Dit-Picard et al. 2012; Ohta et al. 2013; Boyden et al. 2012; Wilson et al. 2001). These genes play an important role in initiating signalling pathways that control the activity of the NCC (Na^+/Cl^- ion cotransporter expressed in the distal tubule) and NKCC2 (Na^+/K^+/$2Cl^-$ cotransporter 2; expressed in the thick ascending limb of the loop of Henle) in the kidney. Activating these cotransporters leads to increased sodium reabsorption whilst at the same time reducing potassium excretion, thus resulting in the combined effects of hyperkalaemia, volume expansion and hypertension (Padmanabhan et al. 2015). To date, although a *WNK4* mutation has been identified by genome-wide association analysis and associated with Burmese hypokalaemic periodic paralysis (Gandolfi et al. 2012), no studies have explored *WNK* mutations and any potential association with hypertension in either the cat or the dog.

Finally, Liddle's syndrome is an autosomal dominant condition where hypertension is associated with low concentrations of both aldosterone and renin. Mutations in the β or γ subunits of the epithelial sodium channel (ENaC; SCN1B, SCNN1G expressed in the cortical connecting tubule) lead to altered half-life of ENaC, therefore, facilitating sodium retention and volume expansion and hypertension. As becomes apparent from understanding the pathophysiological mechanism, these individuals respond well to amiloride or triamterene which block sodium movement through ENaC but do not respond to aldosterone receptor antagonists such as spironolactone.

In human medicine, primary aldosteronism is recognised in ~2% of the general population, ~10% of the hypertensive population and up to 25% of individuals with resistant hypertension (see Chap. 4). These individuals collectively demonstrate hypokalaemia, elevated aldosterone concentrations and low plasma renin activity. Bilateral adrenal hyperplasia is the most common cause of primary aldosteronism; however, less common causes include unilateral adrenal hyperplasia and adenoma of the aldosterone producing adrenal cortex (APA). In approximately 5% of individuals with APA, a Mendelian form of primary aldosteronism is identified (Padmanabhan et al. 2015). Differentiation is important in human medicine because individuals with an APA benefit from adrenalectomy whilst those with bilateral adrenal gland hyperplasia are usually managed medically. An exome sequencing study identified that up to 40% of APA exhibits a gain of function somatic mutation in *KCNJ5*, a potassium channel responsible for membrane depolarization and increasing aldosterone secretion (Scholl and Lifton 2013). Mutations in other genes associated with aldosterone secretion have also been identified including *ATP1A1*, encoding for the α1-subunit of Na^+/K^+ ATPase, *ATP2B3*, encoding a plasma membrane Ca^{2+}ATPase 3 and *CACNA1D*, encoding an L-type Ca^{2+} channel ($Ca_v1.3$) (Scholl et al. 2013; Beuschlein et al. 2013; Fernandes-Rosa et al. 2015; Azizan et al. 2013).

In feline hypertension, there has been some work undertaken to explore the role of the renin-angiotensin-aldosterone system in the pathogenesis of hypertension. Whilst there is heterogeneity within the hypertensive feline population, overall cats with hypertension tend to have significantly higher plasma aldosterone concentrations with proportionally suppressed plasma renin activity (Jensen et al. 1997; Jepson et al. 2014). It has, therefore, been hypothesised that the elevated aldosterone concentrations could be contributing to the pathogenesis of hypertension in cats, although the driving factors behind this increased production remain to be elucidated given the overall trend towards either variable or lower plasma renin activity. A small case series reported that bilateral adrenal gland hyperplasia was commonly identified on histopathology of adrenal glands from geriatric cats with both kidney disease and systemic hypertension. The study hypothesised that bilateral adrenal hyperplasia may be underpinning the development of hypertension in some cats (Javadi et al. 2005). However, a study evaluating adrenal gland morphology at post-mortem examination in 67 cats (30 normotensive and 37 hypertensive) identified that 97% had evidence of bilateral adrenal gland hyperplasia (Keele et al. 2009). Prevalence of this histopathological change was, therefore, not associated with the hypertensive state, although functional analysis and differential gene expression were not explored (Keele et al. 2009). It is interesting to speculate that further exploration of the genes involved in aldosterone synthesis and secretion would be of interest in feline hypertension, particularly, evaluation for somatic mutations within the adrenal glands given the tendency for hyperaldosteronism in hypertensive cats and the role of *CACNA1D*. The cat exhibits a comparatively marked anti-hypertensive response to calcium channel blockers such as amlodipine besylate when compared to clinical efficacy in either the dog or the human, representing therefore an interesting species difference (Huhtinen et al. 2015; Elliott et al. 2001).

Interesting to veterinary medicine, nine genes, *RET, VHL, SDHA, SDHB, SDHC* and *SDHD, KIF1Bbeta, PHD2, SDHAF2*, have been identified which when mutated can result in the development of pheochromocytoma (Iacobone et al. 2011; Opocher and Schiavi 2010; Welander et al. 2011). Pheochromocytoma and paragangliomas are rare tumours in the dog and even scarcer in the cat, and to date no familial associations have been reported (see Chap. 4) (Calsyn and Green 2010; Mai et al. 2015; Gilson et al. 1994; Patnaik et al. 1990). A previous study in dogs evaluated

SDHA, SDHB, SDHC and *SDHD* encoding for the succinate dehydrogenase subunits in pheochromocytoma ($n = 6$) and paraganglioma tissue ($n = 2$) and identified a somatic mutation in *SDHD* in exon 4 that was not present in normal tissue from this dog, and two dogs with pheochormocytomas were heterozygous for this mutation (Holt et al. 2014). These preliminary data suggest that somatic genetic mutations may underlie tumourigenesis in dogs with pheochromocytomas and paragangliomas, similar to humans, and further study in this area would be interesting (Gilson et al. 1994).

6.2 Hypertension as a Complex Disease Trait

Despite the existence of rare monogenic disorders, it is now considered that BP and hypertension in the general population are complex polygenetic traits with many environmental and lifestyle factors that also contribute to the risk of development of hypertension. Nevertheless, given the clear association of genes identified in Mendelian hypertension with pathways of BP regulation, these genes have been of interest in the exploration of genetic associations with BP in the general population. A more extensive understanding of the role that genetics plays in BP regulation and hypertension, however, was difficult to pursue until the advent of single-nucleotide polymorphism (SNP) chip arrays (Munroe et al. 2013; Ehret and Caulfield 2013). A SNP is a variation that is constituted by two, occasionally more, different possible nucleotide bases (alleles) at the same genetic position. In humans, development of SNP chip arrays has enabled direct typing and imputation of ~2.5 million variants across the genome in a rapid-throughput and cost-effective manner, allowing the evaluation of tens to hundreds of thousands of individuals and paving the way for even larger meta-analyses to be performed. Such studies are referred to as genome-wide association studies or GWAS. GWAS studies build on the premise that multiple common genetic variants, each typically with very modest effect (<1 mmHg systolic BP and ~0.5 mmHg for diastolic BP) contribute to the variation in population BP (Ehret and Caulfield 2013).

6.2.1 Mendelian Genes and Their Association with Population Blood Pressure

As described above, many of the genes associated with Mendelian forms of hypertension have effects directly related to physiological pathways fundamental to BP regulation, particularly relating to sodium homeostasis. Studies have supported that variants identified in genes from some of the same pathways, e.g. *WNK1* (With no lysine [K^+]), *ROMK* (inward rectifying potassium channel), *SLC12A3* (NCC; Na^+Cl^- cotransporter), *SLC12A1* (NKCC2; $Na^+K^+2Cl^-$ transporter) and *KCNJ1* (potassium voltage gated channel subfamily J member 1), may have effect on BP within the general population (Ji et al. 2008; Tobin et al. 2005, 2008; Newhouse et al. 2005, 2009). One study evaluated variants in 10 key Mendelian genes (*AGT,*

CYP11B1, CYP17A1, HSD11B2, NR3C1, CNN1A, SCNN1B, SCNN1G, WNK1 and *WNK4*) in different European and African ancestry populations. Although no variants from these genes individually were significantly associated with BP, pooling variants from these genes revealed five regions with cumulative association (*AGT, CYP11B1, NR3C2, SCNN1G* and *WNK1*) (Nguyen et al. 2014). These findings potentially support a 'common disease-rare variant' hypothesis (Welling 2014). Variation in these important genes associated with Mendelian hypertension may, therefore, be of interest to blood pressure regulation in the whole population.

6.2.2 Development of Genome-Wide Association Studies and Exploration of Population Blood Pressure

The first GWAS exploring BP as a quantitative trait using a 'common disease: common variant' hypothesis was published in 2007 as part of the Wellcome Consortium exploring seven common diseases: bipolar disorder, coronary artery disease, Crohn's disease, type-1 and type-2 diabetes, rheumatoid arthritis and hypertension (Burton et al. 2007). This was a case-control study design including 14,000 cases, 2000 from each of the disease conditions of interest and a common group of 3000 controls. Although significant associations were identified for the other disease traits, no significant associations were identified for BP (Burton et al. 2007). Since that time a large number of GWAS have been published with ever increasing sample sizes and number of SNP variants explored in different populations (Burton et al. 2007; Wain et al. 2011; Ganesh et al. 2013; Tragante et al. 2014; Johnson et al. 2011; Ehret et al. 2011; Kato et al. 2011; Levy et al. 2009; Newton-Cheh et al. 2009a; Padmanabhan et al. 2010). The majority of these studies have chosen to explore BP as a quantitative trait rather than dichotomisation to hypertension/normotension, particularly given the challenges associated with determining binary outcome and the changing criteria for the diagnosis of hypertension (Levy et al. 2009; Padmanabhan et al. 2010; Salvi et al. 2012). Significant associations in early studies were reported with SNPs in *STK39* (serine threonine kinase; diastolic BP) (Wang et al. 2009) upstream of *CDH13* (adhesion glycoprotein T-cadherin; systolic BP) (Wang et al. 2009) and *ATP2B1* (ATPase calcium transporting, plasma membrane 1; systolic and diastolic BP) (Wang et al. 2009). However, apart from the variant associated with *ATP2B1*, replication proved challenging, suggesting that the effects could represent false-positive results or are specific to the populations that were studied. In order to improve the power of discovery, collaborative efforts resulted in publication of meta-analyses which were more prolific in terms of identification of significant loci and also included much larger-scale validation within their study design for any significant loci (Wain et al. 2011; Ehret et al. 2011; Kato et al. 2011; Levy et al. 2009; Newton-Cheh et al. 2009a). Studies have also explored ancestral differences using a GWAS approach and considering that genetic variants which affect BP may not have the same constant effect across individuals of European, East Asian, South Asian and African origins (Ehret et al. 2011; Liu et al. 2011). A further approach that has been utilised is the development of bespoke genotyping arrays.

These have been designed to include dense coverage of variants within proximity to genes considered more likely to have functional cardiovascular significance and based on *a priori* knowledge from previously published meta-analyses and consortia working on metabolic, cardiovascular and blood pressure related traits, e.g. Cardio Metabochip (Ehret et al. 2016), and have also been designed to explore specifically the exome (i.e. variants from protein coding genes), e.g. Exome Chip (Liu et al. 2016; Surendran et al. 2016). This approach proved successful, each study evaluating >300,000 individuals and collectively identifying >100 novel BP loci (Ehret et al. 2016; Liu et al. 2016; Surendran et al. 2016). Today over 150 loci have been associated with BP traits although some remain to be validated in independent studies (Munroe et al. 2013; Ehret et al. 2016; Liu et al. 2016; Surendran et al. 2016).

A number of challenges have arisen from the advent of GWAS exploring BP and hypertension in human medicine. One problem is that many of the SNPs that have been identified and associated with BP are found in non-coding regions of the genome and are, therefore, not intragenic and indeed may not even be within the region of association of a gene that could plausibly be linked by current biological knowledge to BP regulation (Ehret et al. 2011). It should be remembered that a specific association with a variant from a GWAS study informs about a region or locus based on linkage disequilibrium (non-random association of alleles in a given region of the genome inherited together) that may be associated with the trait of interest. This locus may extend over many thousands of base-pairs and encompass many genes; determining an actual causal variant is, therefore, very challenging. Ultimately all variants at a given locus need to be evaluated and their functional effects explored (Munroe et al. 2013). In addition, it has been recognised that many of the identified variants and loci may have association not only with BP or hypertension but with other medical conditions where the link between the associations is not immediately apparent. For example, *SH2B3* locus has been associated with BP, celiac disease, myocardial infarction and type-1 diabetes mellitus (Ehret et al. 2011; Gudbjartsson et al. 2009; Smyth et al. 2008). Our ability to find genetic associations is expanding rapidly. A wealth of data is now available in relation to the human genome and both common and rare variants, for example, through projects such as the 1000 Genome Project which now characterises >38 million SNPs, 1.4 million short insertions and deletions (indels) and >14,000 deletions and which will allow the potential to detect new associations (Sudmant et al. 2015; Auton et al. 2015).[3] Furthermore, work from projects such as ENCyclopedia of DNA Elements (ENCODE) have brought about the realisation that it is not just the intragenic regions of DNA that are important (Kellis et al. 2014).[4] The ENCODE project identified many important sites of epigenetic regulation of gene expression, regions of DNA accessible by regulatory proteins, positions for RNA binding, sites of histone modification and transcription factor binding sites which might all have an impact on gene regulation and expression and also provided cell

---

[3] 1000 Genome project; http://www.internationalgenome.org/about/
[4] ENCODE; https://www.genome.gov/10005107/the-encode-project-encyclopedia-of-dna-elements/

specific data that could be of value to cardiovascular and, therefore, blood pressure research (Kellis et al. 2014). From a GWAS perspective, these sites could also have relevance and highlight that GWAS that focus only on genes in proximity to loci may be narrowing the discovery field. Further data have arisen from integration of GWAS data with expression data to provide information about functional variation, e.g. expression Quantitative Trait Loci (eQTL) (Russo et al. 2018). If a loci is identified within ≤1 megabase pairs (Mb) from the gene encoding the transcript, it is called *cis*-eQTL whilst if it is more distant it is called *trans*-eQTL. It is possible for disease susceptibility to be regulated by genes controlled by *trans*-eQTL and, therefore, extended distances of the genome may need to be searched for these associations. It has been estimated that ~30% of mammalian genes are under the control of eQTLs and that eQTLs are very likely to contribute to complex disease phenotypes (Russo et al. 2018). From a functional perspective it has been shown that human genes identified in hypertension GWAS, when conserved in rodent models, may act as *cis* and *trans*-eQTL in multiple tissues (Langley et al. 2013).

A second limitation is how can the data that have been collected be extrapolated to the clinical dilemma of diagnosis and management of hypertension. From human GWAS evaluating hypertension, two variants have been identified with direct translation to a gene of interest relating to regulation of BP: firstly, a variant in the promoter region of *UMOD* encoding for uromodulin which has been linked to a novel role in renal sodium homeostasis and secondly, a variant in the promotor region of *eNOS* encoding for nitric oxide synthase, which is important in the regulation of vascular tone (Salvi et al. 2012, 2013). From studies evaluating the quantitative trait BP, only one pathway has been reliably linked to BP regulation for a SNP in the natriuretic peptide gene (*NPPA/B*) (Newton-Cheh et al. 2009b; Arora et al. 2013). Nevertheless, advances in the analysis of GWAS results and pathway and network analysis give the potential to identify novel biological pathways that may be implicated in BP regulation and prove to be therapeutic targets. From a clinical perspective, however, despite the number and impact of identified BP loci, 'genetic profiling' has not been extrapolated to determine individual BP or hypertension risk and is not at a stage where it can supersede serial measurement of BP for risk prediction and diagnosis. It is interesting to speculate that in the future pharmacogenomics could play a role in determining an individual's most effective anti-hypertensive therapeutic regimen but with currently available data we are far from this point.

Even with the BP loci identified to date there is a continued discrepancy between the estimated heritability of BP as outlined at the beginning of this chapter and the extent to which the genetic loci explain the variance in BP (~2–3%). This has led to consideration as to whether there are still many common variants yet to be identified, each accounting for a small proportional change in BP or whether there are further rare variants with proportionally larger effect that remain to be discovered and which it is not possible to discover with GWAS methodologies (Russo et al. 2018). Newer methodologies such as exome chips and the use of whole-genome sequencing with next-generation sequencing (NGS) provide the opportunity for greater discovery power (Russo et al. 2018).

6.2.3 Genetics of Blood Pressure in the Dog and the Cat

To date there are no published studies that have explored genetic associations with BP and hypertension in the dog, and very limited data are available for the cat. Nevertheless, hypertension is recognised in both species which is of interest given that domesticated dogs and cats share the same environment, exhibit the same phenotypic manifestations of hypertension and show an age-related increase in BP as in humans. This brings potential in terms of exploring and comparing genetic associations with BP between domestic species and man. Our genomic knowledge in relation to the dog and the cat compared to the human is limited. Limitations in terms of the genome build and knowledge of SNPs, indels and deletions make potential interpretation of GWAS challenging. However, for breed-specific and monogenic conditions, GWAS have proved fruitful and more complex disease traits begin to be explored (Skinner and Manuscript 2012; Holden et al. 2018; Mellersh and Ostrander 1997; Boyko 2011). Better evaluation, in particular of the feline genome through projects such as the 99 Lives Cat Genome Sequencing Initiative and the opportunity for whole-genome sequencing, will improve our ability to evaluate genetic associations in the cat (Gandolfi and Alhaddad 2015; Montague et al. 2014; Pontius et al. 2007; Mullikin et al. 2010).[5]

Based on previously published human studies, a candidate gene approach was adopted to explore SNPs within *UMOD* encoding for uromodulin (Tamm-Horsfall protein) in the cat and to look for associations with systolic BP (Jepson et al. 2016). In human medicine, the *UMOD* locus has been linked with chronic kidney disease, blood pressure and hypertension (Devuyst and Pattaro 2017; Seidel and Scholl 2017; Graham et al. 2014; Padmanabhan et al. 2014; Trudu et al. 2013.) In relation to hypertension, uromodulin plays a modulatory role in the activity of TAL sodium–potassium–chloride cotransporter (NKCC2), therefore, impacting sodium and volume homeostasis (Trudu et al. 2013; Scolari et al. 2015). In cats, four SNPs in *UMOD* which were all in linkage disequilibrium were found to be significantly associated with SBP in a population of geriatric cats, although validation of this finding in an independent cohort of cats has not been performed (Jepson et al. 2016). This study has, however, provided proof of concept that some of the genetic associations identified in humans may be of importance in our veterinary patients. Further work is required to explore the role of the cat as a model for hypertension in humans.

References

Arora P, Wu C, May Khan A, Bloch DB, Davis-Dusenbery BN, Ghorbani A et al (2013) Atrial natriuretic peptide is negatively regulated by microRNA-425. J Clin Invest 123(8):3378–3382

Auton A, Abecasis GR, Altshuler DM, Durbin RM, Bentley DR, Chakravarti A et al (2015) A global reference for human genetic variation. Nature 526(7571):68–74

---

[5] 99 Lives Cat Genome Sequencing Initiative: http://felinegenetics.missouri.edu/99lives

Azizan EAB, Poulsen H, Tuluc P, Zhou J, Clausen MV, Lieb A et al (2013) Somatic mutations in ATP1A1 and CACNA1D underlie a common subtype of adrenal hypertension. Nat Genet 45 (9):1055–1060

Barroso I, Gurnell M, Crowley VE, Agostini M, Schwabe JW, Soos MA et al (1999) Dominant negative mutations in human PPARgamma associated with severe insulin resistance, diabetes mellitus and hypertension. Nature 402(6764):880–883

Beuschlein F, Boulkroun S, Osswald A, Wieland T, Nielsen HN, Lichtenauer UD et al (2013) Somatic mutations in ATP1A1 and ATP2B3 lead to aldosterone-producing adenomas and secondary hypertension. Nat Genet 45(4):440–444

Bochud M, Bovet P, Elston RC, Paccaud F, Falconnet C, Maillard M et al (2005) High heritability of ambulatory blood pressure in families of East African descent. Hypertension 45(3):445–450

Boyden LM, Choi M, Choate KA, Nelson-Williams CJ, Farhi A, Toka HR et al (2012) Mutations in kelch-like 3 and cullin 3 cause hypertension and electrolyte abnormalities. Nature 482 (7383):98–102

Boyko AR (2011) The domestic dog: man's best friend in the genomic era. Genome Biol 12(2):216

Burton PR, Clayton DG, Cardon LR, Craddock N, Deloukas P, Duncanson A et al (2007) Genome-wide association study of 14,000 cases of seven common diseases and 3,000 shared controls. Nature 447(7145):661–678

Calsyn JDR, Green RA (2010) Contralateral adrenocortical adenoma in a cat. J Am Anim Hosp Assoc 46(1):36–42

Cerame B, New M (2000) Hormonal hypertension in children: 11beta-hydroxylase deficiency and apparent mineralocorticoid excess. J Pediatr Endocrinol Metab 13(9):1537–1547

Devuyst O, Pattaro C (2017) The UMOD locus: insights into the pathogenesis and prognosis of kidney disease. J Am Soc Nephrol 29(3):713–726

Dobrian AD, Schriver SD, Lynch T, Prewitt RL (2003) Effect of salt on hypertension and oxidative stress in a rat model of diet-induced obesity. Am J Physiol Ren Physiol 285(4):F619–F628

Ehret GB, Caulfield MJ (2013) Genes for blood pressure: an opportunity to understand hypertension. Eur Heart J 34(13):951–961

Ehret GB, Munroe PB, Rice KM, Bochud M, Johnson AD, Chasman DI et al (2011) Genetic variants in novel pathways influence blood pressure and cardiovascular disease risk. Nature 478 (7367):103–109

Ehret GB, Ferreira T, Chasman DI, Jackson AU, Schmidt EM, Johnson T et al (2016) The genetics of blood pressure regulation and its target organs from association studies in 342,415 individuals. Nat Genet 48(10):1171–1184

Elliott J, Barber PJ, Syme HM, Rawlings JM, Markwell PJ (2001) Feline hypertension: clinical findings and response to antihypertensive treatment in 30 cases. J Small Anim Pract 42 (3):122–129

Fernandes-Rosa FL, Giscos-Douriez I, Amar L, Gomez-Sanchez CE, Meatchi T, Boulkroun S et al (2015) Different somatic mutations in multinodular adrenals with aldosterone-producing adenoma. Hypertension 66(5):1014–1022

Fuentes RM, Notkola IL, Shemeikka S, Tuomilehto J, Nissinen A (2000) Familial aggregation of blood pressure: a population-based family study in eastern Finland. J Hum Hypertens 14 (7):441–445

Gandolfi B, Alhaddad H (2015) Investigation of inherited diseases in cats: genetic and genomic strategies over three decades. J Feline Med Surg 17(5):405–415

Gandolfi B, Gruffydd-Jones TJ, Malik R, Cortes A, Jones BR, Helps CR et al (2012) First WNK4-Hypokalemia Animal Model Identified by Genome-Wide Association in Burmese Cats. PLoS One 7(12):e53173

Ganesh SK, Tragante V, Guo W, Guo Y, Lanktree MB, Smith EN et al (2013) Loci influencing blood pressure identified using a cardiovascular gene-centric array. Hum Mol Genet 22 (8):1663–1678

Geller DS, Farhi A, Pinkerton N, Fradley M, Moritz M, Spitzer A et al (2000) Activating mineralocorticoid receptor mutation in hypertension exacerbated by pregnancy. Science 289 (5476):119–123

Gilson SD, Withrow SJ, Wheeler SL, Twedt DC (1994) Pheochromocytoma in 50 dogs. J Vet Intern Med 8(3):228–232

Goldblatt BYH, Lynch J, Ramon F, Ph D, Summerville WW (1934) Studies on experimental hypertension. The production of persistent elevation of systolic blood pressure by means of renal ischaemia. J Exp Med 59(3):347–379

Graham LA, Padmanabhan S, Fraser NJ, Kumar S, Bates JM, Raffi HS et al (2014) Validation of uromodulin as a candidate gene for human essential hypertension. Hypertension 63(3):551–558

Gudbjartsson DF, Bjornsdottir US, Halapi E, Helgadottir A, Sulem P, Jonsdottir GM et al (2009) Sequence variants affecting eosinophil numbers associate with asthma and myocardial infarction. Nat Genet 41(3):342–347

Hansson JH, Nelson-Williams C, Suzuki H, Schild L, Shimkets R, Lu Y et al (1995) Hypertension caused by a truncated epithelial sodium channel γ subunit: genetic heterogeneity of Liddle syndrome. Nat Genet 11(1):76–82

Havlik RJ, Garrison RJ, Feinleib M, Kannel WB, Castelli WP, Mcnamara PM (1979) Blood pressure aggregation in families. Am J Epidemiol 110(3):304–312

Holden LA, Arumilli M, Hytönen MK, Hundi S, Salojärvi J, Brown KH et al (2018) Assembly and analysis of unmapped genome sequence reads reveal novel sequence and variation in dogs. Sci Rep 8(1):1–11

Holt DE, Henthorn P, Howell VM, Robinson BG, Benn DE (2014) Succinate dehydrogenase subunit D and succinate dehydrogenase subunit B mutation analysis in canine phaeochromocytoma and paraganglioma. J Comp Pathol 151(1):25–34

Hottenga J-J, Boomsma DI, Kupper N, Posthuma D, Snieder H, Willemsen G et al (2005) Heritability and stability of resting blood pressure. Twin Res Hum Genet 8(5):499–508

Huhtinen M, Derré G, Renoldi HJ, Rinkinen M, Adler K, Aspegrén J et al (2015) Randomized placebo-controlled clinical trial of a chewable formulation of amlodipine for the treatment of hypertension in client-owned cats. J Vet Intern Med 29(3):786–793

Iacobone M, Schiavi F, Bottussi M, Taschin E, Bobisse S, Fassina A et al (2011) Is genetic screening indicated in apparently sporadic pheochromocytomas and paragangliomas? Surgery 150(6):1194–1201

Imai G, Satoh T, Kumai T, Murao M, Tsuchida H, Shima Y et al (2003) Hypertension accelerates diabetic nephropathy in Wistar fatty rats, a model of type 2 diabetes mellitus, via mitogen-activated protein kinase cascades and transforming growth factor-beta1. Hypertens Res 26 (4):339–347

Javadi S, Djajadiningrat-Laanen SC, Kooistra HS, Van Dongen AM, Voorhout G, Van Sluijs FJ et al (2005) Primary hyperaldosteronism, a mediator of progressive renal disease in cats. Domest Anim Endocrinol 28(1):85–104

Jensen J, Henik R, Brownfield M, Armstrong J (1997) Plasma renin activity and angiotensin I and aldosterone concentrations in cats with hypertension associated with chronic renal disease. Am J Vet Res 58(5):535–540

Jepson RE, Syme HM, Elliott J (2014) Plasma renin activity and aldosterone concentrations in hypertensive cats with and without azotemia and in response to treatment with amlodipine besylate. J Vet Intern Med 28(1):144–153

Jepson RE, Warren HR, Syme HM, Elliott J, Munroe PB (2016) Uromodulin gene variants and their association with renal function and blood pressure in cats: a pilot study. J Small Anim Pract 57 (11):580–588

Ji W, Foo JN, O'Roak BJ, Zhao H, Larson MG, Simon DB et al (2008) Rare independent mutations in renal salt handling genes contribute to blood pressure variation. Nat Genet 40(5):592–599

Johnson T, Gaunt TR, Newhouse SJ, Padmanabhan S, Tomaszewski M, Kumari M et al (2011) Blood pressure loci identified with a gene-centric array. Am J Hum Genet 89(6):688–700

Kato N, Takeuchi F, Tabara Y, Kelly TN, Go MJ, Sim X et al (2011) Meta-analysis of genome-wide association studies identifies common variants associated with blood pressure variation in east Asians. Nat Genet 43(6):531–538

Keele SJ, Smith KC, Elliott J, Syme HM (2009) Adrenocortical morphology in cats with chronic kidney disease (CKD) and systemic hypertension. J Vet Intern Med 23(6):1328

Kellis M, Wold B, Snyder MP, Bernstein BE, Kundaje A, Marinov GK et al (2014) Defining functional DNA elements in the human genome. Proc Natl Acad Sci U S A 111(17):6131–6138

Kimura S, Mulfins JJ, Bunnemann B, Metzger R, Hilgenfeldt U, Zimmermann F et al (1992) High blood pressure in transgenic mice carrying the rat angiotensinogen gene. EMBO J 1(3):821–827

Krege JH, John SWM, Langenbach LL, Hodgin JB, Hagaman JR, Bachman ES et al (1995) Male–female differences in fertility and blood pressure in ACE-deficient mice. Nature 375:146

Kupper N, Willemsen G, Riese H, Posthuma D, Boomsma DI, De Geus EJC (2005) Heritability of daytime ambulatory blood pressure in an extended twin design. Hypertension 45(1):80–85

Lafferty AR, Torpy DJ, Stowasser M, Taymans SE, Lin JP, Huggard P et al (2000) A novel genetic locus for low renin hypertension: familial hyperaldosteronism type II maps to chromosome 7 (7p22). J Med Genet 37(11):831–835

Langley SR, Bottolo L, Kunes J, Zicha J, Zidek V, Hubner N et al (2013) Systems-level approaches reveal conservation of trans-regulated genes in the rat and genetic determinants of blood pressure in humans. Cardiovasc Res 97(4):653–665

Lerman LO, Chade AR, Sica V, Napoli C (2005) Animal models of hypertension: an overview. J Lab Clin Med 146(3):160–173

Levy D, Ehret GB, Rice K, Verwoert GC, Launer LJ, Dehghan A et al (2009) Genome-wide association study of blood pressure and hypertension. Nat Genet 41(6):677–687

Lifton RP, Dluhy RG, Powers M, Rich GM, Cook S, Ulick S et al (1992) A chimaeric ll-β-hydroxylase/aldosterone synthase gene causes glucocorticoid-remediable aldosteronism and human hypertension. Nature 355:262

Lifton R, Gharavi A, Geller D (2001) Molecular mechanisms of human hypertension. Cell 104:545–556

Liu C, Li H, Qi Q, Lu L, Gan W, Loos RJF et al (2011) Common variants in or near FGF5, CYP17A1 and MTHFR genes are associated with blood pressure and hypertension in Chinese Hans. J Hypertens 29(1):70–75

Liu C, Kraja AT, Smith JA, Brody JA, Franceschini N, Bis JC et al (2016) Meta-analysis identifies common and rare variants influencing blood pressure and overlapping with metabolic trait loci. Nat Genet 48(10):1162–1170

Louis-Dit-Picard H, Barc J, Trujillano D, Miserey-Lenkei S, Bouatia-Naji N, Pylypenko O et al (2012) KLHL3 mutations cause familial hyperkalemic hypertension by impairing ion transport in the distal nephron. Nat Genet 44(4):456–460

Luft FC, Volhard F, Helios C, Delbrück M, Medicine M (2003) Mendelian forms of human hypertension and mechanisms of disease. Clin Med Res 1(4):291–300

Maass PG, Aydin A, Luft FC, Schächterle C, Weise A, Stricker S et al (2015) PDE3A mutations cause autosomal dominant hypertension with brachydactyly. Nat Genet 47(6):647–653

Mai W, Seiler GS, Lindl-bylicki BJ, Zwingenberger AL (2015) CT and MRI features of carotid body paragangliomas in 16 dogs. Vet Radiol Ultrasound 56(4):374–383

Mellersh CS, Ostrander EA (1997) The canine genome. Adv Vet Med 40:191–216

Montague MJ, Li G, Gandolfi B, Khan R, Aken BL, Searle SMJ et al (2014) Comparative analysis of the domestic cat genome reveals genetic signatures underlying feline biology and domestication. Proc Natl Acad Sci U S A 111(48):17230–17235

Mullikin JC, Hansen NF, Shen L, Ebling H, Donahue WF, Tao W et al (2010) Light whole genome sequence for SNP discovery across domestic cat breeds. BMC Genomics 11(1):406

Munroe PB, Barnes MR, Caulfield MJ (2013) Advances in blood pressure genomics. Circ Res 112 (10):1365–1379

New MI, Levine LS (1980) Hypertension of Childhood with Suppressed Renin. Endocr Rev 1 (4):421–430

Newhouse SJ, Wallace C, Dobson R, Mein C, Pembroke J, Farrall M et al (2005) Haplotypes of the WNK1 gene associate with blood pressure variation in a severely hypertensive population from the British genetics of hypertension study. Hum Mol Genet 14(13):1805–1814

Newhouse S, Farrall M, Wallace C, Hoti M, Burke B, Howard P et al (2009) Polymorphisms in the WNK1 gene are asociated with blood pressure variation and urinary potassium excretion. PLoS One 4(4):e5003

Newton-Cheh C, Johnson T, Gateva V, Tobin MD, Bochud M, Coin L et al (2009a) Genome-wide association study identifies eight loci associated with blood pressure. Nat Genet 41(6):666–676

Newton-Cheh C, Larson MG, Vasan RS, Levy D, Bloch KD, Surti A et al (2009b) Association of common variants in NPPA and NPPB with circulating natriuretic peptides and blood pressure. Nat Genet 41(3):348–353

Nguyen KH, Pihur V, Ganesh SK, Rakha A, Richard S, Hunt SC et al (2014) Effects of rare and common blood pressure gene variants on essential hypertension: results from the FBPP, CLUE ARIC studies. Circ Res 112(2):318–326

Ohta A, Schumacher F-R, Mehellou Y, Johnson C, Knebel A, Macartney TJ et al (2013) The CUL3-KLHL3 E3 ligase complex mutated in Gordon's hypertension syndrome interacts with and ubiquitylates WNK isoforms: disease-causing mutations in KLHL3 and WNK4 disrupt interaction. Biochem J 451(1):111–122

Okamoto K, Aoki K (1963) Development of a strain of spontaneously hypertensive rats. Jpn Circ J 27(3):282–293

Oldham PD, Pickering G, Fraser Roberts JA, Sowry GSC (1960) The nature of essential hypertension. Lancet 275(7134):1085–1093

Opocher G, Schiavi F (2010) Genetics of pheochromocytomas and paragangliomas. Best Pract Res Clin Endocrinol Metab 24(6):943–956

Padmanabhan S, Melander O, Johnson T, Di Blasio AM, Lee WK, Gentilini D et al (2010) Genome-wide association study of blood pressure extremes identifies variant near UMOD associated with hypertension. PLoS Genet 6(10):e1001177

Padmanabhan S, Newton-Cheh C, Dominiczak AF (2012) Genetic basis of blood pressure and hypertension. Trends Genet 28(8):397–408

Padmanabhan S, Graham L, Ferreri NR, Graham D, McBride M, Dominiczak AF (2014) Uromodulin, an emerging novel pathway for blood pressure regulation and hypertension. Hypertension 64(5):918–923

Padmanabhan S, Caulfield M, Dominiczak AF (2015) Genetic and molecular aspects of hypertension. Circ Res 116(6):937–959

Page IH (1960) The mosaic theory of hypertension. In: Bock KD, Cottier PT (eds) Essential hypertension: an international symposium Berne, June 7th–10th, 1960 Sponsored by CIBA. Springer, Berlin, pp 1–29

Patnaik A, Erlandson R, Lieberman P, Welches C, Marretta S (1990) Extra-adrenal pheochromocytoma (paraganglioma) in a cat. J Am Vet Med Assoc 197(1):104–106

Pickering G (1955) The genetic factor in essential hypertension. Ann Intern Med 43(3):457–464

Pinto YM, Paul M, Ganten D (1998) Lessons from rat models of hypertension: from Goldblatt to genetic engineering. Cardiovasc Res 39(1):77–88

Polderman TJC, Benyamin B, De Leeuw CA, Sullivan PF, Van Bochoven A, Visscher PM et al (2015) Meta-analysis of the heritability of human traits based on fifty years of twin studies. Nat Genet 47(7):702–709

Pontius JU, Mullikin JC, Smith DR, Lindblad-Toh K, Gnerre S, Clamp M et al (2007) Initial sequence and comparative analysis of the cat genome. Genome Res 17:1675–1689

Russo A, Di Gaetano C, Cugliari G, Matullo G (2018) Advances in the genetics of hypertension: The effect of rare variants. Int J Mol Sci 19(3):1–21

Salvi E, Kutalik Z, Glorioso N, Benaglio P, Frau F, Kuznetsova T et al (2012) Genomewide association study using a high-density single nucleotide polymorphism array and case-control design identifies a novel essential hypertension susceptibility locus in the promoter region of endothelial NO synthase. Hypertension 59(2):248–255

Salvi E, Kuznetsova T, Thijs L, Lupoli S, Stolarz-Skrzypek K, D'Avila F et al (2013) Target sequencing, cell experiments, and a population study establish endothelial nitric oxide synthase (eNOS) gene as hypertension susceptibility gene. Hypertension 62(5):844–852

Scholl UI, Lifton RP (2013) New insights into aldosterone-producing adenomas and hereditary aldosteronism: mutations in the K+ channel KCNJ5. Curr Opin Nephrol Hypertens 22(2)

Scholl UI, Goh G, Stölting G, De Oliveira RC, Choi M, Overton JD et al (2013) Somatic and germline CACNA1D calcium channel mutations in aldosterone-producing adenomas and primary aldosteronism. Nat Genet 45(9):1050–1054

Schumacher F-R, Sorrell FJ, Alessi DR, Bullock AN, Kurz T (2014) Structural and biochemical characterization of the KLHL3–WNK kinase interaction important in blood pressure regulation. Biochem J 460(2):237–246

Schuster H, Wienker TF, Bähring S, Bilginturan N, Toka HR, Neitzel H et al (1996) Severe autosomal dominant hypertension and brachydactyly in a unique Turkish kindred maps to human chromosome 12. Nat Genet 13(1):98–100

Scolari F, Izzi C, Ghiggeri GM (2015) Uromodulin: from monogenic to multifactorial diseases. Nephrol Dial Transplant 30(8):1250–1256

Seidel E, Scholl UI (2017) Genetic mechanisms of human hypertension and their implications for blood pressure physiology. Physiol Genomics 49(11):630–652

Shimkets RA, Warnock DG, Bositis CM, Nelson-Williams C, Hansson JH, Schambelan M et al (1994) Liddle's syndrome: Heritable human hypertension caused by mutations in the β subunit of the epithelial sodium channel. Cell 79(3):407–414

Skinner JS, Manuscript A (2012) NIH Public Access 25(2):1–19

Smyth DJ, Plagnol V, Walker NM, Cooper JD, Downes K, Yang JHM et al (2008) Shared and distinct genetic variants in Type 1 diabetes and celiac disease. N Engl J Med 359(26):2767–2777

Snieder H, Harshfield GA, Treiber FA (2003) Heritability of blood pressure and hemodynamics in African- and European-American youth. Hypertension 41(6):1196–1201

Sudmant PH, Rausch T, Gardner EJ, Handsaker RE, Abyzov A, Huddleston J et al (2015) An integrated map of structural variation in 2,504 human genomes. Nature 526(7571):75–81

Surendran P, Drenos F, Young R, Warren H, Cook JP, Manning AK et al (2016) Trans-ancestry meta-analyses identify rare and common variants associated with blood pressure and hypertension. Nat Genet 48(10):1151–1161

Syme H, Barber P, Markwell P, Elliott J (2002) Prevalence of systolic hypertension in cats with chronic renal failure at initial evaluation. J Am Vet Med Assoc 220(12):1799–1804

Tobin MD, Raleigh SM, Newhouse S, Braund P, Bodycote C, Ogleby J et al (2005) Association of WNK1 gene polymorphisms and haplotypes with ambulatory blood pressure in the general population. Circulation 112(22):3423–3429

Tobin MD, Tomaszewski M, Braund PS, Hajat C, Raleigh SM, Palmer TM et al (2008) Common variants in genes underlying monogenic hypertension and hypotension and blood pressure in the general population. Hypertension 51(6):1658–1664

Tragante V, Barnes MR, Ganesh SK, Lanktree MB, Guo W, Franceschini N et al (2014) Gene-centric meta-analysis in 87,736 individuals of European ancestry identifies multiple blood-pressure-related loci. Am J Hum Genet 94(3):349–360

Trudu M, Janas S, Lanzani C, Debaix H, Schaeffer C, Ikehata M et al (2013) Common noncoding UMOD gene variants induce salt-sensitive hypertension and kidney damage by increasing uromodulin expression. Nat Med 19(12):1655–1660

Unseld M, Marienfeld J, Brandt P, Brennicke A (1997) Multilocus linkage of familial hyperkalaemia and hypertension, pseudohypoaldosteronism type II, to chromosomes 1q31-42 and 17p11-q21. Nat Genet 15:57–61

Wain LV, Verwoert GC, O'Reilly PF, Shi G, Johnson T, Johnson AD et al (2011) Genome-wide association study identifies six new loci influencing pulse pressure and mean arterial pressure. Nat Genet 43(10):1005–1012

Walker DJ, Elliott J, Syme HM (2009) Urinary cortisol/cortisone ratios in hypertensive and normotensive cats. J Feline Med Surg 11(6):442–448

Wang Y, O'Connell JR, McArdle PF, Wade JB, Dorff SE, Shah SJ et al (2009) Whole-genome association study identifies STK39 as a hypertension susceptibility gene. Proc Natl Acad Sci U S A 106(1):226–231

Welander J, Söderkvist P, Gimm O (2011) Genetics and clinical characteristics of hereditary pheochromocytomas and paragangliomas. Endocr Relat Cancer 18(6):253–276

Welling PA (2014) Rare mutations in renal sodium and potassium transporter genes exhibit impaired transport function. Curr Opin Nephrol Hypertens 23(1):1–8

Wilson FH, Disse-Nicodeme S, Choate KA, Ishikawa K, Nelson-Willams C, Desitter I et al (2001) Human hypertension caused by mutations in WNK kinases. Science 293(5532):1107–1112

Zuk O, Hechter E, Sunyaev SR, Lander ES (2012) The mystery of missing heritability: Genetic interactions create phantom heritability. Proc Natl Acad Sci U S A 109(4):1193–1198

Part II

Clinical and Pathological Consequences of Hypertension

Hypertension and the Kidney

Jonathan Elliott and Cathy Brown

In human medicine, hypertension is known as the silent killer, associated with catastrophic cardiovascular events that can lead to sudden death, such as stroke and myocardial infarction. The effect hypertension has on the kidney is more subtle but still outwardly clinically silent and so difficult to detect.

As discussed in Chap. 1, the kidney is the long-term regulator of body fluid volume relative to the capacity of the vasculature to accommodate that volume. A number of factors impacted by chronic kidney disease (CKD) (RAAS activation, renal afferent nerve over-activity, reduced production of renalase and phosphate retention leading to bone and mineral disturbances) will lead to raised vascular tone. Both of these (body fluid volume relative to the volume of the vasculature and raised vascular tone) contribute to raised blood pressure.

Autoregulation in the normal kidney protects it against hypertension. Afferent arteriolar constriction is induced in response to increased stretch of the walls of the afferent arteriole, increasing the pressure drop across this vascular bed and preventing a rise in glomerular capillary pressure. With loss of functioning nephrons, afferent arteriolar vasodilation occurs, leaving the glomerular capillary bed exposed to increased perfusion pressure. This glomerular capillary hypertension is exacerbated by local activation of the RAAS system leading to constriction of the efferent arteriole downstream of the glomerular capillary bed.

Glomerular capillary hypertension is silent clinically, except it results in increased loss of plasma protein, primarily albumin, across the glomerular filter. Microalbuminuria is the hallmark of early damage caused by hypertension to the kidney. It is not specific for hypertension. Other factors can damage the integrity of

J. Elliott (✉)
Department of Comparative Biomedical Sciences, Royal Veterinary College, University of London, London, UK
e-mail: jelliott@rvc.ac.uk

C. Brown
University of Georgia College of Veterinary Medicine, Athens, GA, USA
e-mail: cathybro@uga.edu

the glomerular barrier such that albumin translocates more readily. Hyperfiltration and glomerular capillary hypertension also occur with nephron loss in the absence of systemic arterial hypertension, but nevertheless the presence of systemic hypertension exacerbates protein loss across the glomerulus and the associated sequence of events it triggers within the kidney. There is a direct effect of glomerular capillary hypertension on albuminuria and the progressive pathology leading to podocyte effacement, capillary rarefaction and glomerulosclerosis, which can ultimately lead to nephron drop out and obsolescent glomeruli. The vascular responses of the kidney to hypertension can also lead to reduced peritubular capillary flow, causing relative tubular hypoxia, resulting in reduced ability to reclaim filtered proteins through the endocytotic pathways. Thus, there are multiple ways in which hypertension may be linked to proteinuria. The ways in which proteinuria can trigger progressive renal pathology have been reviewed elsewhere (Cravedi and Remuzzi 2013).

The above description of how hypertension impacts on the kidney is taken from human medicine where the three main underlying processes leading to CKD include primary glomerular disease, diabetic nephropathy and hypertensive nephropathy. In each of these, hypertension plays an important role in progressive kidney damage. Much of the detail in the pathological processes summarized above has been gleaned from animal models (rodents and dogs) where renal mass reduction and renal ischaemia in the presence or absence of high-fat diets have been used to model hypertensive and diabetic nephropathy. How much of this is relevant to naturally occurring kidney disease in dogs and cats is considered below.

7.1 Does Hypertension Contribute to Proteinuria in Dogs and Cats: What Is Known from Experimental Models?

Partial renal ablation models have been produced experimentally in both dogs and cats, and micropuncture techniques used demonstrate that hyperfiltration, glomerular capillary hypertension and proteinuria are all demonstrable.

In dogs, a direct relationship between reduction in renal mass and single nephron glomerular filtration rate (SNGFR) was found (Brown et al. 1990). A 74.8% and 86.1% reduction of renal mass (estimated numbers of nephrons) led 4 weeks later to an 86.6% and 127.8% increase in the micropuncture-estimated SNGFR, respectively (see Table 7.1). This resulted in an increase in urine protein to creatinine (UPC) ratio suggesting that the changes in glomerular capillary hydrostatic pressure (associated with a disproportionate decrease in afferent arteriolar resistance compared to efferent arteriolar resistance) and the ultrafiltration coefficient led to greater translocation of protein across the glomerulus. The control dogs had UPCs that were all within the non-proteinuric range (<0.2) whereas the dogs undergoing renal ablation on average had UPCs above 0.5 following 75% renal mass reduction and above 1 for the 86.1% renal mass reduction, which would be classified as proteinuric. Very few morphological changes were seen in glomeruli of the remnant kidneys at this stage except for an increase in glomerular volume related to the reduction in number of glomeruli.

Table 7.1 Data from the literature where single nephron effects of surgical renal reduction have been estimated using micropuncture techniques

	Percentage reduction	SNGFR (nl/min)	Glomerular capillary pressure (mmHg)	Urine protein to creatinine ratio[a]
Dog model[b]				
Control	0	71.0 ± 4.2	63.2 ± 1.9	0.08 ± 0.03
3/4 nephrectomy	74.8	132.5 ± 9.6	73.5 ± 2.0	0.56 ± 0.11
7/8 nephrectomy	86.1	161.8 ± 12.4	77.9 ± 2.2	1.46 ± 0.61
Cat model[c]				
Control	0	28.1 ± 2.8	62.6 ± 1.4	0.065 ± 0.03
3/4 nephrectomy	70.0	56.0 ± 5.9	74.0 ± 1.7	0.31 ± 0.06

SNGFR single nephron glomerular filtration rate
[a]The cat data are estimated from the figure in Brown and Brown (1995)
[b]Brown et al. (1990)
[c]Brown and Brown (1995)

In cats, a similar study examined just one level of renal ablation where one kidney was removed and the vascular supply to half of the remaining kidney was ligated (Brown and Brown 1995). This resulted in a 70% reduction in nephron number but only 36% reduction in kidney weight indicating significant hypertrophy of the remnant kidney when the cats were studied 4–6 weeks later. The increase in the micropuncture estimated SNGFR was 99.3% and was associated with raised glomerular capillary pressure and proteinuria as was found in the dogs (see Table 7.1) again, resulting from a disproportionate decrease in afferent arteriolar resistance relative to efferent arteriolar resistance. The ultrafiltration coefficient also increased, perhaps due to the increased capillary surface area, and contributed to the increased SNGFR and to the loss of protein into the urine. In the case of the cats, following 70% ablation of nephrons, the adaptive responses resulted in an average UPC that would place the cats undergoing renal ablation in the borderline proteinuric category (UPC of 0.2–0.4).

These two experimental studies demonstrate that with acute loss of nephrons the remaining functioning nephrons hypertrophy and hyperfiltrate through dilation of their afferent arterioles as well as changes to the ultrafiltration co-efficient. Dilation of the afferent arteriole would render these animals more susceptible to the effects of hypertension given that higher systemic blood pressure would be transmitted through to the glomerular capillary bed.

Although animals undergoing partial renal ablation to induce CKD experimentally do have increased systemic arterial blood pressure (SABP) in the first few weeks after their surgical preparation, SABP tends to decrease over time from a systolic pressure of around 160 mmHg to pressure of around 140 mmHg after 3 months (Brown et al. 2001, 2003). Additional insults to the kidney are necessary to induce more persistent and severe hypertension, as is seen clinically, in the dog and cat. Examples include wrapping the remnant kidney in cellophane or constricting the renal artery following unilateral nephrectomy (so called renovascular hypertension). Both of these methods lead to significant activation of the RAAS system. The latter has been used in dogs for many years.

In a study involving cats (Mathur et al. 2004), this technique was used and resulted in persistently elevated blood pressure that is damaging to other target organs (e.g. the eye). Cats receiving renal ablation (9/12 of renal mass ablated through arterial ligation) and a renal wrap procedure had 24-h mean systolic blood pressure of around 175 mmHg, which stabilized after about 50 days and continued at this level. By contrast, those receiving an 11/12 renal ablation only had a peak mean 24-h systolic blood pressure of 150 mmHg at 50 days which steadily declined and by 90 days was around 130 mmHg. This was still higher than the control group (with both kidneys intact) whose 24-h mean systolic blood pressure varied a little between 110 and 120 mmHg throughout the study. Cats receiving the renal wrap had significantly higher plasma aldosterone concentrations, plasma renin activity and UPCs when compared to the control and 11/12th nephrectomy group. This study examined the antihypertensive effects of short-term drug administration and unless the cats showed neurological signs of hypertensive encephalopathy or suffered in some other way, they were kept alive for between 140 and 150 days to determine the pathological lesions created by differing levels of hypertension.

Renal pathology in the remnant kidney was more severe in the cats receiving the renal wrap when compared to those receiving the renal ablation only. In fact, when comparing the pathology scores in the kidney removed to create the model within the same group to their remnant kidney scores, the renal ablation group only showed an increase in glomerulosclerosis scores whereas the renal wrap group showed increased in scores of glomerulosclerosis, tubular atrophy and interstitial fibrosis. This study shows that sustained systemic hypertension of a moderate to severe level driven by activation of RAAS is associated with increased proteinuria and more severe renal lesions in glomerular and interstitial compartments. The proteinuria could be driven by systemic hypertension due to the inability of the afferent arteriole to constrict in response to hypertension. It could also be driven by the local RAAS activation leading to efferent arteriolar constriction raising glomerular capillary pressure and reducing peritubular capillary blood flow, thus creating a relative oxygen deficiency for tubules faced with a larger filtered load. Both of these factors appear to have driven more severe pathological changes in the remnant kidneys of these cats, which is likely to indicate more rapid progression of kidney disease occurred in association with the renal wrap model, although this was not confirmed by functional measures.

7.2 Does Hypertension Lead to Increased Proteinuria and Progressive Renal Injury in Canine and Feline Clinical Cases of Hypertension?

The model evidence is clear-cut. Induction of severe hypertension in models of CKD in both the dog and the cat induces more severe proteinuria and results in more marked pathological lesions. Whether this leads to more rapid and progressive loss of kidney function has not been definitively studied in these models. In the next section, data from clinical patients with naturally occurring kidney disease will be reviewed to determine the evidence that systemic hypertension is damaging to the kidney.

The evidence that systemic hypertension is a risk factor for proteinuria in both the cat and dog seems clear from the published data on naturally occurring kidney disease. Syme et al. (2006) studied 136 cats, 28 of which were deemed to be healthy; 94 had azotaemic CKD, of which 28 were hypertensive (SABP > 175 mmHg); and 14 were hypertensive but non-azotaemic. All cats had UPCs and urine albumin to creatinine ratios (UAC) measured at recruitment, and multivariable analysis was used to determine whether age, sex, SABP, plasma creatinine and urine specific gravity were related to proteinuria (UPC and UAC in separate models). Both plasma creatinine and SABP were independently and positively associated with UPC and UAC. Table 7.2 illustrates the findings that if the cats in this study are categorized by the IRIS CKD staging system, then for each stage of CKD, proteinuria is higher in the hypertensive cats than the normotensive cats. This study went on to show that UPC and UAC (in separate models) measured at diagnosis were linearly related to reduced survival independently of both age and plasma creatinine. Figure 7.1 provides an illustration of this relationship using Cox's stepwise regression analysis with stratification based on UPC or UAC. Survival in this study was all-cause mortality with no attempt to define progression of CKD. Hypertension was associated with reduced survival in the univariate analysis but did not stay in the multivariate model as an independent factor, presumably because of its strong association with proteinuria.

Chakrabarti et al. (2012) attempted to study factors influencing progression of CKD in cats diagnosed with naturally occurring CKD. This paper examined cats that were followed longitudinally after a diagnosis of CKD had been made to determine what the risk factors were for them having a 25% of greater increase in their plasma creatinine concentration within a year of diagnosis of their CKD. Although SABP was included in the multivariate analysis having reached the threshold in the univariate analysis, in all the models tested, SABP did not prove to be an independent factor associated with this measure of progression, either when considered as a continuous variable or if the cats were categorized as hypertensive or normotensive. Once again, UPC proved to be associated with reduced survival and it is possible that hypertension is a contributing factor to the occurrence of proteinuria although not an independent factor influencing survival.

In both the above clinical studies hypertensive cats were treated with amlodipine with the aim of reducing their blood pressure to below 160 mmHg. No randomized placebo-controlled clinical trials have been undertaken in cats with amlodipine examining its effects on CKD progression or all-cause mortality. Jepson et al. (2007) looked retrospectively at factors influencing survival of 141 cats treated with amlodipine. Interestingly, neither time-averaged blood pressure nor blood pressure at initial diagnosis was a significant factor influencing all-cause mortality. However, both UPC at initial diagnosis and time-averaged UPC over the survival time were associated with reduced survival. Again, a relationship between blood

Table 7.2 Data showing the effect of hypertension on urine protein excretion in cats

	Variable/IRIS	Control or 1[a]	Control or 2a[a]	2b	3	4
Normotensive	SBP	134 (102–160)	133 (104–147)	140 (101–167)	130 (94–158)	133 (90–164)
	UPC	0.15 (0.07–0.49)	0.18 (0.06–0.25)	0.15 (0.06–4.46)	0.22 (0.06–3.20)	0.65 (0.12–3.64)
	UAC	11 (3–72)	16 (3–63)	16 (2–3065)	35 (2–618)	91 (20–2856)
Hypertensive	SBP	202 (180–223)	199 (178–230)	208 (176–250)	182 (175–215)	179
	UPC	0.27 (0.19–0.43)	0.28 (0.06–0.42)	0.30 (0.08–2.34)	0.49 (0.13–2.03)	0.54
	UAC	69 (4–300)	41 (17–135)	64 (5–2777)	80 (19–509)	173

Cats have been categorized according to the IRIS staging system for CKD. Stage 2 of this system has been divided into two categories: (2a) where plasma creatinine was between ≥140 and <180 μmol/L and (2b) where plasma creatinine concentration was between ≥180 and <250 μmol/L. SBP at initial screening was significantly associated with both log UAC ($P < 0.001$) and log UPC ($P = 0.031$) by multivariable regression analysis
Data taken from Syme et al. (2006)

SBP systolic blood pressure, *UAC* urine albumin to creatinine ratio, *UPC* urine protein to creatinine ratio

[a]Healthy cats that were normotensive were not deemed to have CKD so are not staged according to the IRIS system but are included in this category based on their plasma creatinine concentration for convenience. Hypertensive cats were suspected to have some dysfunction of their kidneys by virtue of the fact that they were persistently hypertensive and so have been categorised as stage 1 or stage 2a

7 Hypertension and the Kidney

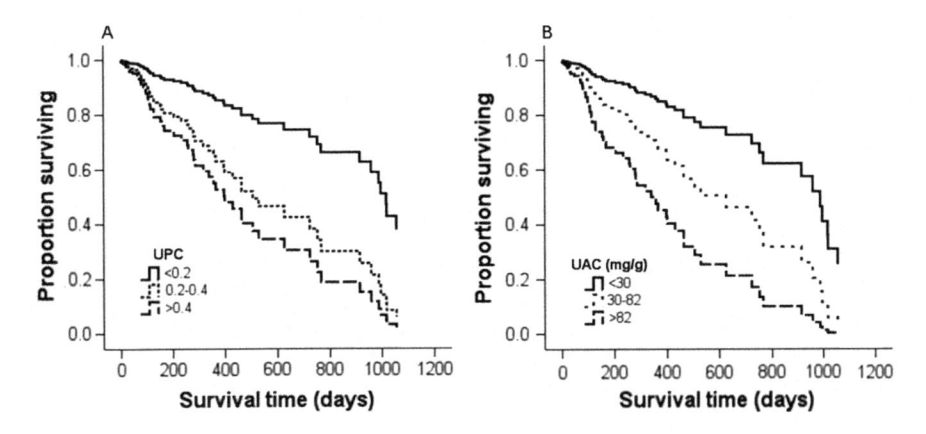

Fig. 7.1 Effect of urine protein and urine albumin to creatinine ratio on all-cause mortality in cats. Survival curves constructed by using Cox's stepwise regression analysis with stratification according to urine protein creatinine (UPC; **a**) and albumin creatinine (UAC; **b**) ratios. The analysis was controlled for the covariates of age and plasma creatinine concentration. The curves are constructed by using data from 126 cats (of which 55 died) who were apparently normal or were suffering from azotaemia and/or systemic hypertension. Reproduced with permission from Syme et al. (2006)

pressure and proteinuria was evident with the cats in the highest quartile for time-averaged blood pressure having the highest UPC values. In addition, treatment with amlodipine reduced UPC as well as blood pressure in cats that had UPC > 0.2 and so were in the borderline or proteinuric categories. It is possible, therefore, that the lack of evidence for an effect of blood pressure on progression of CKD or all-cause mortality in these clinical studies is related to the fact that cats are routinely treated with amlodipine which lowers blood pressure to below 160 mmHg in the majority of cats (Huhtinen et al. 2015).

Comparing the clinical study data with the published model data reviewed above, the following observations are pertinent. The severity of proteinuria observed in the renal wrap model (UPC 1.16 ± 0.3) is higher than most cats with naturally occurring CKD and hypertension at initial diagnosis [Jepson et al. 2007; UPC at initial diagnosis of hypertension was (median and IQR) 0.31 (0.19–0.59); $n = 118$]. This is despite the systolic blood pressure measured in these clinical cases of hypertension being 195 (184–213) mmHg (median and IQR). Even in the highest quartile of blood pressure group the time averaged blood pressure group, although the median blood pressure was close to the value of the renal wrap model group [170 (164–174) mmHg], the time average UPC of this sub-group of clinical cases was still only [0.46 (0.29–0.94)]. Direct comparison of blood pressure between the experimental and clinical studies needs to be made with caution, however, because the clinical measurements are made in a veterinary clinic at a single office visit whereas in the experimental cats blood pressure was continuously measured directly through implantation of telemetry devices. Stress of attending a veterinary clinic and undergoing a physical examination can significantly increase blood pressure

(Belew et al. 1999). Nevertheless, looking at these data together, naturally occurring severe hypertension in the cat does not seem to be associated with such marked proteinuria as seen in experimental feline models. This could be because amlodipine is renoprotective and/or most likely the models of severe hypertension (which involve marked activation of RAAS) are not representative of naturally occurring hypertension in the cat.

In one canine clinical study, blood pressure at diagnosis of CKD was shown to be a risk factor for uraemic death and reduced survival time (Jacob et al. 2003). This study recruited 45 dogs with CKD from the area local to the university where the study was conducted. CKD was confirmed by demonstrating persistence of azotaemia. The majority of these were enrolled into a randomized controlled diet study. Blood pressure was measured using an oscillometric method, and dogs with systolic blood pressure persistently ≥180 mmHg were treated (enalapril, amlodipine or diltiazem) with the aim of reducing systolic blood pressure below 160 mmHg. Based on their blood pressure at recruitment, dogs were assigned to three groups (terciles). The numbers in each group were not equal because at the divisions between terciles; if dogs had the same blood pressure, they were put in the lower blood pressure group. This gave three groups of dogs shown in Table 7.3, which happens to almost match the current IRIS blood pressure categories of normotensive, prehypertensive and hypertensive (and severely hypertensive). Although these allocations were made based on the blood pressure measured at initial diagnosis, follow-up blood pressures (which were not consistently obtained and so time averaged blood pressure could not be used in the analysis) suggested dogs remained in the same categories. This was despite the attempts to treat dogs in the high-

Table 7.3 Clinical parameters from dogs with CKD

Variable	LSBP ($n = 17$)	ISBP ($n = 15$)	HSBP ($n = 13$)	P value
SBP (mmHg)	109–145	147–163	164–217	NA
Age (year)	7.3 ± 5.1	7.8 ± 3.5	8.9 ± 2.8	0.54
Weight (kg)	21.3 ± 13.8	21.7 ± 11.8	19.3 ± 9.9	0.85
BUN (mg/dL)	60 ± 29	70 ± 30	58 ± 17	0.41
Creatinine (mg/dL)	3.2 ± 1.2	3.9 ± 1.8	3.2 ± 0.8	0.20
Phosphorus (mg/dL)	5.2 ± 1.1	4.7 ± 1.8	2.8 ± 2.1	0.25
Total CO_2 (μmol/L)	21.2 ± 2.3	19.7 ± 3.8	21.2 ± 2.8	0.34
Albumin (mg/dL)	3.2 ± 0.3	3.1 ± 0.3	3.0 ± 0.4	0.87
PCV (%)	36.7 ± 8.1	36.4 ± 9.1	39.9 ± 10.3	0.52
Urine P:C ratio	0.80 ± 1.03[a]	1.52 ± 1.04[a]	3.47 ± 3.84[b]	0.009

Data reproduced with permission from Jacob et al. (2003). The original table has been modified to include the blood pressure data from each group of dogs. Data are mean ± SD or range of values.
[a,b]Values with different superscript letters are significantly different from each other as stated in the legend. The P value is given in the final column of the table for the ANOVA that is reported in the paper. The pairwise comparisons and precise P values for comparing each group with the others are not detailed in the original paper

SBP systolic blood pressure, *BUN* blood urea nitrogen, *Urine P:C ratio* urine protein to creatinine ratio

pressure group with the various antihypertensive agents listed above. Eleven of the 13 dogs were treated but only one had their blood pressure reduced below 160 mmHg.

The study examined associations between blood pressure and occurrence of uraemic crisis, death due to CKD and death of any cause. A Cox's proportional hazards model was used to adjust for baseline covariates (age, weight, PCV and serum creatinine, urea nitrogen, phosphorus, total CO_2 and albumin concentrations). Measures of urine protein were not included in this model. Dogs in the highest blood pressure tercile had higher relative risk of each of the outcomes when compared with the dogs in the lowest blood pressure tercile, which was used as the reference group. On average, the relative risk of each of the outcomes occurring was between three and four times higher in the highest blood pressure group when compared to the low pressure group. These dogs were also involved in a study examining the effect of feeding a clinical renal diet on the same outcomes. The researchers adjusted for the effect of feeding the diet within the model and this did not alter the relative risks of the three outcomes and their association with high blood pressure. Figure 7.2 taken from this paper shows the clear difference in the highest blood pressure groups and the other two in the occurrence of renal death. In order to examine the effects of

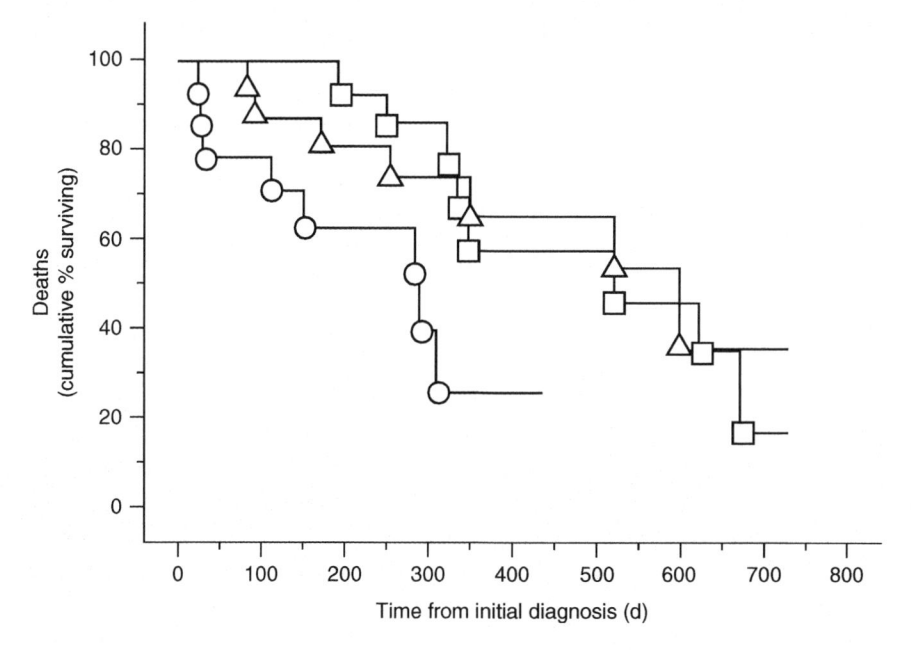

Fig. 7.2 Kaplan Meir survival curves from dogs with CKD illustrating the effects of high, intermediate and low blood pressure on renal mortality. Dogs were grouped according to systolic blood pressure at the time of diagnosis of chronic kidney disease. For this assignment, initial SBP was divided into tertiles and dogs were assigned to a high SBP (circles; initial SBP ranged from 161 to 201 mmHg; $n = 14$); intermediate SBP (squares; 144–160 mmHg; $n = 15$), or low SBP (triangles; initial SBP ranged from 107 to 143 mmHg; $n = 16$) group. Reproduced with permission from Jacob et al. (2003)

blood pressure on decrements in renal function over time, the researchers plotted the reciprocal of serum creatinine against time for the three groups. This showed that the gradient of this line was significantly different in the high pressure group of dogs when compared to the other two groups indicating a faster decline in renal function.

The same group published another paper involving the same 45 dogs but in this study divided the dogs into two groups based on their UPC value at initial diagnosis, a group with dogs with UPCs ≥ 1 ($n = 25$) and a group with UPCs < 1 ($n = 20$) (Jacob et al. 2005). There were dogs in each group that had been randomized to receive a clinical diet and a control diet. The association between proteinuria and outcome was similar regardless of which diet the dogs had been fed. A Cox's proportional hazard model was used to examine the association between the relative risk of uraemic crisis or death from CKD and proteinuria, taking into account a range of other covariates (age; weight; SBP; PCV; and SUN, serum creatinine, inorganic phosphorus, total CO_2 and albumin concentrations). In addition to including SBP as a covariate in this model, the researchers explored the relationship between proteinuria, SBP and uraemic crisis and death by examining the influence of proteinuria (UPC < 1 or UPC ≥ 1) in the three terciles of blood pressure discussed above.

This paper reported a moderate relationship between proteinuria and blood pressure (Spearman's $r = 0.57$; $P < 0.001$). Blood pressure was significantly higher in the proteinuric group compared with the non-proteinuric group (164 ± 23 vs. 139 ± 18 mmHg) at initial diagnosis. The relative risk of uraemic crisis, renal death or all-cause mortality was two to three times higher in the dogs with UPC > 1 (when compared to those with UPC < 1) at each tercile of blood pressure; however, the 95% confidence limits associated with these risks were wide and the influence of proteinuria did not reach statistical significance in this analysis. Proteinuria was associated with an increased risk of uraemic crisis, renal death and all-cause mortality when assessed by Cox's proportional hazards with an unadjusted relative risk of around three for each outcome. When the risk was adjusted for initial blood pressure, the relative risk remained >2 but was no longer statistically significant for uraemic and renal death. However, the adjusted risk remained statistically significant for all-cause mortality when adjusting for blood pressure.

Proteinuria as a continuous variable was also significantly associated with all three adverse outcomes. For every one unit increase in UPC, the risk of the adverse outcome occurring increased by between 60 and 80%. This relationship remained statistically significant adjusting for the initial blood pressure. Reciprocal plots of serum creatinine against time were significantly different between the two proteinuria categories with the UPC ≥ 1 group declining over the 180 days of analysis whereas the UPC < 1 group showed no decline.

These studies taken together show that proteinuria in dogs with CKD is associated with adverse outcomes and progressive deterioration in renal function over time. Blood pressure is related to the severity of proteinuria as it is for cats, suggesting it may be contributing to the glomerular capillary hypertension and loss of protein across the glomerular filtration barrier although it could contribute to proteinuria and progressive renal pathology in other ways. It should be noted that the severity of proteinuria in dogs with CKD is much greater than is seen in cats, and many of the dogs in these clinical studies had levels of protein in their urine which could only result from primary glomerular disease. Seventeen of the 45 dogs in the study had UPCs > 2, and this level of proteinuria is indicative of primary glomerular disease. Primary glomerular disease may have been responsible for the CKD occurring in dogs with lower level proteinuria but renal histopathology was not reported in these papers. Notably, this higher level of proteinuria in dogs with CKD when compared to cats is most likely related to the increased speed of progression of CKD in dogs when compared to cats.

In summary, it is difficult from the clinical data presented to differentiate the effect of hypertension on progressive damage to the kidney from that of proteinuria. The two are associated and it is tempting to suggest that the severity of proteinuria is influenced by hypertension and the damaging effects are related to the proteinuria. Whether proteinuria itself drives progression or is a marker that progression is occurring is much debated (see Cravedi and Remuzzi 2013). Any damaging effects of hypertension on the kidney that are independent of proteinuria are difficult to tease out from these clinical studies. Nevertheless, both need to be considered in the management of CKD in the dog and cat; hence, current best practice guidelines recommend monitoring and treatment of both.

7.3 Renal Pathology Associated with Hypertension in Naturally Occurring Kidney Disease

If hypertension is damaging the kidney directly, one might expect to see classical signs of hypertensive nephropathy described in the human literature (Seccia et al. 2017) in dogs and cats with hypertension. While two papers have been published examining the various clinical factors associated with pathological lesions in the cat kidney (see below), no similar papers have been published comparing the renal pathology of normotensive and hypertensive dogs with CKD. One recent publication involved 47 retired greyhounds (average age 3.7 years), 62% ($n = 29$) of which were hypertensive (SBP > 160 mmHg) (Surman et al. 2012). Serum biochemical evidence of kidney disease was absent in most cases or equivocal in <10% of cases (elevated creatinine accompanied by urine specific gravity > 1.040). Hypertension in these dogs was associated with microalbuminuria. Histopathology (renal biopsy) was available for evaluation from 15 female dogs, 8 categorized as normotensive and 7 as hypertensive. The pathology scores did not differ between the two blood pressure groups although some pathological features consistent with hypertensive renal damage were described in this paper (podocyte effacement, mesangial matrix expansion); these lesions were described as mild and present in both groups, so their relationship to hypertension seems unlikely.

The two cat studies came to different conclusions about hypertension and its association with pathological lesions in cats with CKD. Chakrabarti et al. (2013) found a significant relationship between time-averaged blood pressure and glomerulosclerosis score and the presence of hyperplastic arteriolosclerosis. In this paper, there were 69 cats where blood pressure data were available. Although cats were categorized into normotensive ($n = 35$) and hypertensive ($n = 34$), absolute blood pressure values were available from multiple visits prior to death, and time-averaged blood pressure was calculated using all these measurements. Nine cats from each group were non-azotaemic. Even though antihypertensive treatment (oral amlodipine besylate) was initiated for the hypertensive cats, on average they still had higher time averaged blood pressure than the normotensive cats [136 (119, 146) mmHg vs. 159 (144, 168) mmHg]. The vascular lesions seen in the histopathology undertaken in this study were predominantly hyperplastic arteriolosclerosis (see Fig. 7.3) with only one cat also having hyaline arteriolosclerosis. No cats showed evidence of fibrinoid necrosis. The multivariable model looking for associations between time-averaged systolic blood pressure and renal pathology had an adjusted r^2 value of 0.345 with mean glomerular score and presence of hyperplastic arteriolosclerosis being independently and significantly associated with time-averaged systolic blood pressure. Ten of the 34 hypertensive cats had hyperplastic arteriolosclerosis, whereas this was found in only 1 of the 35 normotensive cats. A scatter plot of the distribution of time-averaged blood pressure for cats with and without hyperplastic arteriolosclerosis is shown in Fig. 7.4.

The mean glomerular score was an index of the severity of glomerulosclerosis as assessed by the mesangial matrix expansion scored on a scale of 0–3 according to the degree of matrix expansion and the proportion of the capillary loops in the glomerulus affected by its expansion. Twenty-five glomeruli were assessed and an average score calculated. Time-averaged blood pressure was directly and significantly related to this mean glomerular score (r^2 0.175; $P < 0.001$) (see Fig. 7.5). However, the matrix expansion associated with hypertension is relatively mild in the cat compared to other species with the highest score indicating >50% of the glomerulus

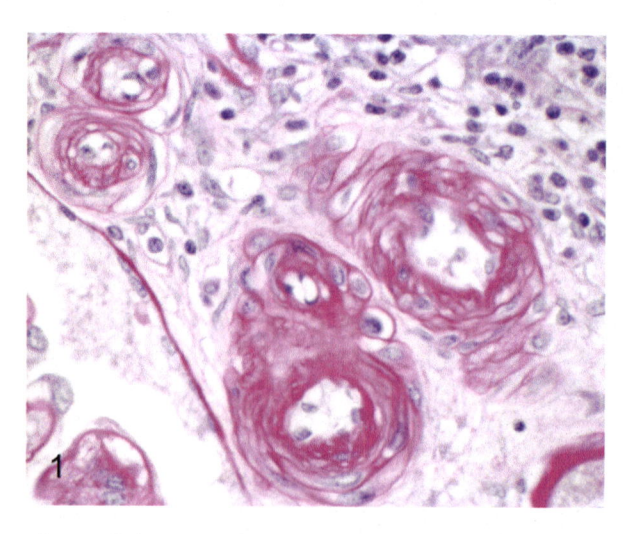

Fig. 7.3 Photomicrograph from a cat demonstrating hyperplastic arteriolosclerosis with medial thickening by concentric layers of basement membrane and smooth muscle cells. Periodic acid-Schiff and haematoxylin. Reproduced with permission from Chakrabarti et al. (2013)

7 Hypertension and the Kidney

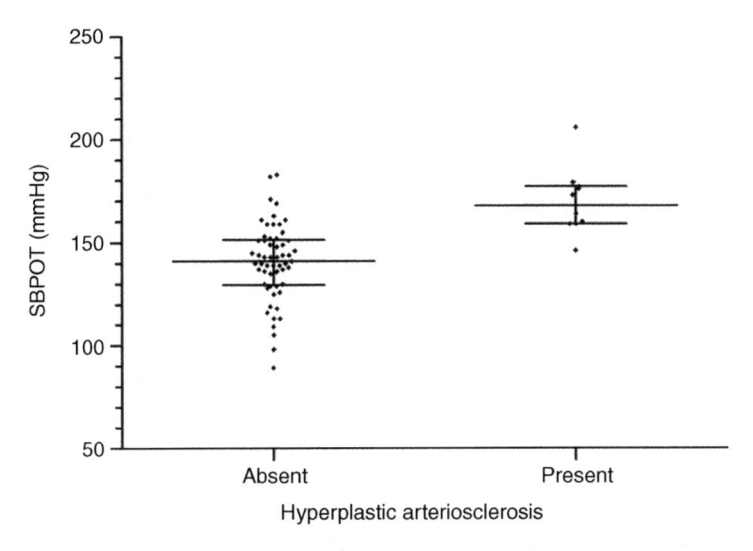

Fig. 7.4 Association of postmortem renal hyperplastic arteriolosclerosis with higher time-averaged systolic blood pressure (SBPOT). The horizontal lines indicate median values and interquartile ranges. Reproduced with permission from Chakrabarti et al. (2013)

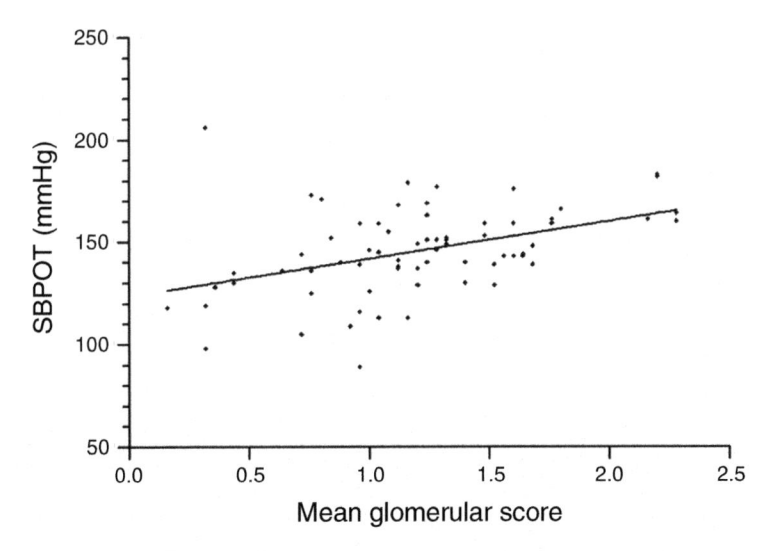

Fig. 7.5 Positive correlation of time-averaged systolic blood pressure (SBPOT) with mean postmortem glomerulosclerosis score in cats with variable renal function. The line of best-fit is drawn on the graph ($r^2 = 0.174$, $P < 0.001$). Reproduced with permission from Chakrabarti et al. (2013)

affected. In human studies it would be not unusual to have glomeruli with between 75 and 100% of the capillary loops affected. The fact that time averaged systolic blood pressure was not related to interstitial fibrosis, the pathological lesion that was related most closely to quantitative assessments of kidney function (e.g. plasma creatinine concentration), suggests hypertension was possibly not driving progressive loss of kidney function. Clearly one of the limitations of this study is that cats were treated for their hypertension so the blood pressure they were exposed to in the immediate period prior to death was lowered by the use of this drug. Specific vascular effects of this drug, in addition to vasodilation, include up-regulation of eNOS in the endothelium of the vasculature (Lenasi et al. 2003) which also may influence connective tissue deposition and proliferation of cells of the walls of blood vessels. Despite these limitations, this paper does show hypertension appears to be associated with relatively mild pathological changes in the kidney and is not associated with those pathological changes that are most related to serum markers of kidney function.

The second paper on kidney pathology in cats appears to contradict the paper by Chakrabarti et al. (2013). McLeland et al. (2015) found vascular lesions, including hyperplastic arteriolosclerosis to be just as common in normotensive cats as they were in hypertensive cats. Indeed, these authors reported hyperplastic arteriolosclerosis to be present in 48% of normotensive cats, whereas Chakrabarti et al. (2013) only found this vascular pathology in 1 of 35 normotensive cats. McLeland et al. (2015) studied 46 cats that underwent post-mortem examination at a veterinary teaching hospital with 33 of these 46 having blood pressure data available. Of these 33, 13 were hypertensive. The absolute blood pressure values for the cats presented in this paper are not provided. The analysis was based on categorical classification of cats as either hypertensive or normotensive. Thus, this paper does not report an association between hypertension and glomerulosclerosis as was reported by Chakrabarti et al. (2013). The reason for the differences between the two papers of renal pathology in cats is difficult to explain. It is possible that the high frequency of vascular lesions in the kidney identified by McLeland et al. is related to causative factors unrelated to hypertension that are absent from the UK but common in the USA.

7.4 Summary

Hypertension occurring in the presence of CKD where the afferent arteriole is dilated and unable to constrict results in increased protein loss in the urine in both dogs and cats. Persistent proteinuria likely damages or is a marker that progressive damage to the kidney is occurring. This is particularly true where the RAAS system is fully activated and is driving glomerular hyperfiltration. Numerous model studies in dogs and a small number in cats support this thinking. Naturally occurring CKD in dogs progresses more rapidly in proteinuric patients (as compared with those that are non-proteinuric) and this progression is exacerbated by hypertension, probably through continued injury to the capillary tuft, acceleration of podocyte loss, increased proteinuria and glomerulosclerosis. The situation in the cat is less clear cut. Hypertension is a risk factor for proteinuria. Renal lesions associated with quite marked hypertension seem relatively mild, and, in the absence of marked systemic activation of the RAAS system, it seems unlikely that hypertension causes marked kidney damage although there is no doubt that hypertension plays a contributory role in progressive CKD through driving more severe proteinuria.

References

Belew AM, Barlett T, Brown SA (1999) Evaluation of the white-coat effect in cats. J Vet Intern Med 13(2):134–142

Brown SA, Brown CA (1995) Single-nephron adaptations to partial renal ablation in cats. Am J Phys 269(5 Pt 2):R1002–R1008

Brown SA, Finco DR, Crowell WA, Choat DC, Navar LG (1990) Single-nephron adaptations to partial renal ablation in the dog. Am J Phys 258(3 Pt 2):F495–F503

Brown SA, Brown CA, Jacobs G, Stiles J, Hendi RS, Wilson S (2001) Effects of the angiotensin converting enzyme inhibitor benazepril in cats with induced renal insufficiency. Am J Vet Res 62(3):375–383

Brown SA, Finco DR, Brown CA, Crowell WA, Alva R, Ericsson GE, Cooper T (2003) Evaluation of the effects of inhibition of angiotensin converting enzyme with enalapril in dogs with induced chronic renal insufficiency. Am J Vet Res 64(3):321–327

Chakrabarti S, Syme HM, Elliott J (2012) Clinicopathological variables predicting progression of azotemia in cats with chronic kidney disease. J Vet Intern Med 26(2):275–281

Chakrabarti S, Syme HM, Brown CA, Elliott J (2013) Histomorphometry of feline chronic kidney disease and correlation with markers of renal dysfunction. Vet Pathol 50(1):147–155

Cravedi P, Remuzzi G (2013) Pathophysiology of proteinuria and its value as an outcome measure in chronic kidney disease. Br J Clin Pharmacol 76:516–523

Huhtinen M, Derré G, Renoldi HJ, Rinkinen M, Adler K, Aspegrén J, Zemirline C, Elliott J (2015) Randomized placebo-controlled clinical trial of a chewable formulation of amlodipine for the treatment of hypertension in client-owned cats. J Vet Intern Med 29(3):786–793

Jacob F, Polzin DJ, Osborne CA, Neaton JD, Lekcharoensuk C, Allen TA, Kirk CA, Swanson LL (2003) Association between initial systolic blood pressure and risk of developing a uremic crisis or of dying in dogs with chronic renal failure. J Am Vet Med Assoc 222(3):322–329

Jacob F, Polzin DJ, Osborne CA, Neaton JD, Kirk CA, Allen TA, Swanson LL (2005) Evaluation of the association between initial proteinuria and morbidity rate or death in dogs with naturally occurring chronic renal failure. J Am Vet Med Assoc 226(3):393–400

Jepson RE, Elliott J, Brodbelt D, Syme HM (2007) Effect of control of systolic blood pressure on survival in cats with systemic hypertension. J Vet Intern Med 21(3):402–409

Lenasi H, Kohlstedt K, Fichtlscherer B, Mülsch A, Busse R, Fleming I (2003) Amlodipine activates the endothelial nitric oxide synthase by altering phosphorylation on Ser1177 and Thr495. Cardiovasc Res 59(4):844–853

Mathur S, Brown CA, Dietrich UM, Munday JS, Newell MA, Sheldon SE, Cartier LM, Brown SA (2004) Evaluation of a technique of inducing hypertensive renal insufficiency in cats. Am J Vet Res 65(7):1006–1013

McLeland SM, Cianciolo RE, Duncan CG, Quimby JM (2015) A comparison of biochemical and histopathologic staging in cats with chronic kidney disease. Vet Pathol 52(3):524–534

Seccia TM, Caroccia B, Calò LA (2017) Hypertensive nephropathy. Moving from classic to emerging pathogenetic mechanisms. J Hypertens 35(2):205–212

Surman S, Couto CG, DiBartola SP, Chew DJ (2012) Arterial blood pressure proteinuria and renal histopathology in clinically healthy retired racing greyhounds. J Vet Int Med 26:1320–1329

Syme HM, Markwell PJ, Pfeiffer D, Elliott J (2006) Survival of cats with naturally occurring chronic renal failure is related to severity of proteinuria. J Vet Intern Med 20(3):528–535

Hypertension and the Heart and Vasculature

8

Amanda E. Coleman and Scott A. Brown

8.1 Introduction

The importance of the heart as a target for end-organ damage in hypertensive disease has long been recognized. The earliest scientific reports detailing myocardial hypertrophy in human patients with renal disease and changes to the "minute arteries of the body" appear in the medical literature of the nineteenth century (Bright 1836; Gull and Sutton 1872). As early as 1913, soon after the ability to non-invasively measure blood pressure and to associate the results of these measurements with pathological outcomes was facilitated by the description of Korotkoff sounds and the sphygmomanometric technique, investigators were able to link elevated blood pressure measurements with clinical signs of left ventricular hypertrophy and cardiac death (Janeway 1913). Approximately 50 years later, in the 1960s, clinical descriptions of cardiac involvement in dogs with naturally occurring systemic arterial hypertension appeared in the veterinary literature (Pirie et al. 1965; Anderson 1968; Anderson and Fisher 1968).

For human patients, systemic arterial hypertension represents a major risk factor for cardiovascular disease and its various manifestations. Compared to normotensives, hypertensive people are at increased risk for all-cause and cardiovascular disease mortality, coronary artery disease, myocardial infarction, heart

A. E. Coleman (✉)
Department of Small Animal Medicine and Surgery, University of Georgia College of Veterinary Medicine, Athens, GA, USA
e-mail: mericksn@uga.edu

S. A. Brown
Department of Small Animal Medicine and Surgery, University of Georgia College of Veterinary Medicine, Athens, GA, USA

Department of Physiology and Pharmacology, University of Georgia College of Veterinary Medicine, Athens, GA, USA
e-mail: sbrown01@uga.edu

failure, atrial fibrillation, stroke and transient ischaemic attack and peripheral vascular disease, with increasing age acting as a major modifier of this effect (Bakris and Sorrentino 2018). Hypertensive people have a two- to threefold greater relative risk for cardiovascular events compared to age-matched normotensives, and even mild increases in systolic blood pressure beyond normal are associated with increased risk (Lewington et al. 2002; Stamler et al. 1993).

Most descriptions of the effects of systemic hypertension on the heart are based upon data generated in experimental (e.g. aortic banding, renal-wrap, genetic) animal models and in spontaneous human disease, none of which faithfully replicate the clinical disease of dogs and cats. The majority of human hypertensives are affected by primary ("essential") hypertension, a heterogeneous condition that is influenced by complex interactions among cardiac, vascular, renal, neurologic and endocrine processes, as well as environmental, genetic and behavioural factors (Bakris and Sorrentino 2018). Furthermore, human hypertension infrequently develops in isolation, instead coexisting with conditions that are also known to be independent risk factors for cardiovascular disease. While the essential hypertension of human beings should be regarded as a pathophysiologic entity that is distinct to the idiopathic or secondary hypertension of companion animals, there remain lessons to be learned from the large amount of information generated by study of these patients.

This chapter describes the structural and functional cardiac and vascular alterations noted with sustained, pathological increases in systemic arterial blood pressure. The clinical relevance of these alterations is considered.

8.2 Pathologic Changes to the Cardiovascular System in Systemic Hypertension

8.2.1 The Heart as a Target Organ

Cardiac Response to Pressure Overload The heart's major phenotypic response to excessive left ventricular afterload, such as that imposed by systemic arterial hypertension, is one of left ventricular hypertrophy. This response is conventionally viewed as an adaptive attempt to normalize excessive left ventricular systolic wall stress and to maintain adequate stroke volume.

In order to understand cardiac responses to systemic arterial hypertension, a brief discussion of ventricular *afterload* is fitting. Ventricular afterload is defined as the sum of external factors that oppose myocardial fibre shortening and, therefore, ventricular ejection. Total ventricular afterload is comprised of both frequency-dependent (i.e. pulsatile) and frequency-independent (non-pulsatile) components (Nichols and Pepine 1982). A major element, *systemic vascular resistance*, represents the load imposed by the peripheral vascular system in a steady-flow (non-pulsatile) state and is determined largely by blood viscosity and arteriolar tone (radius), according to the following relationship (an expression of Poisuelle's Law):

$$\text{Resistance to blood flow} = \frac{8\eta L}{\pi r^4} \quad (8.1)$$

for which η = blood viscosity, L = length of system and r = blood vessel radius (Eq. 8.1).

Afterload is also determined by the resistance to flow imposed by the elastic properties of the aorta (i.e. its "stiffness") and characteristics of pressure waves reflected from downstream portions of the arterial system. Both of these vary over the course of ejection (i.e. are oscillatory, or frequency-dependent), and collectively, the resistance they impose is referred to as *aortic impedance*.

Though an oversimplification that does not appropriately take into account system pulsatility or the effects of changes in ventricular preload or contractility on developed intraventricular pressure, ventricular afterload is often expressed in terms of systolic wall stress (or wall tension). Despite its limitations, the use of wall stress as a measure of ventricular afterload is useful to the understanding of cardiac compensatory responses during pressure-overload states. Based on derivations of the law of Laplace, circumferential wall stress (σ) can be approximated as:

$$\sigma = \frac{Pr}{h} \quad (8.2)$$

for which P = pressure in the left ventricular cavity, r = left ventricular cavity radius and h = left ventricular wall thickness (Eq. 8.2).

Chronic increases in systemic arterial pressure, a cause of ventricular pressure overload, lead to a sustained increase in systolic ventricular wall stress, principally owing to an increase in left ventricular cavity pressure during isovolumic contraction and ventricular ejection periods. In order to move a normal stroke volume into circulation, the heart must invest a greater amount of energy in developing the tension necessary to overcome an increased resistance to ejection. This necessitates an increase in myocardial oxygen demand.

The heart's response to pressure overload stress is one of concentric ventricular hypertrophy, which is characterized by an increase in overall ventricular mass, myocyte size and protein synthesis (Dorn et al. 2003). Ultrastructurally, concentric left ventricular hypertrophy is typified by the addition of sarcomeres in parallel, which leads to widening of the myocyte, and results in an increase in thickness of the ventricular wall with preservation of ventricular cavity size. This increase in ventricular wall thickness serves to mitigate systolic wall stress and myocardial oxygen consumption by directly opposing the increase in intracavitary pressure (Eq. 8.2). Concentric ventricular hypertrophy is classically distinguished from the eccentric ventricular hypertrophy of volume overload, which is characterized by the addition of sarcomeres in series, lengthening of the myocyte and an increase in ventricular cavity size (Dorn et al. 2003).

Conventionally, concentric left ventricular hypertrophy is held to be a protective, adaptive response to increases in ventricular afterload (Grossman et al. 1975). However, recent studies have challenged the view that normalization of wall stress is absolutely necessary for the maintenance of normal cardiac performance (Sano and Schneider 2002; Esposito et al. 2002; Hill et al. 2000), with attention turning to concentric left ventricular hypertrophy as a primary therapeutic target (Frey et al. 2004). Whether the left ventricular hypertrophy of pressure overload is or is not initially protective, it appears to eventually represent a maladaptive response, increasing the risk for detrimental outcomes. As has long been recognized in people with hypertension, the hypertrophy of hypertensive heart disease may transition from a compensated phase, during which cardiac output at rest is maintained, to one of decompensated, overt heart failure. In these cases, the ventricle may or may not be dilated and hypocontractile (heart failure with a reduced ejection fraction or with a preserved ejection fraction, respectively) (Drazner 2011; Frohlich et al. 1992; Meerson 1961).

Gross Pathologic Changes Sustained systemic hypertension may result in detectable alterations in left ventricular geometry, which are exemplified by an increase in left ventricular mass. In human hypertensives, severity of left ventricular hypertrophy is positively (though relatively weakly) correlated to degree of 24-h blood pressure increase (Feola et al. 1998; Levy et al. 1988), and successful antihypertensive treatment is associated with regression of left ventricular hypertrophy and an accompanying decrease in the risk of cardiovascular events (Dahlof et al. 2002), supporting the link between left ventricular hypertrophy and chronic, pathologic increases in blood pressure.

There is substantial variability among untreated hypertensive people in the degree of left ventricular mass increase and phenotypic pattern of hypertrophy (Drazner 2011; Ganau et al. 1992). Systemic arterial hypertension is typically associated with a concentric pattern (i.e. increased relative wall thickness) of left ventricular hypertrophy. However, in human hypertensives, ventricular responses appear to be quite complex. In a landmark echocardiographic study of 165 human patients with essential (primary) hypertension, a concentric pattern of left ventricular hypertrophy was noted in the minority of patients (8–15%, depending on definitions used), with greater percentages exhibiting normal geometry, eccentric hypertrophy or a "concentric remodelling" (i.e. normal left ventricular mass with increased wall thickness) pattern (Ganau et al. 1992). The study's authors postulated that factors known to impact cardiac volume load and stroke volume (e.g. total blood volume, venous return, systemic vascular resistance and innate myocardial properties) are likely modulators of the ventricular response (Ganau et al. 1992). Therefore, in addition to other factors, this heterogeneity of responses may reflect differences in pathophysiology of the hypertension—for example, the relative contributions of renal sodium and water retention versus arteriolar vasoconstriction—among individuals. Since publication of the aforementioned study, additional work has suggested that demographic factors such as race (Kizer et al. 2004), sex (Krumholz et al. 1993) and age (Chahal et al. 2010), concurrent medical and cardiac conditions (e.g. diabetes, obesity, coronary artery disease, valvular heart disease) and genetics (Drazner 2011) may all have an influence on the pattern of ventricular hypertrophy noted in an individual. The influence of differences in neurohormonal activation—for example, whether one's hypertension is characterized by a low-renin or a high-renin state—on the development of concentric or eccentric hypertrophy appears to be controversial, with some work suggesting its importance (du Cailar et al. 2000; Muscholl et al. 1998) and data from a larger study calling this role into question (Velagaleti et al. 2008).

Interindividual variability in degree and patterns of ventricular response has also been documented in naturally occurring feline (Chetboul et al. 2003; Henik et al. 2004) and experimental canine (Koide et al. 1997) hypertension. In a study that described echocardiographic findings in 39 cats with spontaneous systemic hypertension, the most common left ventricular morphology was one of concentric hypertrophy, which was documented in 59% of cases. In the remainder, left ventricular patterns were normal (15%), suggestive of eccentric hypertrophy (13%), or exhibited a focal subaortic interventricular septal bulge (13%) (Chetboul et al. 2003). Factors affecting an individual's ventricular response to systemic hypertension have not been systematically evaluated in cats and dogs, although genetics, age, chronicity of hypertension, pathophysiologic mechanisms underlying the hypertension, superimposition of left-sided valvular insufficiency or anaemia or a combination of these, among other factors, should likely be considered. In the feline study referenced above, with the exception of significantly younger age in cats affected by eccentric hypertrophy, no significant differences in breed, sex, presence of retinopathy or renal failure or systolic blood pressure were identified among the various ventricular response pattern categories (Chetboul et al. 2003).

Cellular and Ultrastructural Alterations Both the cellular and non-cellular components of the myocardium are affected by left ventricular hypertrophy, which is accompanied by an increase in cardiomyocyte size, alterations in extracellular matrix composition and abnormalities of the small intramyocardial coronary arteries. An increase in extracellular structural collagen matrix, including both interstitial and perivascular fibrosis that is characterized by a predominance of fibrillar type I collagen, represents one ultrastructural hallmark of hypertensive heart disease (Frohlich et al. 1992). The increase in cardiomyocyte size, which outpaces concurrent increases in intramyocardial capillary surface area (Tomanek et al. 1982, 1991), leads to decreased capillary density, which may further contribute to the development of interstitial fibrosis by hypoxic signalling. Coronary arteriolosclerosis, characterized by hypertrophy and hyperplasia of the smooth muscle of the tunica media, is also a prominent feature in the hypertensive heart as has been documented in those from cats and dogs with spontaneous systemic hypertension (Littman 1994).

Functional Implications of Hypertensive Cardiac Changes While the concentric hypertrophy of chronic systemic hypertension results in normalization of left ventricular systolic wall stress, this adaptation comes at the expense of normal ventricular functional properties. The hallmark abnormality of the heart exposed to chronic, pathological increases in systemic arterial pressure is diastolic dysfunction, characterized by impaired ventricular relaxation and filling. Even in the normal heart, increases in afterload prolong ventricular relaxation (Gilbert and Glantz 1989; Little 1992). The increase in wall thickness and interstitial collagen content that accompanies left ventricular hypertrophy further slows early, active ventricular relaxation and increases left ventricular stiffness, which predictably alter the diastolic mechanical properties of the heart as evidenced by increases in end-diastolic filling pressure (Hayashida et al. 1997; Munagala et al. 2005; Rouleau et al. 2002). Of several histopathologic factors (i.e. myocyte diameter, percentage fibrosis and myocyte disarray) evaluated in one study of endomyocardial biopsies from human hypertensives, fibrosis severity was the most significant predictor of diastolic dysfunction (Ohsato et al. 1992).

Diastolic alterations have been demonstrated in experimental models of both feline (Wallner et al. 2017) and canine (Munagala et al. 2005) systemic hypertension, including a bilateral renal-wrap model that produces concentric left ventricular hypertrophy and myocardial fibrosis with normal left ventricular volumes after 6–8 weeks. Compared to hearts from sham-operated dogs, those from hypertensive dogs exhibited mean increases in the time constant of left ventricular relaxation [tau, an index of early diastolic function (Bai and Wang 2010)] of up to twice normal (Munagala et al. 2005). Studies of normal dogs have proven the importance of ventricular relaxation in maintaining normal left ventricular filling pressures during exercise (Cheng et al. 1992); therefore, these changes may signal a clinically relevant reduced capability of the hypertrophied canine heart to handle hemodynamic stressors.

The ventricular dysfunction of hypertensive heart disease is not confined to diastole, even in its earliest stages (Schumann et al. 2019; Ayoub et al. 2016). Advanced echocardiographic techniques (e.g. tissue Doppler imaging, speckle tracking echocardiography) have allowed for the early detection of systolic function abnormalities, even as values for conventional echocardiographic parameters, such as ejection fraction, remain apparently normal. With sustained, pathologic increases in afterload, systolic dysfunction may become overt, leading to a dilated, hypocontractile (i.e. "burned out") left ventricle, as has been described in animal models (Meerson 1961) and human hypertensive patients (Drazner 2011). However, progression from concentric hypertrophy to a systolic dysfunction phenotype and heart failure seems to be uncommon in human patients in the absence of precipitating myocardial infarction (Krishnamoorthy et al. 2011; Rame et al. 2004).

Coronary microvascular dysfunction, independent of atherosclerotic coronary artery disease, is a common feature of human hypertensive heart disease (Lazzeroni et al. 2016). The perivascular fibrosis and medial hypertrophy and hyperplasia that characterize intramural coronary arteriolosclerosis occur at the expense of lumen diameter, resulting in reduced coronary vascular reserve, a measure of the ability of the coronary circulation to augment myocardial blood flow during maximal vasodilation. These changes lead to myocardial ischaemia and favour the development of myocardial fibrosis (Lazzeroni et al. 2016). Structural and functional alterations of the coronary circulation also have been documented in experimental models of canine left ventricular hypertrophy. While baseline, resting myocardial blood flow tends to be preserved (Hayashida et al. 1997; Rouleau et al. 2002; Alyono et al. 1986), myocardial oxygen consumption is increased at rest and coronary autoregulation is impaired (Rouleau et al. 2002). In addition, coronary vasodilatory reserve is known to be decreased with hypertrophy of the canine left ventricle (Alyono et al. 1986; Jeremy et al. 1989). Therefore, myocardial demand ischaemia may be induced during periods of hemodynamic stress (e.g. tachycardia) or during abrupt decreases in arterial blood pressure (e.g. with antihypertensive therapies). Canine models of renal hypertension suggest that although capillary growth does not fully compensate for the increase in LV mass seen with chronic (i.e. 7-month) hypertension, coronary angiogenesis does help to mitigate reductions in cardiomyocyte oxygenation (Tomanek et al. 1991).

Finally, left ventricular hypertrophy and fibrosis are associated with electrophysiologic alterations, including an increased susceptibility to spontaneous ventricular arrhythmias and inducible ventricular fibrillation, as have been shown in human hypertensive patients (McLenachan and Dargie 1990) and feline models of left ventricular hypertrophy (Rials et al. 1995, 1996), respectively. Though the mechanism of this increased susceptibility is not fully understood, prolongation of myocyte action potential duration, with associated dispersion of refractoriness or afterpotential development, has been documented consistently (Rials et al. 1995, 1996; Ben-David et al. 1992; Shechter et al. 1989; Kleiman and Houser 1989). Abnormalities of various membrane ion currents, including alterations in channel expression (Ten Eick et al. 1993) and abnormalities of channel function (Kleiman and Houser 1988, 1989), have been noted as contributing factors.

8.2.2 The Systemic Vasculature as a Target Organ

Primary pathologic alterations of the systemic vascular tree in patients with systemic hypertension may contribute to the development and maintenance of systemic hypertension. In addition, they may represent a response to chronic elevations in intraluminal pressure; in other words, the vasculature itself is a target organ.

During hypertensive states, despite arteriolar vasoconstriction and increased systemic vascular resistance, normal tissue perfusion is typically maintained due to local autoregulatory mechanisms that function over a wide range of blood pressure values and which subserve tissue metabolic demands. With chronicity, changes to the architecture of the small systemic arteries and arterioles, including wall thickening due to medial hypertrophic remodelling and arteriolar and capillary rarefaction (Renna et al. 2013), help to maintain relatively normal tissue blood flow despite increases in perfusion pressure. Arteriolosclerosis, the term used to describe remodelling changes to the resistance vessels, features prominently in necropsy studies of dogs and cats with systemic arterial hypertension (Kohnken et al. 2017; Littman et al. 1988; Weiser et al. 1977; Brown et al. 2005) (Fig. 8.1).

Systemic hypertension is more prevalent in older, as compared to younger, cats and dogs. With age, conduit arteries undergo predictable changes that may contribute to an increase in systolic blood pressure, particularly in patients at risk for development of hypertension (Greenwald 2007). In human patients, these arteriosclerotic age-related medial changes, which include loss of elasticity and calcification (and are distinct to the intimal changes of atherosclerotic disease), lead to

Fig. 8.1 Hyperplastic renal arteriolosclerosis, characterized by concentric medial smooth muscle hyperplasia and basement membrane duplication, in a cat with chronic kidney disease and systemic arterial hypertension. Stains: periodic acid-Schiff and haematoxylin. Image provided by Professor Cathy A Brown, University of Georgia

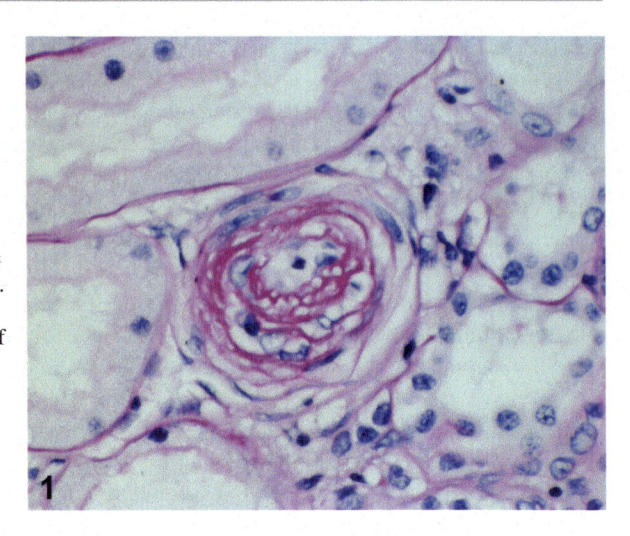

measurable increases in vascular stiffness. During hypertensive states that are characterized by an increase in systemic vascular resistance, ageing-related changes are accelerated and may contribute to further increases in cardiac afterload. Alterations in the elastic properties of the aorta may become a major driving force for systolic blood pressure increases in aged human hypertensive patients. Indeed, it is believed that while an increase in systemic vascular resistance may initiate hypertension, it is the resultant acceleration of vascular ageing and increase in arterial stiffness that propels the progressive increase in systolic blood pressure noted in hypertensive patients with age (Bakris and Sorrentino 2018). As in people, degenerative arteriosclerosis of conduit arteries is a common postmortem finding in older dogs. In two separate necropsy-based studies, aortic arteriosclerotic lesions, characterized by loss of elastic tissue and luminal narrowing, most commonly of the abdominal segment, were noted in 45 of 58 (78%) (Whitney 1976; Valtonen and Oksanen 1972) and 79 of 200 (40%) (Valtonen and Oksanen 1972) randomly selected dogs. Further, age-related decreases in systemic arterial compliance have been documented in aged beagle dogs (mean age, 9.2 years), when compared to younger controls (mean age, 2.3 years), despite similar systemic arterial pressures (Haidet et al. 1996). Age-related changes to the conduit arteries may therefore contribute to a "vicious cycle" in hypertensive animals, similar to that suspected in aged human hypertensives.

In human beings, the degree of age-related increase in the radiographic length of the ascending aorta (i.e. "aortic elongation" or "aortic unfolding"), which has been shown to nearly double between the ages of 20 and 80 years and is attributable to loss of elastic tissue, correlates positively with measures of arterial stiffness, systolic blood pressure and pulse pressure (Sugawara et al. 2008). Aortic redundancy (described by some authors as "undulation", or "tortuosity") is a well-recognized radiographic finding in cats, which may be related to the presence of systemic hypertension, but which appears to be relatively common in older animals

8 Hypertension and the Heart and Vasculature

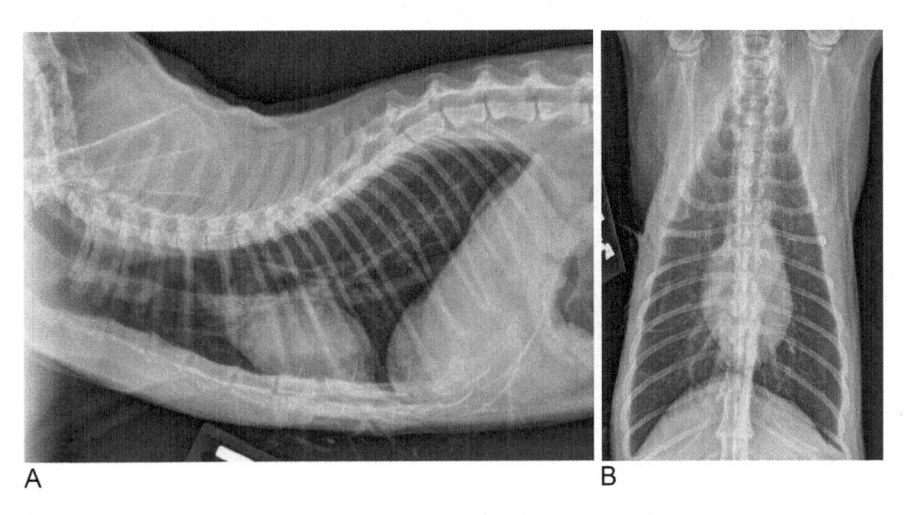

Fig. 8.2 Left lateral (**a**) and ventrodorsal (**b**) thoracic radiographs from a euthyroid, 16-year-old, spayed female domestic shorthair cat with an average indirect systolic arterial blood pressure of 150 mmHg, illustrating the redundant descending thoracic aorta and increased sternal contact ("exaggerated horizontal alignment") of the cardiac silhouette noted commonly in cats of advanced age

(Fig. 8.2). An "undulating, tortuous" thoracic aorta, prominent aortic arch, or both, were present in 7 of 14 hypertensive cats in one study (Littman 1994). Another investigation, which evaluated radiographic findings in a small number of hypertensive cats, documented a greater, though not statistically significant, aortic undulation score in hypertensive, as compared to normotensive, geriatric cats (Nelson et al. 2002). The investigators found a significant relationship between the finding of an undulating aorta and SBP ($p = 0.0097$), but not age. This lack of association with increasing age seemed to contradict an earlier report that suggested otherwise, in which 28% of cats aged 10 years or older had radiographic evidence of a tortuous, redundant aorta, compared to 0% of cats aged 7 years or younger (Moon et al. 1993); however, of note, only cats aged >8 years were included in the study by Nelson and colleagues (Nelson et al. 2002). Although blood pressure was not measured in cats included in the earlier study by Moon et al. (1993), systemic hypertension was considered by the investigators to be unlikely given the absence of left ventricular hypertrophy on contemporaneous echocardiographic examinations. Clearly, the relationship of this radiographic finding to measures of arterial stiffness remains to be determined in cats.

Aortic aneurysm with dissection, reported in several cats (Scollan and Sisson 2014; Wey and Atkins 2000; Gouni et al. 2018) and dogs (Waldrop et al. 2003; Nicolle et al. 2005) with systemic hypertension, is a rare, potentially life-threatening complication of systemic hypertension that may lead to left-sided congestive heart failure or visceral ischaemia due to compression of the lumen of the aorta or its major branches or may result in vessel rupture and sudden death (Patel and Arora 2008). Aortic dissection is thought to be due to a tear in the intimal layer of the vessel, which allows blood from the vessel lumen to move into the plane of a diseased medial layer, creating a false lumen (Patel and Arora 2008). Pre-existing abnormalities of the medial layer are believed to be requisite in atraumatic cases. A recent study of cats with ($n = 11$) and without ($n = 8$) chronic multisystemic arteriolar changes suggestive of systemic hypertension showed that arteriopathy of the vasa vasorum (i.e. the capillary network supplying the medial and adventitial layers of great vessels), which may predispose to aortic aneurysm and dissection in people, is strongly correlated to hypertension status (Kohnken et al. 2017). Further, severe lesions of the vasa vasorum were significantly and strongly correlated to the presence of aortic wall degeneration, suggesting that lesions of this capillary bed, which seem to occur as a response to systemic hypertension, may result in loss of vascular integrity and increase the risk for aortic dissection in hypertensive cats (Kohnken et al. 2017).

8.3 Mechanisms of Cardiovascular Target Organ Damage in Systemic Hypertension

Changes to the heart as a consequence of chronic systemic hypertension may be viewed as a culmination of responses both to hemodynamic stress imposed by an increased cardiac afterload and to the direct tissue effects of several specific mediators (i.e. neurohormones, growth factors, cytokines) (Drazner 2011; Glennon et al. 1995). Additional influences, including age, gender, genetics, dietary salt intake and body composition, factor into the development and magnitude of increases in left ventricular mass in hypertensive people (Bakris and Sorrentino 2018), and it seems reasonable to suspect that one or more of these might be important in hypertensive cats and dogs, as well. The relative contributions of factors leading to ventricular hypertrophy are difficult to dissect. In addition to their direct effects, neurohumoral signals may contribute indirectly to left ventricular hypertrophy via their own load-altering hemodynamic changes or by interactions with other neurohumoral systems.

Increased Cardiac Afterload The hemodynamic effects of increased cardiac afterload are a major factor in the development of left ventricular hypertrophy (Giannattasio and Laurent 2014; Cooper et al. 1985). Increased and sustained mechanical loading of the ventricular myocardium is translated into complex growth signals that result in increased cardiomyocyte size, hypertrophy and hyperplasia of coronary vascular cells and fibroblast stimulation, which lead to distortion of the ventricular microarchitecture (Opie and Gersh 2009). The physical strain imposed on cardiomyocytes during excessive cardiac afterload is detected by biomechanical sensors, which activate several stretch-induced signalling pathways (Sadoshima et al. 1992; Sadoshima and Izumo 1993). Among the most important of these appear to be the calcium-calcineurin pathway, which links stretch-induced increases in intracellular calcium to nuclear transcription and protein synthesis (Maier and Bers 2002; Molkentin et al. 1998), and the intracardiac renin-angiotensin system, although a great deal of pathway redundancy exists (Sadoshima and Izumo 1993). Through its direct effects and by stimulation of local production of angiotensin II, myocardial stretch also signals the release of the growth factor, endothelin, from vascular endothelial cells and stimulates fibroblasts to increase extracellular matrix collagen deposition, leading to interstitial fibrosis.

As is true in human hypertension (Feola et al. 1998; Levy et al. 1988; Lauer et al. 1991; Devereux et al. 1997), echocardiographic studies of hypertensive cats and dogs have documented weak or no correlation between SBP and degree of echocardiographic left ventricular hypertrophy (Chetboul et al. 2003; Lesser et al. 1992; Snyder et al. 2001; Misbach et al. 2011), as well as dissociation between the rate of blood pressure increase and rate of left ventricular hypertrophy development (Morioka and Simon 1981), suggesting that non-hemodynamic (i.e. non-load-altering) factors likely contribute to its development and severity. However, complicating interpretation of these data are limitations inherent to the measurement of blood pressure in veterinary patients, the high prevalence of situational (i.e. "white-coat") increases in blood pressure, and the likelihood that single, in-hospital measurements poorly represent time-averaged blood pressure. Indeed, the degree of left ventricular hypertrophy in human hypertensive patients is more closely correlated to 24-h blood pressure variables derived from ambulatory blood pressure monitoring (Feola et al. 1998) than to in-hospital measurements (Levy et al. 1988). In addition, the results of a recent metanalysis of 12 studies, including a total of 6341 human patients, suggested that left ventricular mass is more strongly associated with central (i.e. aortic), as opposed to peripheral (i.e. brachial), systolic blood pressure measurements, which do not always correspond (Kollias et al. 2016). Even so, experimental canine studies in which identical, known increases in afterload are imposed via controlled, gradual aortic banding reveal heterogeneity in ventricular hypertrophic response patterns among individuals, with responses closely linked to differences in baseline (i.e. pre-loading) left ventricular mass and wall stress (Koide et al. 1997).

Influences of Neurohumoral and Growth Factors Within 30 min of exposure to a direct, non-biomechanical hypertrophic stimulus (e.g. stimulation of alpha-adrenoreceptors, angiotensin II receptors or endothelin receptors), detectable alterations in gene expression have been documented in isolated rat myocardial cells, with induction of early-response genes that favour cell growth (Glennon et al. 1995). Within 12 h, there is demonstrable up-regulation of genes associated with foetal programming, including those encoding contractile and non-contractile proteins, and within 24 h, up-regulation of constitutively expressed genes encoding sarcomeric proteins has been documented (Glennon et al. 1995). The sum total of these effects are general increases in cell protein and RNA content and an increase in cell size. Numerous intracellular signalling pathways mediate the cellular effects of these and other factors. A discussion of these pathways is beyond the scope of this chapter but has been reviewed by others (Frey et al. 2004; Glennon et al. 1995).

Of the known myocardial hypertrophic factors, those belonging to the renin-angiotensin-aldosterone system have been described in the greatest detail.

Angiotensin II Angiotensin II, the major effector of the renin-angiotensin-aldosterone system, acts predominantly via interaction with the angiotensin II subtype 1 (AT_1) receptor to effect both direct and indirect actions on the heart and vasculature. The direct, afterload-increasing effects of arteriolar vasoconstriction, for which activation of arteriolar AT_1 receptors is one stimulus, are discussed above. Recent work, using genetic knockout mice infused with angiotensin II, illustrates the relative importance of these hemodynamic effects in the development of angiotensin II-induced left ventricular hypertrophy (Crowley et al. 2006). Predictably, angiotensin II infusion of wild-type mice leads to an increase in blood pressure, as well as the development of cardiac hypertrophy and fibrosis, while angiotensin II infusion of AT_1 knockout mice (i.e. those with complete absence of the AT_1 receptor) does not produce similar changes (Fig. 8.3). However, in the face of angiotensin II administration, wild-type mice receiving renal transplants from AT_1 receptor knockout donors (i.e. "kidney knock-outs") do not develop statistically significant increases in blood pressure or cardiac hypertrophy and fibrosis. Conversely, if renal AT_1 receptors are provided to AT_1 receptor knockout mice via renal transplantation (i.e. "systemic knock-outs"), angiotensin II infusion is again associated with significant blood pressure increases and cardiac hypertrophy and fibrosis, though these develop more slowly than in wild-type mice. These experiments illustrate the importance of afterload alterations in the development of cardiac target organ damage caused by angiotensin II and the critical role of the kidney in that process.

Additional data support the role of non-hemodynamic, hypertrophic, pro-fibrotic effects of cardiac AT_1 receptor stimulation in hypertension-induced hypertrophy, although these are largely based on in vitro studies using supraphysiologic doses of angiotensin II and in vivo studies of human hypertensives treated with blockers of the renin-angiotensin-aldosterone system. For example, angiotensin II (along with endothelin and transforming growth factor-β1) stimulates the transformation of fibroblasts to myofibroblasts (Berk et al. 2007), which are integral to myocardial collagen deposition. Angiotensin II modulates the activity of the metalloproteinases and their inhibitors, interfering with collagen degradation and "tipping the balance" towards collagen accumulation (Woessner 1991). Further, results from clinical trials suggest that treatment with blockers of the renin-angiotensin-aldosterone system (i.e. angiotensin-converting enzyme inhibitors and angiotensin II receptor blockers) results in a degree of cardiac hypertrophy regression that is independent of blood pressure reduction and exceeds that noted with other classes of antihypertensive agents, when adjusted for degree of blood pressure reduction (Kjeldsen et al. 2002; Mathew et al. 2001; Klingbeil et al. 2003).

Stimulation of AT_1 receptors by angiotensin II causes growth and hypertrophy of smooth muscle cells (Geisterfer et al. 1988), which may contribute to arteriolar remodelling in chronic systemic hypertension. However, when the AT_1 receptor was deleted from smooth muscle cells in one study, severity of hypertension-induced vascular remodelling was not affected, despite a reduction in local vascular oxidative stress (Sparks et al. 2011). This suggested that intraluminal pressure may be the more important mediator of vascular changes during sustained hypertension.

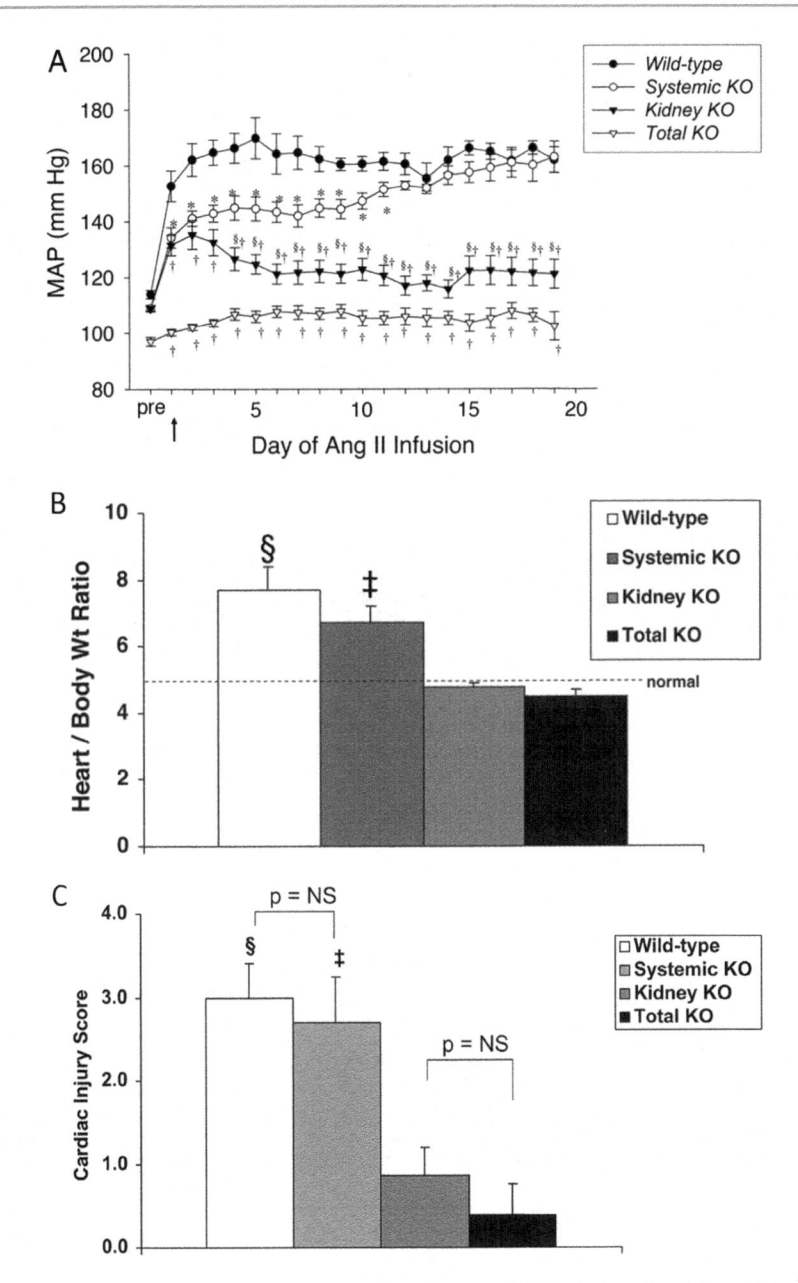

Fig. 8.3 Mean arterial blood pressure (**a**), cardiac hypertrophy (**b**) and cardiac injury (**c**) in mice undergoing kidney cross-transplantation and angiotensin II infusion. In all panels, "systemic KO" refers to mice lacking AT_1 receptors in all tissues *except* the kidney; "kidney KO" denotes mice lacking AT_1 receptors in the kidney only, and "total KO" refers to mice lacking AT_1 receptors in all tissues. (**a**) Daily, 24-h mean arterial blood pressure measurements before ("pre") and during 21 days of angiotensin II infusion. Asterisk, significantly different (SD) vs. wild-type; section sign, SD vs. systemic KO; dagger, SD vs. wild-type. (**b**) Mean heart-to-body weight ratios after

Aldosterone In dogs, aldosterone infusion at a dosage producing systemic hypertension is associated with the development of concentric left ventricular hypertrophy and echocardiographic evidence of diastolic dysfunction (Asemu et al. 2013; Stanley et al. 2013). In studies of people with essential hypertension, treatment with a mineralocorticoid receptor antagonist lowers blood pressure and is associated with regression of left ventricular hypertrophy (Pitt et al. 2003; Sato et al. 2001), suggesting an important role of this hormone in cardiac target organ damage when activation or inappropriate suppression of the renin-angiotensin-aldosterone system is a contributing factor.

Hyperaldosteronism, or the failure of adequate aldosterone suppression in response to sodium retention or in the face of high dietary sodium intake, is important to the pathogenesis of HT in some human patients, particularly those with systemic hypertension that is resistant to standard therapy. In one study, up to 20% of such patients were affected by hyperaldosteronism (Calhoun 2013). Hypertensive patients with primary hyperaldosteronism appear to be at greater risk for target organ damage of the heart and kidneys, when compared to patients with other "types" of systemic hypertension (Milliez et al. 2005), and severity of left ventricular hypertrophy appears to be worse in these patients when compared to age-, BP- and HT duration-matched patients with hypertension of other etiologies (Tanabe et al. 1997).

Data from spontaneously hypertensive cats with chronic kidney disease have suggested a potential role for relative or absolute hyperaldosteronism in the pathogenesis of feline hypertension (Jensen et al. 1997; Jepson et al. 2014), (see also Chap. 1) but associations between aldosterone levels and cardiac target organ damage have not been explored in this species.

8.4 Clinical Implications of Cardiac Target Organ Damage in Hypertensive Veterinary Patients

Increases in echocardiographic indices of left ventricular mass, which take into account both concentric and eccentric patterns of hypertrophy, are an independent risk factor for cardiovascular and all-cause mortality in hypertensive human patients (Levy et al. 1990; Vakili et al. 2001). Some studies further suggest clinical relevance of the pattern of left ventricular hypertrophy, with concentric patterns conferring

Fig. 8.3 (continued) 28 days of Ang II infusion. Mean heart-to-body weight ratio in non-infused wild-type mice, established from previous experiments by the same group of investigators, is represented by the dashed line. Section sign, SD vs. kidney KO and total KO; double dagger, SD vs. kidney KO and total KO. (**c**) Semiquantitative cardiac pathology scores, evaluating severity of cellular infiltrates, myocardial cell injury, vessel wall thickening and fibrosis. Section sign, SD vs. kidney KO and total KO; double dagger, SD vs. kidney KO and total KO). From: Crowley SD, Gurley SB, Herrera MJ, et al. Angiotensin II causes hypertension and cardiac hypertrophy through its receptors in the kidney. Proc Natl Acad Sci U S A 2006;103:17985–17990. Reproduced with permission

greater risk of cardiovascular complications as compared to eccentric and normal patterns, regardless of the presence or absence of coronary artery disease (Verdecchia et al. 1995; Muiesan et al. 2004; Ghali et al. 1998).

To date, studies of spontaneously hypertensive cats and dogs have not been able to detect or quantify a similar increase in risk for patients with left ventricular hypertrophy. A cardiovascular cause of death (or complication at the time of euthanasia) was assigned in 4 of 19 (21%) hypertensive cats dying in one study (Elliott et al. 2001); however, no significant difference in mean survival time was noted between spontaneously hypertensive cats ($n = 39$) with and without echocardiographic evidence of left ventricular hypertrophy in a separate retrospective study with a follow-up period of 12 months and event rate of 47% (Chetboul et al. 2003). Of note, all cats included in the latter study (Chetboul et al. 2003) were treated with oral amlodipine after echocardiographic evaluation. Another study failed to find an increased risk of hypertensive retinopathy in cats with echocardiographic evidence of left ventricular hypertrophy (Sansom et al. 2004). However, all of these studies involve relatively small number of cats and so are likely underpowered to detect these associations.

Prior to the development of effective oral antihypertensive therapies, the most common cause of death in human patients with untreated systemic hypertension was progressive heart failure, experienced by approximately 33% (median SBP, 215–220 mmHg) (Janeway 1913). In contrast, congestive heart failure occurs uncommonly in cats (Littman 1994; Wey and Atkins 2000; Elliott et al. 2001), and only rarely in dogs (Nicolle et al. 2005), with isolated systemic hypertension. More frequently, hypertensive cardiomyopathy (i.e. left ventricular hypertrophy) remains clinically silent unless superimposed on unrelated, underlying cardiac disease or unless a substantial hemodynamic stressor (e.g. administration of large volumes or rapid rates of intravenous fluids, blood transfusion, severe anaemia) challenges the cardiovascular system. In a recent, large, retrospective cohort study that explored survival after hypertension diagnosis in cats, treatment for "cardiac disease" was initiated in only 14 of 282 (5%) cases, although further details regarding disease specifics were not available due to the retrospective nature of the study, which involved accessing case records of primary care practices (Conroy et al. 2018). Species differences in natural history may perhaps be attributable to the relatively short lifespan of dogs and cats, differences in underlying diseases and comorbidities, or a rather low prevalence of coronary artery disease or similar risk factors for heart failure, as compared to human beings. Indeed, data from hypertensive people suggests that the progression from concentric left ventricular hypertrophy to a dilated, failing phenotype is uncommon unless the former is complicated by myocardial infarction (Krishnamoorthy et al. 2011; Rame et al. 2004). Because myocardial infarction is recognized only rarely in cats and dogs, this may explain, in part, interspecies differences in disease progression and heart failure development.

Although fairly uncommon in veterinary patients, clinically relevant complications of hypertensive cardiovascular disease may be life-threatening. Collapse and sudden death, presumed to be cardiovascular in origin, have been reported as the cause of death in hypertensive cats (Littman 1994; Elliott et al. 2001), as has congestive heart failure. As discussed above, aortic aneurysm with dissection is a rare, but potentially dangerous, complication of systemic arterial hypertension (Scollan and Sisson 2014; Wey and Atkins 2000; Waldrop et al. 2003; Nicolle et al. 2005). Superimposition of excessive cardiac afterload on an already diseased heart may increase the risk for serious complications, such as congestive heart failure or cardiac arrhythmia. Finally, the hypertensive heart may be ill-equipped to handle hemodynamic stressors, such as intravascular volume expansion, which would be otherwise accommodated.

8.5 Clinical Cardiovascular Findings in Hypertensive Veterinary Patients

In veterinary patients, the cardiovascular manifestations of systemic hypertension are often identified in apparently asymptomatic dogs and cats for which routine physical examination reveals auscultable abnormalities or in which cardiac evaluation is prompted by the diagnosis of systemic hypertension. Less commonly, hypertensive patients may be presented with clinical signs indicative of cardiovascular complications. Although rare, respiratory signs attributable to cardiogenic pulmonary oedema (or, in cats, pleural effusion) may be present. Epistaxis, which occurs uncommonly in dogs and cats with systemic hypertension, may also be noted as the cardinal presenting complaint (Littman 1994; Maggio et al. 2000; Bissett et al. 2007). In one study of hypertensive dogs, syncope was noted in 2 (6%) of 30 (Misbach et al. 2011).

Several diagnostic modalities, each with differences in sensitivity, specificity and availability, allow for the assessment of cardiac target organ damage in patients with systemic hypertension. Because one of the major goals of antihypertensive therapy is to avoid or slow the progression of target organ damage, cardiac evaluation may be advisable if patient, client, and availability factors allow. In addition, regression of cardiac changes may represent one method by which the clinician can evaluate efficacy of antihypertensive therapy.

Physical Examination Clinically detectable cardiac physical examination abnormalities, including a systolic heart murmur, gallop sound, arrhythmia, dyspnoea or a combination of these, are present in 69–80% of feline (Chetboul et al. 2003; Snyder et al. 2001; Elliott et al. 2001) and 67% of canine (Misbach et al. 2011) hypertensive patients. In one study of cats with chronic kidney disease, the prevalence of cardiovascular physical examination abnormalities (i.e. murmur, arrhythmia or gallop sound) was 65% in those with systemic hypertension, as compared to 35% in those without (Syme et al. 2002). In two studies of cats with systemic hypertension due to various underlying diseases, in which cardiovascular physical examination findings were detailed, a heart murmur and gallop sound were detected in 40–62% and 16–27% of cases, respectively (Chetboul et al. 2003; Elliott et al. 2001). Systolic heart murmurs were reported in 3 of 5 (60%) and 20 of 30 (67%) dogs in two separate studies (Littman et al. 1988; Misbach et al. 2011).

Systolic heart murmurs in hypertensive cats may be caused by mitral valve insufficiency or dynamic right ventricular outflow tract obstruction (Rishniw and Thomas 2002; Sampedrano et al. 2006), with an intensity that may be exacerbated by concurrent anaemia. A diastolic murmur due to aortic insufficiency is rarely auscultated in hypertensive cats. In hypertensive dogs, systolic heart murmurs are typically associated with the presence of mitral valve regurgitation (Misbach et al. 2011), while a diastolic heart murmur may be auscultated over the left heart base if substantial aortic insufficiency is present (Cortadellas et al. 2006). Gallop sounds in hypertensive cats are most commonly caused by an audible fourth heart sound, which is associated with decreased left ventricular compliance and coincident with atrial contraction. Additional auscultable abnormalities may include accentuation of the first and second heart sounds (Keene et al. 2015).

Auscultation of a heart murmur in a hypertensive cat or dog does not always signal the presence of hypertension-associated left ventricular hypertrophy (Snyder et al. 2001). Systemic hypertension tends to affect older animals, in which unrelated, structural heart disease (e.g. myxomatous valvular disease) is also relatively common. Further, physiologic heart murmurs are common in cats, particularly at higher heart rates. In contrast, the finding of a gallop sound in a cat with systemic hypertension may be useful; in one study, all hypertensive cats with this physical examination finding had echocardiographic evidence of left ventricular hypertrophy (Chetboul et al. 2003).

Mean heart rate in hypertensive cats is not significantly different from that of normal controls (Snyder et al. 2001). A diagnosis of tachycardia was made in 5 of 30 (17%) hypertensive cats in one study (Elliott et al. 2001) although heart rate cutoffs used to define tachycardia were not described by the investigators. In the same study, arrhythmia leading to pulse deficits was present in 4 of 30 (13%) cases (Elliott et al. 2001).

Echocardiography As expected, the most common abnormality noted on standard (i.e. two-dimensional and M-mode) echocardiographic evaluation in cats and dogs with systemic hypertension is left ventricular hypertrophy. As discussed above, in spontaneously hypertensive cats, the pattern of hypertrophy is most often concentric (Fig. 8.4) but may also be eccentric in nature. When concentric, hypertrophy may be diffuse (i.e. affecting both the interventricular septum and the left ventricular free wall) or segmental. Left atrial enlargement, consistent with elevated left ventricular filling pressures, may also be identified. Table 8.1 provides a summary of selected veterinary studies reporting conventional echocardiographic findings in cats and dogs with naturally occurring systemic hypertension. There are limited published data regarding echocardiographic findings in spontaneously hypertensive dogs, with the majority presented as case reports or series describing hypertension of various endocrine diseases. In one study that described conventional echocardiographic and tissue Doppler imaging abnormalities in 30 hypertensive dogs, 47% (14/30) had diffuse concentric left ventricular hypertrophy, which was symmetric in most cases (10/14; 72%) and asymmetric with predominance of septal thickening in the remainder (4/14; 28%) (Misbach et al. 2011).

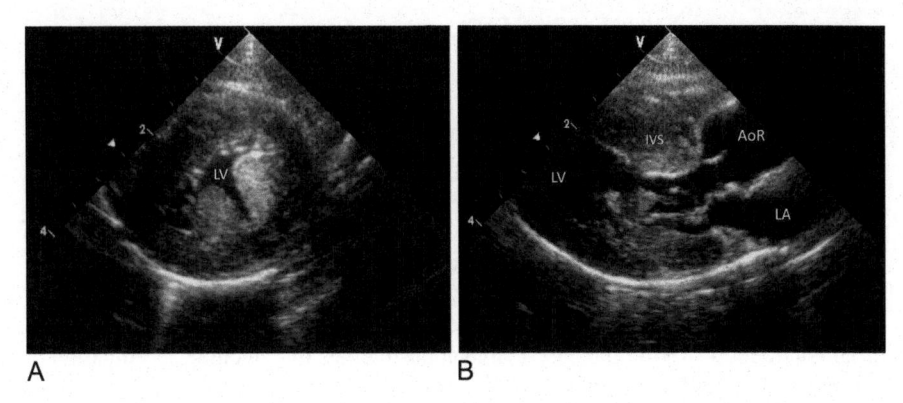

Fig. 8.4 Two-dimensional short- (**a**) and long- (**b**) axis echocardiographic images taken from a hypertensive (average indirect systolic blood pressure, 220 mmHg) 12-year-old neutered male domestic long-haired cat, presented for acute blindness due to hypertensive retinopathy. Each image was taken at end-diastole from a right parasternal window. There is generalized concentric hypertrophy of the left ventricular wall and dilatation of the aortic root. *LV* left ventricle, *LA* left atrium, *AoR* aortic root, *IVS* interventricular septum. Images courtesy of Gregg Rapoport, University of Georgia College of Veterinary Medicine, Athens, GA, USA

The incidence of subclinical left ventricular hypertrophy or other cardiac abnormalities in veterinary hypertensive patients is not well defined, as large, prospective, longitudinal echocardiographic screening studies, which compare findings to those from age-matched normotensive controls, have not been reported. In relatively small cross-sectional studies, echocardiographic evidence of hypertensive cardiomyopathy was present in 48–85% of feline (Chetboul et al. 2003; Henik et al. 2004; Littman 1994; Snyder et al. 2001) and 47–74% of canine (Misbach et al. 2011; Cortadellas et al. 2006) patients with systemic hypertension, although differences in the definitions used to diagnose systemic hypertension and to designate echocardiographic abnormalities complicate interpretation of these estimates. As a minimum, these studies suggest that abnormalities of left ventricular geometry are common in hypertensive dogs and cats. However, given that echocardiographic abnormalities may be absent in some hypertensive dogs and cats, and that when present, the type and degree are variable, echocardiography should not be considered an adequate screening tool for detection of hypertension.

The relatively high prevalence of idiopathic, primary hypertrophic cardiomyopathy (HCM) in the general cat population, estimated at 14.7–15.5% of apparently healthy cats (Paige et al. 2009; Payne et al. 2015), deserves mention when considering the ventricular hypertrophy of hypertensive patients. Because HCM results in structural and functional changes that are generally indistinguishable from those of hypertensive heart disease using echocardiography (Sampedrano et al. 2006), one or more of the cats included in the above-cited studies (i.e. those characterizing left ventricular geometry in hypertension) may have been concurrently affected by these conditions. Few veterinary studies have been designed to evaluate hypertrophy

Table 8.1 Summary of veterinary studies reporting conventional (i.e. two-dimensional and M-mode) echocardiographic findings in cats and dogs with naturally occurring systemic hypertension

References	Species	n	Age (y); mean ± SD (range)	SBP (mmHg); mean ± SD (range)	Concurrent diseases	SBP (mmHg) used for diagnosis of HT	End-diastolic wall thickness used for diagnosis of concentric LVH	Prevalence of LVH	Prevalence of various LVH patterns	IVSd (range; mm)	PWd (range; mm)	Prevalence of LA enlargement	Prevalence of systolic dysfunction
Sampedrano et al. (2006)	Cat	17	13.8 ± 3.8	237 ± 38 (180–300)	–	>160	> 6 mm	100% (cats without LVH excluded)	–	4.1–8.9	3.4–7.8	–	–
Henik et al. (2004)	Cat	75	13.8 ± 3.6 (5–19)	205 ± 29.1 (170–300)	8% (6/75) hyperthyroid; 68% (44/65) azotaemic	≥170	≥ 5.5 mm	48% (36/75)	Concentric only; IVS +PW in 32% (24/75); IVS only in 9% (7/75); PW only in 7% (5/75)	3.0–9.0	3.0–10.0	28% (19/67)	8% (6/75)
Chetboul et al. (2003)	Cat	39	12.7 ± 3.5	239 ± 38	8% (3/39) hyperthyroid; 62% (32/52) azotaemic	>170	> 6 mm	85% (33/39)	Concentric in 59% (23/39); eccentric in 13% (5/39); basilar IVS bulge only in 13% (5/39)	3.9–14.0	4.0–10.9	–	–
Nelson et al. (2002)	Cat	15	12.4 ± 3.2	217.7 ± 25.6	20% (3/15) hyperthyroid; 87% (13/15) "renal disease"	>180	–	–	–	–	–	–	–
Snyder et al. (2001)	Cat	19	13.6 ± 4.1 (6–18)	216 ± 25 (175–275)	0% hyperthyroid; 26% (5/19) azotaemic	>170	> 6 mm	74% (14/19)	Concentric only: IVS + PW in 16% (3/19); IVS	4.2–7.4	3.8–9.4	–	–

(continued)

Table 8.1 (continued)

References	Species	n	Age (y); mean ± SD (range)	SBP (mmHg); mean ± SD (range)	Concurrent diseases	SBP (mmHg) used for diagnosis of HT	End-diastolic wall thickness used for diagnosis of concentric LVH	Prevalence of LVH	Prevalence of various LVH patterns	IVSd (range; mm)	PWd (range; mm)	Prevalence of LA enlargement	Prevalence of systolic dysfunction
Littman (1994)	Cat	12	–	–	–	>160	–	83% (10/12)	only in 5% (1/19); PW only in 53% (10/19)	–	–	–	–
Misbach et al. (2011)	Dog	30	9.6 ± 3.4 (0.9–13.0)	198 ± 25 (167–260)	63% (19/30) CKD; 16% (5/30) adrenal gland disorders	>160	> Maximum expected by allometric scaling method	47% (14/30)	Concentric only: IVS +PW in 47% (14/30)	−24.4 to 45.9% increase from expected maximum values	−44.8 to 37.7% increase from expected maximum values	0%	–
Cortadellas et al. (2006)	Dog	43	–	150–179 (n = 18); ≥180 (n = 25)	100% Leishmania-positive	≥ 180 or ≥150 with TOD	> 15% above upper limit of weight-based reference range	74% (32/43)	Concentric only: IVS +PW in 35% (15/43); PW only in 40% (17/43)	–	–	–	–
Littman et al. (1988)	Dog	5	8.6 ± 1.8 (2–12)	218 ± 18.3	80% (4/5) "renal disease"	–	–	4/5 (80%)	–	–	–	–	–

– not reported or not applicable, SBP systolic arterial blood pressure, HT hypertension, LV left ventricle, LVH left ventricular hypertrophy, IVS_d end-diastolic interventricular septum wall thickness, PW_d end-diastolic left ventricular posterior wall thickness, $LVID_d$ end-diastolic left ventricular internal diameter, LA left atrium, CKD chronic kidney disease

regression with control of systemic hypertension, which might shed some light on this issue. In one that reported serial echocardiographic examinations in amlodipine-treated hypertensive cats ($n = 14$), significantly fewer cats had left ventricular hypertrophy after at least 3 months of therapy, at which time 93% of cases were considered normotensive (i.e. SBP < 170); however, left ventricular hypertrophy was still present in 6 of 14 (43%) (Snyder et al. 2001). It is possible that one or more of these cats may have been affected by HCM, which would not be expected to respond to treatment with amlodipine, although there are other important potential explanations for failure to regress, as discussed in greater detail below.

Echocardiographic indicators of diastolic dysfunction may precede overt left ventricular hypertrophy, as has been demonstrated in studies of human (Dreslinski et al. 1981), experimental canine (Douglas and Tallant 1991) and naturally occurring canine (Misbach et al. 2011) and feline (Sampedrano et al. 2006) hypertension. In addition, advanced ultrasonographic techniques have allowed for identification of myocardial functional abnormalities that are otherwise undetected using conventional echocardiographic modalities. Tissue Doppler imaging, which provides a sensitive and non-invasive method for evaluation of myocardial motion, has been used in spontaneously hypertensive cats and dogs to document diastolic and systolic function abnormalities in both hypertrophied and non-hypertrophied myocardial segments (Misbach et al. 2011; Sampedrano et al. 2006). These studies suggest that cardiac target organ damage may be more prevalent than previously thought. However, whether these changes are clinically relevant, or signal an increased risk for the development of clinically relevant complications, remains unexplored in dogs and cats.

In the absence of other causes (e.g. left ventricular outflow tract stenosis with post-stenotic alterations), findings of aortic root dilatation (Fig. 8.4) and aortic valve insufficiency suggest the possibility of systemic arterial hypertension. In one study, echocardiographic measurements of the proximal ascending aorta (i.e. distal to the sinuses of Valsalva) were significantly greater in cats with systemic arterial hypertension as compared to those from normotensive geriatric cats (Nelson et al. 2002). These measurements were significantly ($p = 0.0001$) and positively correlated to systolic BP, but not age, and the authors advocated indexing the proximal ascending aorta diameter to aortic annular diameter as a method of differentiating hypertensive cats from normal geriatric cats (Nelson et al. 2002). Aortic insufficiency is noted infrequently in hypertensive cats but seems to be much more common in hypertensive dogs. In one study, mild or moderate aortic valve regurgitation, associated with dilatation of the proximal aorta and an apparently normal aortic valve in all cases, was documented in 33% of 30 dogs (Misbach et al. 2011).

Electrocardiography Electrocardiographic signs of left ventricular hypertrophy, including increased R wave amplitude, increased QRS complex duration and leftward mean electrical axis deviation, may be present in dogs and cats with systemic hypertension (Tilley 1992). However, the sensitivity and specificity of electrocardiography for the detection of left ventricular hypertrophy is poor. Six-lead electrocardiographic criteria of left ventricular hypertrophy (i.e. increased R wave amplitude, leftward mean electrical axis deviation in the frontal plane or both) were fulfilled in only 4 of 32 (12.5%) hypertensive dogs with echocardiographic evidence of left ventricular hypertrophy in one study (Cortadellas et al. 2006). A recent investigation of cats with left ventricular hypertrophy due to systemic hypertension with ($n = 7$) or without ($n = 2$) hyperthyroidism, hyperthyroidism alone ($n = 3$) or idiopathic hypertrophic cardiomyopathy ($n = 23$) found that among the 12 electrocardiographic variables evaluated, only duration of the QT interval and heart rate corrected QT interval were somewhat useful predictors of left ventricular hypertrophy (Romito et al. 2018). Using ROC analysis, the optimal cutoff for QT interval duration (170 ms) was shown to have a sensitivity and specificity of 48% and 91%, respectively, for this purpose. The case is only marginally better for human patients, in whom 12-lead electrocardiography has a sensitivity of 50–60% for the detection of left ventricular hypertrophy, depending on the criteria applied (Pewsner et al. 2007).

The prevalence of specific arrhythmias in veterinary hypertensive patients has been reported only sporadically. Atrial or ventricular premature complexes were identified in 3 of 13 (23%) hypertensive cats for which an electrocardiogram was performed in one study (Littman 1994), but the indication for performing this test was not described by the authors. In another study of hypertensive cats ($n = 15$), ventricular premature complexes were documented in 5 (33%), whereas ectopy was noted in no healthy geriatric control cat ($n = 15$) (Nelson et al. 2002). In a study describing electrocardiographic findings in cats with left ventricular hypertrophy due to systemic hypertension, hyperthyroidism or idiopathic hypertrophic cardiomyopathy, the presence of an arrhythmia of any kind on a 2-min recording had a sensitivity and specificity of 29% and 100%, respectively, for the identification of concentric left ventricular hypertrophy (Romito et al. 2018). Importantly, the comparator group was echocardiographically and clinically normal cats; no cats with systemic illness or other cardiac diseases were included.

Thoracic Radiography Radiographic evidence of cardiomegaly was noted in 9 of 14 (64%) cats with systemic hypertension in one early study (Littman 1994). As mentioned above, an "undulating, tortuous" thoracic aorta, prominent aortic arch or both were present in 7 of 14 hypertensive cats in one study (Littman 1994). Another investigation, which evaluated radiographic findings in a small number of hypertensive cats, documented a greater, though not statistically significant, aortic undulation score in hypertensive, as compared to normotensive, geriatric cats (Nelson et al. 2002).

Compared to echocardiography, radiography is insensitive for the detection of cardiomegaly, particularly when mild or moderate in severity. Additionally, this modality does not allow for assessment of left ventricular function or wall thickness. For these reasons, thoracic radiography is not reliable as a screening tool for cardiac target organ damage.

8.6 Expected Response to Treatment of Hypertensive Cardiac Target Organ Damage

As is the rationale in almost all persistently hypertensive patients, antihypertensive treatment is indicated to reverse or limit ongoing cardiac target organ damage and to prevent future damage. The road from hypertension to hypertrophy is not one-way, and regression is expected in many, though not all, cases following normalization of blood pressure (Drazner 2011). In people, regression of left ventricular hypertrophy and fibrosis in response to antihypertensive therapy is associated with improved ventricular diastolic function (Wachtell et al. 2002; Diez et al. 2002) and cardiovascular outcomes (Okin et al. 2004; Devereux et al. 2004). Interestingly, these effects are independent of blood pressure reduction (Wachtell et al. 2002; Okin et al. 2004; Devereux et al. 2004). When adjusted for treatment duration and degree of blood pressure reduction, regression of left ventricular hypertrophy appears to be most pronounced in response to antagonism of the renin-angiotensin-aldosterone system and with calcium channel blockade, as compared to diuretics and beta-adrenergic blockers (Klingbeil et al. 2003).

There is need for prospective, longitudinal studies of treated hypertensive cats and dogs that focus specifically on regression of cardiac target organ damage and its prognostic implications. One study that evaluated hypertensive cats ($n = 14$) before and after oral amlodipine treatment found that significantly fewer had left ventricular hypertrophy after at least 3 months of therapy (Snyder et al. 2001). However, this echocardiographic finding was still present in 6 of 14 (43%) at follow-up. Although 93% of cases were considered normotensive at the time of recheck echocardiography, the systolic blood pressure cutoff used to define normotension (i.e. <170 mmHg) was greater than has been advocated as an ideal therapeutic target more recently (Acierno et al. 2018). Further, as has been suggested in human patients, 24-h blood pressure monitoring may more accurately assess blood pressure control than single or multiple in-hospital measurements. Therefore, lack of adequate, sustained blood pressure control may have contributed to the lack of regression in all cases in the cited study. As mentioned above, it is also possible that one or more of these cats may have been affected by HCM, which would not be expected to respond to treatment with amlodipine.

8.7 Conclusions

Through mechanisms that include the hemodynamic effects of increased cardiac afterload, direct cellular influences and activation of and interactions with various neurohumoral systems, systemic hypertension is able to affect phenotypic changes to the heart that most commonly include concentric left ventricular hypertrophy and which result in altered diastolic and systolic myocardial function. Although typically subclinical, the cardiovascular alterations noted in response to chronic, pathological increases in systemic arterial blood pressure may increase the affected animal's risk for life-threatening complications, particularly in patients with pre-existing heart disease or in those faced with additional hemodynamic stressors. Among other questions requiring answers, additional work is needed to define risk factors (beyond degree of blood pressure increase) for left ventricular mass increases in hypertensive dogs and cats, the optimal anti-hypertensive approach for reducing the risk of cardiovascular complications and the prognostic value of cardiac hypertrophy regression.

References

Acierno MJ, Brown S, Coleman AE et al (2018) ACVIM consensus statement: guidelines for the identification, evaluation, and management of systemic hypertension in dogs and cats. J Vet Intern Med 32:1803–1822

Alyono D, Anderson RW, Parrish DG et al (1986) Alterations of myocardial blood flow associated with experimental canine left ventricular hypertrophy secondary to valvular aortic stenosis. Circ Res 58:47–57

Anderson LJ (1968) Arterial disease in canine interstitial nephritis. J Pathol Bacteriol 95:47–53

Anderson LJ, Fisher EW (1968) The blood pressure in canine interstitial nephritis. Res Vet Sci 9:304–313

Asemu G, O'Connell KA, Cox JW et al (2013) Enhanced resistance to permeability transition in interfibrillar cardiac mitochondria in dogs: effects of aging and long-term aldosterone infusion. Am J Physiol Heart Circ Physiol 304:H514–H528

Ayoub AM, Keddeas VW, Ali YA et al (2016) Subclinical LV dysfunction detection using speckle tracking echocardiography in hypertensive patients with preserved LV ejection fraction. Clin Med Insights Cardiol 10:85–90

Bai X, Wang Q (2010) Time constants of cardiac function and their calculations. Open Cardiovasc Med J 4:168–172

Bakris G, Sorrentino M (2018) Hypertension: a companion to Braunwald's heart disease. Elsevier Saunders, Philadelphia

Ben-David J, Zipes DP, Ayers GM et al (1992) Canine left ventricular hypertrophy predisposes to ventricular tachycardia induction by phase 2 early after depolarizations after administration of BAY K 8644. J Am Coll Cardiol 20:1576–1584

Berk BC, Fujiwara K, Lehoux S (2007) ECM remodeling in hypertensive heart disease. J Clin Invest 117:568–575

Bissett SA, Drobatz KJ, McKnight A et al (2007) Prevalence, clinical features, and causes of epistaxis in dogs: 176 cases (1996–2001). J Am Vet Med Assoc 231:1843–1850

Bright R (1836) Tabular view of the morbid appearances occurring in one hundred cases in connection with albuminous urine. Guys Hosp Rep 1:380–400

Brown CA, Munday JS, Mathur S et al (2005) Hypertensive encephalopathy in cats with reduced renal function. Vet Pathol 42:642–649

Calhoun DA (2013) Hyperaldosteronism as a common cause of resistant hypertension. Annu Rev Med 64:233–247

Chahal NS, Lim TK, Jain P et al (2010) New insights into the relationship of left ventricular geometry and left ventricular mass with cardiac function: a population study of hypertensive subjects. Eur Heart J 31:588–594

Cheng CP, Igarashi Y, Little WC (1992) Mechanism of augmented rate of left ventricular filling during exercise. Circ Res 70:9–19

Chetboul V, Lefebvre HP, Pinhas C et al (2003) Spontaneous feline hypertension: clinical and echocardiographic abnormalities, and survival rate. J Vet Intern Med 17:89–95

Conroy M, Chang YM, Brodbelt D et al (2018) Survival after diagnosis of hypertension in cats attending primary care practice in the United Kingdom. J Vet Intern Med 32:1846–1855

Cooper G, Kent RL, Uboh CE et al (1985) Hemodynamic versus adrenergic control of cat right ventricular hypertrophy. J Clin Invest 75:1403–1414

Cortadellas O, del Palacio MJ, Bayon A et al (2006) Systemic hypertension in dogs with leishmaniasis: prevalence and clinical consequences. J Vet Intern Med 20:941–947

Crowley SD, Gurley SB, Herrera MJ et al (2006) Angiotensin II causes hypertension and cardiac hypertrophy through its receptors in the kidney. Proc Natl Acad Sci U S A 103:17985–17990

Dahlof B, Devereux RB, Kjeldsen SE et al (2002) Cardiovascular morbidity and mortality in the Losartan intervention for endpoint reduction in hypertension study (LIFE): a randomised trial against atenolol. Lancet 359:995–1003

Devereux RB, Roman MJ, de Simone G et al (1997) Relations of left ventricular mass to demographic and hemodynamic variables in American Indians: the Strong Heart Study. Circulation 96:1416–1423

Devereux RB, Wachtell K, Gerdts E et al (2004) Prognostic significance of left ventricular mass change during treatment of hypertension. JAMA 292:2350–2356

Diez J, Querejeta R, Lopez B et al (2002) Losartan-dependent regression of myocardial fibrosis is associated with reduction of left ventricular chamber stiffness in hypertensive patients. Circulation 105:2512–2517

Dorn GW 2nd, Robbins J, Sugden PH (2003) Phenotyping hypertrophy: eschew obfuscation. Circ Res 92:1171–1175

Douglas PS, Tallant B (1991) Hypertrophy, fibrosis and diastolic dysfunction in early canine experimental hypertension. J Am Coll Cardiol 17:530–536

Drazner MH (2011) The progression of hypertensive heart disease. Circulation 123:327–334

Dreslinski GR, Frohlich ED, Dunn FG et al (1981) Echocardiographic diastolic ventricular abnormality in hypertensive heart disease: atrial emptying index. Am J Cardiol 47:1087–1090

du Cailar G, Pasquie JL, Ribstein J et al (2000) Left ventricular adaptation to hypertension and plasma renin activity. J Hum Hypertens 14:181–188

Elliott J, Barber PJ, Syme HM et al (2001) Feline hypertension: clinical findings and response to antihypertensive treatment in 30 cases. J Small Anim Pract 42:122–129

Esposito G, Rapacciuolo A, Naga Prasad SV et al (2002) Genetic alterations that inhibit in vivo pressure-overload hypertrophy prevent cardiac dysfunction despite increased wall stress. Circulation 105:85–92

Feola M, Boffano GM, Procopio M et al (1998) Ambulatory 24-hour blood pressure monitoring: correlation between blood pressure variability and left ventricular hypertrophy in untreated hypertensive patients. G Ital Cardiol 28:38–44

Frey N, Katus HA, Olson EN et al (2004) Hypertrophy of the heart: a new therapeutic target? Circulation 109:1580–1589

Frohlich ED, Apstein C, Chobanian AV et al (1992) Medical progress—the heart in hypertension. N Engl J Med 327:998–1008

Ganau A, Devereux RB, Roman MJ et al (1992) Patterns of left ventricular hypertrophy and geometric remodeling in essential hypertension. J Am Coll Cardiol 19:1550–1558

Geisterfer AA, Peach MJ, Owens GK (1988) Angiotensin II induces hypertrophy, not hyperplasia, of cultured rat aortic smooth muscle cells. Circ Res 62:749–756

Ghali JK, Liao Y, Cooper RS (1998) Influence of left ventricular geometric patterns on prognosis in patients with or without coronary artery disease. J Am Coll Cardiol 31:1635–1640

Giannattasio C, Laurent S (2014) Central blood pressure. Manual of hypertension of the European society of hypertension, 2nd edn. CRC Press, London, pp 257–268

Gilbert JC, Glantz SA (1989) Determinants of left ventricular filling and of the diastolic pressure-volume relation. Circ Res 64:827–852

Glennon PE, Sugden PH, Poole-Wilson PA (1995) Cellular mechanisms of cardiac hypertrophy. Br Heart J 73:496–499

Gouni V, Papageorgiou S, Debeaupuits J et al (2018) Aortic dissecting aneurysm associated with systemic arterial hypertension in a cat. Schweiz Arch Tierheilkd 160:320–324

Greenwald SE (2007) Ageing of the conduit arteries. J Pathol 211:157–172

Grossman W, Jones D, McLaurin LP (1975) Wall stress and patterns of hypertrophy in the human left ventricle. J Clin Invest 56:56–64

Gull WW, Sutton HG (1872) On the pathology of the morbid state commonly called chronic Bright's disease with contracted kidney ("arterio-capillary fibrosis."). Med Chir Trans 55:273–330.1

Haidet GC, Wennberg PW, Finkelstein SM et al (1996) Effects of aging per se on arterial stiffness: systemic and regional compliance in beagles. Am Heart J 132:319–327

Hayashida W, Donckier J, Van Mechelen H et al (1997) Diastolic properties in canine hypertensive left ventricular hypertrophy: effects of angiotensin converting enzyme inhibition and angiotensin II type-1 receptor blockade. Cardiovasc Res 33:54–62

Henik RA, Stepien RL, Bortnowski HB (2004) Spectrum of M-mode echocardiographic abnormalities in 75 cats with systemic hypertension. J Am Anim Hosp Assoc 40:359–363

Hill JA, Karimi M, Kutschke W et al (2000) Cardiac hypertrophy is not a required compensatory response to short-term pressure overload. Circulation 101:2863–2869

Janeway TIC (1913) A clinical study of hypertensive cardiovascular disease. Arch Intern Med 12:755–798

Jensen J, Henik RA, Brownfield M et al (1997) Plasma renin activity and angiotensin I and aldosterone concentrations in cats with hypertension associated with chronic renal disease. Am J Vet Res 58:535–540

Jepson RE, Syme HM, Elliott J (2014) Plasma renin activity and aldosterone concentrations in hypertensive cats with and without azotemia and in response to treatment with amlodipine besylate. J Vet Intern Med 28:144–153

Jeremy RW, Fletcher PJ, Thompson J (1989) Coronary pressure-flow relations in hypertensive left ventricular hypertrophy. Comparison of intact autoregulation with physiological and pharmacological vasodilation in the dog. Circ Res 65:224–236

Keene BW, Smith FWK, Tilley LP et al (2015) Rapid interpretation of heart and lung sounds: a guide to cardiac and respiratory auscultation in dogs and cats, 3rd edn. Elsevier Saunders, St. Louis, MO

Kizer JR, Arnett DK, Bella JN et al (2004) Differences in left ventricular structure between black and white hypertensive adults: the hypertension genetic epidemiology network study. Hypertension 43:1182–1188

Kjeldsen SE, Dahlof B, Devereux RB et al (2002) Effects of losartan on cardiovascular morbidity and mortality in patients with isolated systolic hypertension and left ventricular hypertrophy: a Losartan Intervention for Endpoint Reduction (LIFE) substudy. JAMA 288:1491–1498

Kleiman RB, Houser SR (1988) Calcium currents in normal and hypertrophied isolated feline ventricular myocytes. Am J Phys 255:H1434–H1442

Kleiman RB, Houser SR (1989) Outward currents in normal and hypertrophied feline ventricular myocytes. Am J Phys 256:H1450–H1461

Klingbeil AU, Schneider M, Martus P et al (2003) A meta-analysis of the effects of treatment on left ventricular mass in essential hypertension. Am J Med 115:41–46

Kohnken R, Scansen BA, Premanandan C (2017) Vasa Vasorum Arteriopathy: relationship with systemic arterial hypertension and other vascular lesions in cats. Vet Pathol 54:475–483

Koide M, Nagatsu M, Zile MR et al (1997) Premorbid determinants of left ventricular dysfunction in a novel model of gradually induced pressure overload in the adult canine. Circulation 95:1601–1610

Kollias A, Lagou S, Zeniodi ME et al (2016) Association of central versus brachial blood pressure with target-organ damage: systematic review and meta-analysis. Hypertension 67:183–190

Krishnamoorthy A, Brown T, Ayers CR et al (2011) Progression from normal to reduced left ventricular ejection fraction in patients with concentric left ventricular hypertrophy after long-term follow-up. Am J Cardiol 108:997–1001

Krumholz HM, Larson M, Levy D (1993) Sex differences in cardiac adaptation to isolated systolic hypertension. Am J Cardiol 72:310–313

Lauer MS, Anderson KM, Levy D (1991) Influence of contemporary versus 30-year blood pressure levels on left ventricular mass and geometry: the Framingham Heart Study. J Am Coll Cardiol 18:1287–1294

Lazzeroni D, Rimoldi O, Camici PG (2016) From left ventricular hypertrophy to dysfunction and failure. Circ J 80:555–564

Lesser M, Fox PR, Bond BR (1992) Assessment of hypertension in 40 cats with left-ventricular hypertrophy by Doppler-Shift sphygmomanometry. J Small Anim Pract 33:55–58

Levy D, Anderson KM, Savage DD et al (1988) Echocardiographically detected left ventricular hypertrophy: prevalence and risk factors. The Framingham Heart Study. Ann Intern Med 108:7–13

Levy D, Garrison RJ, Savage DD et al (1990) Prognostic implications of echocardiographically determined left ventricular mass in the Framingham Heart Study. N Engl J Med 322:1561–1566

Lewington S, Clarke R, Qizilbash N et al (2002) Age-specific relevance of usual blood pressure to vascular mortality: a meta-analysis of individual data for one million adults in 61 prospective studies. Lancet 360:1903–1913

Little WC (1992) Enhanced load dependence of relaxation in heart failure. Clinical implications. Circulation 85:2326–2328

Littman MP (1994) Spontaneous systemic hypertension in 24 cats. J Vet Intern Med 8:79–86

Littman MP, Robertson JL, Bovee KC (1988) Spontaneous systemic hypertension in dogs: five cases (1981–1983). J Am Vet Med Assoc 193:486–494

Maggio F, DeFrancesco TC, Atkins CE et al (2000) Ocular lesions associated with systemic hypertension in cats: 69 cases (1985–1998). J Am Vet Med Assoc 217:695–702

Maier LS, Bers DM (2002) Calcium, calmodulin, and calcium-calmodulin kinase II: heartbeat to heartbeat and beyond. J Mol Cell Cardiol 34:919–939

Mathew J, Sleight P, Lonn E et al (2001) Reduction of cardiovascular risk by regression of electrocardiographic markers of left ventricular hypertrophy by the angiotensin-converting enzyme inhibitor ramipril. Circulation 104:1615–1621

McLenachan JM, Dargie HJ (1990) Ventricular arrhythmias in hypertensive left ventricular hypertrophy. Relationship to coronary artery disease, left ventricular dysfunction, and myocardial fibrosis. Am J Hypertens 3:735–740

Meerson FZ (1961) On the mechanism of compensatory hyperfunction and insufficiency of the heart. Cor Vasa 3:161–177

Milliez P, Girerd X, Plouin PF et al (2005) Evidence for an increased rate of cardiovascular events in patients with primary aldosteronism. J Am Coll Cardiol 45:1243–1248

Misbach C, Gouni V, Tissier R et al (2011) Echocardiographic and tissue Doppler imaging alterations associated with spontaneous canine systemic hypertension. J Vet Intern Med 25:1025–1035

Molkentin JD, Lu JR, Antos CL et al (1998) A calcineurin-dependent transcriptional pathway for cardiac hypertrophy. Cell 93:215–228

Moon ML, Keene BW, Lessard P et al (1993) Age-related-changes in the feline cardiac Silhouette. Vet Radiol Ultrasound 34:315–320

Morioka S, Simon G (1981) Echocardiographic evidence for early left-ventricular hypertrophy in renal hypertensive dogs. Am J Cardiol 47:478–478

Muiesan ML, Salvetti M, Monteduro C et al (2004) Left ventricular concentric geometry during treatment adversely affects cardiovascular prognosis in hypertensive patients. Hypertension 43:731–738

Munagala VK, Hart CY, Burnett JC Jr et al (2005) Ventricular structure and function in aged dogs with renal hypertension: a model of experimental diastolic heart failure. Circulation 111:1128–1135

Muscholl MW, Schunkert H, Muders F et al (1998) Neurohormonal activity and left ventricular geometry in patients with essential arterial hypertension. Am Heart J 135:58–66

Nelson L, Reidesel E, Ware WA et al (2002) Echocardiographic and radiographic changes associated with systemic hypertension in cats. J Vet Intern Med 16:418–425

Nichols WW, Pepine CJ (1982) Left ventricular afterload and aortic input impedance: implications of pulsatile blood flow. Prog Cardiovasc Dis 24:293–306

Nicolle AP, Carlos Sampedrano C, Fontaine JJ et al (2005) Longitudinal left ventricular myocardial dysfunction assessed by 2D colour tissue Doppler imaging in a dog with systemic hypertension and severe arteriosclerosis. J Vet Med A Physiol Pathol Clin Med 52:83–87

Ohsato K, Shimizu M, Sugihara N et al (1992) Histopathological factors related to diastolic function in myocardial hypertrophy. Jpn Circ J 56:325–333

Okin PM, Devereux RB, Jern S et al (2004) Regression of electrocardiographic left ventricular hypertrophy during antihypertensive treatment and the prediction of major cardiovascular events. JAMA 292:2343–2349

Opie LH, Gersh BJ (2009) Drugs for the heart, 7th edn. Saunders/Elsevier, Philadelphia

Paige CF, Abbott JA, Elvinger F et al (2009) Prevalence of cardiomyopathy in apparently healthy cats. J Am Vet Med Assoc 234:1398–1403

Patel PD, Arora RR (2008) Pathophysiology, diagnosis, and management of aortic dissection. Ther Adv Cardiovasc Dis 2:439–468

Payne JR, Brodbelt DC, Luis FV (2015) Cardiomyopathy prevalence in 780 apparently healthy cats in rehoming centres (the CatScan study). J Vet Cardiol 17(Suppl 1):S244–S257

Pewsner D, Juni P, Egger M et al (2007) Accuracy of electrocardiography in diagnosis of left ventricular hypertrophy in arterial hypertension: systematic review. BMJ 335:711

Pirie HM, Mackey JL, Fisher EW (1965) The relationship between renal disease and arterial lesions in the dog. Ann N Y Acad Sci 127:861–873

Pitt B, Reichek N, Willenbrock R et al (2003) Effects of eplerenone, enalapril, and eplerenone/enalapril in patients with essential hypertension and left ventricular hypertrophy: the 4E-left ventricular hypertrophy study. Circulation 108:1831–1838

Rame JE, Ramilo M, Spencer N et al (2004) Development of a depressed left ventricular ejection fraction in patients with left ventricular hypertrophy and a normal ejection fraction. Am J Cardiol 93:234–237

Renna NF, Heras NDL, Miatello RM (2013) Pathophysiology of vascular remodeling in hypertension. Int J Hypertens 2013:808353

Rials SJ, Wu Y, Ford N et al (1995) Effect of left ventricular hypertrophy and its regression on ventricular electrophysiology and vulnerability to inducible arrhythmia in the feline heart. Circulation 91:426–430

Rials SJ, Wu Y, Pauletto FJ et al (1996) Effect of an intravenous angiotensin-converting enzyme inhibitor on the electrophysiologic features of normal and hypertrophied feline ventricles. Am Heart J 132:989–994

Rishniw M, Thomas WP (2002) Dynamic right ventricular outflow obstruction: a new cause of systolic murmurs in cats. J Vet Intern Med 16:547–552

Romito G, Guglielmini C, Mazzarella MO et al (2018) Diagnostic and prognostic utility of surface electrocardiography in cats with left ventricular hypertrophy. J Vet Cardiol 20:364–375

Rouleau JR, Simard D, Blouin A et al (2002) Angiotensin inhibition and coronary autoregulation in a canine model of LV hypertrophy. Basic Res Cardiol 97:384–391

Sadoshima J, Izumo S (1993) Mechanical stretch rapidly activates multiple signal transduction pathways in cardiac myocytes: potential involvement of an autocrine/paracrine mechanism. EMBO J 12:1681–1692

Sadoshima J, Jahn L, Takahashi T et al (1992) Molecular characterization of the stretch-induced adaptation of cultured cardiac cells. An in vitro model of load-induced cardiac hypertrophy. J Biol Chem 267:10551–10560

Sampedrano CC, Chetboul V, Gouni V et al (2006) Systolic and diastolic myocardial dysfunction in cats with hypertrophic cardiomyopathy or systemic hypertension. J Vet Intern Med 20:1106–1115

Sano M, Schneider MD (2002) Still stressed out but doing fine: normalization of wall stress is superfluous to maintaining cardiac function in chronic pressure overload. Circulation 105:8–10

Sansom J, Rogers K, Wood JL (2004) Blood pressure assessment in healthy cats and cats with hypertensive retinopathy. Am J Vet Res 65:245–252

Sato A, Takane H, Saruta T (2001) High serum level of procollagen type III amino-terminal peptide contributes to the efficacy of spironolactone and angiotensin-converting enzyme inhibitor therapy on left ventricular hypertrophy in essential hypertensive patients. Hypertens Res 24:99–104

Schumann CL, Jaeger NR, Kramer CM (2019) Recent advances in imaging of hypertensive heart disease. Curr Hypertens Rep 21:3

Scollan K, Sisson D (2014) Multi-detector computed tomography of an aortic dissection in a cat. J Vet Cardiol 16:67–72

Shechter JA, O'Connor KM, Friehling TD et al (1989) Electrophysiologic effects of left ventricular hypertrophy in the intact cat. Am J Hypertens 2:81–85

Snyder PS, Sadek D, Jones GL (2001) Effect of amlodipine on echocardiographic variables in cats with systemic hypertension. J Vet Intern Med 15:52–56

Sparks MA, Parsons KK, Stegbauer J et al (2011) Angiotensin II type 1A receptors in vascular smooth muscle cells do not influence aortic remodeling in hypertension. Hypertension 57:577–585

Stamler J, Stamler R, Neaton JD (1993) Blood pressure, systolic and diastolic, and cardiovascular risks. US population data. Arch Intern Med 153:598–615

Stanley WC, Cox JW, Asemu G et al (2013) Evaluation of docosahexaenoic acid in a dog model of hypertension induced left ventricular hypertrophy. J Cardiovasc Transl Res 6:1000–1010

Sugawara J, Hayashi K, Yokoi T et al (2008) Age-associated elongation of the ascending aorta in adults. JACC-Cardiovasc Imaging 1:739–748

Syme HM, Barber PJ, Markwell PJ et al (2002) Prevalence of systolic hypertension in cats with chronic renal failure at initial evaluation. J Am Vet Med Assoc 220:1799–1804

Tanabe A, Naruse M, Naruse K et al (1997) Left ventricular hypertrophy is more prominent in patients with primary aldosteronism than in patients with other types of secondary hypertension. Hypertens Res 20:85–90

Ten Eick RE, Zhang K, Harvey RD et al (1993) Enhanced functional expression of transient outward current in hypertrophied feline myocytes. Cardiovasc Drugs Ther 7(Suppl 3):611–619

Tilley LP (1992) Essentials of canine and feline electrocardiography: interpretation and treatment, 3rd edn. Lea & Febiger, Philadelphia

Tomanek RJ, Searls JC, Lachenbruch PA (1982) Quantitative changes in the capillary bed during developing, peak, and stabilized cardiac hypertrophy in the spontaneously hypertensive rat. Circ Res 51:295–304

Tomanek RJ, Wessel TJ, Harrison DG (1991) Capillary growth and geometry during long-term hypertension and myocardial hypertrophy in dogs. Am J Phys 261:H1011–H1018

Vakili BA, Okin PM, Devereux RB (2001) Prognostic implications of left ventricular hypertrophy. Am Heart J 141:334–341

Valtonen MH, Oksanen A (1972) Cardiovascular disease and nephritis in dogs. J Small Anim Pract 13:687–697

Velagaleti RS, Gona P, Levy D et al (2008) Relations of biomarkers representing distinct biological pathways to left ventricular geometry. Circulation 118:2252–2258, 2255p, following 2258

Verdecchia P, Schillaci G, Borgioni C et al (1995) Adverse prognostic-significance of concentric remodeling of the left-ventricle in hypertensive patients with normal left-ventricular mass. J Am Coll Cardiol 25:871–878

Wachtell K, Bella JN, Rokkedal J et al (2002) Change in diastolic left ventricular filling after one year of antihypertensive treatment: the Losartan Intervention For Endpoint Reduction in Hypertension (LIFE) study. Circulation 105:1071–1076

Waldrop JE, Stoneham AE, Tidwell AS et al (2003) Aortic dissection associated with aortic aneurysms and posterior paresis in a dog. J Vet Intern Med 17:223–229

Wallner M, Eaton DM, Berretta RM et al (2017) A feline HFpEF model with pulmonary hypertension and compromised pulmonary function. Sci Rep 7:16587

Weiser MG, Spangler WL, Gribble DH (1977) Blood pressure measurement in the dog. J Am Vet Med Assoc 171:364–368

Wey AC, Atkins CE (2000) Aortic dissection and congestive heart failure associated with systemic hypertension in a cat. J Vet Intern Med 14:208–213

Whitney JC (1976) Some aspects of pathogenesis of canine arteriosclerosis. J Small Anim Pract 17:87–97

Woessner JF Jr (1991) Matrix metalloproteinases and their inhibitors in connective tissue remodeling. FASEB J 5:2145–2154

Hypertension and the Eye

Elaine Holt

9.1 Introduction

Chronically sustained increases in blood pressure cause injury to tissues; the rationale for the treatment of systemic hypertension (SH) in dogs and cats is the prevention of this injury (Brown et al. 2007). Frequently the eye is the first organ to manifest clinically apparent complications of SH in dogs and cats (Dubielzig et al. 2010) and, the vast majority of hypertensive cats have been reported to have either ocular or cardiac abnormalities at the time of first diagnosis (Elliott et al. 2001). Ocular disease has also been reported to be the most specific sign associated with SH in cats (Chetboul et al. 2003), and in dogs, the presence of ≥ 1 type of ocular lesion has been shown to have moderate sensitivity and specificity for identification of the disease (LeBlanc et al. 2011). Reports of the prevalence of ocular injury in dogs and cats with systemic hypertension vary, but this has been shown to be as high as 100% in cats (Littman 1994; Turner et al. 1990; Maggio et al. 2000; van de Sandt et al. 2003; Chetboul et al. 2003; Van Boxtel 2003; Littman et al. 1988; Bovee et al. 1989; Sansom and Bodey 1997; Jacob et al. 2003).

Ocular disease and its relationship to SH was originally described in humans by Lieberich in 1859, and a similar link between blood pressure and ocular pathology was later demonstrated experimentally in dogs and monkeys (Goldblatt et al. 1934; Garner et al. 1975). Hyareh first elucidated the pathophysiologic mechanisms for ocular involvement with SH and described clinical findings in human patients and animal models (Hayreh et al. 1986a, b). Much of our knowledge concerning the pathophysiology of the disease in dogs and cats is based on clinical observations as there are few experimental studies in domestic species (Turner et al. 1990; Stiles 1994; Littman 1994; Sansom et al. 1994; Crispin and Mould 2001; Maggio et al. 2000; Kobayashi et al. 1990; Morgan 1986).

E. Holt (✉)
University of Pennsylvania School of Veterinary Medicine, Philadelphia, PA, USA
e-mail: eholt@upenn.edu

Systemic hypertension fundamentally impacts on ocular blood flow. Differences in the blood supply and the anatomy of the retina, choroid and optic nerve result in distinctive ocular manifestations of the disease (Hayreh et al. 1986c; Crispin and Mould 2001). Knowledge of the mechanisms that exist to regulate blood flow to the retina and choroid is integral to understanding the pathophysiology of ocular disease associated with SH. A schematic diagram of the anatomy and vascular beds of the ocular fundus is presented in Fig. 9.1, together with normal histology of the feline eye in Fig. 9.2.

9.2 Ocular Blood Flow

The internal maxillary artery is the main blood supply to the eye and orbit in domestic species, and this gives rise to the external ophthalmic artery. This is in contrast to primates, where the entire microcirculation of the globe is supplied by the internal ophthalmic artery. An anastomosis of the external and internal ophthalmic arteries in the dog gives rise to both long and short posterior ciliary arteries which supply the vascular beds of the retina and choroid. The internal ophthalmic artery also provides the blood supply to the optic nerve (Wong and Macri 1964; Glenwood and Edward 2013) (Fig. 9.3).

The inner layers of the neurosensory retina receive their nutrition from the retinal vascular system. Blood vessels are excluded from the outer retina, which includes the photoreceptors, where their presence could compromise vision. Consequently, the high-energy needs of the rods and cones are met by diffusion of glucose and oxygen from the adjacent vascular choroid (Alder et al. 1990; Braun et al. 1995; Linsenmeier

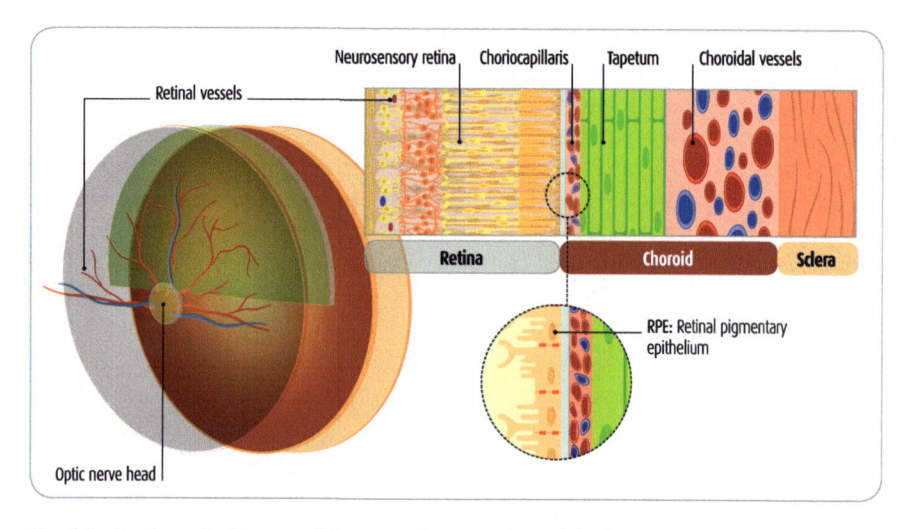

Fig. 9.1 A schematic diagram of the general organization of the fundus

9 Hypertension and the Eye

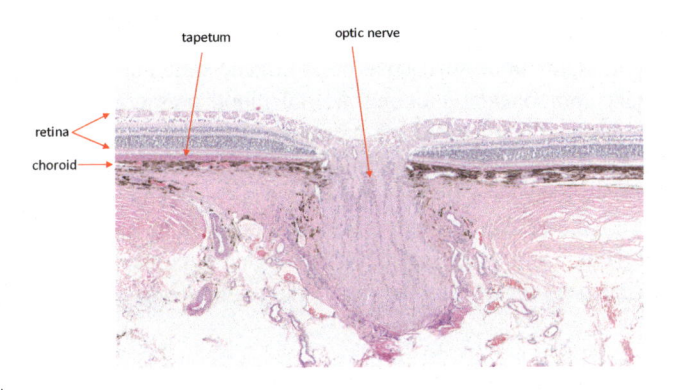

Fig. 9.2 Normal feline fundus—H&E section (Reproduced with permission from EJ Scurrell)

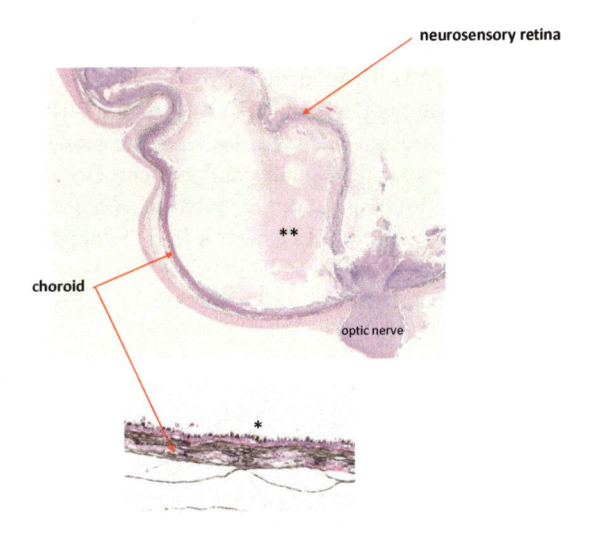

Fig. 9.3 Cat with pathological retinal detachment, hypertrophy of the RPE (asterisk) and subretinal effusion (double asterisk) +/− mild subretinal haemorrhage (Reproduced with permission from EJ Scurrell)

1986). The choroidal capillaries, collectively known as the choriocapillaris, are highly fenestrated, which allows leakage of plasma proteins and fluid into the choroidal interstitial space. In the cat, the throughput of blood from the choriocapillaris is 40–50 times that of the retinal circulation, which exposes the outer retina to near-arterial levels of oxygen (Bill et al. 1980; Alm and Bill 1972). Only 20% of the oxygen consumed by the retina is delivered through the retinal circulation.

9.2.1 Autoregulation

Autoregulation is a well-documented feature of the retinal vascular bed in a number of species, including the cat (Alm and Bill 1972). It ensures a relatively stable level of blood flow to the eye in the presence of changes in ocular perfusion pressure (mean blood pressure minus intraocular pressure) and varied metabolic demand. The most important component regulating arterial blood pressure to the retina is the change in the luminal diameter of the precapillary arterioles (Hayreh et al. 1986b). As blood pressure increases, an intrinsic myogenic and contractile response to stretch of smooth muscle in the blood vessel wall results in vascular constriction (Jepson 2011). Autoregulation is not generally thought to be a feature of the choroid vascular bed (Nickla and Wallman 2010), although a possible myogenic or vasomotor response has been reported in the rabbit and pigeon (Kiel and Shepherd 1992).

9.2.2 Blood-Retinal Barrier

A blood-retinal barrier (BRB) also protects the eye from the systemic circulation and is essential for maintaining retinal homeostasis. The BRB comprises tight junctions both between the non-fenestrated endothelial cells of the retinal capillaries (inner BRB) and between the retinal pigment epithelial cells (outer BRB). The inner BRB mediates highly selective diffusion of molecules from the blood to the retina; the outer BRB regulates the movement of solutes and nutrients from the choroid to the sub-retinal space. The blood-retinal barrier is most permeable at the optic nerve head, where substances from the choroid can pass into the optic nerve (Glenwood and Edward 2013). Failure of the BRB results in leakage of fluids and proteins from vessels and subsequent retinal, choroidal and optic nerve damage.

9.3 Pathophysiology

The effect of SH on the eye has been extensively investigated in non-human primates through experimental studies (Keyes 1937; Garner et al. 1975; Hayreh et al. 1986b; Maggio et al. 2000), and experimental hypertension-induced retinopathy has also been described in the dog (Keyes 1937). Although specific knowledge regarding the pathophysiology of SH in domestic species is lacking, there appears to be a good comparison to non-human primates.

With SH, sustained retinal arteriolar constriction results in occlusion and ischaemic necrosis of blood vessels, subsequent obliteration or focal dilation of retinal blood vessels. Increased vascular permeability causes failure of autoregulation and decompensation of the BRB, which results in leakage of fluids and proteins from vessels and subsequent tissue damage. Fibrinoid necrosis of the choroidal arteries leads to choroidal ischaemia, which allows fluid to accumulate between the neurosensory retina and the retinal pigment epithelium (RPE) (Hayreh et al. 1986b), causing localized separation of the neurosensory retina from the RPE. Vascular occlusion follows ischaemia and, ultimately, recanalization of the choroidal vessels. Ischaemia also causes destruction and disruption of the RPE, which is a common feature in hypertensive cats (Crispin and Mould 2001). With separation of the neurosensory retina from the underlying RPE, the photoreceptors are detached from their blood supply, the choriocapillaris (Fig. 9.4). Consequently, the

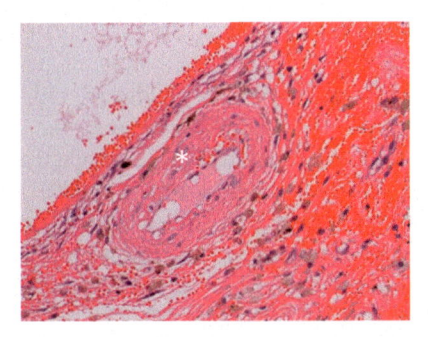

Fig. 9.4 Feline iris. H&E ×200. Hyperplastic arteriolosclerosis with accumulation of erythrocytes and fibrin in the wall of affected blood vessel indicating endothelial cell damage (asterisk). Thickening of vessel wall has resulted in extreme narrowing of the blood vessel lumen (Reproduced with permission from EJ Scurrell)

metabolism of the retina is lowered, and atrophy of the photoreceptors eventually occurs. The hypoxic retina secretes VEGF, which promotes neo-vascular proliferation in the iris, ciliary body and optic nerve head. Pre-iridal fibrovascular membranes are associated with the development of glaucoma and distortion of the iris profile (Dubielzig et al. 2010). Rarely, in advanced or severe cases of SH, ischaemia leads to axonal oedema (papilledema), and, chronically, optic nerve atrophy can develop.

9.4 Pathology

In a clinical study involving cats with ocular disease and SH, the ocular histopathology was reported in four cats following euthanasia for a variety of reasons. Lesions were bilateral, with the least severe changes reported as microscopic serous retinal detachments adjacent to the optic disc and retinoschisis.

Arteriolosclerosis is the primary lesion seen on histopathology in eyes with chronic, sustained high blood pressure. Both hyaline and hyperplastic forms of arteriolosclerosis occur, and both result in arteriole luminal narrowing. Lesions are seen in the arterioles of the retina, choroid and, less frequently, the iris. With hyaline degeneration, PAS-positive plasma proteins enter the damaged endothelial cell blood vessel lining, which, with time, appears as an amorphous, brightly eosinophilic, thickened wall (Fig. 9.4). Although more commonly associated with acute and/or severe SH in humans, the hyperplastic form of arteriolosclerosis is also documented in the eyes of dogs and cats. Myocyte-like cell proliferation in the damaged blood vessel wall creates an 'onion-skin' effect. Haemorrhage and fibrin deposition are also distinctive features of the diseased retina in domestic species, with focal retinal destruction and RPE disruption; sub-retinal fluid and focal, multifocal or total bullous retinal detachments are also common findings (Dubielzig et al. 2010; Maxie 2015). The most severe changes were widespread sub-retinal and intra-retinal haemorrhage, marked degeneration of inner retinal layers and retinal arteriolar thickening with fibrinoid necrosis; choroidal vascular changes were less marked (Sansom et al. 1994).

9.5 Ocular Manifestations of Systemic Hypertension

Hypertensive retinopathy, hypertensive choroidopathy and hypertensive optic neuropathy first described by Hayreh et al. (1986b) are the three recognized manifestations of SH in humans and reflect the fundatmental differences in the ocular vascular beds. Hypertensive retinopathy is most common with a reported prevalence of 2–14% in non-diabetic adults over age 40 years (Wong et al. 2007). A variety of grading systems have been proposed in people in an attempt to classify the severity of ocular lesions, although no consensus or universally accepted system currently exists (Keith et al. 1974; Wong and Mitchell 2004). There is evidence that more gradual or insidious changes in blood pressure primarily impact the retinal vascular system, while acute and severe increases in blood pressure predominantly affect the choroid (de Venecia and Jampol 1984; Villagrasa and Cascales 2000).

The funduscopic findings seen with SH have been described in the cat by Crispin et al. Fluorescein angiography has also been performed in dogs to identify the origin of fundus changes and to detect the earliest lesions associated with SH (Villagrasa and Cascales 2000; Crispin and Mould 2001).

9.5.1 Hypertensive Retinopathy

- *Narrowing of retinal arterioles*—localized or generalized narrowing, attenuation or even absence of arterioles results in variations in the calibre of arterioles and a beaded appearance. Early in the course of the disease, retinal oedema can obscure vessels causing a 'pseudo narrowing' effect. Retinal veins are often unaffected although venous nipping can be seen where an arteriole crosses the venule resulting in compression and bulging of the venule on either side (Fig. 9.5). Retinal vessels in the pigmented, non-tapetal region of the fundus may appear white because of arteriolosclerosis (Crispin and Mould 2001).
- *Retinal oedema*—the result of transudate from terminal retinal arterioles or from sub-retinal oedema that diffuses into adjacent tissues; oedema can be generalized or localized (Fig. 9.6). Chronic oedema may result in cystoid degeneration and sub-retinal fibrosis (Maggio et al. 2000).
- *Retinal haemorrhages*—varied in appearance from discrete, dot-like intraretinal haemorrhages to large pre-retinal haemorrhages; some haemorrhages result from ruptured aneurysms. "Spoke-like" peripapillary haemorrhages at the periphery of the optic nerve head may also be present (Fig. 9.7).
- *Retinal vascular tortuosity*—the result of bullous or flat retinal detachments, as well as a consequence of arteriolosclerosis (Maggio et al. 2000; Crispin and Mould 2001).

9 Hypertension and the Eye

Fig. 9.5 Congestion of retinal vasculature with retinal detachment in a 13 year old MN dog with systemic hypertension. The asterisk indicates arteriovenous nicking—an arteriole crossing a venule resulting in compression and bulging of the venule on either side (Courtesy of J Sansom, Animal Health Trust)

Fig. 9.6 Multifocal retinal oedema (asterisk) with areas of detachment throughout the retina of a hypertensive cat (Courtesy of J Mould)

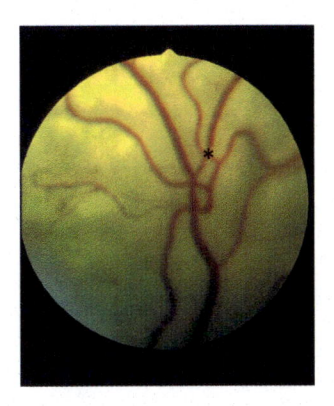

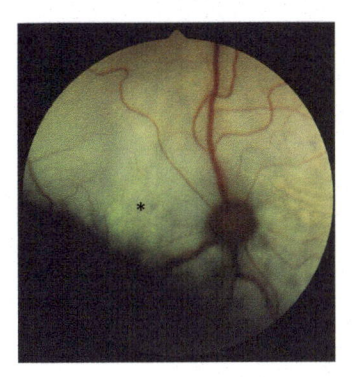

9.5.2 Hypertensive Choroidopathy

- *Sub-retinal oedema*—Villagrasa and Cascales identified choroidal ischaemia as the earliest angiographic indicator of SH in dogs in their study and peripapillary sub-retinal oedema as the earliest funduscopic lesion (Villagrasa and Cascales 2000).
- *Retinal detachment*—the separation of the neurosensory retina from the RPE because of fluid or cellular debris (Alder et al. 1990) (Fig. 9.8). Bullous retinal separation may be the earliest observable funduscopic change preceding retinal detachment in cats (Sansom et al. 1994). Retinal detachment can be partial or complete, and sub-retinal fluid serous or with variable amounts of haemorrhage (Maggio et al. 2000, Sansom et al. 2004. With chronic retinal separation, irreversible dialysis or disinsertion of the retina can result (Fig. 9.9) (Crispin and Mould 2001; Elliott et al. 2001; Komáromy et al. 2004).

9.5.3 Hypertensive Optic Neuropathy

- *Optic disc oedema*—is the primary manifestation of hypertensive optic neuropathy in people. By the time our patients are evaluated, the optic disc is often

Fig. 9.7 Punctate intra-retinal haemorrhages and peripapillary haemorrhage. There is also an inferior retinal detachment. English Springer Spaniel with adrenal tumour (Courtesy of J Mould)

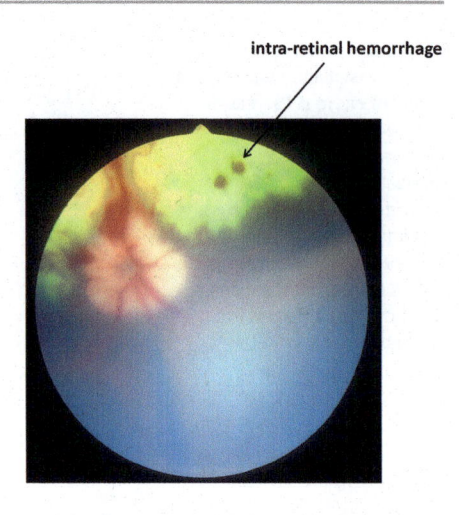

Fig. 9.8 Serous retinal detachment in a hypertensive cat. The entire superior region of the retina is separated from the RPE by serous fluid (asterisk); there is also an inferior detachment (double asterisk) (Courtesy of J Mould)

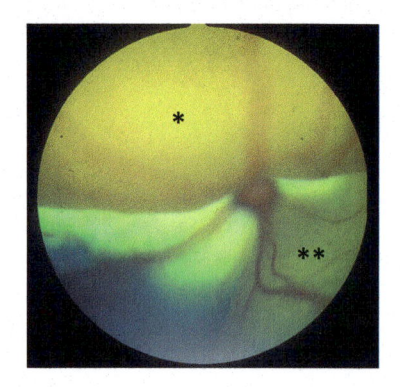

obscured by retinal oedema, retinal detachment or haemorrhage. In the cat, the recessed optic nerve head also means that papilledema is not as obvious in this species (Komáromy et al. 2004).

- *Optic nerve atrophy*—usually only seen in chronic disease along with advanced retinal degeneration.

Several investigators have proposed that the choroid may be more important than the retina in the development of hypertensive ocular disease (Sansom et al. 1994; Maggio et al. 2000; Crispin and Mould 2001; LeBlanc et al. 2011; Young et al. 2018). Consequently, hypertensive chorioretinopathy, rather than hypertensive retinopathy, may be the more appropriate clinical diagnosis in animals with systemic hypertension (Young et al. 2018).

In addition to fundus abnormalities, ocular lesions reported with SH can involve the anterior segment of the eye (cornea, anterior chamber, iris, pupil, lens) as well as the vitreous body (Locke 1992); multiple lesions may also often present in the same eye (Morgan 1986; Bodey et al. 1996). The specific lesions present will depend on the severity and duration of the high blood pressure.

9 Hypertension and the Eye

Fig. 9.9 Retinal disinsertion (retinal dialysis) in a dog with systemic hypertension. There is marked sub-retinal and choroidal haemorrhage (white asterisk). The detached retina hangs in folds over the optic nerve head; the tapetal region of the fundus appears brighter without the overlying retina (Courtesy of J Sansom, Animal Health Trust)

Fig. 9.10 Complete hyphaema OD in a hypertensive cat. The pupil is obscured by blood; only the temporal region of the iris is visible. Focal posterior synechiae (white asterisk) and hyphaema OS (Courtesy of J Mould)

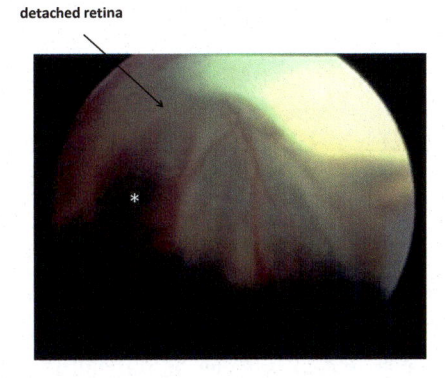

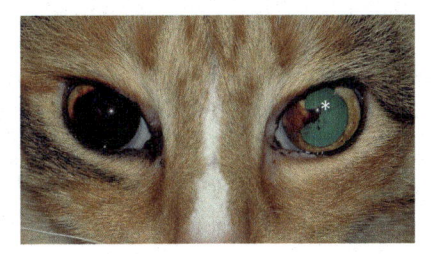

Anterior segment abnormalities reported with SH include hyphaema, anterior uveitis, cataract and corneal ulcer (Bovee et al. 1989; Maggio et al. 2000; Littman et al. 1988; Littman 1994; Wolfer et al. 1998; Ginn et al. 2013) (Fig. 9.10). Vitreous haemorrhage, in association with hyalitis (inflammation of the vitreous), may also be present (Sansom and Bodey 1997; Stiles 1994; Komáromy et al. 2004; Young et al. 2018).

9.6 The Clinical Picture

Owners seldom observe ocular abnormalities in their pet until they are very advanced, and, consequently, early fundus changes, such as intraretinal oedema and intraretinal haemorrhage, are rarely seen in clinical patients (Littman et al. 1988; Maggio et al. 2000). Small, serous retinal detachments may not cause clinically detectable impairment of vision but serious visual deficits can result from perivasculitis (linear, white sheathing around retinal vessels), sub-retinal oedema, retinal vascular tortuosity, retinal detachment, retinal degeneration and papilloedema (Sansom and Bodey 1997; Locke 1992; Wolfer et al. 1998; Komáromy et al. 2004; Ginn et al. 2013; Morgan 1986; Sansom et al. 1994; Maggio et al. 2000; Elliott et al. 2001). With large volumes of sub-retinal fluid, blindness is common. Sudden blindness, or rapidly progressing vision loss, is often the initial presenting complaint because of the late stage of the disease at which pets are presented for evaluation. Vision loss is because of hyphaema, intraocular haemorrhage, retinal detachment or glaucoma (Morgan 1986; Littman et al. 1988; Henik et al. 1997; Sansom et al. 1994; Littman 1994; Stiles 1994; Maggio et al. 2000; Elliott et al. 2001; Chetboul et al. 2003; Komáromy et al. 2004; Ginn et al. 2013; Elliott and Barber 1998).

Blindness on initial examination is reported in 9–83% of cats with SH (Elliott et al. 2001; Littman 1994; Maggio et al. 2000; Sansom et al. 1994; Chetboul et al. 2003), and the duration of blindness prior to presentation, ranging from 1 day to 2 years (Maggio et al. 2000; Littman 1994). One reason for the wide difference in numbers the numbers reported between studies that information regarding vision loss usually relies on owner observation. Even in a clinical situation, visual behaviour in cats can be difficult to assess, and is particularly challenging if the vision deficit is unilateral. Young et al. (2018) found 61% ($n = 54$) of cats with hypertensive chorioretinopathy to be blind in both eyes on presentation, and the duration of blindness less than 2 weeks in 52% of eyes and greater than 2 weeks in 23% of eyes. Blindness however was not reported in any cat in a study of hypertensive animals with either chronic kidney disease or untreated hyperthyroidism (Kobayashi et al. 1990). As cats in this study were initially evaluated for non-ocular reasons, it is possible that their blood pressure prior to joining the study was not sufficient to cause significant target organ damage (TOD). Blindness has also been reported as the initial presenting complaint in hypertensive dogs but only in isolated case reports (Bovee et al. 1989; Sansom and Bodey 1997; Wolfer et al. 1998; Ginn et al. 2013). The presenting complaint was not recorded in the only published retrospective study of hypertensive dogs with ocular lesions (LeBlanc et al. 2011).

Other ocular signs reported on initial presentation in hypertensive animals include a red eye (conjunctival hyperemia, chemosis, episcleral congestion), changes in iris colour, changes in size or responsiveness of pupils, a cloudy eye (anterior uveitis, cataract) and a painful eye (corneal ulcer, glaucoma) (Bovee et al. 1989; Maggio et al. 2000; Sansom and Bodey 1997; Littman et al. 1988; Littman 1994; Young et al. 2018; Komáromy et al. 2004; Wolfer et al. 1998; Ginn et al. 2013). When the cause of the hypertension is secondary, non-ocular signs related to the underlying disease may be the initial complaint, e.g. weight loss, recumbency, laboured breathing, diarrhoea, nasal discharge or epistaxis, polyuria, polydipsia and neurologic abnormalities have been reported (Littman 1994; Young et al. 2018).

9.6.1 Signalment

There appears to be an increased risk of hypertensive chorioretinopathy with age (LeBlanc et al. 2011; Young et al. 2018; Sansom et al. 1994). Cats ≥ 10 years of age in one study were seven times more likely to develop hypertensive retinopathy than younger cats (Sansom et al. 2004). Brown et al. also suggested it is reasonable to screen cats and dogs ≥ 10 years of age for SH since conditions that cause secondary hypertension are more prevalent in older animals (Brown et al. 2007).

A gender predilection for hypertensive ocular disease has also been reported, with female cats significantly overrepresented in one study (Young et al. 2018). Female and neutered cats were also considered to be at an increased risk for hypertensive retinopathy in a study by Sansom et al. (2004), although in an earlier study by the same author of cats with hypertensive ocular disease, no gender predisposition was found (Sansom et al. 1994). Maggio et al., found that more female than male cats (43/69) had hypertensive ocular lesions, although this study lacked a control population (Maggio et al. 2000). A gender predilection for hypertensive ocular disease has not been reported in dogs.

9.6.2 Ocular Examination

A complete ocular examination is essential in animals suspected of SH, and accurate identification of ocular lesions is important for both diagnosis and monitoring of the disease. A thorough anterior segment examination, as well as funduscopic examination, including indirect (see Chap. 12) and direct ophthalmoscopy, is recommended. Recognizing early signs of hypertensive ocular disease is key if the functional integrity of the eye is to be preserved; however, early fundus changes can be open to misinterpretation and subjective analysis. Large retinal detachments can be readily viewed directly through the pupil without the aid of an ophthalmoscope (Fig. 9.11). The segments of detached retina balloon anteriorly and appear as a greyish-membranous curtain within which blood vessels may be visible. With retinal dialysis or disinsertion, the retina is seen to hang in folds over the optic nerve head; dorsally the exposed tapetum is diffusely hyper-reflective (Fig. 9.9). Indirect ophthalmoscopy has been shown to lack sensitivity for detecting subtle retinal lesions compared to other techniques in humans (Aryan et al. 2005; Klein et al. 1984). This author feels that direct ophthalmoscopy is more useful in dogs and cats if small, focal lesions (e.g. retinal oedema or retinal haemorrhage) are to be correctly identified, although direct ophthalmoscopy has been reported to correlate poorly with blood pressure measurements in human patients with mild to moderate hypertension, with significant inter and intra-observer variability (Dimmitt et al. 1989).

Fig. 9.11 Complete, serous retinal detachment OU in a hypertensive cat. Retroillumination from the tapetum facilitates visualization of the detached retina, which balloons anteriorly and appears as a greyish-membranous curtain with blood vessels

Digital fundus photography, for the purposes of documenting and monitoring retinal lesions, has become more popular with rapid advances in technology for capturing and sharing images. Affordable alternatives to conventional fundus cameras have become increasingly available (Pirie and Pizzirani 2011), and several smartphone-assisted ocular photographic techniques have been described both in humans and in animals (Haddock et al. 2013; Balland et al. 2017). Regardless of device, fundus photography is inherently difficult in animals where patient movement and the presence of a tapetum can mean that even when utilizing a specialized fundus camera, many minor camera adjustments need to be made in order to prevent over- or underexposure and obtain a consistent image (McMullen et al. 2013). Automated image analysis offers promise to improve diagnostic accuracy of retinal imaging, and a computer-based algorithm used in human medicine has recently been investigated in cats with systemic hypertension for objective diagnosis of retinal vasculature changes (Cirla et al. 2019).

Ocular ultrasound may be necessary to evaluate retinal integrity, especially if significant intraocular haemorrhage is present or retinal detachment is suspected; Electroretinography can also be useful to assess retinal function. Fluorescein angiography, which allows detailed evaluation of retinal and choroidal vascular beds, has been recommended by some authors, although it is rarely practical in a clinical situation (Crispin and Mould 2001).

Accurately assessing visual behaviour in dogs and cats can be difficult, and it is particularly challenging if the vision deficit is unilateral. A menace response, or reflex, is a blink in response to a visual threat and is not a robust test for the presence or absence of vision. Vision assessment in clinical patients should include tests that demonstrate the animal's ability to track objects and navigate an obstacle course, although test results can be difficult to interpret in cats.

9.6.3 Prevalence of Ocular Lesions

Although information is presented here, comparing prevalence rates of ocular disease across studies can be misleading, since animals are evaluated at different stages of disease. There is also often inadequate follow-up, and the ability of some authors to assess subtle ocular lesions may be limited.

Cats Retinal detachment is commonly reported and is often bilateral (Littman 1994; Sansom et al. 1994, 2004; Maggio et al. 2000). A retrospective study of cats with hypertensive chorioretinopathy, identified 90% ($n = 158$) of eyes with retinal detachment, over half of which were described as complete, and half of affected cats had retinal detachment in both eyes (Young et al. 2018). Partial to complete retinal detachment was also found in 62% ($n = 86$) of eyes diagnosed with hypertensive retinopathy, with approximately one-third of cats examined having both retinal detachment and intraocular haemorrhage; retinal vascular tortuosity was described as common in this study (Maggio et al. 2000). Other authors report only a few cats with retinal detachment as a result of SH (Elliott et al. 2001; Chetboul

Fig. 9.12 Marked intra-retinal haemorrhage in a hypertensive cat (Courtesy of J Mould)

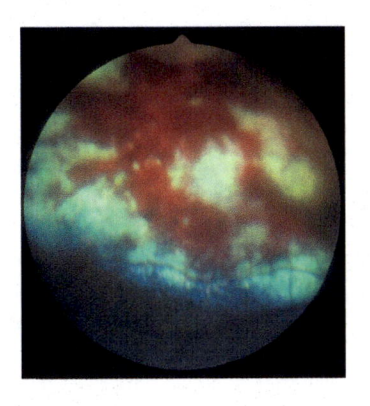

et al. 2003; Stiles 1994) with the degree and type of detachment (e.g. serous or haemorrhagic) not always recorded.

Retinal haemorrhages are also a frequent finding (Chetboul et al. 2003; Elliott et al. 2001; Stiles 1994; Morgan 1986; Littman 1994) and were present in 41% ($n = 78$) of eyes in a retrospective study of cats with hypertensive chorioretinopathy (Fig. 9.12) (Young et al. 2018). Intraocular haemorrhage, in the form of hyphaema and/or vitreous haemorrhage, has been reported in 22–28% of hypertensive cats (Elliott et al. 2001; Young et al. 2018; Littman 1994; Sansom et al. 1994), and retinal oedema and foci of intraretinal serous exudates were the most consistent findings in 66% ($n = 8$) of cats with hypertension secondary to CKD. Tapetal hyperreflectivity, a sign of retinal degeneration, has been reported in 19% ($n = 34$) (Young et al. 2018) and 9% ($n = 13$) of eyes of hypertensive cats on initial examination (Maggio et al. 2000). Although the regions of hyper-reflectivity were described as large in the study by Maggio et al., the degree of retinal degeneration is often not recorded. Interestingly, retinal vascular tortuosity does not appear to occur consistently in hypertensive cats (Stiles 1994).

Dogs The prevelence of ocular lesions in dogs with SH has been infrequently reported, with specific ocular findings mostly described in case reports or case series (Littman et al. 1988; Bovee et al. 1989; Jacob et al. 2003; Ginn et al. 2013; Sansom and Bodey 1997), LeBlanc et al. (2011) found sixty-two percent ($n = 26$) of hypertensive dogs had more than one ocular lesion presumed to be indicative of the hypertension, with over 50% having more than two lesions (LeBlanc et al. 2011). Retinal haemorrhage was the most common lesion ($n = 17$), with retinal detachment present in ten dogs. Although neither was considered significant when compared to normotensive dogs in the study, it was however, noted by the authors that some normotensive dogs with ocular lesions were being treated for SH at the time of examination, which may suggest that blood pressure was not well controlled in these animals. Retinal oedema ($n = 5$) and tortuous retinal vessels ($n = 2$) were considered specific for hypertension in the patient population, as both were found only in hypertensive dogs.

9.6.4 Differential Diagnosis for Hypertensive Ocular Lesions

Most of the ocular lesions that have been described are suggestive of, but not specific for SH and consideration of alternative causes for these lesions is important particularly when blood pressure measurements suggest only a mild or moderate risk of TOD. Violette et al., recently investigated the causes of pre-retinal haemorrhages (PRH) in dogs. In dogs with PRH, SH, diabetes mellitus and multiple meyeloma were found to be statistically over represented but other serious sytemic diseases were also identified including, immune mediated thrombocytopenia, leptospirosis, metastic neoplasia and thromboembolic diease (Violette and Ledbetter 2018).

Other possible causes of intraocular haemorrhage, retinal vascular tortuosity and retinal detachment include coagulopathies, vasculitis, uveitis and neoplastic syndromes, specifically, plasma cell myeloma and associated hyperviscosity syndrome, lymphosarcoma, polycythemia vera and systemic lupus erythematosus (Ginn et al. 2013; LeBlanc et al. 2011; Morgan 1986). Infectious diseases in cats that can result in chorioretinitis (posterior uveitis) include toxoplasmosis, feline infectious peritonitis, feline immunodeficiency virus, feline leukaemia virus, cryptococcosis, blastomycosis, histoplasmosis, candidiasis and coccidioidomycosis. Less likely causes of chorioretinitis in cats include tuberculosis, periarteritis nodosa, ethylene glycol intoxication and hyperviscosity syndrome (Komáromy et al. 2004).

9.6.5 Blood Pressure

A classification system has been developed based on the perceived relative risk of developing TOD in dogs and cats (Brown et al. 2007; further refined by Acierno et al. 2018), but the degree of SH required to produce ocular signs is not known. In experimental animal models, neither the time of onset nor the peak severity of the ocular lesions shows a correlation with measured blood pressure (Hayreh et al. 1986c). In clinical studies some cats do not appear to develop ocular disease for several months following a diagnosis of SH, while others develop ocular disease simultaneously with onset (Stiles 1994). Elliott et al. reported no significant difference between the systolic blood pressure measurements of hypertensive cats presenting with retinal detachment and those hypertensive cats that did not present with acute onset blindness (Elliott et al. 2001). It may be that fluctuations in blood pressure, rather than absolute or incremental changes, are more damaging to the vessels of the eye than sustained high pressures, and the ocular vasculature may be able to adapt to some extent if the increase in blood pressure is gradual before the threshold for damage is exceeded.

A range of blood pressure measurements associated with ocular injury in cats has been published (Morgan 1986; Wolfer et al. 1998; Turner et al. 1990; Henik 1997). An increased risk of ocular disease when systolic blood pressure (SBP) exceeds 180 mmHg has been suggested (Komáromy et al. 2004; Sansom et al. 1994), and SBP has been found to be significantly higher in hypertensive cats with retinopathies

(262 ± 34 mmHg) than in hypertensive cats without ocular disease (221 ± 34 mmHg)(Chetboul et al. 2003). The protocols for measuring blood pressure in clinical studies often differ. When measured by the Doppler technique, a mean SBP of 219 mmHg was reported in 24 cats with retinal haemorrhages and/or retinal detachments (Littman 1994) and a mean SBP of 238 mmHg (range 170–300 mmHg) in 69 cats with hypertensive retinopathy (Maggio et al. 2000). Using an oscillometric monitor, SBP values ≥168 mmHg were recorded in 90% ($n = 48$) of hypertensive cats with ocular disease, and the presence of hypertensive retinopathy was found to be significantly associated with increases in SBP, diastolic blood pressure (DBP) and mean arterial pressure (MAP) (Sansom et al. 2004). SBP was also reported to be better at predicting the likelihood of ocular hypertensive disease than DBP or MAP.

In dogs with hypertensive ocular disease, a wide range of SBP (173–265 mmHg) has been reported (Littman et al. 1988; Sansom and Bodey 1997; Jacob et al. 2003; Bovee et al. 1989; Ginn et al. 2013). This may in part be because the breeds of dogs involved in studies often vary and it is known that substantial interbreed differences in blood pressure exist. (Brown et al. 2007). In one study, a median SBP of 197 mmHg was found in 62% ($n = 42$) of dogs with confirmed SH, although no significant difference in SBP was found between hypertensive dogs with ocular lesions and those without (LeBlanc et al. 2011). In the same study however, hypertensive dogs with retinal haemorrhages were found to have a significantly higher SBP when compared to hypertensive dogs without retinal haemorrhages. Jacob et al. only detected retinal lesions in 3/14 dogs with SBP >180 mmHg and CKD (Jacob et al. 2003). This led the authors to suggest that ocular lesions may not be a reliable indicator of high SBP, although the absence of retinal lesions in some dogs could have been related to intermittent reductions in SBP associated with antihypertensive treatment.

9.6.6 Concurrent Disease

As secondary hypertension is the most common category of high blood pressure in dogs and cats (Brown et al. 2007), concurrent clinical disease is often present in animals with hypertensive ocular disease. It is also possible that the susceptibility of the eye to injury from SH is influenced by concurrent disease (Jepson 2011).

Cats Concurrent clinical disease was identified in 40% ($n = 36$) of hypertensive cats with ocular lesions, specifically, chronic kidney disease ($n = 20$), hyperthyroidism ($n = 16$), hypertrophic cardiomyopathy ($n = 3$) and neoplasia ($n = 3$), with some cats having more than one disease (Young et al. 2018). Primary hyperaldosteronism and diabetes mellitus have also been reported in cats with hypertensive ocular lesions (Maggio et al. 2000; Littman 1994).

All hypertensive cats with ocular signs were considered to have renal pathology in one study (Littman 1994), and most cats had CKD in a study where 61% ($n = 33$) of cats with hypertensive retinopathy had concurrent disease (Sansom et al. 2004).

One or more abnormal findings for renal function were also reported in 78% ($n = 54$) of cats with hypertensive retinopathy, with 32% of cats having CKD prior to the diagnosis of hypertension and onset of ocular disease (Maggio et al. 2000). In the same study, hyperthyroidism was diagnosed in only five cats with hypertensive ocular lesions, and it has been suggested by several authors that ocular disease is infrequent in cats with SH associated with hyperthyroidism (Littman 1994; Sansom et al. 1994, 2004). Only one of three cats with SH and hyperthyroidism had ocular signs (retinal arterial tortuosity and focal retinal degeneration) related to SH, compared to 80% ($n = 12$) of cats with chronic renal failure (Stiles 1994). Even in the absence of SH, hyperthyroidism does not seem to be a frequent cause of ocular abnormalities in cats (van der Woerdt and Peterson 2000; Williams et al. 2013). The relationship between feline hyperthyroidism and hypertension is discussed further in Chap. 5.

Dogs Concurrent clinical disease reported in dogs with hypertensive ocular changes include arteriosclerosis, hypothyroidism, acute and chronic kidney disease, hypertensive encephalopathy, cardiac disease and hypercholesterolaemia (Rubin 1975; Gwin et al. 1978; Littman et al. 1988; Sansom and Bodey 1997; Jacob et al. 2003). An association with sudden acquired retinal degeneration (SARD) has also been proposed, but it is unclear whether this represented a direct causal relationship or was linked to a concurrent diagnosis of hyperadrenocorticism (Carter et al. 2009; Auten et al. 2018). Diabetes mellitus is associated with SH in the dog (Brown et al. 2007), but no significant association was detected between the presence of retinopathy and systolic hypertension in 11 dogs with spontaneous diabetes mellitus followed over a 2-year period (Herring et al. 2014). The absence of retinopathy in these dogs may have been because the SH was considered mild with regard to the risk of developing TOD in all but two animals.

LeBlanc et al. found that, 75% ($n = 32$) of dogs with hypertensive ocular lesions had concurrent systemic disease: renal disease ($n = 11$), diabetes mellitus ($n = 5$), iatrogenic hypertension (PPA induced) ($n = 5$), hyperadrenocorticism ($n = 3$), pheochromocytoma ($n = 2$), congestive heart failure ($n = 1$) and neoplasia ($n = 1$) (LeBlanc et al. 2011) (Fig. 9.13).

9.7 Treatment

The goals of treatment for patients with SH are to control the underlying disease if present and to decrease the magnitude, severity and likelihood of TOD (SBP <150 mmHg and/or DBP >95 mmHg) (Brown et al. 2007; Sansom and Bodey 1997). Antihypertensive treatment recommendations are described elsewhere (Henik et al. 1997; Brown et al. 2007; see Chaps. 12 and 13), but if TOD has already occurred, additional therapeutic strategies suited to the specific organ may be appropriate. The successful management of animals with ocular manifestations of SH depends on early recognition of ocular changes (Crispin and Mould 2001).

Fig. 9.13 Bullous retinopathy in an 11-month-old English Cocker Spaniel with juvenile nephropathy and renal failure; systolic blood pressure 210 mmHg. Focal retinal oedema is present throughout (asterisk). Both eyes had similar retinal changes (Courtesy of J Sansom, Animal Health Trust)

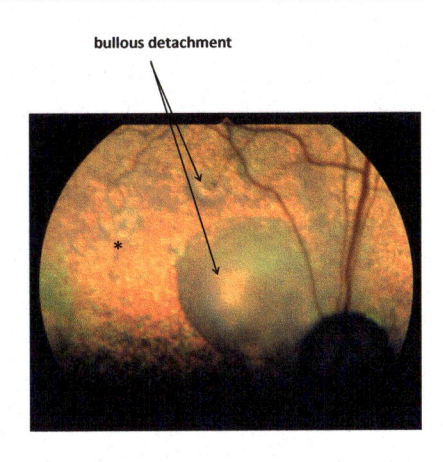

When retinal detachment is present, prompt treatment to reattach the retina is critical in order to preserve as much retinal function as possible and increase the likelihood that vision will return. The retina will only successfully reattach if the elevated blood pressure is reduced. The resolution or improvement of chorioretinal lesions, intraocular haemorrhage or hyphaema also depends on appropriate systemic antihypertensive therapy.

If secondary ocular disease is present, such as glaucoma and anterior uveitis, additional treatment is required. Topical corticosteroids (e.g. 1% prednisolone acetate q 6–8 h) are the mainstay of treatment for anterior uveitis; a topical mydriatic/cycloplegic (e.g. 0.5% tropicamide, 1% atropine) may be needed to control iridocyclospasm or maintain pupil mobility (Maggio et al. 2000). A carbonic anhydrase inhibitor (e.g. dorzolamide HCl 2% in cats, or dorzolamide combined with timolol in dogs) every 8–12 h is a good choice for managing early stages of glaucoma.

After initiation of treatment, follow-up examinations should be performed regularly until there is improvement of ocular signs and/or a decrease in blood pressure is achieved. Fundus changes reported in dogs and cats following retinal reattachment include early optic disc atrophy, retinal degeneration, regions of tapetal hyporeflectivity and retinal vascular attenuation (Crispin and Mould 2001; Komáromy et al. 2004; Sansom et al. 1994). Once the animal has stabilized, monthly or bi-monthly ocular and especially fundus examinations are recommended. It has been suggested that animals with serious or rapidly progressing ocular disease, regardless of the magnitude of blood pressure, should have the eye re-evaluated 1–3 days following initial diagnosis and then at intervals of 7–10 days (Brown et al. 2007).

Cats Calcium channel blockers (CCB) are usually the first choice of treatment for cats with hypertension and ocular lesions. Amlodipine besylate is used as the sole agent if no underlying cause for the SH is identified (Henik et al. 1997; Snyder 1998; Elliott et al. 2001; Young et al. 2018). An ACE inhibitor (ACEi) or angiotensin receptor blocker can be added if proteinuria or albuminuria is present or if blood pressure remains uncontrolled after an increase in the initial dose of CCB (Stepien 2011; Acierno et al. 2018). Since the authorisation of the angiotensin receptor blocker, telmisartan as an anti-hypertensive agent in the cat, this can also be considered as a first-line agent (Glaus et al. 2018). Maggio et al. reported that sixty-nine percent ($n = 18$) of cats with ocular lesions showed improvement in one or both eyes when treated with amlodipine besylate, Eyes that had retinal detachment prior to treatment, had either substantially less sub-retinal fluid, or partial to complete retinal reattachment, or both following treatment. Eyes with retinal oedema or haemorrhage but no detachment prior to treamtent, also had partial to complete resolution of retinal lesions (Maggio et al. 2000). In the same study, cats that were not treated with amlodipine but treated instead with either beta-blocking agents (alone or in combination with ACEi) or diltiazem had only partial and often temporary improvement of ocular signs. Of fourteen cats with hypertensive ocular disease that were treated with diuretics and systemic corticosteroids, and monitored regularly, eyes that responded favourably to treatment were those that had been diagnosed on presentation with bullous retinopathy, partial retinal detachment or intraocular haemorrhage. However, eyes with total retinal detachment showed poor response to treatment (Sansom et al. 1994).

Infrequent follow up, especially repeat ocular examinations, is a limitation to the findings of these and other clinical studies. Amlodipine besylate was used once daily as the first-line treatment to reduce blood pressure in hypertensive cats, 60% ($n = 18$) of which had ocular abnormalities including complete vision loss, but no description of the outcome of ocular lesions or visual status was reported in this study (Elliott et al. 2001). In retrospective studies which report response of hypertensive ocular lesions to treatment, it is also not always clear if a single drug or combination of drugs was used, nor whether the drug dose given to animals was consistent, or if a poor response to therapy was because chronic, subclinical ocular disease existed prior to treatment (Young et al. 2018).

In an experimental study of cats with coexistent systemic hypertension and renal insufficiency, oral administration of amlodipine produced a significant antihypertensive effect and fewer cats in the treated group had ocular lesions when compared to an equal number of cats in the placebo group. The difference in the prevalence of ocular lesions was statistically significant, and results of this study suggest that treatment of hypertension early in the course of the disease could prevent ocular damage (Mathur et al. 2002).

Dogs ACE Inhibitors or angiotensin receptor blockers are usually recommended as the initial drug of choice for dogs with SH (Brown et al. 2007; Acierno et al. 2018); however, information about the specific response of ocular lesions to antihypertensive therapy is inconclusive (Bovee et al. 1989; Sansom and Bodey 1997; Ginn et al. 2013; Wolfer et al. 1998). One dog with end-stage ocular disease and SH did not show improvement, whereas the retina reattached and the blood pressure returned to normal in two dogs with CKD when the only treatment was diet change (Littman et al. 1988). LeBlanc et al. reported that dogs receiving antihypertensive drugs, whether these drugs were effective or not at controlling blood pressure, were not uniformly spared from ocular TOD (LeBlanc et al. 2011).

9.8 Prognosis

The prognosis for vision in animals with ocular TOD depends on multiple factors. With widespread retinal detachment, the duration of the detachment is likely the most important factor influencing return of vision (Young et al. 2018; Erickson et al. 1983). The physiologic state of the retina and the degree of retinal degeneration present at the time of retinal detachment also both play a role in determining visual recovery (Komáromy et al. 2004; Sansom et al. 1994). Furthermore, a poor prognosis for return of vision has been suggested in cats with haemorrhagic detachment compared to serous retinal detachment (Sansom et al. 1994), and in the rabbit, when sub-retinal fluid is haemorrhagic or exudative, or the retina is more elevated (i.e. bullous detachment), retinal degeneration has been shown to occur more rapidly (Glatt and Machemer 1982; Erickson et al. 1983). Recently, sex has been suggested to be a clinically relevant factor with regard to visual prognosis in cats with SH; female cats in one study were four times less likely than male cats to have a menace response at the time of last examination (Young et al. 2018).

In cats with experimental retinal detachment, the recovery of the reattached retina is reported to be optimal if it is detached for less than a week. Permanent histologic changes are evident in the feline retina as early as 3 days following detachment, and photoreceptor synaptic terminals are no longer detectable histologically by 50 days. In most species studied experimentally, marked degeneration of photoreceptor outer segments, cystoid oedema of inner retinal layers, dedifferentiation and hypertrophy of RPE cells occur within the first week following detachment (Erickson et al. 1983; Anderson et al. 1983).

The resolution of retinal lesions, rate of retinal reattachment and long-term outcome for vison have been evaluated in clinical studies of cats with SH. Fifty percent ($n = 14$) of cats with ocular signs considered to be 'early' (multifocal retinal oedema or haemorrhage) had either partial or complete resolution of retinal lesions or no progression of ocular disease following treatment to reduce blood pressure. In the same study, vision was not maintained in cats with large areas of retinal degeneration following retinal reattachment, even though there was improvement in vision initially (Maggio et al. 2000). Similarly, Littman noted areas of retinal degeneration in hypertensive cats that remained blind in spite of retinal reattachment and resolution of retinal haemorrhages (Littman 1994).

Young et al. also reported that eyes with retinal detachment that were visual prior to treatment had a good prognosis for maintaining vision. Following treatment, 93% ($n = 41$) of visual eyes were reported to have vision as assessed by the presence of a menace response (Young et al. 2018) and 57% ($n = 76$) of blind eyes regained some vision during the treatment period, with half of the eyes regaining vision in less than 3 weeks. Reattachment of the retina also occurred within 3 weeks in 29% ($n = 46$) of eyes and by day 60 in an additional 22% of eyes. When comparing eyes that had been blind for less than 2 weeks at presentation, and those that had been blind for greater than 2 weeks, there was no significant difference in the menace response, which led the authors to conclude that the duration of blindness prior to presentation did not appear to affect the final visual status of cats. One of the inherent limitations

Fig. 9.14 Confirmed systemic hypertension in a dog. Parasagittal section of a formalin-fixed globe showing complete retinal detachment (asterisk) and marked intraocular haemorrhage (double asterisk) including hyphaema (Reproduced with permission from EJ Scurrell)

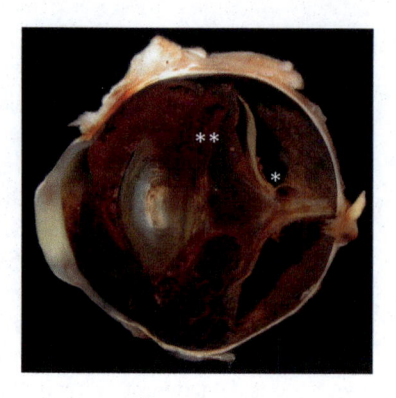

of this retrospective study is that visual outcome was only assessed by the presence of a menace response which is often the only practical way of assessing vision in cats in a clinical setting (Young et al. 2018). Return of vision following retinal reattachment has also been reported in dogs with SH, with the length of time to return of vision ranging from within 2 days to 4 weeks (Wolfer et al. 1998; Ginn et al. 2013; Sansom and Bodey 1997). Howver, long-term follow-up of dogs is lacking and often the number of animals examined is small.

Successful reattachment of the retina appears to be common in clinical patients with SH, especially if eyes are visual before treatment. In an attempt to assess retinal function following retinal reattachment, ERG was performed in a hypertensive cat, blind from bilateral retinal detachment and intraretinal haemorrhages. The cat had no retinal function on presentation, as determined by ERG; 11 days following treatment the cat was visual, the retina had reattached, and the blood pressure was reduced. An ERG performed on day 25 showed evidence of retinal function in both eyes, and by day 78 the ERG showed improvement in one eye although not in the other; the cat remained visual, but the authors did not comment on whether the ERG was normal (Komáromy et al. 2004).

Retinal reattachment in dogs and cats does not necessarily engender visual success long term. Animals with hypertensive ocular disease can also remain blind because extensive vitreous or pre-retinal haemorrhage rarely resolves. Glaucoma can also develop acutely from complete hyphaema or chronically from the development of pre-iridal fibrovascular membranes (Bovee et al. 1989) (Fig. 9.14).

9.9 Conclusion

The effects of sustained systemic hypertension in animals are directly visible in the eye and careful sequential ocular examinations, including fundoscopy, are recommended for both diagnosis and monitoring of the disease. Retinal detachment is common in dogs and cats, and retinal reattachment is only possible with reduction in blood pressure. The duration of retinal detachment is the most important factor affecting visual recovery following treatment for systemic hypertension, and the degree and type of retinal detachment also influences the success of reattachment. Retinal reattachment does not necessarily imply visual success in hypertensive animals as retinal degeneration is often advanced by the time animals are first evaluated. Animals with hypertensive ocular disease can also remain blind because of extensive vitreous or pre-retinal haemorrhage or from glaucoma which can develop acutely from complete hyphaema or chronically from the development of pre-iridal fibrovascular membranes.

The prevalence of ocular injury with systemic hypertension has been reported to be as high as 100% in cats, and routine screening of animals at risk for the disease provides the opportunity for prompt and early treatment. The degree of systemic hypertension required to produce ocular injury is not known; however, implementation of treatment before SBP is allowed to reach severe levels (>180 mmHg) should offer protection to the eye, reduce the risk for blindness and afford the best prognosis for vision long term.

References

Acierno MJ, Brown S, Coleman AE, Jepson RE, Papich M, Stepien RL, Syme HM (2018) ACVIM consensus statement: guidelines for the identification, evaluation, and management of systemic hypertension in dogs and cats. J Vet Intern Med 32(6):1803–1822

Alder VA, Ben-Nun J, Cringle SJ (1990) PO2 profiles and oxygen consumption in cat retina with an occluded retinal circulation. Invest Ophthalmol Vis Sci 31:1029–1034

Alm A, Bill A (1972) The oxygen supply to the retina. II. Effects of high intraocular pressure and of increased arterial carbon dioxide tension on uveal and retinal blood flow in cats. A study with radioactively labelled microspheres including flow determinations in brain and some other tissues. Acta Physiol Scand 84:306–319

Anderson DH, Stern WH, Fisher SK et al (1983) Retinal detachment in the cat: the pigment epithelial-photoreceptor interface. Invest Ophthalmol Vis Sci 24:906–926

Aryan HE, Ghosheh FR, Jandial R, Levy ML (2005) Retinal hemorrhage and pediatric brain injury: etiology and review of the literature. J Clin Neurosci 12:624–631

Auten CR, Thomasy SM, Kass PH et al (2018) Cofactors associated with sudden acquired retinal degeneration syndrome: 151 dogs within a reference population. Vet Ophthalmol 21:264–272

Balland O, Russo A, Isard PF et al (2017) Assessment of a smartphone-based camera for fundus imaging in animals. Vet Ophthalmol 20(1):89–94

Bill A, Törnquist P, Alm A (1980) Permeability of the intraocular blood vessels. Trans Ophthalmol Soc UK 100:332–336

Bodey AR, Michell AR, Bovee KC et al (1996) Comparison of direct and indirect (oscillometric) measurements of arterial blood pressure in conscious dogs. Res Vet Sci 61:17–21

Bovee KC, Littman MP, Crabtree BJ, Aguirre G (1989) Essential hypertension in a dog. J Am Vet Med Assoc 195:81–86

Braun RD, Linsenmeier RA, Goldstick TK (1995) Oxygen consumption in the inner and outer retina of the cat. Invest Ophthalmol Vis Sci 36:542–554

Brown S, Atkins C, Bagley R et al (2007) Guidelines for the identification, evaluation, and management of systemic hypertension in dogs and cats. J Vet Intern Med 21:542–558

Carter RT, Oliver JW, Stepien RL, Bentley E (2009) Elevations in sex hormones in dogs with sudden acquired retinal degeneration syndrome (SARDS). J Am Anim Hosp Assoc 45:207–214

Chetboul V, Lefebvre HP, Pinhas C et al (2003) Spontaneous feline hypertension: clinical and echocardiographic abnormalities, and survival rate. J Vet Intern Med 17:89–95

Cirla A, Drigo M, Ballerini L, et al (2019) VAMPIRE® fundus image analysis algorithms: validation and diagnostic relevance in hypertensive cats. Vet Ophthalmol 1–9

Crispin SM, Mould JR (2001) Systemic hypertensive disease and the feline fundus. Vet Ophthalmol 4:131–140

de Venecia G, Jampol LM (1984) The eye in accelerated hypertension. II. Localized serous detachments of the retina in patients. Arch Ophthalmol 102:68–73

Dimmitt SB, West JN, Eames SM et al (1989) Usefulness of ophthalmoscopy in mild to moderate hypertension. Lancet 1:1103–1106

Dubielzig RR, Ketring KL, McLellan GJ, Albert DM (2010) Veterinary ocular pathology E-book: a comparative review. Elsevier, London, pp 370–372

Elliott J, Barber PJ (1998) Feline chronic renal failure: clinical findings in 80 cases diagnosed between 1992 and 1995. J Small Anim Pract 39:78–85

Elliott J, Barber PJ, Syme HM et al (2001) Feline hypertension: clinical findings and response to antihypertensive treatment in 30 cases. J Small Anim Pract 42:122–129

Erickson PA, Fisher SK, Anderson DH et al (1983) Retinal detachment in the cat: the outer nuclear and outer plexiform layers. Invest Ophthalmol Vis Sci 24:927–942

Garner A, Ashton N, Tripathi R et al (1975) Pathogenesis of hypertensive retinopathy. An experimental study in the monkey. Br J Ophthalmol 59:3–44

Ginn JA, Bentley E, Stepien RL (2013) Systemic hypertension and hypertensive retinopathy following PPA overdose in a dog. J Am Anim Hosp Assoc 49:46–53

Glatt H, Machemer R (1982) Experimental subretinal hemorrhage in rabbits. Am J Ophthalmol 94:762–773

Glaus TM, Elliott J, Herberich E et al (2018) Efficacy of long-term oral telmisartan treatment in cats with hypertension: results of a prospective European clinical trial. J Vet Intern Med 33:413–422

Glenwood G, Edward M (2013) Chapter 3: Physiology of the eye. In: Gelatt KN, Gilger BC, Kern TJ (eds) Veterinary ophthalmology. Wiley, Ames, IA, pp 181–183

Goldblatt H, Lynch J, Hanzal RF, Summerville WW (1934) Studies on experimental hypertension: i. the production of persistent elevation of systolic blood pressure by means of renal ischemia. J Exp Med 59:347–379

Gwin RM, Gelatt KN, Terrell TG, Hood CI (1978) Hypertensive retinopathy associated with hypothyroidism, hypercholesterolemia, and renal-failure in a dog. J Am Anim Hosp Assoc 14:200–209

Haddock LJ, Kim DY, Mukai S (2013) Simple, inexpensive technique for high-quality smartphone fundus photography in human and animal eyes. J Ophthalmol 2013:518479

Hayreh SS, Servais GE, Virdi PS (1986a) Fundus lesions in malignant hypertension. VI. Hypertensive choroidopathy. Ophthalmology 93:1383–1400

Hayreh SS, Servais GE, Virdi PS (1986b) Fundus lesions in malignant hypertension. V. Hypertensive optic neuropathy. Ophthalmology 93:74–87

Hayreh SS, Servais GE, Virdi PS et al (1986c) Fundus lesions in malignant hypertension. III. Arterial blood pressure, biochemical, and fundus changes. Ophthalmology 93:45–59

Henik RA (1997) Systemic hypertension and its management. Vet Clin North Am Small Anim Pract 27:1355–1372

Henik RA, Snyder PS, Volk LM (1997) Treatment of systemic hypertension in cats with amlodipine besylate. J Am Anim Hosp Assoc 33:226–234

Herring IP, Panciera DL, Werre SR (2014) Longitudinal prevalence of hypertension, proteinuria, and retinopathy in dogs with spontaneous diabetes mellitus. J Vet Intern Med 28:488–495

Jacob F, Polzin DJ, Osborne CA et al (2003) Association between initial systolic blood pressure and risk of developing a uremic crisis or of dying in dogs with chronic renal failure. J Am Vet Med Assoc 222:322–329

Jepson ER (2011) Feline systemic hypertension: classification and pathogenesis. J Feline Med Surg 13:25–34

Keith NM, Wagener HP, Barker NW (1974) Some different types of essential hypertension: their course and prognosis. Am J Med Sci 268:336–345

Keyes JEL (1937) Experimental hypertension. IV. Clinical and pathologic studies of the eyes: a preliminary report. Arch Ophthalmol 17:1040

Kiel JW, Shepherd AP (1992) Autoregulation of choroidal blood flow in the rabbit. Invest Ophthalmol Vis Sci 33:2399–2410

Klein BE, Davis MD, Segal P et al (1984) Diabetic retinopathy. Assessment of severity and progression. Ophthalmology 91:10–17

Kobayashi DL, Peterson ME, Graves TK et al (1990) Hypertension in cats with chronic renal failure or hyperthyroidism. J Vet Intern Med 4:58–62

Komáromy AM, Andrew SE, Denis HM et al (2004) Hypertensive retinopathy and choroidopathy in a cat. Vet Ophthalmol 7:3–9

LeBlanc NL, Stepien RL, Bentley E (2011) Ocular lesions associated with systemic hypertension in dogs: 65 cases (2005–2007). J Am Vet Med Assoc 238:915–921

Liebreich R (1859) Ophthalmoskopischer befund bei morbus Brightii. Albrecht von Graefes Arch Ophthalmol 5:265–8. Cited by Hayreh S et al (1986) Fundus lesions in malignant hypertension III Arterial blood pressure, biochemical, and fundus changes. Ophthalmology 93:45–59

Linsenmeier RA (1986) Effects of light and darkness on oxygen distribution and consumption in the cat retina. J Gen Physiol 88:521–542

Littman MP (1994) Spontaneous systemic hypertension in 24 cats. J Vet Intern Med 8:79–86

Littman MP, Robertson JL, Bovée KC (1988) Spontaneous systemic hypertension in dogs: five cases (1981–1983). J Am Vet Med Assoc 193:486–494

Locke LC (1992) Ocular manifestations of hypertension. Optom Clin 2:47–76

Maggio F, DeFrancesco TC, Atkins CE et al (2000) Ocular lesions associated with systemic hypertension in cats: 69 cases (1985–1998). J Am Vet Med Assoc 217:695–702

Mathur S, Syme H, Brown CA et al (2002) Effects of the calcium channel antagonist amlodipine in cats with surgically induced hypertensive renal insufficiency. Am J Vet Res 63:833–839

Maxie G (2015) Jubb, Kennedy & Palmer's pathology of domestic animals: 3-volume set. Saunders, St. Louis, MO, pp 472–473

McMullen R, Millichamp N, Pirie C (2013) Chapter 12: Ophthalmic photography. In: Gelatt KN, Gilger BC, Kern TJ (eds) Veterinary ophthalmology. Wiley, Ames, IA, pp 760–762

Morgan R (1986) Systemic hypertension in four cats: ocular and medical findings. J Am Anim Hosp Assoc 22:615–621

Nickla DL, Wallman J (2010) The multifunctional choroid. Prog Retin Eye Res 29:144–168

Pirie CG, Pizzirani S (2011) Anterior and posterior segment photography: an alternative approach using a dSLR camera adaptor. Vet Ophthalmol 14:1–8

Rubin LF (1975) Ocular manifestations of arteriosclerosis and hypertension in dogs. Trans Am Coll Vet Ophthalmol 6:55–68

Sansom J, Bodey A (1997) Ocular signs in four dogs with hypertension. Vet Rec 140:593–598

Sansom J, Barnett KC, Dunn KA et al (1994) Ocular disease associated with hypertension in 16 cats. J Small Anim Pract 35:604–611

Sansom J, Rogers K, Wood JLN (2004) Blood pressure assessment in healthy cats and cats with hypertensive retinopathy. Am J Vet Res 65:245–252

Snyder PS (1998) Amlodipine: a randomized, blinded clinical trial in 9 cats with systemic hypertension. J Vet Intern Med 12:157–162

Stepien RL (2011) Feline systemic hypertension: diagnosis and management. J Feline Med Surg 13:35–43

Stiles J (1994) The prevalence of retinopathy in cats with systemic hypertension and chronic renal failure or hyperthyroidism. J Am Anim Hosp Assoc 30:564–572

Turner JL, Brogdon JD, Lees GE, Greco DS (1990) Idiopathic hypertension in a cat with secondary hypertensive retinopathy associated with a high-salt diet. J Am Anim Hosp Assoc 26:647–651

Van Boxtel SA (2003) Hypertensive retinopathy in a cat. Can Vet J 44:147–149

van de Sandt RROM, Stades FC, Boevé MH, Stokhof AA (2003) Arterial hypertension in the cat. A pathobiologic and clinical review with emphasis on the ophthalmologic aspects. Tijdschr Diergeneeskd 128:2–10 van der Woerdt A, Peterson ME (2000) Prevalence of ocular abnormalities in cats with hyperthyroidism. J Vet Intern Med 14:202–203

Villagrasa M, Cascales MJ (2000) Arterial hypertension: angiographic aspects of the ocular fundus in dogs. A study of 24 cases. Eur J Compan Anim Pract 10:177–190

Violette NP, Ledbetter EC (2018) Punctate retinal hemorrhage and its relation to ocular and systemic disease in dogs: 83 cases. Vet Ophthalmol 21:233–239

Williams TL, Elliott J, Syme HM (2013) Renin-angiotensin-aldosterone system activity in hyperthyroid cats with and without concurrent hypertension. J Vet Intern Med 27:522–529

Wolfer J, Grahn B, Arrington K (1998) Diagnostic ophthalmology. Bilateral, bullous retinal detachment. Can Vet J 39:57–58

Wong VG, Macri FJ (1964) Vasculature of the cat eye. Arch Ophthalmol 72:351–358

Wong TY, Mitchell P (2004) Hypertensive retinopathy. N Engl J Med 351:2310–2317

Wong TY, Wong T, Mitchell P (2007) The eye in hypertension. Lancet 369:425–435

Young WM, Zheng C, Davidson MG, Westermeyer HD (2018) Visual outcome in cats with hypertensive chorioretinopathy. Vet Ophthalmol 22(2):161–167

Hypertension and the Central Nervous System

10

Kaspar Matiasek, Lara Alexa Matiasek, and Marco Rosati

10.1 Introduction

The first paper on arterial hypertension[1] causing dysfunction of the central nervous system (CNS) was published in 1928 by Oppenheimer and Fishberg. These scholars introduced the term *hypertensive encephalopathy* (HE) and used it to describe a life-threatening acute complication of elevated blood pressure (Oppenheimer and Fishberg 1928). Since then, the medical literature has featured a plethora of reports and studies on hypertensive CNS lesions, most of which comprise clinical descriptions and experiments in laboratory rodents. Indeed, the latter appear to reflect the brain lesions seen in humans (Nag and Robertson 2005). On the other hand, hypertensive rats do not mirror the natural history of the human condition in the same way as dogs and cats with long-standing spontaneous hypertension do. The mechanisms underlying the pathophysiology of CNS involvement are multifactorial and far from completely understood. Based on our current understanding, hypertension evokes direct hypertensive lesions and predisposes individuals to encephalovascular accidents[2] that require additional triggers.

---

[1] Hereinafter, arterial hypertension will simply be referred to as *hypertension*.
[2] Most papers and textbooks refer to cerebrovascular accidents. This anatomical prefix implies involvement of the forebrain alone, excluding the brainstem and cerebellum in a strict sense. The authors therefore use the term *encephalovascular* instead whenever the situation involves all brain areas. Similarly, we use the term *encephalic* in situations where *cerebral* gives the false impression that the relevant feature is restricted to the forebrain (e.g. cerebral blood flow).

K. Matiasek (✉) · M. Rosati
Section of Clinical and Comparative Neuropathology, Centre for Clinical Veterinary Medicine, Ludwig-Maximilians-Universität München, Munich, Germany
e-mail: kaspar.matiasek@neuropathologie.de

L. A. Matiasek
Anicura Kleintierklinik, Babenhausen, Germany

Direct hypertensive lesions include the following: (1) *acute (de novo) encephalopathy* without preceding target organ damage (TOD) of the brain and its vasculature, (2) *acute breakthrough of chronic CNS-TOD* (Eberhardt 2018) and (3) *chronic arterial small-vessel disease* (CASVD) (Liu et al. 2018; Pantoni 2010).

Both acute forms converge to HE, accounting for up to 20% of hypertensive emergencies in humans. Clinically, HE presents as forebrain-predominant encephalopathy with the dynamics of a vaso-disruptive brain disorder (Eberhardt 2018). CASVD, on the other hand, causes insidious loss of brain cells (also called *hypertensive neurodegeneration*). This process is due to chronic malnutrition of brain tissue and has been associated with progressive cognitive decline in humans (Dunn and Nelson 2014; Liu et al. 2018).

No such stratification is available yet for spontaneous hypertensive lesions in the canine or feline brain. Despite the high prevalence of hypertension in dogs and cats in veterinary practice (Acierno et al. 2018; Brown et al. 2007) and between 29% (Maggio et al. 2000) and 46% (Littman 1994) of hypertensive cats with ocular TOD showing neurological signs, reports on CNS involvement are remarkably sparse and refer only to the acute presentation of HE (Church et al. 2019; Jacob et al. 2003; Littman 1994; Maggio et al. 2000; O'Neill et al. 2013).

The lack of systematic studies on the subject impedes awareness of this condition and leaves the practitioner with significant uncertainty about its incidence and relevance. Clinical diagnosis of hypertensive CNS changes, on the other hand, is considered an important determinant of encephalovascular accidents, of risk levels in hypertensive crises and of long-term outcomes in human patients. Therefore, hypertensive CNS changes should ideally be translated ad hoc into therapeutic considerations and clinical management of the patient (Acierno et al. 2018; Eberhardt 2018).

Early antihypertensive treatment of HE-affected dogs and cats may reverse neurological signs and, furthermore, prevent progression of brain lesions and of other forms of TOD (Kyles et al. 1999; O'Neill et al. 2013). Remission, however, very much depends on the stage and reversibility of hypertensive tissue changes. In general, haemodynamic, early pervasive and metabolic disturbances are expected to be reversible with treatment, while advanced vascular remodelling and vaso-disruptive lesions can only be silenced and functionally compensated for by the patient.

10.2 Basic Vascular Physiology of the CNS: Blood Supply and Autoregulation

Due to continuous ion transport and neurotransmitter responses, the brain has a high metabolic rate that demands approximately 15% of resting cardiac output and consumes as much as 20% of total blood oxygen and glucose at the peak of its activity (Miller et al. 2018). As the capacity of the brain to store energy is limited, execution and maintenance of neurological functions is dependent on a continuous blood flow that overcomes intracranial vascular resistance and senses and adapts to regional metabolic needs (Dunn and Nelson 2014). To achieve the required blood supply to the brain, large extracranial arteries connect to a set of intracranial collaterals in a species-specific manner. In the cat, the bulk of blood flow employs the maxillary artery and its *rete mirabile* and, to a lesser extent, the pharyngeal artery. Both of these vessels are derived from the external carotid artery. The maxillary and pharyngeal arteries feed the blood bilaterally into the circle of Willis at the base of the middle and rostral cranial fossa, from which they supply the basilar artery with blood flowing rostrocaudally. This anatomy differs significantly from that of the canine brain, which, similar to the human brain, receives most of the forebrain blood supply from communicating branches of the internal carotid arteries, while the midbrain and hindbrain are supported from the caudal direction through the vertebrobasilar system.

The circle of Willis then releases the paired rostral, middle and caudal cerebral arteries and the rostral cerebellar artery. The basal forebrain structures are supported by a set of paramedian striate and choroidal arteries and their branches. Similarly, there are paramedian arteries to supply the brainstem from the basilar artery. Laterally, the latter releases the caudal cerebellar and labyrinthine arteries to supplement the caudal half of the cerebellum and the dorsolateral brainstem.

Across species, the arteries must maintain an average encephalic blood flow (EBF) of 50 ml per 100 g brain tissue per minute. Due to its predominance regarding energy-demanding activities, the grey matter requires to be supplemented by 70 ml/100 g/min, while the white matter can cope with 20 ml/100 g/min (Strandgaard and Paulson 1990).

Conforming to the Monro-Kellie doctrine of the encased brain, the arterial pulse does not merely translate into venous blood flow. Instead, the diameter increase of blood vessels during pulsatile dilation also triggers the drainage of extravascular interstitial fluids. Therefore, one (larger) fraction of the haemodynamic force continues towards the veins as blood flow, while another (smaller) fraction is absorbed by the cerebrospinal fluid (CSF) system and triggers the effusion of CSF from the ventricles and intracranial subarachnoid spaces (Fig. 10.1). This transfer of pressure from one fluid compartment to the next is only possible because of the elasticity of blood vessels, a patent craniospinal CSF outflow pathway, the relative rigidity of the brain parenchyma and the invariable volume of the cranial vault. The net exchange of blood volume per time unit is reflected by encephalic perfusion, while encephalic perfusion pressure (EPP) represents the driving force.

There are two important dynamic modifiers of brain perfusion: (1) intracranial pressure (ICP) and (2) the vascular tone (and luminal dimension) of intracranial arterioles.

ICP may be elevated haemodynamically by adaptive vasodilation (e.g. as a response to $PaCO_2$) and transient gravitational/positional venostasis. ICP also increases significantly with brain oedema, mass lesions and hydrocephalus, which increase intracranial vascular resistance by extramural compression of blood vessels under pathological conditions.

In the healthy brain, perfusion relies on sophisticated vasomotor control that distributes arterial blood according to actual metabolic needs and avoids hypotensive

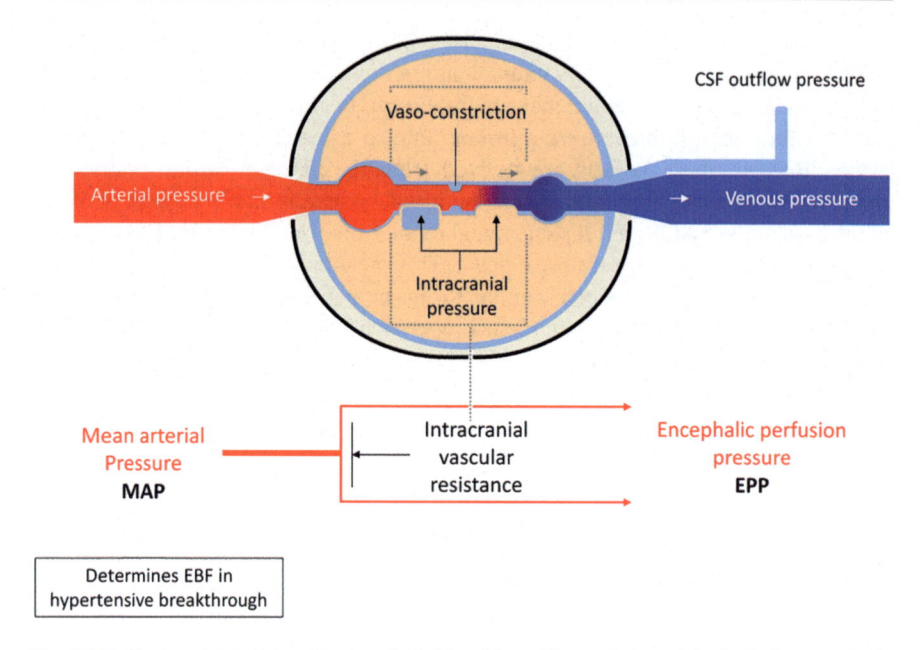

Fig. 10.1 Factors determining the encephalic blood flow. Net perfusion of the brain [= encephalic blood flow (EBF)] results from encephalic (brain) perfusion pressure (EPP). The effective EPP equals the mean arterial pressure (MAP) minus, firstly, local opposing forces by vasoconstriction, luminal stenosis and external pressure onto the vessel walls from interstitial fluid accumulation or tissue gain, all of which determine the intracranial vascular resistance, and, secondly, minus resistance of the extracranial veins. The MAP is defined as diastolic pressure plus 1/3 pulse pressure. Hence, a certain fraction of the MAP intra-cranially is absorbed by the tissue as well as by the CSF. The remaining pressure translates into the venous outflow pressure. Note that extracranial venous resistance, in particular that of the jugular vein, also affects the EBF and CSF outflow pressure. Under physiological conditions, blood vessels are capable of luminal adaptation to local metabolic demands, which is called autoregulation. In hypertension, this fine tuning becomes successively impaired until the MAP drives EBF on its own

stasis as well as hypertensive brain damage (Czosnyka et al. 2003). Within the MAP safety range of 50–160 mmHg (Nag and Robertson 2005), vasogenic autoregulation is the primary regulator of EPP. In this way, all regions receive sufficient glucose and oxygen despite circadian and activity-related blood pressure variation (Rivera-Lara et al. 2017).

If systemic arterial pressure exceeds the normotensive range, two mechanisms protect the brain from harmful effects. The first is vasoconstriction of pial arteries and sub-pial arterioles following sympathetic regulation via nerve fibres from the superficial cervical ganglion (Edvinsson and Krause 2002; Hamel 2006). The main function of this mechanism is to attenuate excessive blood flow to the parenchyma. Once the so-called penetrating arteries leave the meninges to enter the parenchyma amidst Virchow-Robin spaces, sympathetic nerves are stripped off. These vessels now directly respond to media distension, metabolic signals and transmitters from surrounding neurons and astrocytes (Dunn and Nelson 2014).

Myogenic vasoconstriction is a key mechanism of prevention in all vascular territories, both at the brain surface and in the deep regions served by the transpial arterial system and deep perforators, wherever a transmural pressure gradient is detected. Abrupt pressure changes of 10–25 mmHg lead to immediate and effective local vasoconstriction (Osol and Halpern 1985). In experimental video analyses, the response lagged only 0.25 s behind the stimulus (Halpern et al. 1984). This type of autoregulation is very effective but requires healthy vascular smooth muscle cells (also called mediacytes) within a compliant vessel wall.

Under physiological conditions, myogenic vasoconstriction is reversed if the incoming blood pressure drops to normal values. Alternatively, glycolytic intermediates of anaerobic metabolism, potentially toxic ion and neurotransmitter concentrations, radicals, phospholipid degradation products, nitric oxide release and changes in pH enforce elevated perfusion rates to protect the brain from metabolic injury.

10.3 Pathophysiological Mechanisms of Chronic Hypertension in the CNS

Although myogenic responses are extremely helpful in buffering hypertensive episodes, over-employment of the system leads to significant complications. One problem for brain perfusion and treatment likewise is that in chronic hypertension, myogenic control resets pressure to an increased level (i.e. right shift) by up to 40 mmHg (Bor-Seng-Shu et al. 2012). After being triggered, however, myogenic contraction is effective over a wide range of pressure changes (Dunn and Nelson 2014; Nag and Robertson 2005). For this reason, global and regional resting EBF remain sufficient, and fluctuations remain clinically silent over a certain period. After some time, when decompensation finally occurs, resting EBF becomes significantly reduced as well (Fujishima et al. 1995). In hypertensive people, this reduction is most evident in the thalamus and striatum, followed by the cerebral cortex, reflecting the order of vulnerability to cerebrovascular accidents. The vulnerability of deep cerebral nuclei suggests that the threshold between protective and exaggerated vasoconstriction is lower in deep perforating arteries than in superficial transpial arteries of cerebral cortex and the vertebrobasilar supply system (Fujishima et al. 1995).

Apart from sustained myogenic vasoconstriction, it has been documented that chronic shear stress and circulating angiotensin-II can induce the release of pro-constrictive thromboxane-A_2 and endothelin-1 from the distended endothelium (Golding et al. 2002). These factors either evoke or facilitate additive contraction of vascular smooth muscle cells by activation of dihydropyridine-sensitive Ca^{2+} channels and thereby lead to sustained vasospasms that respond poorly to the natural vasorelaxant, nitric oxide (Bkaily et al. 2000; Yoshida et al. 1994). In addition, angiotensin-II, among other factors, compromises nitric oxide-dependent arteriolar dilation in response to increased neural activity and thereby disrupts another important axis of neuroprotection (Dunn and Nelson 2014). The consequences of failure of vasodilation are insidious and predominantly neurodegenerative in nature (Dunn and

Nelson 2014). In addition to malnutrition, neurons and glial cells are exposed to metabolic (waste) stress as well as oxidative and nitrosative stress, and such a milieu favours neuroinflammation. Subsequent cell loss lowers neurological capacities, especially in individuals with reduced compensatory capabilities, such as elderly patients and those suffering from concurrent neurodegenerative conditions (Fujishima et al. 1995).

As hypertension persists, vascular narrowing gains two structural and, therefore, less reversible components that increase the risk of transient ischaemic attacks, infarction and CNS haemorrhage significantly, namely, (1) medial hyperplasia and (2) hypertrophic vascular remodelling (Fig. 10.2).

Upon exaggerated contractile activity, smooth muscle cells of pial and parenchymal arteries and arterioles initially adapt to the increased workload by cellular

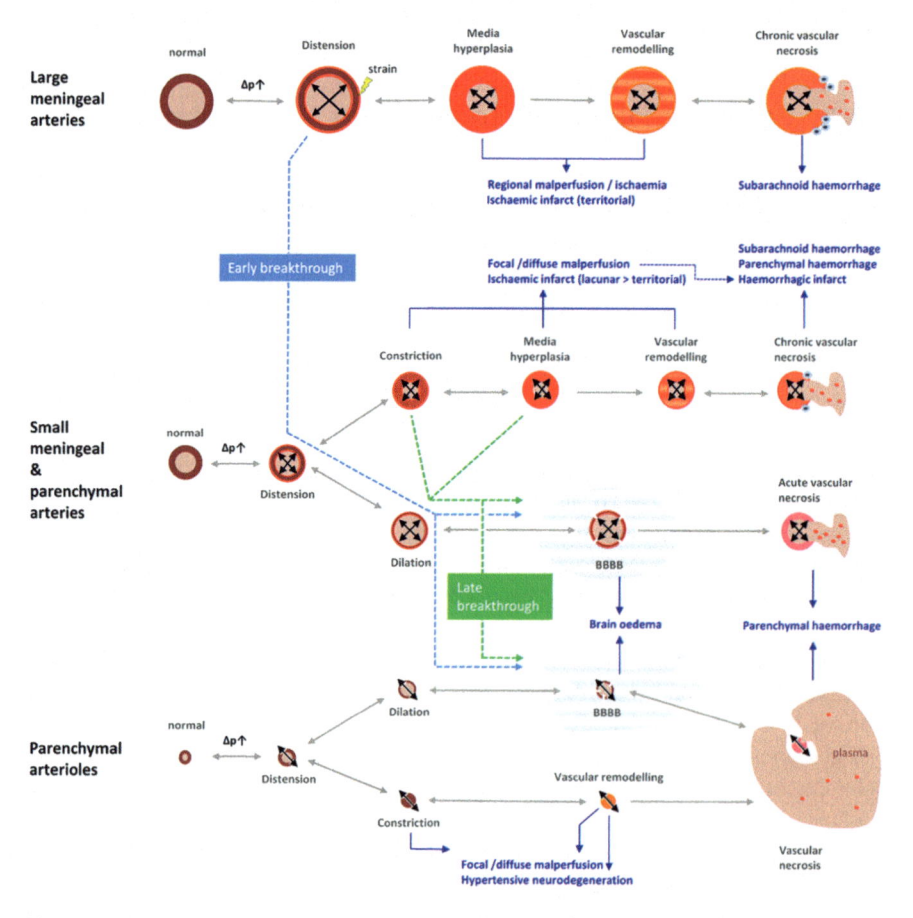

Fig. 10.2 Summarized consequences of hypertension for brain vasculature and parenchyma. Schematic diagram summarizing the pathophysiologic pathways of adaptive and maladaptive vascular changes including decompensation and associated tissue changes. Details are provided in the text

hypertrophy (Owens et al. 1981), followed by dedifferentiation, cell-cycle activation and proliferation (Touyz et al. 2018). If vascular distension and mitogenic stimulation persist, medial hyperplasia escapes control. Exaggerated thickening of the smooth muscle layer (i.e. arteriosclerosis) increases the wall-to-lumen ratio (Fig. 10.3) and the rigidity of the vessel walls (Touyz et al. 2018), while its sensitivity to constrictor agents is reduced (Pires et al. 2013). Cell proliferation is accompanied by deposition of the inelastic extracellular matrix components laminin, fibronectin and collagen IV (Nag and Robertson 2005). In particular, in the large supplying arteries (the circle of Willis and the vessels proceeding from it), nutrition of the thickened vascular wall becomes increasingly compromised. At a certain point, this nutrient deficiency leads to medial degeneration and formation of rigid, non-contractile fibro-collagenous replacement tissue and so called "onion-skinning" (Church et al. 2019). From the luminal side, hypertensive micro-lesions of the endothelium evoke plasma insudation, sub-endothelial fibroplasia and disruption of the elastic lamina. Both intimal and medial remodelling decrease the luminal diameter significantly (Church et al. 2019; Nag and Robertson 2005) until exhaustive dilation comes into effect (Fig. 10.4). The impact of vascular narrowing on the blood supply of CNS tissue may be compensated as long as redistribution of flow direction among arterial leptomeningeal anastomoses is functional. If, however, blood flow in the main nutritive arteries (rostral/medial/caudal cerebral and rostral/caudal cerebellar) and their branches is critically diminished, large territorial infarcts can occur.

In a similar way, spinal cord infarction can be caused by hypertensive spinal artery changes. Of these, the ventral spinal artery and its branches are most commonly affected. They usually present with concentric fibro-intimal remodelling and/or hyaline degeneration, as described in case series in cats (Rylander et al. 2014).

Medial proliferation and remodelling also affects small parenchymal arteries between 200 and 800 µm in diameter, in particular, deep perforators such as the (lenticulo)striate artery (Lammie 2005). Vascular wall changes in animals, as opposed to those in humans, lack the atheromatous component with the possible exception of hypertensive dogs with Cushing's disease. Narrowing of these perforators segregates with the risk of lacunar infarction, particularly in deep grey and white matter (Fig. 10.5). As vascular lumina within infarcted areas are often still patent, tissue necrosis has been interpreted to arise from post-stenotic low-flow states (Lammie 2005).

Smaller, often incomplete and clinically silent infarcts are associated with eccentric collagenous and hyaline sclerosis of arteries and arterioles between 40 and 100 µm in diameter (Lammie 2005). Again, basal nuclei represent a common lesion site in humans and animals in general, whereas dogs and cats show more frequent involvement of the hindbrain.

Increased fragility and rigidity of the degenerated vascular wall predisposes patients to rhexis bleeding if an enduring or sudden increase in transmural pressure occurs. The highest fatality rates are encountered in complete rupture of parenchymal arteries during hypertensive crisis (>200 mmHg). In humans, hypertensive

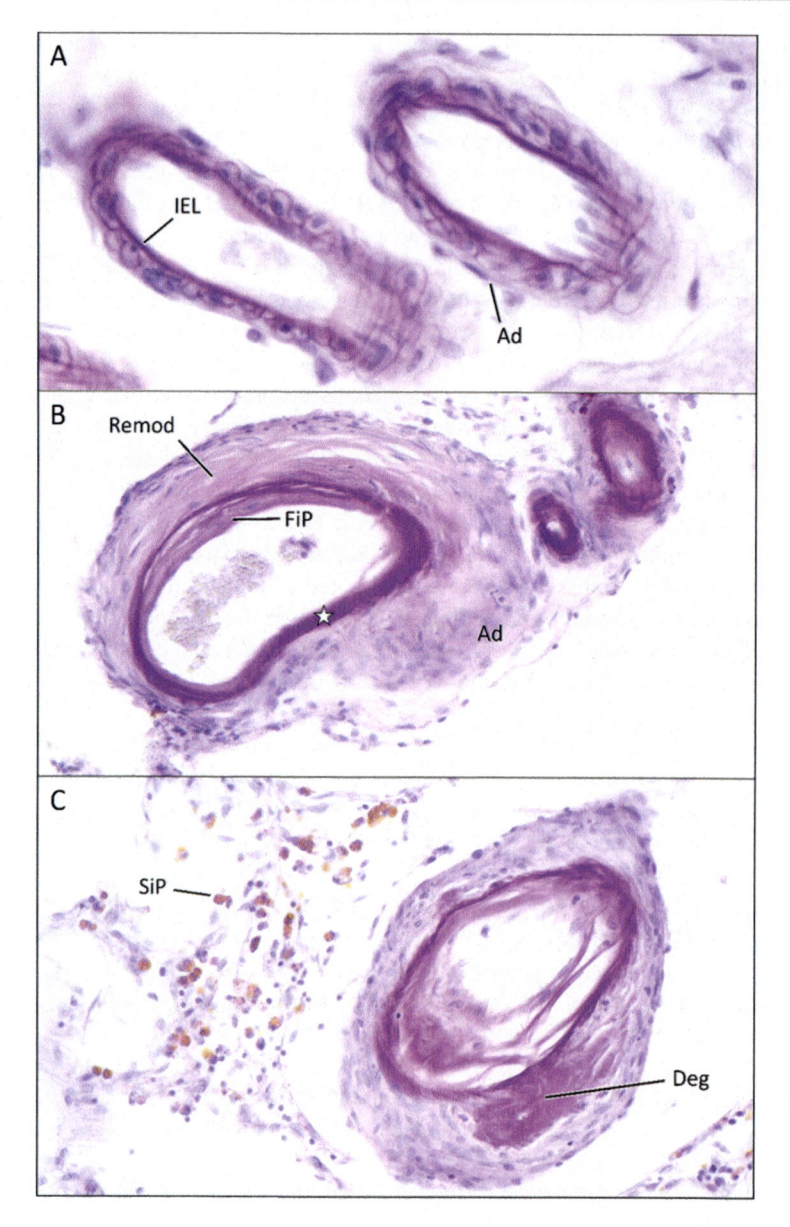

Fig. 10.3 Effects of increased blood pressure on the integrity of meningeal arteries. In unaffected arteries (**a**) the endothelium intimately resides on a delicate, PAS-positive internal elastic lamina (IEL) and the adventitia of small arteries (Ad) consists of a simple sheath of fibroblast like spindle cells. With sustained hypertension, as in this 15-year-old female domestic shorthair cat with chronic kidney disease and hyperthyroidism (**b**), endothelial injury leads to subendothelial fibrointimal proliferation (FiP), thus abluminal displacement, granular disintegration and broadening of the internal elastic lamina (asterisk). Further abluminal layers of the media show fibroblastic remodelling (Remod) that appears contiguous with concentric fibrocollagenous proliferation of

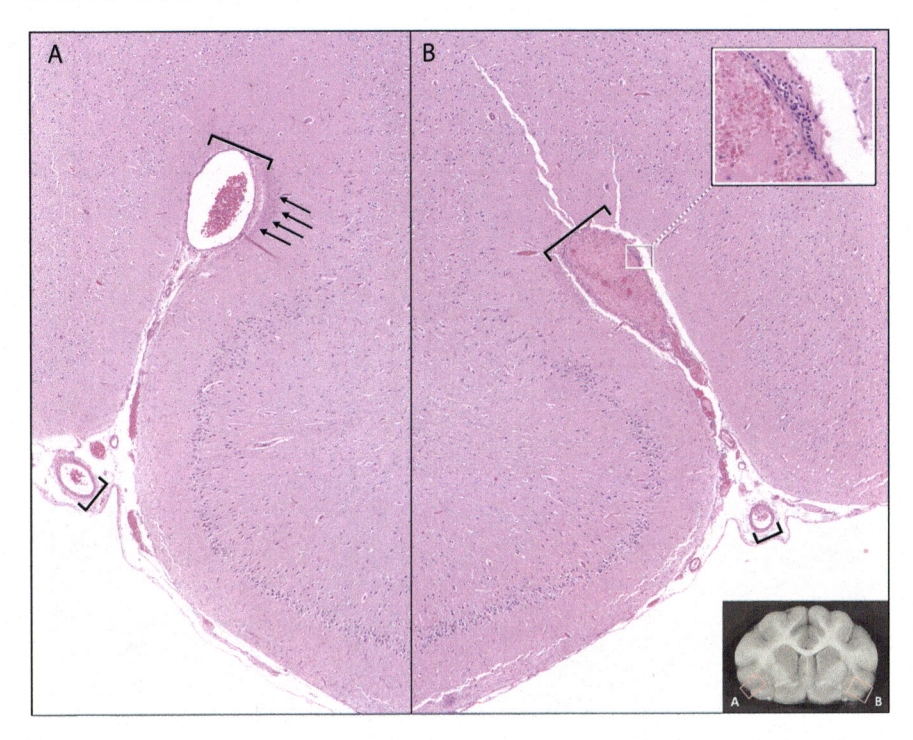

Fig. 10.4 Sulcus base vasodilation in a hypertensive Maine Coon (11-year-old, male, castrated) with chronic kidney disease and hyperthyroidism. Forced vasodilation often is seen first in "exhausted" less muscular blood vessels at the depth of sulci (**a**, **b**: large square brackets) while preceding arterial segments at the surface are capable of keeping their vascular tone up for longer time periods (**a**, **b**: small square brackets). This holds true in particular for basal sulci such as the depicted lateral rhinal sulci. For anatomical orientation, see lower inset. Compression of and leakage of plasma contents into adjacent brain tissue may give rise to a focal glial response (**a**: arrows). Prolonged physical disruption of the vessel wall, on the other hand, can lead to focal perivascular/meningeal inflammation (**b**: upper inset). Histological stain (**a**, **b**): *HE* haematoxylin-eosin

parenchymal haemorrhage (HPBH) most often occurs in the deep parts of the cerebral hemispheres, including basal nuclei, and, less frequently, in the pons and cerebellum (McCarron et al. 2005).

In small animals that died from suspected HPBH, the hindbrain and, in particular, the cerebellopontine angle (Fig. 10.5) appear to be affected more often than other regions. However, a clinical history of hypertension is often missing, and, as in

Fig. 10.3 (continued) the adventitia (Ad). With further advanced stages (**c**), malsupply of the media leads to degeneration of the vessel wall (Deg). Chronic seepage of blood into the subarachnoid space from this or other blood vessels, nearby, is seen as siderophage (SiP) accumulation. Histological stain (**a–c**): *PAS* periodic acid-Schiff histochemistry. Annotations: *Ad* adventitia, *Deg* media degeneration, *FiP* fibrointimal proliferation, *Remod* media remodelling, *SiP* siderophages

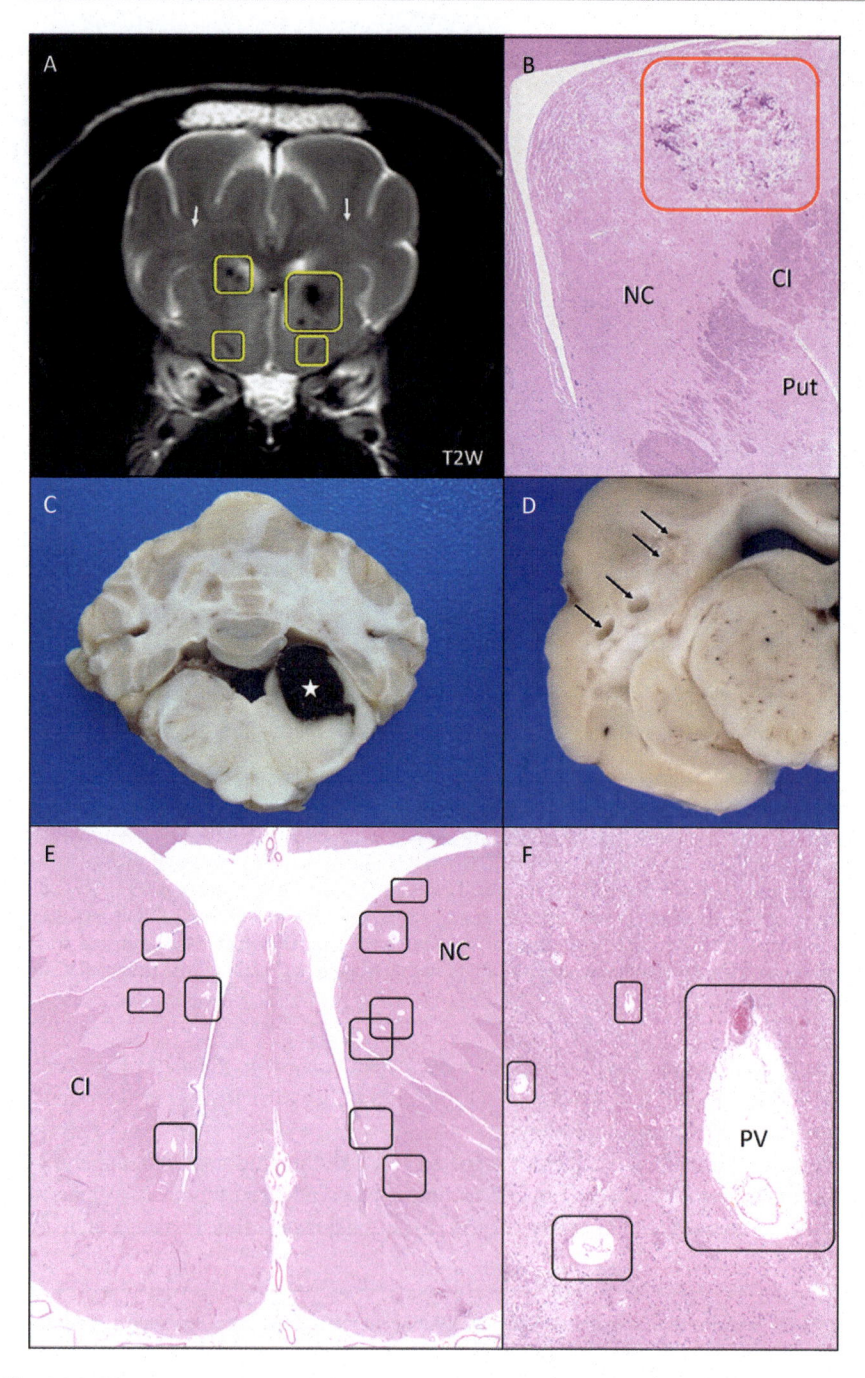

Fig. 10.5 Tissue consequences of hypertension (related images in Fig. 10.8). Corresponding to Fig. 10.8b, c, haemorrhages are most commonly confined to the deep penetrator system of arteries. These frequently are associated with lacunar infarcts as seen in the basal nuclei in this dog (**a**: yellow frame; same dog as in Fig. 10.8) and cat (**b**: red frame; domestic shorthair, 19-year-old,

humans (McCarron et al. 2005), the ruptured blood vessel can rarely be seen in histological sections to show definitively that hypertensive damage of the vascular wall had occurred.

Hypertensive haemorrhages may also evolve from meningeal arteries. It appears, however, that concentric adventitial fibroplasia prevents extensive bleeding. Instead, there is continuous seepage of small amounts of blood into the subarachnoid space (Fig. 10.6), which may evoke vasospasm of neighbouring small arteries and arterioles and thereby cause or aggravate malperfusion of superficial cortical areas (Frontera et al. 2009). The disrupted endothelial surface also attracts platelets to adhere and shows expression of the adhesion molecules ICAM-1 and PECAM-1, as well as the pro-inflammatory cytokines IL-1α, IL-6 and IL-8 (Suzuki et al. 2001).

Notably, all disorders associated with meningeal arterial wall necrosis and rupture are accompanied by significant inflammatory changes (Chalouhi et al. 2013; Suzuki et al. 2001). Depending on the stage, inflammatory infiltrates range from purely polymorphonuclear to pyogranulomatous to purely lymphoid populations and may be diagnostically misleading, particularly if the cells are found in CSF cytology of clinical cases (Fig. 10.6).

10.4 Acute HE: Breaking Through and Breaking Down

Brain infarcts and HPBH may lead to acute or per-acute deterioration in hypertensive patients characterized by more (infarcts) or less (diffuse haemorrhage) focal neurological signs. Classic HE, on the other hand, resembles the clinical manifestation of diffuse hyperperfusion syndrome. Therefore, an abrupt increase in systolic blood pressure by 30 mmHg may thwart the attempt of naïve vasoconstrictors to keep the EPP within physiological limits. In 'hypertension-experienced' vasculature, forced vasodilation has been recognized as an exhaustion phenomenon at sustained systolic pressures >160 mmHg in animals and >200 mmHg in humans (Church et al. 2019; Brown et al. 2007; Nag and Robertson 2005; O'Neill et al. 2013). In most affected

Fig. 10.5 (continued) female, spayed; CKD). Rupture of pre-damaged larger arteries, on the other hand, can lead to lethal parenchymal (**c**: asterisk; Whippet, 9-year-old, male; hypertension of unknown origin) or subarachnoid (not shown) haemorrhages with intracranial pressure rise and widespread malperfusion. In very long-standing cases, clinically mild or even silent lacunar infarcts arise mainly again in basal nuclei, thalamus or subcortical white matter, where they end up as CSF-filled cavities (**d**: arrows; Poodle, 6-year-old, female spayed; CKD). The depicted multiplicity also is referred to as status lacunaris. It ought not to be mistaken as status cribrosus, a rather typical sequela of long-term hypertension characterized by massive concentric enlargement of perivascular spaces (surrounding dilated veins and small arteries likewise) without evidence of tissue necrosis (**e**, **f**: frames; domestic shorthair cat, 13-year-old, female; hyperthyroidism). Histological stain (**b**, **e**, **f**): HE. Landmarks and annotations: *CI* internal capsule, *NC* caudate nucleus, *Put* putamen, *PV* perivascular space of Virchow-Robin. (MRI scans have been kindly made available by PD Dr. Thomas Flegel DACVIM (Neurology) & DECVN and Dr. Shenja Loderstedt DECVN, University of Leipzig, Germany)

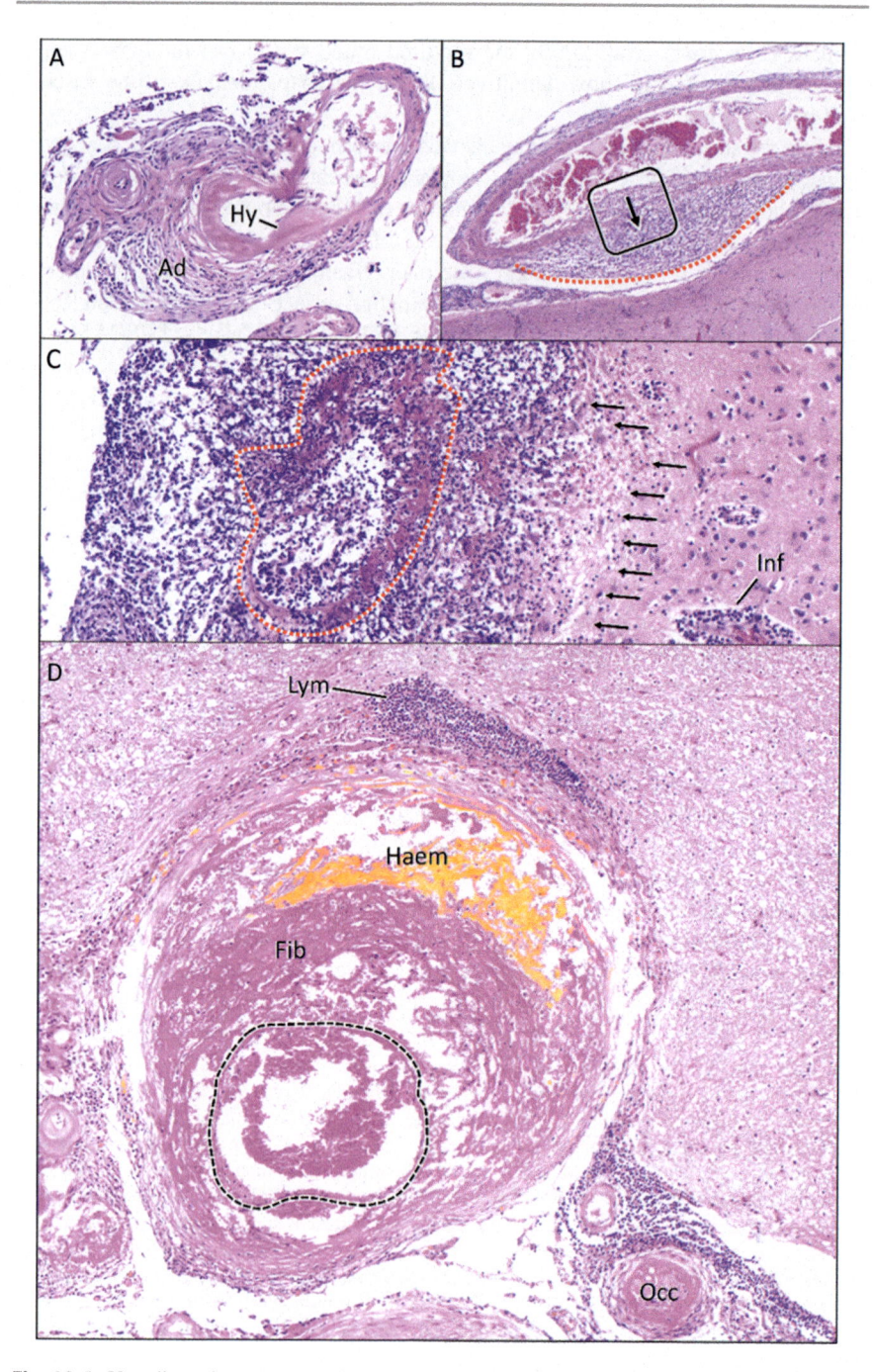

Fig. 10.6 Vasodisruption causes meningitis as seen in these hypertensive cats with chronic kidney disease. With hyaline remodelling (**a**: Hy; domestic shorthair cat, 15-year-old male, castrated) and malsupply, arterial vessel becomes rigid and, hence, vulnerable to shearing forces caused by hydrostatic pressure. As a consequence, arterial wall necrosis and rupture may occur focally (**b**:

vascular territories, the EPP then approaches the MAP. The unbuffered arterial pressure, supported by local metabolic factors, forces segmental dilation of the deep parenchymal arterioles, increases Starling forces (hydrostatic pressure in capillaries), induces blood-brain barrier breakdown (BBBB) and causes fibrinoid necrosis of microvessels <40 μm in diameter (Lammie 2005) (Fig. 10.7).

The pathological equivalents of BBBB are vasogenic oedema and leakage of plasma proteins that appear hyperintense (brighter than normal brain tissue) on water-sensitive T2-weighted and fluid-attenuated inversion recovery (FLAIR) images, iso- to hypointense (equal to or darker than normal brain tissue) on T1-weighted images, variably contrast enhancing on T1-weighted imaging and variably hyperintense on diffusion-weighted imaging (DWI) and on apparent diffusion coefficient (ADC) maps that both trace the movement of extracellular fluids (O'Neill et al. 2013) compared to purely ischaemic lesions (Schwartz 2002).

Both in humans and animals, free fluid is most evident in the subcortical white matter of the parietal and occipital lobes, followed by the semi-oval centre of the frontal lobe (Eberhardt 2018; O'Neill et al. 2013) (Fig. 10.8). This distribution is attributable to the drop in sympathetic innervation of the meningeal arteries in the caudal/posterior direction (Hamel 2006). The recognizable spatial pattern and response to antihypertensive treatment have given rise to the name *posterior reversible encephalopathy syndrome* (PRES) in humans (Hinchey et al. 1996; Hinchey 2008; Lamy et al. 2004). PRES is the most frequent hypertension-associated syndrome in humans with sudden dizziness and headaches, although PRES occasionally can result from other disorders without blood pressure changes (Lamy et al. 2004).

As with dogs and cats, in which HE is overrepresented in kidney disease (Church et al. 2019; O'Neill et al. 2013), PRES seems to affect patients with renal hypertension disproportionately often (Hinchey 2008), which may indicate an additive effect of metabolic and hormonal imbalances and toxic agents associated with azotaemic kidney disease. Furthermore, PRES and associated phenomena have been recognized in human and feline renal transplant recipients (Hinchey et al. 1996; Kyles et al. 1999; Tejani 1983).

Fig. 10.6 (continued) frame; same cat as in **a**) or affect the vessel concentrically (**c**: dashed line, domestic shorthair cat, 14-year-old, female). Necrosis and leakage likewise trigger inflammation that might be held in place if still confined by the adventitia (**b**: dashed line). If the process breaks through, it may spread diffusely to the subarachnoid space diffusely and cause necro-inflammatory reactions and gliosis in the adjacent superficial brain parenchyma (**c**: arrows) and remote inflammatory cell cuffs (**c**: Inf). In large arteries, as the basilar artery seen in (**d**) (domestic shorthair cat, 14-year-old, female, spayed), endothelial disruption may cause insudation and intramural bleeding with fibrin clots (Fibr) and blood pigment, histologically seen as brown siderophages (outer fringe) and the iron-free yellow haematin (Haem). Other smaller arteries undergo occlusive lipohyalinosis (Occ). Again tissue necrosis and leakage caused inflammation. As actual tissue necrosis has occurred, the composition is predominantly lymphocytic (Lym) at this stage. Histological stain (**a–d**): HE. Annotations: *Ad* adventitia, *Fib* fibrin insudation/clot, *Haem* haematin, *Hy* hyaline degeneration, *Inf* inflammatory cell cuff, *Lym* lymphocytic infiltrate, *Occ* occluded artery

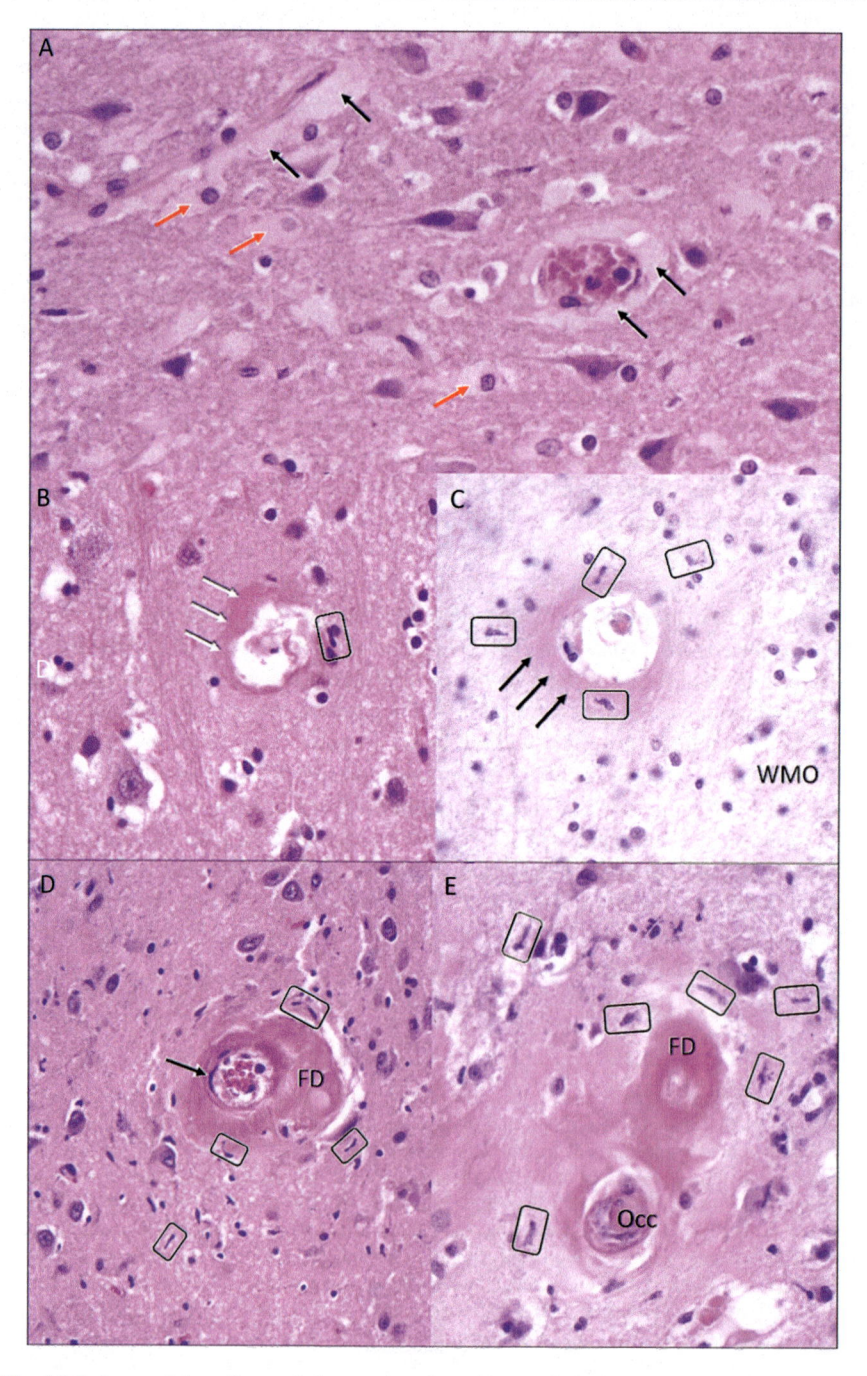

Fig. 10.7 Intra-axial small vessel changes in a dog with protein-losing nephropathy (**a**) and three cats (**b–e**) with chronic kidney disease. Acute breakthrough causes blood-brain barrier breakdown with leakage of plasma (**a**; German shepherd dog, 7-year-old, female-spayed; protein losing nephropathy), which in grey matter is immediately taken up by astrocytic end feet (black arrows),

Notably, the distribution of subcortical white matter oedema often starts bilaterally asymmetrical (Fig. 10.8) and may progress to a symmetrical fashion (Church et al. 2019). Perivascular astrocytes frequently show clasmatodendrosis mimicking or progressing into gemistocytic reaction or even Alzheimer type II-like astrocytes (Church et al. 2019). Moreover, perivascular microgliosis is seen (Fig. 10.7).

Grey matter oedema has been documented in ischaemic lesions of the caudate nuclei and thalamus on MRI (O'Neill et al. 2013) and, likewise, on histological sections (Fig. 10.7). Although human patients show early involvement of the cerebral cortex on MRI (Rykken and Mckinney 2014), no such changes have been reported to date in dogs or cats (O'Neill et al. 2013). This absence, however, might be a question of timing, as there is an increase in the expression of the water-channel protein aquaporin-4 in hypertensive feline cortices, independent of seizure activity, suggesting that grey matter oedema is rapidly re-absorbed and removed by astrocytes (unpublished observation). The shift from extracellular space to astrocytic cytoplasm (Fig. 10.7) may correspond to MRI features suggestive of cytotoxic oedema (Rykken and Mckinney 2014) and resembles a rescue phenomenon to attenuate disturbed homeostasis and increased pressure in affected grey matter.

Acute hypertensive BBBB has been modelled extensively by vascular occlusion and application of the pressure-altering agents angiotensin-II, norepinephrine and epinephrine (Nag and Robertson 2005). Three endothelial mechanisms were identified. First, stressed endothelial cells lose their negative surface charge (glycocalyx) and attract protein anions, which are taken up by endocytosis. Secondly, proteins taken up and those bound to the luminal surface are carried to the abluminal surface via caveolae and transendothelial channels. The abluminal release of these proteins results in the typical bland perivascular plasma extravasation centred on histologically preserved capillary endothelium (Lammie 2005). If the blood pressure remains elevated, disruption of tight junctions results in excessive pericellular leakage of water and proteins. If restricted to a few vessels, such leakage can lead to focal necrosis and cavitation (Nag and Robertson 2005). If larger areas are affected, progressive volume gain leads to mass shifting and a global intracranial pressure rise, both of which constitute life-threatening emergencies (Brown et al. 2005; Eberhardt 2018). With the cerebral hemispheres being preferentially affected, untreated brain oedema may force a descending transtentorial herniation, which

Fig. 10.7 (continued) redistributed through the astrocytic syncytium and impresses as cellular post-resorption oedema (red arrows). Continuation of leakage passes by the end feet into the perivascular grey (**b**: white arrows; domestic shorthair cat, 11-year-old female, spayed) and white matter [**c**: black arrows; Persian cat, 15-year-old male, castrated (same as **e**)], where the fluids remain in extracellular space as white matter oedema. The pervasive effect triggers an inflammatory response by perivascular microglial cells (black frames, **b** through **e**). Enduring stress to the vessel walls further leads to fibrinoid degeneration (**d, e**: FD) with successive destruction of the endothelium (**d**: arrow; domestic shorthair cat, 16-year-old, female; CKD and diabetes mellitus) and vascular occlusion (Occ). Histological stain (**a–e**): HE. Annotations: *FD* fibrinoid (also: fibrinohyaline) degeneration, *Occ* vascular occlusion, *WMO* white matter oedema

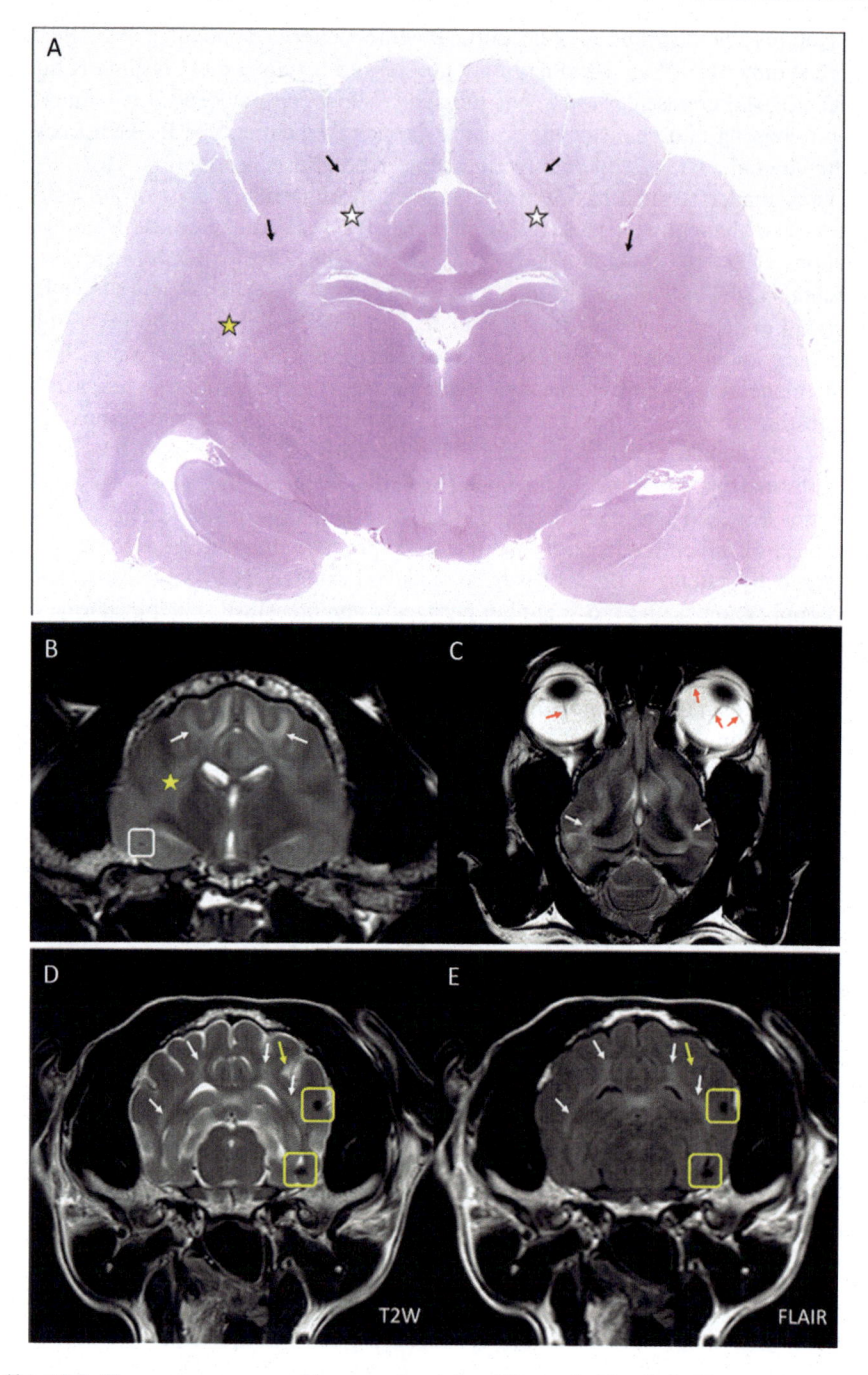

Fig. 10.8 Tissue consequences of hypertension (related images in Fig. 10.5). Net uptake of fluid gives rise to the typical fronto-parieto-occipital subcortical oedema (**a**: white asterisks; domestic shorthair cat, 13-year-old male, castrated; hyperaldosteronism) that variably extends into the thalamic radiation (**a**: yellow asterisk), semi-oval centre and fasciculus subcallosus. Histologically

translates into caudal displacement of the cerebellum, vermal *foramen magnum* herniation, and, finally, compression of the medullary respiratory control centre and death (Brown et al. 2005; Church et al. 2019).

10.5 Hypertension and Seizures

Cerebrovascular disorders are a major risk factor for late-onset epilepsy in elderly people (Ferlazzo et al. 2016). It is therefore not surprising that hypertension in mid-life adults inflicts a hazard ratio for epilepsy of 1.26 (1.05–1.51, CI 95%) (Johnson et al. 2018) only that seizure development via vascular lesions reflects one single setting of epileptogenesis in hypertensive patients only. In addition to structural epilepsy caused by epileptogenic scars after stroke, cortical microinfarcts and (inhibitory) de-afferentiation, it has been shown that dysregulation of the renin-angiotensin system can give rise to so-called unprovoked seizures on its own (Gasparini et al. 2019). Increased angiotensin concentrations appear to lower the neuronal excitation threshold via AT_1 receptors either directly or via interference with dopaminergic and GABAergic circuits (Tchekalarova and Georgiev 2005). As proof of principle, the pro-excitatory effect of angiotensin II has been reversed successfully by the AT_1 receptor antagonist losartan in an epileptic rat mode (Pereira et al. 2010). The interdependence gains further credence by a retrospective study that established hypertension as an independent risk factor (OR: 1.57) for development of unprovoked seizures in a cohort of human epilepsy patients after adjustment for cerebrovascular diseases (Ng et al. 1993).

It is further assumed that the vasogenic oedema may be directly related to seizure development in about two-thirds of human PRES patients. As hypertension represents just one of several causes of PRES in people, the causative association is less than clear (Gasparini et al. 2019; Ng et al. 1993).

Among our companion animals, seizures appear to be more common in hypertensive cats, even though firm epidemiological data are missing. Our own

Fig. 10.8 (continued) white matter oedema frequently is accompanied by a grey-white interface oedema (**a**: black arrows). On brain scans of hypertensive cats (**b, c**: domestic shorthair, 12-year-old, female; hyperaldosteronism) and dogs (**d, e**: cross-breed, 11-year-old, male; hypertension of unknown origin), brain oedema on MRI usually presents as hyper-intensities on T2 (**b–d**: white arrows) and FLAIR (**e**: white arrows). Similar to PRES in people, the intensity extends into the semi-oval centre (**b**: yellow asterisk), thalamic/optic radiation and occipital subcortical white matter (**c**: white arrows). Dilation of deep lateral rhinal sulcus arteries occasionally becomes evident as small, seemingly temporal signal voids (**b**: white frame). Moreover, horizontal/dorsal scans may depict concurrent hypertensive retinal detachment (**c**: red arrows). Among vascular accidents, MRI picks up post-resorptive cavities of ischaemic infarcts (**d, e**: yellow arrows) and the signal voids caused by iron accumulation in hypertensive brain haemorrhages (yellow frames). Histological stain (**a**): HE. (MRI scans have been kindly made available by PD Dr. Thomas Flegel DACVIM (Neurology) & DECVN and Dr. Shenja Loderstedt DECVN, University of Leipzig, Germany)

preliminary investigations in a group of 36 cats with hypertension due to CKD and/or hyperthyroidism revealed an overall seizure frequency of 25%. Furthermore, ongoing postmortem studies in epileptic cats with fatal outcome provide straightforward pathological features of hypertensive encephalopathy in about 15% of cases. However, hypertension may have contributed to vascular lesions that were seen on postmortem in another 10% of epileptic cats. Notably the oedema pattern across epileptic and non-epileptic cats with hypertensive encephalopathy does not differ significantly. It remains therefore questionable whether the PRES-like vasogenic oedema in feline hypertensive breakthrough implicates an individual risk of seizure development.

Prospective clinical studies are required to establish prevalence and risk of epilepsy in hypertensive dogs and cats further and to separate the effects of structural brain damage and neuromodulation by angiotensin.

10.6 Clinical Presentation and Diagnosis of Hypertensive Brain Lesions in Dogs and Cats

As there is sufficient evidence to prove a causal relationship between neurological dysfunction/brain lesions and sustained hypertension in dogs and cats, any neurological deterioration in hypertensive pet patients—and in particular those with kidney disease—must be considered a consequence of poor blood pressure control.

Unfortunately, the clinical presentation of hypertensive brain lesions is non-specific. To date, we have no clear understanding of whether or how long-standing hypertensive vasoconstriction and neuroinflammation might lead to neurodegeneration, cognitive decline and behavioural changes or how they influence coordination of movement in dogs and cats.

Acute neurological complications manifest as focal vascular accidents, vestibular disease (head tilt, circling and falling, vestibular ataxia and nystagmus) and other neuroophthalmological phenomena, diffuse forebrain signs, such as changes in mentation (obtundation, stupor, coma), altered behaviour and seizures or a combination of these signs. Clinical and laboratory signs of the underlying disease or other TOD, such as hyphaema and retinal detachment [especially in cats (Littman 1994; Maggio et al. 2000)], as well as biomarkers of kidney disease and cardiac dysfunction, might be helpful indicators of an association with hypertension but do not, on their own, justify the risk of hypotensive crisis and distributive shock upon a premature decision to render antihypertensive treatment.

In a clinical setting, the most reliable tool for clinical diagnosis of hypertension-induced brain lesions is MRI (Table 10.1). This modality shows characteristic bilateral asymmetric ill-defined T2-weighted hyperintensities of subcortical white matter, particularly in the parietal and occipital lobes in HE. Vascular accidents, on the other hand, may present as T2-hyperintense ischaemic lesions or haemorrhages causing signal voids on gradient-echo imaging. Moreover, MRI allows the grading of subsequent mass effects.

Table 10.1 Hypertensive brain lesions most frequently depicted on MRI scans in dogs and cats

Tissue change	Signal characteristics	Preferred areas/spatial characteristics
Vasogenic white matter oedema (PRES-like)	*Intensity*[a] ↑ T2W, T2W FLAIR, T2*W, DWI, ADC →–↓ T1W, T1W FLAIR ↗ T1W post contrast *Borders* (a) Ill-defined margin to normal WM (b) Well-defined margin to adjacent GM	Corona radiata > semi-oval centre > internal capsule Parietal, occipital > frontal > temporal (multifocal-diffuse, bilateral, often asymmetric)
Cytotoxic grey matter oedema and actual ischaemic necrosis (infarcts)	*Intensity*[b] ↑ T2W, T2W FLAIR, DWI[c] →–↓ T1W, T1W FLAIR ↑ T1W post contrast ↓ ADC[c] *Borders* (a) Ill-defined margin to normal GM (b) Fairly well-defined margin to adjacent WM	Parietal cortex, caudate nucleus > other basal nuclei, parahippocampal cortex > paraflocculus, > thalamus, brainstem (focal-multifocal, unilateral or asymmetric)
Parenchymal haemorrhages	*Intensity*[b] ↘–↓ T1W, T1W FLAIR, T2W, T2W FLAIR ↓↓: T2*W → T1W post contrast *Borders* Well-defined	Parietal cortex, caudate nucleus, paraflocculus > thalamus, brainstem Extensive haemorrhage occasionally in cerebellopontine angle. (focal-oligofocal, unilateral or asymmetric)
Inactive lacunar infarcts	*Intensity*[a,b] ↑ T2W ↓ T1W, T1W FLAIR, T2W FLAIR → T1W post contrast *Borders* Well-defined	Parietal cortex, basal nuclei, subcortical white matter (parietal, temporal) (focal-oligofocal, unilateral or asymmetric)

WM white matter, *T1W* T1 weighted, *T2W* T2 weighted, *FLAIR* fluid attenuated inversion recovery, *GM* grey matter, *DWI* diffusion weighted imaging, *ADC* apparent diffusion coefficient, *PRES* posterior reversible encephalopathy syndrome

↗–↑: Mildly to markedly hyperintense; →: Isointense; ↘–↓: Mildly to markedly hypointense

[a] Compared to unaffected WM
[b] Compared to unaffected grey matter
[c] Diffusion maps may change with combined vasogenic and cytotoxic oedema

Blood pressure measurement is a standard procedure in vascular accidents. The characteristic spatial pattern of brain oedema and BBBB in suspected HE should prompt clinicians to assess blood pressure. Any increase in blood pressure from a level that poses a moderate risk of TOD (160–179/100–119 mmHg) to one that poses a high risk of TOD ($\geq$180/120 mmHg) (Brown et al. 2007) might support the diagnosis. However, it may be necessary to repeat the measurements; in the case of normotensive episodes, they should ideally be performed by arterial catheterization and should comply with current consensus guidelines (Acierno et al. 2018) (see Chap. 2).

Therefore, it is worth considering that ICP increases, encephalovascular accidents and other intracranial diseases may affect systemic blood pressure. Any ICP rise can trigger the Cushing reflex, characterized by bradycardia and hypertension through stimulation of the sympathetic nervous system (Wan et al. 2008). In the early stages of ischaemic infarcts, disrupted neurogenic control of cardiac action and vascular tone, baroreflex dysregulation, sympathetic hyperactivity and distress can contribute to hypertensive crisis, while after several days, patients may show a decrease of 20–30/10 mmHg (Dunn and Nelson 2014).

The diagnosis of HE may also be supported by transcranial Doppler sonographic evaluation of intracranial blood flow and vascular resistance. This technique can reliably measure the perfusion of specific arteries in humans and dogs. More sophisticated indices (resistive index, pulsatility index, ratio of systolic to diastolic mean velocity) mirror the mixed impact of autoregulation and ICP on EBF. Unfortunately, the algorithms are not yet sufficiently validated for application in domestic animals; it is difficult to obtain reliable values in cats, and there are only a few experts currently undertaking this procedure (Sasaoka et al. 2018).

Once the neurological problem has been associated with hypertension, animals should be examined for underlying disease, including kidney disease, hyperthyroidism, Cushing's disease, diabetes mellitus, primary hyperaldosteronism and pheochromocytoma (Kent 2009; O'Neill et al. 2013). In aged animals, the presence of multiple conditions predisposing to hypertension occurring concomitantly is not uncommon.

There is also a considerable fraction of animals—up to 20% in cats—that might suffer from idiopathic hypertension without abnormalities in complete blood counts, serum biochemistry, urinalysis or extended test panels (Acierno et al. 2018; Brown et al. 2007).

10.7 Treatment and Prognosis

Both encephalovascular accidents in hypertensive patients and HE warrant immediate and aggressive treatment and monitoring in a 24-h care unit (Acierno et al. 2018; Brown et al. 2007). In the first stage, proper treatment includes control of ICP, hypertension and brain perfusion. In chronic hypertension, vasogenic autoregulation of brain and kidney perfusion has adapted to higher blood pressures; thus, abrupt antihypertensive treatment may result in critical hypoperfusion of the brain, which becomes even worse if ICP is increased (Acierno et al. 2018; Brown et al. 2007; Dunn and Nelson 2014). A recommended practice, therefore, is to decrease the systolic blood pressure by approximately 10% over the first hour and by another 15% over the next few hours (Elliott 2003), followed by a gradual decrease until physiological pressures have been reached (Acierno et al. 2018; Brown et al. 2007).

While it is beyond the scope of this chapter to suggest specific medications (see Chaps. 12 and 13), a case series on HE in dogs and cats presents an empirical regimen that circumvented the risk of hypotensive complication by oral use of the calcium channel blocker amlodipine besylate (0.1 mg/kg PO q 12 h) in renal hypertension and hypertension of unknown causes (Kent 2009; O'Neill et al. 2013). Dogs received this treatment in combination with the angiotensin-converting enzyme inhibitor, enalapril (0.5 mg/kg PO q 24 h) (O'Neill et al. 2013). All animals improved within 72 h, some within a few hours after the first treatment, and all animals returned to normal neurological states (O'Neill et al. 2013). Another study in cats with HE upon kidney transplantation reported an effective response to subcutaneous injection of 2.5 mg hydralazine per cat. Out of the entire cohort of 21 cats affected by postoperative hypertension, one animal failed to respond to hydralazine, and one became hypotensive (Kyles et al. 1999).

Given that 24-h blood pressure monitoring is available, intravenous treatment with other antihypertensive drugs should be considered to abolish further neurological emergencies. In that case, focus also needs to be placed on ICP increase, brain oedema, brain haemorrhage and seizure control. As the first two entail diuretic therapies and the use of hypertonic agents, their potential to increase blood pressures transiently requires special attention. Hypertensive patients, especially those with increased ICP, are at risk for anaesthetic accidents during possible brain surgery for evacuation of haematoma.

Given that 25–45% of PRES cases in humans show persistent MRI lesions and 10–25% of patients have persistent neurological deficits (Heo et al. 2016), the outcome of treated HE in dogs and cats appears comparatively optimistic. Unfortunately, however, systematic studies in pets are still lacking, and the number of published cases with follow-up is too small an evidence base on which to confidently recommend management protocols.

10.8 Conclusion

High metabolic demands, a sophisticated vasculature and limited space for expansion render the CNS a preferential site of direct and indirect hypertensive damage. The involvement of the CNS in a hypertensive disorder entails high morbidity and a considerable risk of fatality if left untreated. As neurological complications are non-specific, evaluation of blood pressure and TOD of other organs, particularly the eye, should be included in the work-up of any neurological patient.

References

Acierno MJ, Brown S, Coleman AE et al (2018) ACVIM consensus statement: guidelines for the identification, evaluation, and management of systemic hypertension in dogs and cats. J Vet Intern Med 32:1803–1822

Bkaily G, Shbaklo H, Taoudi-Benchekroun M et al (2000) Nitric oxide relaxes the vascular smooth muscle independently of endothelin-1- and U46619-induced intracellular increase of calcium. J Cardiovasc Pharmacol 36:S110–S116

Bor-Seng-Shu E, Kita WS, Figueiredo EG et al (2012) Cerebral hemodynamics: concepts of clinical importance. Arq Neuropsiquiatr 70:352–356

Brown CA, Munday JS, Mathur S et al (2005) Hypertensive encephalopathy in cats with reduced renal function. Vet Pathol 42:642–649

Brown S, Atkins C, Bagley R et al (2007) Guidelines for the identification, evaluation, and management of systemic hypertension in dogs and cats. J Vet Intern Med 21:542–558

Chalouhi N, Hoh BL, Hasan D (2013) Review of cerebral aneurysm formation, growth, and rupture. Stroke 44:3613–3622

Church ME, Turek BJ, Durham AC (2019) Neuropathology of spontaneous hypertensive encephalopathy in cats. Vet Pathol 56(5):778–782

Czosnyka M, Smielewski P, Czosnyka Z et al (2003) Continuous assessment of cerebral autoregulation: clinical and laboratory experience. Acta Neurochir Suppl 86:581–585

Dunn KM, Nelson MT (2014) Neurovascular signaling in the brain and the pathological consequences of hypertension. Am J Physiol Heart Circ Physiol 306:H1–H14

Eberhardt O (2018) Hypertensive crisis and posterior reversible encephalopathy syndrome (PRES). Fortschr Neurol Psychiatr 86:290–300

Edvinsson L, Krause DN (2002) Cerebral blood flow and metabolism. Lippincott, Williams & Wilkins, Philadelphia

Elliott WJ (2003) Management of hypertension emergencies. Curr Hypertens Rep 5:486–492

Ferlazzo E, Gasparini S, Beghi E, Sueri C, Russo E, Leo A, Labate A, Gambardella A, Belcastro V, Striano P, Paciaroni M, Pisani LR, Aguglia U (2016) Epilepsy in cerebrovascular diseases: review of experimental and clinical data with meta-analysis of risk factors. Epilepsia 57(8):1205–1214

Frontera JA, Fernandez A, Schmidt JM et al (2009) Defining vasospasm after subarachnoid hemorrhage: what is the most clinically relevant definition? Stroke 40:1963–1968

Fujishima M, Ibayashi S, Fujii K et al (1995) Cerebral blood flow and brain function in hypertension. Hypertens Res 18:111–117

Gasparini S, Ferlazzo E, Sueri C, Cianci V, Ascoli M, Cavalli SM, Beghi E, Belcastro V, Bianchi A, Benna P, Cantello R, Consoli D, De Falco FA, Di Gennaro G, Gambardella A, Gigli GL, Iudice A, Labate A, Michelucci R, Paciaroni M, Palumbo P, Primavera A, Sartucci F, Striano P, Villani F, Russo E, De Sarro G, Aguglia U (2019) Hypertension, seizures, and epilepsy: a review on pathophysiology and management. Neurol Sci 40(9):1775–1783

Golding EM, Marrelli SP, You J et al (2002) Endothelium-derived hyperpolarizing factor in the brain: a new regulator of cerebral blood flow? Stroke 33:661–663

Halpern W, Osol G, Coy GS (1984) Mechanical behavior of pressurized in vitro prearteriolar vessels determined with a video system. Ann Biomed Eng 12:463–479

Hamel E (2006) Perivascular nerves and the regulation of cerebrovascular tone. J Appl Physiol 100:1059–1064

Heo K, Cho KH, Lee MK et al (2016) Development of epilepsy after posterior reversible encephalopathy syndrome. Seizure 34:90–94

Hinchey JA (2008) Reversible posterior leukoencephalopathy syndrome: what have we learned in the last 10 years? Arch Neurol 65:175–176

Hinchey J, Chaves C, Appignani B et al (1996) A reversible posterior leukoencephalopathy syndrome. N Engl J Med 334:494–500

Jacob F, Polzin DJ, Osborne CA et al (2003) Association between initial systolic blood pressure and risk of developing a uremic crisis or of dying in dogs with chronic renal failure. J Am Vet Med Assoc 222:322–329

Johnson EL, Krauss GL, Lee AK, Schneider ALC, Dearborn JL, Kucharska-Newton AM, Huang J, Alonso A, Gottesman RF (2018) Association between midlife risk factors and late-onset epilepsy: results from the atherosclerosis risk in communities study. JAMA Neurol 75(11):1375–1382

Kent M (2009) The cat with neurological manifestations of systemic disease. Key conditions impacting on the CNS. J Feline Med Surg 11:395–407

Kyles AE, Gregory CR, Wooldridge JD et al (1999) Management of hypertension controls postoperative neurologic disorders after renal transplantation in cats. Vet Surg 28:436–441

Lammie AG (2005) Small vessel disease. In: Kalimo H (ed) Cerebrovascular diseases. ISN Neuropath Press, Basel

Lamy C, Oppenheim C, Meder JF et al (2004) Neuroimaging in posterior reversible encephalopathy syndrome. J Neuroimaging 14:89–96

Littman MP (1994) Spontaneous systemic hypertension in 24 cats. J Vet Intern Med 8:79–86

Liu Y, Dong YH, Lyu PY et al (2018) Hypertension-induced cerebral small vessel disease leading to cognitive impairment. Chin Med J 131:615–619

Maggio F, Defrancesco TC, Atkins CE et al (2000) Ocular lesions associated with systemic hypertension in cats: 69 cases (1985–1998). J Am Vet Med Assoc 217:695–702

McCarron MO, Cohen NR, Nicoll JAR (2005) Parenchymal brain hemorrhage. In: Kalimo H (ed) Cerebrovascular diseases. ISN Neuropath Press, Basel

Miller JB, Suchdev K, Jayaprakash N et al (2018) New developments in hypertensive encephalopathy. Curr Hypertens Rep 20:13

Nag S, Robertson DM (2005) The brain in hypertension. In: Kalimo H (ed) Cerebrovascular diseases. ISN Neuropath Press, Basel

Ng SK, Hauser WA, Brust JC, Susser M (1993) Hypertension and the risk of new-onset unprovoked seizures. Neurology 43(2):425–428

O'Neill J, Kent M, Glass EN et al (2013) Clinicopathologic and MRI characteristics of presumptive hypertensive encephalopathy in two cats and two dogs. J Am Anim Hosp Assoc 49:412–420

Oppenheimer B, Fishberg A (1928) Hypertensive encephalopathy. Arch Intern Med 42:264–278

Osol G, Halpern W (1985) Myogenic properties of cerebral blood vessels from normotensive and hypertensive rats. Am J Phys 249:H914–H921

Owens GK, Rabinovitch PS, Schwartz SM (1981) Smooth muscle cell hypertrophy versus hyperplasia in hypertension. Proc Natl Acad Sci USA 78:7759–7763

Pantoni L (2010) Cerebral small vessel disease: from pathogenesis and clinical characteristics to therapeutic challenges. Lancet Neurol 9:689–701

Pereira MG, Becari C, Oliveira JA, Salgado MC, Garcia-Cairasco N, Costa-Neto CM (2010) Inhibition of the renin-angiotensin system prevents seizures in a rat model of epilepsy. Clin Sci (Lond) 119(11):477–482

Pires PW, Dams Ramos CM, Matin N et al (2013) The effects of hypertension on the cerebral circulation. Am J Physiol Heart Circ Physiol 304:H1598–H1614

Rivera-Lara L, Zorrilla-Vaca A, Geocadin RG et al (2017) Cerebral autoregulation-oriented therapy at the bedside: a comprehensive review. Anesthesiology 126:1187–1199

Rykken JB, Mckinney AM (2014) Posterior reversible encephalopathy syndrome. Semin Ultrasound CT MR 35:118–135

Rylander H, Eminaga S, Palus V et al (2014) Feline ischemic myelopathy and encephalopathy secondary to hyaline arteriopathy in five cats. J Feline Med Surg 16:832–839

Sasaoka K, Nakamura K, Osuga T et al (2018) Transcranial Doppler ultrasound examination in dogs with suspected intracranial hypertension caused by neurologic diseases. J Vet Intern Med 32:314–323

Schwartz RB (2002) Hyperperfusion encephalopathies: hypertensive encephalopathy and related conditions. Neurologist 8:22–34

Strandgaard S, Paulson OB (1990) Hypertension: pathophysiology, diagnosis, and management. In: Laragh JH, Brenner BM (eds) Hypertensive disease and the cerebral circulation. Raven Press, New York, pp 399–416

Suzuki K, Masawa N, Takatama M (2001) The pathogenesis of cerebrovascular lesions in hypertensive rats. Med Electron Microsc 34:230–239

Tchekalarova J, Georgiev V (2005) Angiotensin peptides modulatory system: how is it implicated in the control of seizure susceptibility? Life Sci 76(9):955–970

Tejani A (1983) Post-transplant hypertension and hypertensive encephalopathy in renal allograft recipients. Nephron 34:73–78

Touyz RM, Alves-Lopes R, Rios FJ et al (2018) Vascular smooth muscle contraction in hypertension. Cardiovasc Res 114:529–539

Wan WH, Ang BT, Wang E (2008) The Cushing response: a case for a review of its role as a physiological reflex. J Clin Neurosci 15:223–228

Yoshida M, Suzuki A, Itoh T (1994) Mechanisms of vasoconstriction induced by endothelin-1 in smooth muscle of rabbit mesenteric artery. J Physiol 477:253–265

Part III

Pharmacology and Therapeutic Use of Antihypertensive Drugs

Pharmacology of Antihypertensive Drugs

11

Jonathan Elliott and Ludovic Pelligand

11.1 Introduction

Chapter 1 of this book discussed the pathophysiology of hypertension, focussing on that associated with chronic kidney disease (CKD) since this is the most common scenario faced by veterinary practitioners where antihypertensive treatment is required. The take home message from Chap. 1 is that CKD contributes or exacerbates hypertension through:

- Dysregulation of the sympathetic nervous system (renal afferent nerves and baroreceptors)
- Activation of the renin-angiotensin-aldosterone system (RAAS; locally within the kidney)
- Dysfunction of endothelial cell function (various mechanisms, e.g. oxidative stress, RAAS activation, asymmetric dimethylarginine accumulation)

Thus the major control points at which blood pressure regulation is disturbed in the hypertensive patient are the regulation of peripheral resistance and regulation of circulating fluid volume driven by overactivity of the sympathetic nervous system and the RAAS and exacerbated by endothelial cell dysfunction (Fig. 1.10 summarises the interactions between these systems in the pathophysiology of hypertension in CKD).

Thus it would be logical to assess the selection of antihypertensive agents with this understanding in mind as the ideal antihypertensive agent would normalise vascular tone (whatever is driving the increase in vascular tone) and circulating fluid volume and correct endothelial cell dysfunction.

J. Elliott (✉) · L. Pelligand
Department of Comparative Biomedical Sciences, Royal Veterinary College, University of London, London, UK
e-mail: jelliott@rvc.ac.uk; lpelligand@rvc.ac.uk

In this chapter we will focus on the pharmacology of drugs inhibiting the two main systems that are thought to be dysfunctional in hypertension secondary to CKD, namely the sympathetic nervous and the RAAS systems. Non-competitive inhibition of the vascular actions of both noradrenaline and angiotensin II can be achieved by blocking calcium entry into vascular smooth muscle cells as both of these mediators require calcium to enter the cell to mediate tonic vasoconstriction. Calcium channel blockers will be discussed in some depth because of the experience of their use in veterinary medicine (particularly amlodipine).

11.2 Antihypertensive Drug Targets

11.2.1 Sympathetic Nervous System

In human medicine overactivity of the sympathetic nervous system is well documented in hypertensive patients (Grassi et al. 2015; see Chap. 1). Although this is not well studied in veterinary patients, it seems likely to be the case also. The history of antihypertensive drugs for human use started in the 1950s with identifying ways of inhibiting sympathetic nerve activity. The first drugs used were inhibitors of ganglionic transmission (e.g. hexamethonium), but these agents blocked transmission at both the sympathetic and parasympathetic ganglia with all the side effects that parasympathetic blockade causes (dry mouth, constipation).

The noradrenergic neuron blocking drugs, such as guanethidine, gave selectivity for the sympathetic system. These drugs are taken up into sympathetic nerve endings on the transporters that return noradrenaline to the nerve (neuronal uptake or uptake 1), become concentrated in the vesicles and block the fusion of the storage vesicles with the nerve terminal membrane, thus preventing neurotransmission. Eventually these drugs damage sympathetic nerves, and the mode of action meant that reflex activation of the sympathetic system could not over-ride the blockade leading to postural hypotension, diarrhoea and impotence.

As pharmacology developed as a discipline and scientific discovery of how neurotransmission was regulated advanced, new drugs were discovered with greater selectivity of action and more controlled effects on the sympathetic nervous system resulting in fewer clinical side effects. Prior to the development of effective ways of inhibiting the renin-angiotensin-aldosterone system, antagonists at adrenoceptors were used to treat hypertension. Beta-blockers are cardioprotective, reducing the detrimental effects of overstimulation of the heart by the sympathetic nervous system, and they are also mildly antihypertensive. The concept of how beta-blockers work as antihypertensives has developed over the years. Blockade of postsynaptic peripheral beta-adrenoceptors is probably not as important as blockade of pre-synaptic beta-receptors (activation of which enhances the release of noradrenaline from sympathetic nerve endings; Rosen et al. 1990) and central actions to reduce central drive for sympathetic nerve activity (Grassi 2016). Thus, the ability of beta-blockers to penetrate into the central nervous system may be important for their antihypertensive actions. In human medicine, non-selective beta-blockers increased the risk of asthmatic attacks, and in the 1960s the discovery of β_1 and β_2 subtypes of adrenoceptor allowed the development of cardioselective beta-blockers. In human patients beta-blockers are not generally first-line drugs for the management of essential hypertension but are used in patients with angina or heart failure secondary to their hypertension. In veterinary patients, beta-blockers have been used infrequently as antihypertensive agents, and the range of different drugs available have not been assessed in a systematic way.

Drugs with peripheral actions to inhibit sympathetic tone to the vasculature (alpha-adrenoceptor blocking drugs) were the next development in antihypertensive medication. Phentolamine and phenoxybenzamine are non-selective alpha-adrenoceptor antagonists, the former competitive reversible and the latter a competitive but irreversible antagonist (alkylate receptors), which were used as antihypertensive agents in the 1970s but led to marked reflex tachycardia (Grassi 2016). The development of more selective alpha$_1$-adrenoceptor antagonists (such as prazosin) meant the postsynaptic receptors on the smooth muscle of arterioles and venules could be blocked, but the pre-synaptic alpha$_2$ receptors on the nerve terminals were left unoccupied; hence normal feedback control of noradrenaline release from the nerve terminals remained intact. This meant excessive stimulation of the heart was reduced. Nevertheless, any peripherally acting drug that lowers blood pressure leads to increased sympathetic nerve activity as the baroreceptors and other sensory mechanisms that are part of the blood pressure control system try to keep blood pressure at the hypertensive level. The longer the duration of action of the drug and the slower its onset of action, the less marked the reflex response to the blood pressure lowering effect will be. Doxazosin is a long-acting (once-daily dosing) alpha$_1$-adrenoceptor antagonist used in human medicine. However, in addition to central activation of sympathetic nerve activity, postural hypotension and impotence are both unwanted side effects of these drugs.

The use of these competitive antagonists at alpha$_1$ adrenoceptors alone is insufficient to treat hypertension generally because other mediators (predominantly produced by activation of RAAS) are important in the pathophysiology of hypertension. Increased activity of the sympathetic nervous system centrally also increases activation of the RAAS system which further upregulates the activity of the sympathetic nervous system. Thus, in human medicine these drugs are used in combination with inhibitors of RAAS and/or calcium channel blockers to achieve appropriate control of blood pressure. The exception to this would be in cases of phaeochromocytoma where a catecholamine-producing tumour is responsible for the hypertension. Here removal of the tumour is the treatment of choice, but the use of adrenoceptor antagonists (alpha- and beta-blockers) will be indicated to control blood pressure until this can be achieved (see Chap. 4).

Drugs that act centrally to down-regulate sympathetic nerve activity would be ideal for use in the management of hypertension provided they were free of any serious side effects resulting from their central action. It is now thought that some of the antihypertensive effect of beta-blockers results from a central sympathoinhibitory action, and it certainly seems that beta-blockers, unlike alpha-blockers, are not associated with increased sympathetic nerve activity in the periphery (Rosen et al. 1990).

Clonidine, an alpha$_2$-receptor agonist, was the first centrally acting drug identified to lower sympathetic nerve activity and blood pressure, suggesting that alpha$_2$-adrenoceptors in the nucleus tractus solitarius and the locus ceruleus, areas of the brain involved in regulating blood pressure, reduced sympathetic nerve activity. However, the alpha$_2$-adrenoceptor agonists with central action have marked sedative effects and other unpleasant side effects which meant they could not be tolerated therapeutically (Vongpatanasin et al. 2011). These side effects are due to central activation of alpha$_2$ receptors, whereas it has been suggested that the antihypertensive action of clonidine can be explained by its action at an imidazoline binding site in the rostroventrolateral medulla. Moxonidine is a centrally acting drug that also reduces sympathetic nerve activity via its effects on imidazoline receptors and lacks the sedative action of clonidine and other alpha$_2$-agonists (Edwards et al. 2012).

In summary, there is a long history in human medicine of the development of sympathoinhibitory drug for the treatment of hypertension. The peripherally acting drugs tend to be limited by side effects and by the fact they upregulate sympathetic nervous activity centrally, which in turn leads to activation of the renin-angiotensin-aldosterone system (RAAS). Drugs acting centrally which selectively reduce sympathetic nerve activity associated with blood pressure regulation are likely to be the most useful clinically in assisting in the control of hypertension associated with CKD in dogs and cats. However, the use of drugs interacting with the sympathetic nervous system to treat hypertension in dogs and cats has received much less attention than in human medicine.

11.2.2 The Renin-Angiotensin-Aldosterone System

As discussed above, drugs which inhibit the activation or activity of the renin-angiotensin-aldosterone system (RAAS) are important antihypertensive agents because this is a key blood pressure regulating system in hypertension associated with CKD (see Chap. 1). The complexity of RAAS has emerged over the last 40 years (see Putnam et al. 2012) such that there are now thought to be:

- Multiple metabolites of angiotensin I which have biological activity.
- Multiple enzymatic pathways by which these metabolites can be formed, some of which do not involve the classical angiotensin-converting enzyme (ACE-1).
- The metabolites of angiotensin I can act on a number of different receptors, some of which contribute to hypertension whilst others are counter-regulatory, and their activation lowers blood pressure.

Figure 11.1 depicts the current state of knowledge of the metabolic pathways involved in the generation of peptides from angiotensinogen and angiotensin I. Angiotensinogen is the precursor of angiotensin I and is a 58-kDa protein produced predominantly by the liver, although the kidney, adipose tissue and

11 Pharmacology of Antihypertensive Drugs

Angiotensinogen

Prorenin → Renin
Renin + (P)RR
Prorenin + (P)RR

↓

Angiotensin I —ACE2→ **Angiotensin 1–9**
Asp-Arg-Val-Tyr-Ile-His-Pro-Phe-His-Leu   Asp-Arg-Val-Tyr-Ile-His-Pro-Phe-His

ACE ↓   ACE ↓

Angiotensin II —ACE2→ **Angiotensin 1–7**
Asp-Arg-Val-Tyr-Ile-His-Pro-Phe   Asp-Arg-Val-Tyr-Ile-His-Pro

Aminopeptidase A ↓   ↘ AT1R / AT2R   ↘ Mas

Angiotensin III
Arg-Val-Tyr-Ile-His-Pro-Phe

Aminopeptidase M ↓

Angiotensin IV
Val-Tyr-Ile-His-Pro-Phe

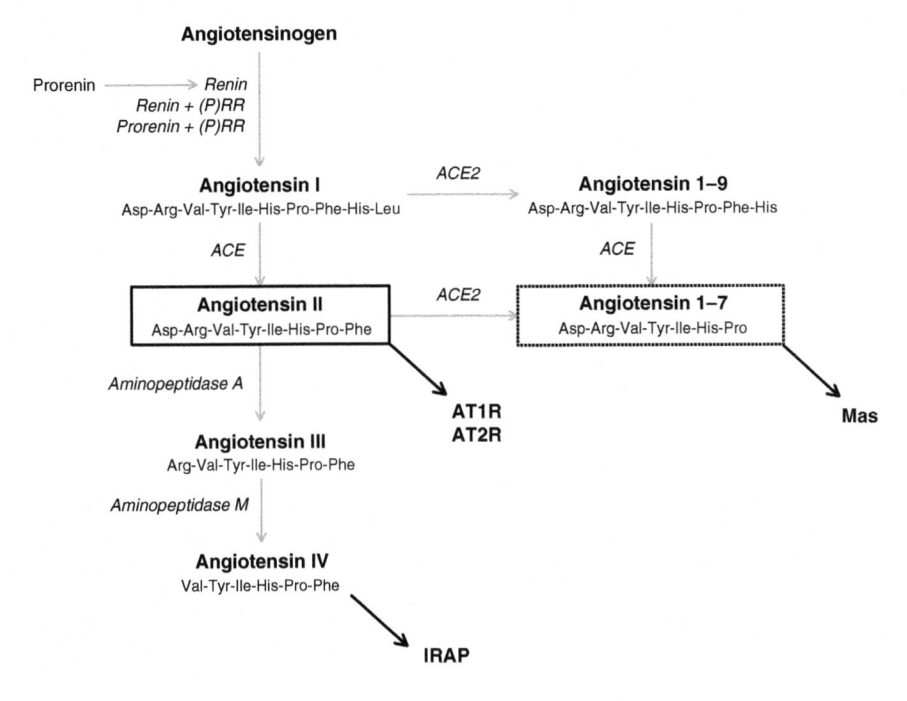

↘ IRAP

Kelly Putnam et al. Am J Physiol Heart Circ Physiol
2012;302:H1219-H1230

Fig. 11.1 Components of the renin-angiotensin system (RAS). The precursor peptide, angiotensinogen, is cleaved by renin to form the decapeptide angiotensin I. The catalytic activity of renin increases when bound to the (pro)renin receptor [(P)RR], and furthermore, the otherwise inactive prorenin can become catalytically active when bound to the (P)RR. The dipeptidase angiotensin-converting enzyme (ACE) cleaves angiotensin I to form the octapeptide angiotensin II (ANG II), the central active component of this system. ANG II can be catabolised by angiotensin-converting enzyme 2 (ACE2) into angiotensin-(1–7) [ANG-(1–7)], another active peptide of this system which typically opposes the actions of ANG II. ANG II can also be cleaved into smaller fragments, such as angiotensin III and angiotensin IV by aminopeptidases A and M, respectively. Most effects of ANG II are mediated by the angiotensin type 1 receptor (AT1R); however, ANG II can also bind to the angiotensin type 2 receptor (AT2R), which generally exhibits opposing effects to those at the AT1R. ANG-(1–7) acts via the Mas receptor, and angiotensin IV can bind to the insulin-regulated aminopeptidase receptors (IRAP). Reproduced with permission from Putnam et al. (2012)

neuronal tissue (glial cells) are amongst the other tissues capable of secreting this protein or other proteins from which angiotensin II can be derived. The importance of some of these pathways and the factors involved in regulating the enzymes required to generate the metabolites of angiotensin I in health and disease remain to be determined in human medicine and have not been investigated in veterinary patients. Realisation that this system is far more complicated than was once thought is important when interpreting the clinical effects of drugs that interfere with RAAS.

Table 11.1 Angiotensin peptide receptor properties

	AT_1 receptor	AT_2 receptor	Mas receptor
Expression	Always expressed (density influenced by sex hormones)	Expressed at times of stress or injury	Factors influencing Mas receptor expression not well characterised
Activator	Angiotensin II	Angiotensin II	Angiotensin 1–7
Effects of receptor activation			
Vascular tone	Vasoconstriction	Vasodilation (via NO)	Vasodilation (via NO)
Endothelial function	Inhibits endothelial cell function	Improves endothelial cell function	Improves endothelial cell function
Sodium homeostasis	Sodium retention (direct action on proximal tubule and aldosterone secretion)	Sodium excretion	Sodium excretion
Sympathetic nerve activity	Increased	No effect	No effect
Inflammation	Pro-inflammatory	Anti-inflammatory	Anti-inflammatory
Cell proliferation (smooth muscle)	Mediates cell growth and proliferation	Inhibits cell growth and proliferation	Inhibits cell growth and proliferation
Connective tissue deposition	Pro-fibrotic	Anti-fibrotic	Anti-fibrotic

The action of ACE-2 enzyme on angiotensin I and angiotensin II to form Ang_{1-9} and Ang_{1-7}, respectively, is important as the latter can activate the Mas receptor which counterbalances some of the detrimental actions of AT_1 receptor activation (see below). The ACE-2 enzyme has 400-fold higher affinity for angiotensin II compared to angiotensin I and is not inhibited by any of the conventional ACE inhibitors used therapeutically.

Angiotensin II is the major physiologically active metabolite formed by the RAAS pathway and its classical actions that lead to increased blood pressure are mediated by activation of the AT_1 receptor (see Table 11.1). The physiological significance of activation of AT_2 and Mas receptors remains to be determined, but their actions seem to counterbalance excessive activation of AT_1 receptors (Santos et al. 2018).

11.2.2.1 Non-ACE Pathways for Angiotensin II Formation

The classical pathway depicted in Fig. 11.1 shows circulating angiotensinogen (produced by the liver) being acted upon by the enzyme renin (secreted by the kidney) to form angiotensin I which is then acted upon by tissue ACE-1, the classical plasma membrane-bound converting enzyme which generates angiotensin II within the extracellular space. This is the accepted endocrine system, known to be an important regulator of blood pressure, extracellular fluid volume via regulating sodium homeostasis (see Chap. 1). There is evidence that alternative precursors other than angiotensinogen for angiotensin I formation are expressed by some tissues, including cardiac myocytes. These include the dodecapeptide Ang_{1-12} and the larger peptide Ang_{1-25}. An intracrine system is proposed whereby these peptides are synthesised and converted intracellularly into angiotensin II via intracellular chymase enzymes (Ahmad et al. 2014; Reyes et al. 2017). The chymase enzymes do not contribute to extracellular generation of angiotensin II because of the abundance of serine protease inhibitors which circulate in plasma. Angiotensin II generation intracellularly can lead to activation of nuclear receptors for this peptide and to the release of angiotensin II from the cell, producing local activation of cell surface receptors within the organ responsible for its genesis. Most evidence for an intracrine system generating angiotensin II exists for cardiac myocytes (Reyes et al. 2017). The relevance of this intracrine system to veterinary clinical patients remains to be determined. The importance of this local intracrine system is that it will be resistant to the two major approaches we have adopted to inhibiting RAAS activation in veterinary medicine, namely, ACE inhibition and angiotensin receptor blockade: the former because conventional ACE is not involved in the generation of angiotensin II and the latter because angiotensin receptor blockers do not seem to inhibit the intracrine or intracellular action of angiotensin II generated by chymase (Ferrario and Mullick 2017).

11.2.2.2 ACE Inhibitors

ACE-1 is a carboxypeptidase, cleaving the C-terminal pair of amino acids from peptides. This enzyme converts angiotensin I to angiotensin II and is also responsible for the breakdown of bradykinin to its inactive metabolite. Thus the pharmacodynamic effects of ACE inhibitors are secondary to reducing the production of angiotensin II and potentiating the action of bradykinin. The structure of this enzyme has been comprehensively studied, and captopril was the first drug designed by structural analysis of the active site of this enzyme to bind to and block the active site. The active site of the enzyme has a zinc atom in a pocket which normally attracts the carboxyl group of the peptide adjacent to the peptide bond which is cleaved. Captopril possesses a sulphydryl group which forms an ionic bond with the zinc atom. Unfortunately, although a very effective ACE inhibitor, captopril proved to have a number of important side effects (immunological), most of which were related to the sulphydryl group.

Thus, the next generation of ACE inhibitors was produced which did not possess the sulphydryl group. Enalapril was the first to be authorised in veterinary medicine in the 1990s followed by benazepril, ramipril and imidapril. The metabolites of these drugs differ in their potency as inhibitors of ACE-1 as evaluated by their binding affinity for the enzyme and by their lipophilicity, hence their ability to diffuse across cell membranes and into transcellular fluids (e.g. cerebrospinal fluid) that are only crossed by highly lipophilic molecules.

11.2.2.3 Pharmacokinetics of ACE Inhibitors

Benazeprilat, enalaprilat, ramiprilat and imidaprilat are the active principles of the ACE-1 inhibitors currently licensed in small animals. As their oral bioavailability is low, somewhere between 2% and 13%, these drugs are formulated as esters (prodrugs: benazepril, enalapril, ramipril and imidapril) to increase absorption by passive diffusion through the GI tract. Prodrugs are subsequently activated by the liver or at peripheral sites, as all prodrugs are not necessarily converted through a first passage through the liver (first-pass effect). Absorption is not affected by the presence of food.

Peak prodrug concentration occurs 30–40 min after administration. After a single oral dose of 0.5 mg/kg benazeprilat, a C_{max} of 37.6 ng/mL is achieved at T_{max} 1.25 h in dogs and 77.0 ng/mL at T_{max} 2 h in cats. The plasma concentration-time curve of ACEis classically has a biphasic decay past T_{max}. For benazeprilat, the initial fast phase half-life is 1.7 and 2.4 h, whilst the terminal phase is 19 and 29 h in dogs and cats, respectively. After repeated administration of a drug with terminal half-life of around 20 h, accumulation should become evident when measuring plasma concentrations, but this is not the case with ACEi. This unexpected behaviour cannot be explained by conventional compartmental modelling but is explainable with a physiologically based model that considers a complex binding mechanism. Benazeprilat is extensively protein bound (85–90%). Indeed, ACEis bind non-specifically and non-saturably to albumin and also bind its targeted biophase (ACE). ACE exists as a soluble fraction circulating in blood (cleaved from ACE anchored in the endothelium) and a more extensive fraction still attached to the endothelium (especially in the pulmonary vascular bed) that influences the disposition of ACEi. The fraction of ACEi bound to endothelial ACE is non-measurable but acts as a reservoir that releases the drug slowly in the circulation (which is measurable). Therefore, the first decay phase represents the elimination of the free drug (liver or kidney), and the late decay phase reflects the release of benazeprilat that was bound to ACE, mainly in the tissue pool (Toutain et al. 2000; Toutain and Lefebvre 2004). The strength of the bound portion is modelled with a Michaelis-Menten model, controlled by ACE capacity (B_{max}, property of the animal species with polymorphism described) and ACEi affinity (K_d, property of the drug) between ACEi and ACE (release from the reservoir). This model accounts for a non-linear pharmacokinetics at low concentrations and accumulation of benazeprilat with repeated dosing due to progressive saturation of ACE.

In comparison, C_{max} values for ramiprilat were 12.1 ng/mL (day 1) and 17.7 ng/mL (day 8), respectively, for an average T_{max} of 1 h after repeated administration of the clinically licensed dose in dogs (0.125 mg/kg/day). The concentrations then declined in a biphasic pattern with biological half-lives of 0.5 and 8.9 h (day 1) and 0.7 and 10.5 h (day 8) on average, respectively.

Benazeprilat and ramiprilat are cleared by both the liver and kidney. The excretion pathway of benazeprilat is primary hepatic in cats (85% biliary/15% renal) and is more balanced in dogs (54% biliary/46% renal). In contrast, enalaprilat (hydrophilic drug) is predominantly cleared by the kidney (95%) in dogs and cats.

Reduction in GFR in dogs did increase exposure to enalaprilat but not benazeprilat or ramiprilat in dogs (Lefebvre et al. 1999, 2006).

11.2.2.4 Pharmacokinetic/Pharmacodynamic (PK/PD) Relationship of ACEi

Initial investigations of the PK/PD relationship of ACEi in dogs adopted ACE inhibition as a pharmacodynamic end point (Toutain and Lefebvre 2004). Their PK/PD approach predicted that any dose of benazepril equal or in excess of 0.25 mg/kg would give similar duration and intensity of effect on ACE inhibition. Starting at low doses (0.03 mg/kg/day), progressive saturation of the pool of ACE, occurred achieving complete inhibition but only after several days. A dose of 0.125 mg/kg/day achieves total inhibition after 2 days. ACE inhibition as biomarker was a highly sensitive surrogate for the RAAS, but the effects on the rest of the system (renin activity, hormone concentrations, electrolyte excretion and blood pressure) were not considered, and these conclusions would need to be backed up by clinical studies.

Mochel et al. (2013) used non-linear mixed effect modelling to describe the periodicity of the RAAS, urinary electrolyte excretion and blood pressure with a cosine function. The parameters estimated were the mesor (average value over time), amplitude (up and down swings) and the acrophase (time period in a cycle). Circadian variations in RAAS and blood pressure were supported by model analysis; it revealed small but significant 5–7 mmHg amplitude swing in blood pressure in parallel to the RAAS periodicity in non-treated healthy dogs. BP does not dip at night in dogs as it is the case in humans but rises during the night time period. There is also a post-prandial decrease in renin activity, suggesting that different feeding times could impact the chronobiology of the renin cascade. The same team (Mochel et al. 2014) demonstrated that the timing of food intake appears pivotal to the circadian organisation of the renin cascade, with food intake leading to a significant but small reduction of BP that could be related to the decreased activity of renin and the secretion of vasodilatory gut peptides.

In a placebo-controlled study, Mochel et al. (2015) comprehensively studied the effect of administering benazepril at a dose of 0.34 mg/kg, superimposed onto aforementioned RAAS circadian variations. In this study, plasma benazeprilat concentrations were measured concomitantly with plasma renin activity (PRA), angiotensin-II (AII) and aldosterone (ALD) concentrations (but not ACE activity). The non-linear saturable pharmacokinetic model was combined with an integrated pharmacodynamic model to describe the effect of ACE-bound benazeprilat not only on angiotensin II concentrations (IC_{50} = 17.9 ng/mL, I_{max} = 1) but also the cascading effect of the predicted angiotensin II concentrations on both ALD concentration and PRA (benazepril-induced interruption of the angiotensin II-renin negative feedback loop). It was essential to model circadian RAAS variations (cosine function) during both periods of the crossover to avoid overestimation of the effect of benazepril on angiotensin II and ALD formation and underestimation of its effect on RA. Despite long-lasting inhibition of ACE (Toutain and Lefebvre 2004), benazeprilat triggered a marked fall in angiotensin II and ALD but for a much shorter period of time. This is consistent with earlier observations in canine and human patients and is due to other enzymes (chymase, cathepsin G) contributing to the conversion of angiotensin I into angiotensin II. This demonstrates the possible limitation of ACE as a sole biomarker for model-based determination of benazepril dose and advocates for higher doses when the whole cascade is considered.

Finally, the chronobiology of ACEi administration should be considered. Assuming that drug efficacy is maximum when the peak effect time of the drug and the peak of the underlying biological rhythm are synchronised, optimised efficacy with bedtime dosing in dogs has been suggested (Mochel and Danhof 2015).

(a) Antihypertensive Action in Dogs ACE inhibitors are authorised for clinical use in dogs in the treatment of congestive heart failure. Administration of ACE inhibitors to healthy dogs under experimental conditions leads to a very small reduction in blood pressure in dogs fed a 'normal diet' (c10 mmHg; see Lefebvre and Toutain 2004). Administration of a sodium-restricted diet increases the hypotensive response to ACE inhibitors as the maintenance of blood pressure is more dependent on RAAS activation under these conditions, but even so the blood pressure reduction seen is still only around 20 mmHg. Models of CKD in dogs, induced by renal mass reduction, have been used to demonstrate the systemic and glomerular antihypertensive action of enalapril (Brown et al. 2003) and benazepril (Mishina and Watanabe 2008) although the dose-response relationship of the antihypertensive action was not explored in these studies. The remnant kidney model (7/8 or 11/12th reduction) leads to activation of the RAAS system and is associated with an increase of between 20 and 35 mmHg systolic, mean and diastolic arterial pressure. Brown et al. (2003) administered enalapril for 6 months to dogs that had had 11/12ths of their renal mass surgically removed. The initial oral dose rate was 1 mg/kg twice daily starting 10 days after surgery. The dose was reduced after 2 weeks to 0.5 mg/kg twice daily for the remainder of the 6 month period. Indirect blood pressure measurement and exogenous creatinine clearance were measured every 3 months, and micropuncture studies were undertaken at the end of the 6 month study period. Compared to placebo, the enalapril-treated groups had significantly lower (by between 22 and 24 mmHg) mean and diastolic arterial blood pressures at 3 months. The difference in systolic blood pressure at 3 months was 17 mmHg but this did not reach statistical significance. Blood pressure tended to decrease in both groups over the first 3 months of the study compared to the pre-treatment values and stabilised in the placebo group at 6 months. The arterial blood pressure in the enalapril-treated group appeared to increase after 3 months, such that there was no statistical difference between the two groups 6 months after surgery. However, direct measurement of blood pressure during micropuncture studies after 6 months did show a significant difference in mean systemic arterial blood pressure of about 24 mmHg (lower in the enalapril-treated group), and micropuncture studies showed that the glomerular capillary pressure was also 12–13 mmHg lower. Despite these haemodynamic changes, there was no difference in the single nephron glomerular filtration rate between groups. Nevertheless, these effects of enalapril treatment were accompanied by lower proteinuria and reduced severity of glomerular and tubular lesions. Interpretation of the structural data from this study is complicated by the significantly worse tubular lesion and infiltrate scores in the placebo group at the start of the study.

Mishina and Watanabe (2008) also examined the effect of ACE inhibition on the hypertension associated with surgical renal mass reduction in dogs. They undertook a 7/8th nephrectomy and, at least 1 month later, administered benazpril at a dose rate of 2 mg/kg orally once daily for 2 weeks. This treatment lowered arterial blood pressure by about 10–15 mmHg during this period. On cessation of dosing, the blood pressure values returned to the pre-treatment levels suggesting the model was stable in terms of the degree of hypertension.

There are limited randomised controlled clinical trials in dogs examining the effect of ACE inhibitors on systemic arterial blood pressure. In one study of biopsy-proven glomerulonephritis, dogs treated with enalapril for 6 months (0.5 mg/kg daily or twice daily) showed a mild reduction in systolic blood pressure (-12.8 ± 27.3 mmHg; $n = 16$) which was significantly greater than the change in systolic blood pressure seen in the placebo group (5.9 ± 21.5 mmHg; $n = 14$) (Grauer et al. 2000). The pre-treatment systolic blood pressure in the enalapril-treated dogs was 154 ± 25 mmHg and in the dogs that received placebo was 148 ± 25 mmHg.

(b) Antihypertensive Action of ACE Inhibitors in Cats

Benazepril when administered at a daily oral dose of 2.5 mg/cat for 7 days to a group of six healthy male cats had no significant effect on their awake resting blood pressure (systolic pressure 111 ± 3.8 vs. 106.3 ± 3.3 mmHg) (Jenkins et al. 2015). This dosing regimen significantly reduced the rapid pressor response to bolus dose of angiotensin I under anaesthesia by more than 75% when the pressor response test was assessed 4 h after dosing. A cutoff of 75% attenuation of a submaximal pressor response to angiotensin I was chosen as this is used in human medicine to predict the clinical efficacy of RAAS blocking drug (Buchwalder-Csajka et al. 1999). However, this attenuation of the angiotensin I pressor response, although still measurable, was much reduced (and well below 75% attenuation) after 24 h of dosing suggesting that, with this dosing regimen, the effect of benazepril on conversion of angiotensin I to angiotensin II does not persist for 24 h. Similar effects on blood pressure have been reported in cats with reduced renal mass as those reported in the dog with a 30–50 mmHg increase in mean, systolic and diastolic blood pressure after surgery which reduces over the next 6 months (Brown et al. 2001). This study involved four groups of eight cats. Renal mass was reduced by 15/16ths leading to mild systemic hypertension (systolic blood pressure c160 mmHg 1 month after the surgical reduction in renal mass which fell gradually over the next 5 months) and glomerular capillary hypertension (c75 mmHg which is about 10–15 mmHg above normal) 6 months after renal mass reduction. Benazepril (0.25–1 mg/kg once daily by mouth) lowered glomerular capillary pressure back to normal levels and reduced systemic arterial blood pressure by about 10–15 mmHg, which was a consistent difference over the 6 months of the study (see Fig. 11.2). The effect of benazepril on systemic arterial blood pressure was statistically significant but only if the data from one cat (treated with the highest dose of benazepril), with systolic blood pressure that was particularly high (>185 mmHg) and which did not respond to treatment, were excluded. Even on excluding this cat,

Fig. 11.2 Effect of benazepril on systolic arterial blood pressure in cats with experimentally reduced kidney mass (11th/12th nephrectomy model). Reproduced with permission from Brown et al. (2001)

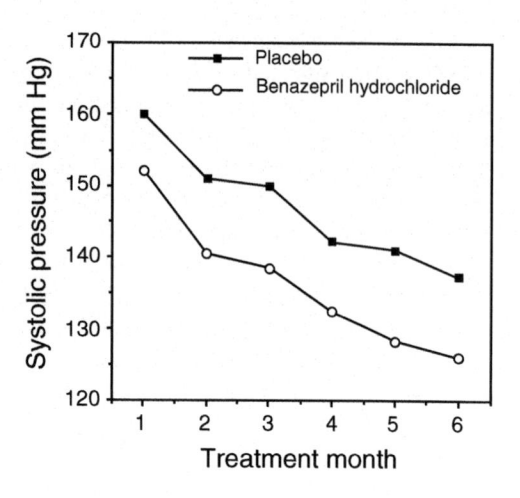

the authors reported no difference in the antihypertensive effect between the different doses used in this study, suggesting the effect was maximal at 0.25 mg/kg once a day although the group sizes and individual cat variability may have been too great to see a dose-dependent effect. Despite this mild systemic antihypertensive effect and more marked reduction in glomerular capillary pressure, no significant reduction in proteinuria or effect on renal lesions was seen with treatment of this model of CKD with benazepril.

Similarly mild antihypertensive action has been reported by clinicians using ACE inhibitors to treat cats with hypertension where experience suggests that ACE inhibitors as a sole treatment generally provide insufficient antihypertensive effect to control blood pressure adequately. No large-scale randomised controlled clinical trials have been undertaken in cats examining the effects of ACE inhibitors on systemic arterial blood pressure.

11.2.2.5 Angiotensin Receptor Blockers

The 'sartan' family of angiotensin receptor blockers (ARBs) are well established as antihypertensive, cardio- and renoprotective agents in human medicine. Peptide ligands for the AT_1 receptor were first produced in the 1970s but were not very effective clinically due to lack of oral bioavailability and a tendency to have partial agonist effects. Losartan was the first non-peptide AT_1 receptor antagonist to be developed by computer modelling of the structure of the receptor binding site. It was approved by the FDA for use in human medicine in 1995. The newest sartan, azilsartan, was approved for human clinical use in 2010. The relative binding affinities of eight ARBs for the AT_1 receptor reported by Miura et al. (2011) was irbesartan > olmesartan > candesartan > EXP3174 (losartan metabolite) > eprosartan > telmisartan > valsartan > losartan. All have kd values in the 2–10 nM range, and all show selectivity for the AT_1 receptor over the AT_2 receptor of 10,000- to 30,000-fold. In vitro experiments show all of the sartans, except losartan and eprosartan, cause insurmountable inhibition of AT_1 receptor activation by angiotensin II. However, none of these drugs bind irreversibly to the receptor, and the insurmountable inhibition is likely to be due to their tight binding and slow dissociation from the receptor. Azilsartan appears to have the higher affinity for the human AT_1 receptor than olmesartan, telmisartan, valsartan and irbesartan and dissociates from the receptor extremely slowly (Ojima et al. 2011). Whether this gives azilsartan any clinical therapeutic advantage is debatable.

Both the AT_1 receptor and the AT_2 receptor are members of the G protein-coupled receptor (GPCR) super family. GPCRs can have constitutive activity in the absence of the activating ligand, particularly when highly expressed in tissues. Some of the existing ARBs have been shown to act as inverse agonists at the AT_1 receptor, thus inhibiting constitutive activity of this receptor. This may not be a property of all ARBs. For example, in mutant receptor systems showing a high level of constitutive activity, olmesartan and valsartan proved to be more effective inverse agonists than losartan (Miura et al. 2006). The importance of inverse agonism to the clinical therapeutic effect of these drugs remains to be seen. However, it may be relevant to the long-term adaptations to hypertension as AT_1 receptors appear to be upregulated (Yasuda et al. 2008) and constitutively activated by myocyte stretch (Zou et al. 2004). The molecular mechanisms by which ARBs interact with AT_1 receptors and the chemical structures that promote inverse agonism are the subject of much research to develop new AT_1 receptor ligands, including those that are 'biased agonist' and favour the receptor activation of β-arrestin signalling pathway instead of the Gq protein activation as these may have additional benefits in cardiovascular and renal disease (Takezako et al. 2017).

Unlike ACE inhibitors, administration of an ARB does not prevent the formation of angiotensin II. Thus, blockade of the AT_1 receptor leaves angiotensin II and its metabolites available to activate the AT_2 and Mas receptors, which may have beneficial effects in cardiovascular and renal disease and may contribute to their antihypertensive effects (see Fig. 11.1 and Table 11.1). As the AT_1 receptor is involved in feedback inhibition of renin secretion by the juxtaglomerular apparatus in the kidney, ARBs lead to increases in renin secretion (by blocking this negative feedback loop). This means that ARB administration will lead to increased angiotensin II formation. With the AT_1 receptors blocked, activation of AT_2 receptors by angiotensin II will increase. The importance of AT_2 receptor activation in bringing about the antihypertensive effects of ARBs has not been investigated in human medicine but is a theoretical advantage of ARBs over ACE inhibitors.

Some ARBs have other pharmacological actions that may be relevant to their beneficial clinical effects. Some of the drugs (e.g. irbesartan and telmisartan) are partial agonists at the peroxisome proliferator-activated receptor-γ (PPARγ) regulating cellular glucose and fatty acid metabolism. Whether this effect is of clinical relevance and leads to insulin sensitising effects at clinical dose rates has been questioned (Miura et al. 2011). However, it has been proposed that ARBs with this property are more likely to be protective against vascular dementia. Ischaemia to the brain leads to upregulation of AT_2 receptors. Their activation is neuroprotective, and their activation involves the PPARγ pathway (Horiuchi and Mogi 2011). Thus, it has been suggested that ARBs with partial PPARγ agonist activity are most likely to protect against stroke and vascular dementia (Towfighi and Ovbiagele 2008).

(a) Effects of ARBs on Blood Pressure in Dogs

Knowledge of the pharmacology of ARBs in veterinary species has resulted from the use of dogs as an experimental model for human drug development. The cardiovascular pharmacology of losartan, the first non-peptide ARB, was extensively studied in the dog during its development for human use in the 1990s. Spontaneously hypertensive dogs at the University of Pennsylvania bred by Professor Bovee were used (Bovee et al. 1991), and administration of losartan was shown to lead to mild dose-dependent reductions in blood pressure. The Penn hypertensive dogs had resting blood pressures of 163 ± 8.7 and 102 ± 2.6 mmHg (systolic and diastolic, respectively). Intravenously administered losartan (0.3 to 30 mg/kg) dose dependently reduced systolic pressures by 10–15 mmHg at 45–75 min following administration. At this time the highest dose of losartan tested (30 mg/kg) inhibited the pressor response to exogenously administered bolus doses of angiotensin I and II by 80–85%. Christ et al. (1994) demonstrated that the dog formed very little of the active metabolite EXP 3174. Studies (controlled or uncontrolled) of the effect of ARBs on the resting blood pressure of normal dogs or in dogs with renal hypertension or naturally occurring chronic kidney disease are lacking in the literature as most experimental studies involving dogs and ARBs have focused on models of heart disease (e.g. ischaemic heart disease). This makes comparison with the studies reported above on dogs treated with ACE inhibitors (many of which are authorised for use in dogs to treat cardiac disease) not possible.

(b) Effects of ARBs on Cats

One ARB, namely, telmisartan, has been developed for use in cats, and so there is a good understanding of the pharmacology of this drug in cats. Telmisartan was authorised for use in Europe in cats with chronic kidney disease for the management of proteinuria in 2013. In 2018 a marketing authorisation was obtained for the treatment of systemic hypertension in cats in Europe and for the control of systemic hypertension in cats in the USA.

(i) In Vitro Pharmacology

The effects of telmisartan on the responses of isolated rings of cat uterine artery were studied in vitro using wire myography (see Fig. 11.3; Berhane et al. 2009). Telmisartan proved to be a potent insurmountable inhibitor of angiotensin II in this system at concentrations of 10 nM and above. At 1 nM a parallel rightward shift in the concentration-response curve is seen with no reduction in the maximum response supporting a competitive interaction. The apparent dissociation constant calculated from this experiment for telmisartan was 0.68 nM.

(ii) Pharmacokinetics of Telmisartan in the Cat

In all species, telmisartan binds extensively (around 98 to 99%) and reversibly to plasma albumin and α1-acid glycoprotein. Despite the high protein binding, due to its lipophilic weak acid nature (pKas 3.5, 4.1 and 6.0), telmisartan is indeed the most lipophilic of all sartans and crosses biological membranes very readily, which could support a high volume of distribution. There are no published studies of the in vivo

11 Pharmacology of Antihypertensive Drugs

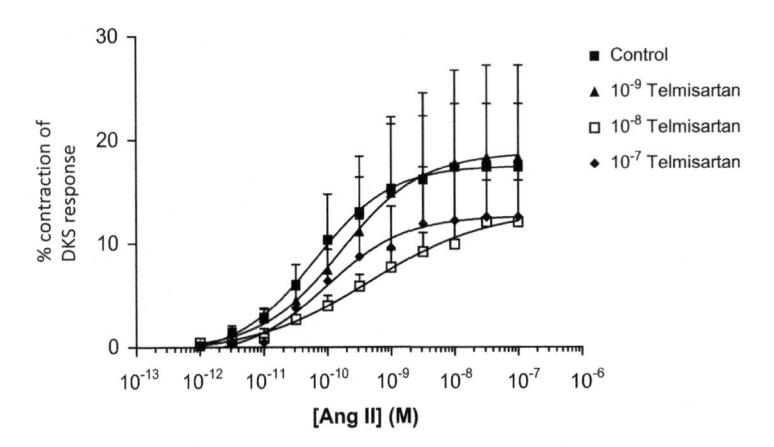

Fig. 11.3 The effect of telmisartan on angiotensin II-mediated contractile response of feline isolated uterine arteries. Uterine arteries were incubated with 10^{-9}, 10^{-8} or 10^{-7} M telmisartan or equal volumes of vehicle (control) for 30 min, and cumulative concentration-response curve to Ang II was obtained ($n = 6$). Data from an abstract presented at the European Association of Veterinary Pharmacology and Toxicology, Leipzig 2009

pharmacokinetics of telmisartan in the cat. Here the data on file with Boehringer Ingelheim used to gain the product authorisation for the cat is referred to (with the permission of the company).

Initial pharmacokinetic study of a pilot formulation of telmisartan was described in four healthy cats after intravenous and oral administration of 1 mg/kg dose (Boehringer Ingelheim, data on file). It confirmed a medium to high clearance (44.5 ml/kg/min), and a large volume of distribution was (8.9 L/kg). Oral absorption was rapid (T_{max} 26 min), and bioavailability was 33%.

The effect of feeding on the disposition of telmisartan was investigated in a pivotal crossover study including 12 cats administered with a daily dose of 1 mg/kg for 5 days. Fasting (for about 19 h) did not alter significantly the overall drug exposure, but fasting was associated with earlier T_{max} (21 vs. 32 min) and a C_{max} that was more than twice than that recorded when food was administered (169 vs 75 ng/mL). Elimination half-life was not different (5 h fasted vs 6 h with food). The estimation of the terminal phase was obscured by the presence of 'multiple peaks' on the plasma concentration-time curve that correspond to enterohepatic cycling (ERH), complicating the evaluation of the terminal slope.

The pharmacokinetics of the final formulation of telmisartan was confirmed in a pivotal study following single and chronic oral administration of 0.5 (single dose), 1 (once-daily administration for 21 days) and 3 mg/kg (single dose) to 12 cats (6 male and 6 female). Following rapid oral absorption (Tmax 0.5 h), exposures were dose proportional (linear pharmacokinetics), and no significant sex effect was seen. Terminal elimination HL was 7.3 to 8.6 h. After 21 days of dosing, modest accumulation was observed: mean AUC accumulation ratios (mean of individual ratios) of 1.2 for males and 1.4 for females were observed. C_{trough} plasma concentrations (just before next administration) were in most cases below the lower limit of quantification of the assay (LLOQ).

Lipophilic drugs such as telmisartan require transformation to promote their excretion. There is in vitro and in vivo evidence of telmisartan conjugation via UDP glucuronosyltransferases, yielding an inactive 1-O-acylglucuronide metabolite efficiently excreted in the bile. In an in vitro study using pooled liver microsomes, telmisartan was effectively glucuronidated in cats and more effectively than in rats, dogs and humans (higher V_{max}, similar K_m) (Ebner et al. 2013). Analysis of samples from the pivotal PK study revealed that telmisartan 1-O-acylglucuronide was generated in vivo as a major metabolite (concentration in the range of 21% of parent compound).

(iii) In Vivo Pharmacodynamics Effects of Telmisartan

Telmisartan was examined for its effects on resting conscious cat blood pressure and for its ability to inhibit the pressor responses to angiotensin I (Jenkins et al. 2015). This crossover study compared three different doses of telmisartan (0.5, 1 and 3 mg/kg) with the therapeutic dose of benazepril (2.5 mg/cat), two doses of irbesartan (6 and 10 mg/kg) and a single dose of losartan (2.5 mg/kg). All drugs were administered orally and daily for 7 days before assessments were made on resting blood pressure (measured by telemetry). The responses to angiotensin I infusion were made under anaesthesia. There was a 7 day washout period between each drug administration of this 8 period study.

Only telmisartan significantly reduced resting systolic arterial blood pressure at all doses tested. The 3 mg/kg dose lowered systolic arterial blood pressure from 111.1 ± 3.8 to 88.9 ± 6.7 mmHg, a 22.2 mmHg reduction. Telmisartan proved the most effective drug in blocking the effects of exogenously administered angiotensin I at the doses selected. Losartan was the least effective drug perhaps suggesting that cats, like dogs, do not produce the active metabolite of this drug very effectively. This confirms the finding reported by Mathur et al. (2004) where losartan (4 mg/kg) also proved ineffective in lowering blood pressure in cats with experimental hypertension, including a model that activates RAAS (renal ablation and renal wrap model). Jenkins et al. (2015) showed irbesartan to be intermediate to telmisartan and losartan at the two doses tested, neither reducing the submaximal pressor response of angiotensin I (control response of 30–35 mmHg) by more than 75%. The high dose of telmisartan tested (3 mg/kg) reduced the pressor response of all doses angiotensin I to below 11 mmHg (whereas the response to the highest dose of angiotensin I in the placebo cats was 138 mmHg). It is possible that these studies under-estimated the effects of irbesartan and losartan in the cat if the rate of absorption of these drugs was slower than telmisartan and benazepril and that 90 min post-dosing did not coincide with the peak blood levels.

Despite the reported relatively short half-life of telmisartan in the cat (see above), the effect of telmisartan was maintained for at least 24 h after dosing (attenuation of angiotensin I was the same 24 h after dosing when compared to 90 min post-dosing). This contrasts with benazepril which attenuated the angiotensin pressor response by only 31.2% after 24 h compared with a 78.9% reduction seen with telmisartan (3 mg/kg) at 24 h post-dosing.

An additional study examined the effect of 14 days of oral dosing with telmisartan on the indirect blood pressure (Doppler method) of healthy adult male and female experimental cats fed a standard maintenance diet (Coleman et al. 2018). Twenty

11 Pharmacology of Antihypertensive Drugs

eight cats were block randomised to receive seven different dosing regimens (including placebo) for 14 days. The intention was to expose each group of 4 cats to three of the dosing regimens so that there would have been 12 cats per dosing regimen. However, in the analysis a period effect was evident despite the 7 day washout between each period, and the paper presents the data from the first period only giving just four cats per group. The dosing regimens were placebo, 1, 1.5, 2 and 3 mg/kg once daily (morning) and 1 and 1.5 mg/kg dosed every 12 h. Telmisartan was dosed from an oral proprietary preparation (Semintra™) containing 4 mg/ml telmisartan. Blood pressure was recorded daily during the 14 day dosing period approximately 3 h after dosing. Mean baseline systolic blood pressure was remarkably similar between each of the seven groups (127–130 mmHg). Blood pressures measured in week 1, week 2 and week 1 + 2 were averaged and statistically analysed for each group. The daily mean systolic blood pressures measured in each group are presented in Fig. 11.4. As can be seen, for all dose groups the mean SBP tended to fall throughout the 14 day period and the duration of dosing may not have been long enough to determine the maximum effect of each dose.

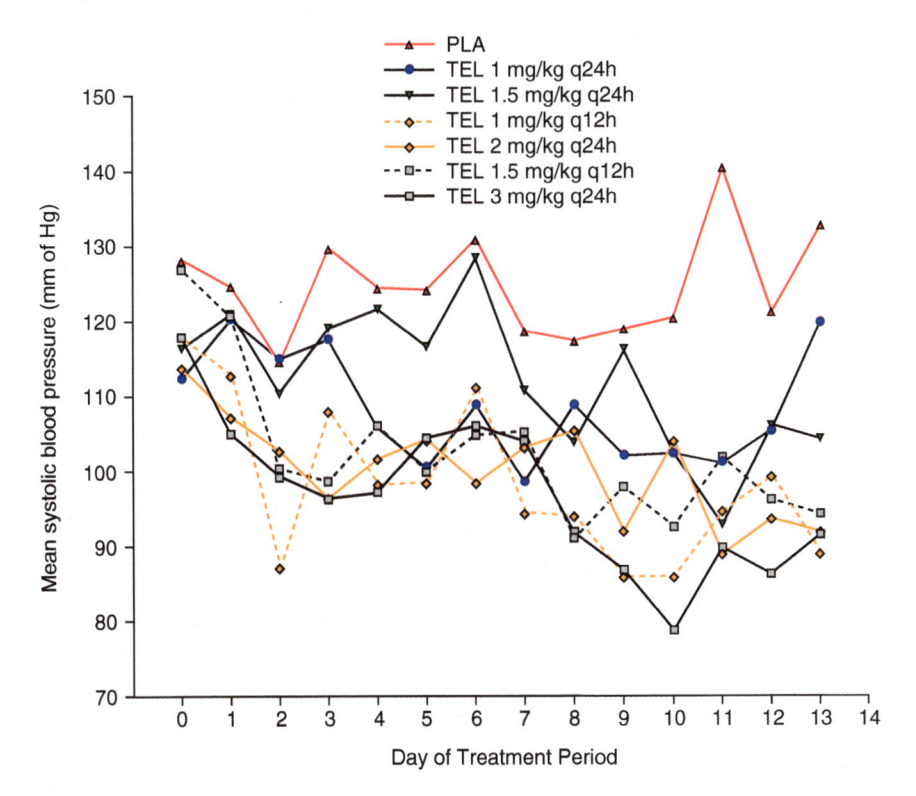

Fig. 11.4 Effect of telmisartan on blood pressure in normal cats. Mean indirectly measured systolic arterial blood pressure in 28 awake, clinically normal cats ($n = 4$ cats/treatment group) administered various dosages of telmisartan (TEL) or placebo (PLA) orally for 14 days. All values were obtained 3 ± 1 h following the morning dose of TEL or PLA. For ease of interpretation, error bars are not included. Reproduced with permission from Coleman et al. (2018)

When the average blood pressures over the 2 week period are considered, telmisartan lowered blood pressure at all doses tested except 1 and 1.5 mg/kg daily when compared to placebo. If the average blood pressures from week 2 only are considered, all dose groups had significantly lowered blood pressure compared to the placebo group. The reduction was about 25 mmHg for the low doses (1 and 1.5 mg/kg once daily) and 30–40 mmHg for the higher doses (2 and 3 mg/kg once daily and 1 and 1.5 mg/kg q12h). By contrast, the blood pressure of the cats receiving placebo changed by only 7 mmHg over the second week of the study. The degree of blood pressure reduction is greater than expected in normal healthy cats fed a sodium-replete diet where activation of RAAS should play a minimal part in maintaining systolic blood pressure. The experience of administering benazepril to conscious healthy cats resulted in very little change in blood pressure (Jenkins et al. 2015). Indeed, in a previous laboratory study involving cats chronically instrumented with telemetry to measure blood pressure continuously, lisinopril had much less of an effect on resting blood pressure in the conscious cat, even when combined with feeding of a sodium-restricted diet (Brown et al. 1997). This study was designed to determine whether blood pressure varied with the light-dark cycle. A difference of approximately 3 mmHg was found (126.2 ± 10.8 vs. 123.5 ± 9.3 mmHg in the hours of light and dark, respectively). Reducing sodium intake from 65 to 35 mg/kg/day had no discernible effect on blood pressure. Lisinopril treatment (2.5 mg/day for 3 weeks) in combination with the lower sodium intake caused a small but significant reduction in mean arterial pressure (c10 mmHg) which persisted throughout the 24 h inter-dosing period and was accompanied by an increase in heart rate. By contrast, Coleman et al. (2018) recorded systolic blood pressures of below 80 mmHg in some cats treated with telmisartan although none were reported to show signs of hypotension. The marked effect of telmisartan on resting blood pressure in cats when fed a sodium-replete diet might suggest a blood pressure lowering action in addition to antagonism of angiotensin II at the AT_1 receptor.

Two randomised controlled clinical trials have been reported in cats to study the clinical effect of telmisartan. The first was a comparison of the effect of telmisartan (1 mg/kg daily for 6 months; $n = 112$) with benazepril (0.5–1.0 mg/kg daily for 6 months; $n = 112$) in cats with chronic kidney disease and urine protein-to-creatinine ratio (UPC) ≥ 0.2 (Sent et al. 2015). The end point was reduction in proteinuria, and the study was designed to assess non-inferiority of telmisartan to benazepril which had been shown to reduce proteinuria compared to placebo in a pivotal clinical trial (the BENRIC study; King et al. 2006). In this first clinical trial, telmisartan proved non-inferior to benazepril (confidence interval, −0.035 to 0.268) based on the weighted time-averaged UPC. Indeed, telmisartan treatment significantly reduced UPC compared to baseline after 180 days, whereas benazepril treatment did not. Blood pressure was not reported as an efficacy end point in this study.

The second randomised controlled clinical trial involving telmisartan in cats was the European pivotal field efficacy and safety study for the EMA authorisation for telmisartan as an antihypertensive treatment for cats (Glaus et al. 2019). This was a placebo-controlled multicentre study involving cats with persistently elevated blood pressure (>160 mmHg but <200 mmHg; measured by Doppler method) regardless of the underlying cause of hypertension. Cats were randomised in a 1:2 ratio to receive placebo or telmisartan (2 mg/kg daily) for 28 days (efficacy phase), and those cats receiving telmisartan continued on the drug for a further 120 days (safety phase). The predefined co-primary efficacy end points were that telmisartan reduced systolic blood pressure to a significantly greater extent than did placebo at 14 days of treatment and that the population mean systolic blood pressure reduction for telmisartan-treated cats was greater than 20 mmHg at 28 days. Cats with systolic arterial blood pressure > 200 mmHg after 14 days were excluded from the study for welfare reasons but their data were carried forward in the assessment of efficacy at 28 days. The intention-to-treat population after 14 days of treatment consisted of 285 cats (189 receiving telmisartan and 96 receiving placebo). The change in systolic blood pressure from baseline was significantly larger in the telmisartan-treated cats (-19.2 (-22.4 to -16.0) mmHg) when compared to the placebo cats (-8.8 (-12.4 to -5.3) mmHg; $P < 0.001$). In the intention-to-treat population, mean reduction in systolic blood pressure in the telmisartan-treated group at 28 days was -24.5 (-27.9 to -21.0) mmHg and so exceeded 20 mmHg reduction with the lower 95% confidence limit, thus meeting the second primary efficacy end point. For comparison, the mean reduction in blood pressure in the placebo group was -10.8 (-14.4 to -7.1). In the intention-to-treat analysis, data from 2 out of 189 telmisartan and 1 out of 96 placebo cats were carried forward into the analysis with these cats having systolic blood pressure > 200 mmHg at 14 days.

If the population was divided into cats with pre-treatment blood pressure $\geq$ 180 mmHg and compared to those with blood pressure between 160 and 179 mmHg pre-treatment, a similar proportionate reduction in blood pressure (14.1 vs. 13.3% reduction) was seen in both groups treated with telmisartan according to the study protocol (see Table 11.2), suggesting telmisartan is equally effective when used to treat moderately and severely hypertensive cats (based on the risk of target

Table 11.2 Response of hypertensive cats to telmisartan

	SABP at baseline:			
	Between **160–179 mmHg**		Between **180–200 mmHg**	
	Telmisartan Baseline mSABP: **171.9 mmHg** ($n = 98$)	**Placebo** Baseline mSABP: **170.9 mmHg** ($n = 56$)	**Telmisartan** Baseline mSABP: **188.4 mmHg** ($n = 76$)	**Placebo** Baseline mSABP: **188.5 mmHg** ($n = 32$)
Visit	Mean SABP reduction compared to baseline at respective visit			
Day 14	16.7 ($n = 98$)	8.8 ($n = 56$)	22.3 ($n = 76$)	9.5 ($n = 32$)
Day 28	22.9 ($n = 94$)	11.9 ($n = 55$)	26.6 ($n = 71$)	10.9 ($n = 32$)

Data are taken from Glaus et al. (2019) and consist of the per protocol population divided according to their pre-treatment blood pressure into hypertensive and severely hypertensive groups (according to the ACVIM classification of hypertension)

organ damage according to the ACVIM consensus statement on hypertension; Acierno et al. 2018). However, the efficacy of telmisartan in cats with blood pressure exceeding 200 mmHg (a not uncommon finding in veterinary practice) remains to be determined. The data obtained from the safety component of the study suggests the control of blood pressure persists for 120 days and the level of adverse events seen during this period was acceptable. An equivalent clinical trial was undertaken in the USA with a very similar design. The results of the US trial were very similar to the European trial (Coleman et al. 2019).

This study (and the US equivalent study[1]) demonstrated that telmisartan is safe and effective as an antihypertensive agent when dosed at 2 mg/kg once daily. As no increase in dose was allowed in this study for cats that did not meet the predefined efficacy criteria, the benefit in those cats that do not respond adequately (to currently accepted blood pressure targets) to 2 mg/kg of increasing the dose remains to be determined. In addition, this clinical trial did not assess proteinuria in response to telmisartan. The target blood pressure in human patients treated for hypertension tends to be lower for the more proteinuric patients as this gives greater renoprotection (Peterson et al. 1995). Such an approach has not been studied in veterinary medicine although proteinuria has been shown to be a risk factor for all-cause mortality in hypertensive cats treated with amlodipine (Jepson et al. 2007).

The current suggested minimal target post-treatment systolic blood pressure is <160 mmHg with the optimal target being <140 mmHg (Acierno et al. 2018) based on the observational study that shows cats with systolic blood pressure > 140 mmHg are at increased risk of developing hypertension often associated with target organ damage as they age, particularly if they have chronic kidney disease (Bijsmans et al. 2015).

11.2.2.6 Mineralocorticoid Receptor Antagonists

Mineralocorticoid receptor antagonists have been used since the 1960s as potassium-sparing diuretic drugs when spironolactone was approved for use in human medicine. At that time it was used as a way of counteracting the hyperaldosteronism and subsequent potassium loss that limits the diuretic effects of thiazides and furosemide. The affinity of spironolactone for the mineralocorticoid receptor appears to be in the low nanomolar concentration range, about tenfold higher than its affinity for the androgen receptor. This contrasts with the drug's affinity for the progesterone and glucocorticoid receptors which are in the low micromolar and high nanomolar range respectively.

The classical actions of spironolactone as a potassium-sparing diuretic occur through its competitive binding of the mineralocorticoid receptor. Aldosterone acts on tissues by interacting with the mineralocorticoid receptor (MR). This receptor is found in highest density in epithelial tissues involved in transporting sodium and potassium (collecting tubule and collecting duct of the kidney, sweat glands and

---

[1] Dosing regimen in the US study was 1.5 mg/kg twice daily for 14 days followed by 2 mg/kg once daily.

colonic epithelium) where the MR co-localises with an enzyme called 11-β-hydroxysteroid dehydrogenase type II (11-β-HSDII) which is responsible for inactivating cortisol (by metabolising it to cortisone) and preventing it activating the MR. Interaction of aldosterone with the MR leads to an increase in density of apical membrane sodium channels and increased activity of the basolateral sodium-potassium ATPase (sodium pumps). Potassium and hydrogen ion transport across the apical membrane of these epithelial cells also results from MR activation. Most of the effects of aldosterone activation of the MR result from genomic effects (resulting from the interaction of aldosterone with its cytoplasmic MR, which migrates to the nucleus and alters DNA synthesis) and require new protein synthesis taking hours to come into full effect but starting after 20–60 min.

Non-genomic actions of aldosterone have been reported through a membrane-bound receptor where the effects occur within minutes rather than hours of receptor activation. In the kidney, for example, increased activity of Na/H exchanger stimulated by aldosterone is very rapid, being mediated through cell surface MRs. Non-genomic effects have also been shown to occur in vascular smooth muscle where activation of surface MRs seems to reduce calcium entry through voltage-dependent calcium channels. MRs have more recently been identified in non-epithelial tissues, particularly, in the heart, vasculature and brain, further supporting a role for aldosterone in regulating function of these tissues (see Jaisser and Farman 2016). Nevertheless the physiological importance of the non-genomic actions of aldosterone remains unclear.

From its introduction into human medicine in the 1960s, spironolactone was viewed as a potassium-sparing diuretic. In 1999, the seminal RALES study (Pitt et al. 1999) changed this view and demonstrated effects of spironolactone, the explanation for which could not be through its renal actions. The rationale for the investigators conducting the RALES study was the hypothesis that aldosterone breakthrough occurs in the face of ACE inhibitor treatment and that addition of an aldosterone receptor antagonist would prove beneficial. The identification and characterisation of mineralocorticoid receptors in tissues that are not sodium-transporting epithelia, such as cardiac myocytes, vascular smooth muscle and neuronal cells (baroreceptors), has proved a very active area of basic research stimulated by the seminal clinical observations of the RALES study. When a low dose of spironolactone (25 mg/day) was added to conventional therapy for heart failure (furosemide, ACE inhibitor and digoxin), this resulted in a 30% (18–40%) reduction in all-cause mortality. Death due to cardiac causes (progression of cardiac failure and sudden cardiac death) were both significantly reduced (by more than 30%) as were hospitalisations due to cardiac causes. The benefit was seen regardless of age of patient, or cause of cardiac failure (ischaemic or non-ischaemic; low or normal ejection fraction). The trial which involved 1663 patients was stopped at the 24 month interim analysis because spironolactone treatment had proven so beneficial.

Follow-up studies supported the hypothesis that part of the beneficial effect of spironolactone was explained by its anti-fibrotic effect as those patients with above the population median circulating concentration of the cardiac fibrosis biomarker pro-collagen type III amino-terminal peptide (PIIINP) benefited the most from spironolactone treatment (Zannad et al. 2000). Further smaller studies demonstrated that spironolactone also improved endothelial cell function when added to conventional treatment of human heart failure patients (Farquharson and Struthers 2000). The results of Farquharson and Struthers (2000) suggested that aldosterone breakthrough led to increased conversion of angiotensin I to angiotensin II in the vasculature of heart failure patients, despite them receiving ACE inhibitor therapy. Further experimental studies demonstrated that hyperaldosteronism leads to enhanced desensitisation of baroreceptors (Wang 1994), a phenomenon that can be reversed by spironolactone, lowering centrally mediated sympathetic drive in heart failure patients.

More selective mineralocorticoid receptor antagonists have been developed for human medicine, principally eplernerone which lacks the anti-androgen and progesterone-mimetic effects of spironolactone. Similar results have been obtained in a large clinical trial (Zannad et al. 2011) where eplernerone has reproduced the effect of spironolactone when added to conventional treatment for heart failure (including ARBs as well as ACE inhibitors).

Rationale for Use of Aldosterone Receptor Blockers with ACE Inhibitors

The concept that angiotensin II is *the* stimulus for aldosterone production is an oversimplification. Certainly, following the introduction of an ACE inhibitor it is possible to document a reduction in circulating aldosterone concentrations. However, plasma aldosterone concentration tends to increase over time despite continued ACE inhibition, and the term 'aldosterone breakthrough' has been used to describe this phenomenon, the mechanism of which is poorly understood but is observed to occur in c35% (range 10–53% depending on the definition) of human patients treated with ARBs or ACE inhibitors (Bomback and Klemmer 2007). Plasma potassium concentration and ACTH are additional stimuli for aldosterone secretion from the adrenal gland zona glomerulosa. Aldosterone may be produced from sites other than the adrenal gland where angiotensin II may not control its production. It is also possible that the oxidative stress that occurs in tissues, particularly when stimulated by angiotensin II in the heart failure state, reduces the metabolism of cortisol by the enzyme 11-β-HSD type II which prevents cortisol from accessing and stimulating the mineralocorticoid receptor. All of these mechanisms may contribute to the 'aldosterone breakthrough' phenomenon although the latter would not explain a return of plasma aldosterone concentrations to pre-treatment levels. The precise mechanism(s) involved has not been determined. The favoured explanation is alternative pathways for the formation of angiotensin II from angiotensin I (e.g. chymase enzymes). However, this explanation would suggest the phenomenon should be more common in patients treated with ACE inhibitors compared to ARBs which does not appear to be the case. Indeed, small studies involving human patients suggest that patients given direct renin inhibitors alone demonstrate a similar prevalence of aldosterone breakthrough to those given ARBs or a combination of ARBs and direct renin inhibitors (Bomback et al. 2012).

Use of Aldosterone Receptor Antagonists in Veterinary Patients

Spironolactone has been developed as a treatment for dogs with heart failure secondary to mitral valve disease where it is added to conventional therapy (ACE inhibitors and diuretics). A combined product containing both spironolactone and benazepril has also been developed for use in dogs. The use of spironolactone to control blood pressure in dogs or cats has not been studied in controlled clinical trials despite the fact that mineralocorticoid receptor antagonists are increasingly being used in human medicine to treat 'drug-resistant' hypertension (Narayan and Webb 2016).

Aldosterone breakthrough from inhibition following treatment with ACE inhibitors can be demonstrated in naturally occurring heart failure in dogs (Häggström et al. 1996), thus strengthening the rationale for using spironolactone in canine heart failure patients in combination with ACE inhibitors, although the body of published literature addressing this issue is very small. Ames et al. (2016) demonstrated the phenomenon of aldosterone breakthrough subacutely in experimental dogs given an ACE inhibitor (benazepril or enalapril) when the RAAS system had been stimulated by the administration of furosemide or amlodipine. Finally, Ames et al. (2017) studied 39 dogs on treatment for myxomatous heart disease and estimated that 32% of dogs with signs of congestive heart failure and 30% with no signs of heart failure showed aldosterone breakthrough using a definition based on a reference interval for urinary aldosterone-to-creatinine ratio.

The development of a mineralocorticoid receptor antagonist for registration for use in dogs has meant the pharmacology of this drug has been studied in detail in the dog. To determine the optimal dose of spironolactone to use in dogs, a pre-clinical experimental study was conducted to examine the natriuretic effect and its relationship to plasma concentrations of active metabolites of spironolactone. Pharmacokinetic and pharmacodynamics modelling was used to identify the optimal dose to take forward into clinical trials for dogs with naturally occurring heart failure.

Spironolactone is rapidly absorbed from the GI tract and circulates in plasma highly bound to proteins (>90%). Spironolactone is difficult to measure due to its degradation after freeze/thaw cycles. Rapid liver biotransformation generates two main active metabolites, one that retains the sulphur group (7α-methyl-spironolactone) and canrenone (sulphur-free) that is the most important metabolite in dogs. Hence, canrenone is routinely measured as a surrogate for spironolactone. The metabolites showed linear kinetics over the spironolactone dose range from 1 to 4 mg/kg. The oral bioavailability of spironolactone is 50% in fasting conditions and improves to 80–90% with food. In radiolabelled studies, 70% of the dose is recovered from faces and 20% in urine. After multiple oral doses of 2 mg/kg for 10 days, steady-state conditions are reached by day 2 and no accumulation is observed. Mean C_{max} of 382 and 94 µg/L are achieved for 7α-thiomethyl-spironolactone and canrenone, after 2 and 4 h respectively (EMA - Veterinary Medicines 2007).

The initial dose recommended for dogs with heart failure (2 mg/kg/day) had been based on an allometric approach. Spironolactone does not have any significant diuretic or natriuretic effect at doses ranging from 1 to 8 mg/kg/day in laboratory dogs without hyperaldosteronism (Jeunesse et al. 2007). Intramuscular injection of

3 μg aldosterone/kg in healthy dogs achieves a transient state of hyperaldosteronism representative of concentrations reported in dogs with congestive heart failure (Guyonnet et al. 2010). This model was suitable for spironolactone dose determination by pharmacokinetic and pharmacodynamics modelling. In a 5 period crossover study, 15 dogs received doses of either no aldosterone (control) or aldosterone with 0, 0.8, 2 and 8 mg/kg spironolactone. Urine was collected over 6 h to measure the effect of aldosterone and spironolactone on the ratio of Na^+/K^+ excreted in urine. Measured plasma concentrations of canrenone increased proportionally with dose. The relationship between dose and reversal of aldosterone's effect on Na^+/K^+ excretion was modelled with a non-linear E_{max} model. The dose that achieved half the maximal effect (ED_{50}) was 1 mg/kg. The effect of aldosterone on Na^+/K^+ ratio was virtually completely reversed (88%) with a dose of 2 mg/kg, thus confirming the candidate dose that was taken forward into pivotal clinical trials for dogs with naturally occurring heart failure.

Clinical Evidence of Efficacy and Safety of Spironolactone in Dogs

One multicentre large clinical trial has been conducted for registration purposes in dogs with mitral valve disease (Bernay et al. 2010), and the safety data resulting from this trial has been published in a peer-reviewed paper (Lefebvre et al. 2013). The multicentre randomised controlled clinical trial was conducted in Europe and enrolled 214 dogs with mitral valve disease. Data from 212 dogs were included in the final analysis (intention to treat) and 190 dogs in the per protocol analysis. The composite primary end point was cardiac-related death or euthanasia because of mitral regurgitation or severe worsening of mitral regurgitation (defined as the need to introduce unauthorised medication or to increase furosemide >10 mg/kg to prevent congestive heart failure). Most of the cases included were in International Small Animal Cardiac Health Council classification (ISACHC) class II (advanced heart disease with either mild heart failure or where heart failure has been controlled with cardiac therapy) [$n = 190$] with the remainder in class III (advanced heart failure with clinical signs including respiratory distress, marked ascites and profound exercise intolerance).

To be included dogs had to have confirmation of mitral valve disease by echocardiography and identification of heart enlargement on thoracic radiography (vertebral heart score > 10.5). They also had to be showing at least three of the following six clinical signs, including one of cough, syncope and dyspnoea and one of decreased activity, decreased mobility and altered demeanour. Spironolactone (2 mg/kg daily orally with food) or placebo was added to conventional therapy (ACE inhibitor (benazepril, ramipril, enalapril or imidapril) plus furosemide and digoxin if needed).

Dogs were recruited by both specialist centres and first-opinion practices and had a somewhat unusual design with two different routes into the final 12 month common study period. The first route lasted 2 months and required the dogs to be receiving furosemide during this period. The second route lasted 3 months, and the dogs were not allowed furosemide treatment for at least the first 5 days of this study period. Cases reached the end points, were lost or withdrawn during these two routes into the common study period, such that 179 dogs completed these initial studies. Of these, only 123 were entered into the 12 month common study period (some owners declined to continue into the next phase of the trial), and 79 dogs completed the full 12 months. The 56 cases where the owners declined to continue in the trial were evenly distributed between the spironolactone- and placebo-treated groups. The consort diagram in the publication defines the reasons for withdrawal of cases at each stage.

Of the 212 cases in the intention-to-treat population, 39 cases reached the composite primary end point (11 in the spironolactone-treated group and 28 in the placebo group), 38 were withdrawn for other reasons (26 in the spironolactone group and 12 in the placebo group), 56 were not enrolled in the 12 month study because the owner declined (30 in the spironolactone group and 26 in the placebo group) and 79 completed the study (35 in the spironolactone group and 44 in the control group). Thus the event rate was low in this trial with only 18.3% of the intention-to-treat population reaching the predefined composite end point. Nevertheless, the event rate was 10.8% in the spironolactone-treated group and 25.4% in the placebo group. The hazard ratio of the treatment effect was 0.45 (0.22 to 0.90; $P = 0.023$). The wide 95% confidence limits on the hazard ratio is a reflection of the low event rate. Nevertheless, these data suggest spironolactone reduced the risk of an event happening by 55% (10–78%). The event rate found in this study is lower than that found in other large clinical trials involving dogs with mitral valve disease, primarily because the majority of dogs enrolled in this study were at the early clinical stages of the disease (190/212 in ISACHC stage 2). Indeed, because less than 50% of the population reached an end point, a median survival time could not be calculated for this study. Factors that negatively impacted on survival included furosemide treatment (yes or no). The effect of spironolactone treatment was evident whether or not furosemide had been used in the treatment regimen.

The Bernay study has been criticised for its complicated design and low event rate (Kittleson and Bonagura 2010). The enrolment of cases without assessment by clinical specialists may have led to dogs without clinical signs related to mitral valve disease being included in the study which could well have contributed to the low event rate seen. The complicated design of the study with two different routes and the relatively high withdrawal rate at entry to the 12 month common study are also less than ideal. Nevertheless, classification of cardiac-related deterioration and death were clearly defined end points, and the study had a randomised placebo-controlled design with appropriate blinding suggesting that, despite these limitations, an effect of spironolactone on progression of cardiac disease can be seen. The magnitude of that effect remains somewhat uncertain (based on the very wide 95% confidence limits identified by this study. The study was designed at a time prior to the widespread use of pimobendan in the management of dogs with mitral valve disease. Further studies with higher event rates are needed to determine the magnitude of the effect of spironolactone with more confidence and whether this is evident in cases treated with pimobendan. No attempt was made to measure blood pressure in this study and the effect of spironolactone on blood pressure in this context and its contribution (if any) to the effect seen remains unknown.

The safety aspects of use of spironolactone in dogs in placebo-controlled clinical trials have been critically assessed, including the data from the Bernay study, and reported in a separate publication (Lefebvre et al. 2013). Combining spironolactone with ACE inhibitors theoretically increases the risk of hyperkalaemia occurring. Lefebvre et al. (2013) found no difference in the frequency of adverse events between the dogs treated with spironolactone and placebo. Analysis of adverse event by systemic organ class found no difference between the frequency and number of these between the two groups. The risk of renal dysfunction or hyperkalaemia was the same between the spironolactone- and placebo-treated groups. As the majority of dogs enrolled in these studies had mild heart failure (ISACHC class 2), the safety of spironolactone in dogs with severe heart failure remains to be determined.

Clinical Evidence of Efficacy and Safety of Spironolactone in Cats

The clinical evidence for efficacy of spironolactone in cats consists of two relatively small clinical studies in cats with hypertrophic cardiomyopathy. The first (MacDonald et al. 2008) involved a colony of Maine Coon cats. Cats were included in the study if they had left ventricular wall thickness in diastole of >6 mm and decreased early lateral mitral annular velocity (Em) or summated early and late mitral annular velocity (EAsum) measured by pulsed wave tissue Doppler imaging echocardiography (markers of diastolic dysfunction). The study had a randomisation design that paired cats based on the Em/EAsum measurements to ensure these were equivalent between the two groups as these were of particular interest over time to determine whether spironolactone treatment improved diastolic function through an anti-fibrotic effect. Cats received either spironolactone (2 mg/kg q12 h orally) or placebo for 4 months. Although blood pressure was measured at entry to the study (to ensure systolic pressure measured by Doppler was between 100 and 160 mmHg), blood pressure measurements whilst on treatment were not reported in the paper.

Twenty-six cats met the entry criteria and were randomised so there were 13 per group. Of these five had pre-treatment plasma aldosterone concentrations above the laboratory reference interval (388 pg/ml), one in the spironolactone group and four in the placebo group. No significant treatment-related cardiac effects could be identified over the 4 month period of the study. Four of the spironolactone-treated cats developed severe ulcerative facial dermatitis 2.5 months after starting the drug. These lesions resolved within 4–6 weeks of cessation of spironolactone and commencement of amoxicillin-clavulanate and prednisolone treatment and did not recur over the subsequent 12 months. One spironolactone-treated cat developed non-regenerative anaemia associated with myeloid dysplasia, and six of the cats had mild increases in blood urea nitrogen after 2 months of treatment with one cat having serum creatinine outside of the reference range. As expected, plasma aldosterone concentration increased markedly in cats treated with spironolactone. Thus, this pilot study found no evidence of a beneficial effect on hypertrophic cardiomyopathy of 4 months of treatment on this relatively small group of cats and raised concerns about the relatively high prevalence (30%) of what is likely to be drug-related acute severe facial dermatitis which resolved on cessation of drug treatment.

A second clinical trial in cats with heart disease has been published (James et al. 2018). Although this was a randomised controlled clinical trial involving 20 cats, interpretation of the results is hampered by the cardiac disease apparently being worse at entry to the study in the placebo-treated cats. This pilot study found no safety issues and is suggested as the basis for further clinical investigation of spironolactone in cats given that the high prevalence of drug-induced skin reactions was not seen in any of the nine cats treated in this study. All cats entered into this study were seen by a specialist cardiologist and were assessed as having congestive heart failure secondary to cardiomyopathy with evidence of left ventricular remodelling and left atrial enlargement on echocardiography. Cats with kidney disease (IRIS stage 3 and above), hyperthyroidism, systemic hypertension and non-cardiac systemic disease or those that were pregnant or lactating were excluded. The dose rate of spironolactone used in this study was 2–4 mg/kg once daily by oral administration. Cats had to be concomitantly treated with ACE inhibitors and furosemide from entry to the study. The use of clopidogrel was allowed at the investigator's discretion. Use of pimobendan, antidysrhythmic drugs and aspirin was not permitted.

The study was scheduled to last 60 weeks with the primary end point defined as death or euthanasia due to cardiac causes. The baseline characteristics of the cats at entry to the study showed that the nine cats treated with spironolactone had lower blood pressure, better body condition score, fewer other electrocardiographic abnormalities and lower left atrium-to-aortic ratio (1.9 ± 0.3 vs. 2.5 ± 0.6) when compared to the placebo group suggesting the two groups were not well matched for some important characteristics that might influence outcome. Although cardiac mortality was higher in cats receiving placebo (with none of these cats completing the 60 month study) than those receiving spironolactone (five of the nine cats completed the study), interpretation of these data is not possible given the differences in baseline characteristics between the two populations. Blood pressure was measured throughout the study and examination of the supplementary data provided on this suggested no reduction in blood pressure was seen following introduction of spironolactone (median (range) of SBP (mmHg) were 120 (85–144) $n = 9$; 121 (106–134) $n = 8$; 132 (112–144) $n = 7$ for visits 1, 2 and 5, respectively). The prevalence of adverse events in the spironolactone-treated cats did not differ from that reported in the placebo cats, so providing reassurance that immunological reactions may not be common in an outbred population of cats although more cats need to be treated under controlled studies to be completely sure of this conclusion.

Spironolactone has not been well evaluated in pre-clinical studies in cats to be sure that the dose rates used in the clinical trials are optimal. The characteristic increase in serum aldosterone concentrations noted by MacDonald et al. (2008) suggests the dose rate of 2 mg/kg twice daily was blocking the mineralocorticoid receptor in vivo, but the extent of inhibition at this dose rate remains to be determined. The finding that plasma aldosterone concentration tends to be normal or high in association with low plasma renin activity in the hypertensive cat (Jepson et al. 2014) and that hypertensive cats tend to have lower plasma potassium ion concentrations (Syme et al. 2002) possibly implicates aldosterone in the pathophysiology of hypertension in cats, suggesting the use of spironolactone as an antihypertensive agent might be logical. However, for many cats good control of hypertension can be achieved with amlodipine (see below) as a single agent; thus spironolactone has not been used as a first-line treatment for hypertension in cats. Indeed, drug-resistant hypertension is relatively uncommon in cats if a post-treatment target blood pressure of 160 mmHg is used (Acierno et al. 2018). Whether this target is appropriate and whether spironolactone could be used in both dogs and cats in combination with ACE inhibitors/ARBs and amlodipine to achieve lower post-treatment blood pressure targets remain to be determined.

11.2.3 Calcium Channel Blockers

Calcium channel blockers are vasodilator drugs which non-competitively antagonise vasoconstriction induced by a range of different hormones, neurotransmitters and paracrine agents, most of which depend on calcium entry via voltage-dependent calcium channels in vascular smooth muscle cell membranes to maintain sustained increase in vascular tone. Thus, this group of drugs acts peripherally to lower systemic vascular resistance (having a predominant effect on the arterial side of the circulation). They lower blood pressure but stimulate increased centrally driven sympathetic nerve activity as a response to the reduction in vascular tone, upregulating both the sympathetic nervous system and the RAAS as a result.

Amlodipine is an example of a voltage-dependent calcium channel blocker or antagonist (also known as a calcium entry blocker) that has been used in human clinical medicine since the late 1980s. Its use in veterinary medicine for the treatment of hypertension in cats was first suggested in the mid-1990s. It is a member of the dihydropyridine group of calcium channel antagonists and binds selectively to voltage-dependent calcium channels of the L-type, favouring the L-type calcium channels found in vascular smooth muscle over those found in cardiac muscle and cardiac nodal tissue. Thus amlodipine predominantly has vascular effects, which is in contrast to verapamil, which is relatively cardioselective, and diltiazem, which acts on both vascular and cardiac muscle.

Like other drugs from this class, amlodipine binds to the α_1 subunit of the L-type calcium channel acting allosterically with the gating mechanism to prevent it from opening in response to depolarisation. It shows preference for the channel in mode 0, in which state the channel does not open in response to depolarisation. This will non-competitively antagonise a range of agonists that stimulate arterial vasoconstriction through receptors that are linked to arterial smooth muscle depolarisation, such as norepinephrine, angiotensin II and endothelin-1. These mediators also activate intracellular second messenger systems (inositol trisphosphate) to stimulate calcium release from intracellular stores, but for many agonists, tonic contraction of arteriolar smooth muscle is dependent on entry of calcium into the cell to replenish these stores, and entry occurs through the voltage-dependent channels. Calcium causes smooth muscle contraction by calmodulin-mediated activation of the enzyme myosin light-chain kinase which phosphorylates the myosin light chain causing it to detach from the actin and allowing muscle shortening to occur. The kinetics of the binding of amlodipine to the L-type calcium channel are slow, meaning that its onset of action is slow. This is a desirable property as rapid reductions in blood pressure, such as seen with nifedipine (which binds to and dissociates rapidly from the calcium channel), lead to stimulation of reflex tachycardia as a result of activation of baroreceptors. Meta-analyses of clinical trials in human medicine have demonstrated a positive benefit of use of long-acting calcium channel blocking drugs (such as amlodipine) over those of intermediate duration of action (Chaugai et al. 2018).

As a class, as discussed above, dihydropyridines are arterial/arteriolar dilators, thus reducing arterial blood pressure, but have little effect on the venous side of the circulation. Other types of smooth muscle are affected by amlodipine (e.g. biliary, urinary and gastrointestinal tracts and uterus). Through selective arterial/arteriolar vasodilation, amlodipine has antihypertensive effects in animals where blood pressure is raised through an increase in peripheral vascular resistance. Thus, the magnitude of the effect of this class of drug on blood pressure will depend on how much vasoconstrictor tone is contributing to the systemic vascular resistance at the time the drug is administered to a given animal. As discussed in Chap. 1, raised vascular resistance is one of the mechanisms by which hypertension secondary to CKD occurs.

Calcium channel blockers have differing effects on different vascular beds. The kidney is a major target organ for hypertension, receiving 20–25% of the cardiac output and having a high-pressure first capillary bed (the glomerular capillary bed) to facilitate the formation of glomerular filtrate. With loss of functioning nephrons, the afferent arterioles of the remaining functioning nephrons dilate to compensate and increase glomerular capillary pressure leading to hyperfiltration and increased load of albumin appearing in the filtrate which is proposed as one mechanism, whereby intrinsic progression of CKD occurs. The ideal antihypertensive agent would not only lower systemic arterial pressure but would also reduce glomerular capillary pressure by dilating the efferent arteriole more than the afferent arteriole. The second-generation calcium channel blockers, like amlodipine, do not have such effects and are thought to preferentially dilate the afferent arteriole over the efferent arteriole. For this reason, in human medicine, amlodipine tends to be used concomitantly with inhibitors of the RAAS (ACE inhibitors, angiotensin receptor blockers or renin inhibitors), particularly in hypertensive patients at risk of renal disease (e.g. diabetic patients) and those with proteinuria. Third-generation calcium channel blockers, such as benidipine (Yao et al. 2006) and lercanidipine (Grassi et al. 2017), also dilate efferent arterioles and potentially are more renoprotective than the second-generation calcium channel blockers.

Most of the pharmacological effects of amlodipine can be attributed to its actions through the L-type calcium channel with the major blood pressure lowering effect being related to this action on arteries and resistance arterioles. However, in vitro data also suggests amlodipine has actions on endothelial cells which increase the generation of nitric oxide. In human patients this is an important effect as endothelial cell dysfunction accompanies hypertension, coronary heart disease and diabetes, all conditions in which amlodipine is used as part of the treatment regimen. In isolated coronary microvessels of dogs, amlodipine was shown to dose dependently increase nitrite production (nitrite is the breakdown product of nitric oxide; Zhang and Hintze 1998). This is an effect that was not seen with diltiazem or verapamil. The dose-dependent effects of amlodipine in this study appeared to be biphasic with initial stimulation occurring between 10^{-10} and 10^{-8} M and further stimulation seen at 10^{-6} to 10^{-5} M. Similar effects were seen in large vessels. These experiments suggested that the mechanism by which amlodipine stimulated NO production involved the bradykinin receptor system as the B_2 receptor antagonist HOE 140 inhibited the effect.

Further studies have shown amlodipine to increase the generation of NO from native endothelial cells, as detected by electron spin resonance spectroscopy, and stimulated an eightfold increase in cyclic GMP levels in cultured endothelial cells (Lenasi et al. 2003), a process that involved changes in phosphorylation of eNOS. Although activation of eNOS involved the B_2 receptor pathway in porcine coronary artery, B_2-independent activation of eNOS was identified in the rat aorta and cultured cells involving PKC. Further work by Batova et al. (2006) showed in subcellular experiments using bovine aortic endothelial cells that amlodipine dose dependently (in the nanomolar concentration range) liberated eNOS from caveolin-rich low-density membranes, potentiating the eNOS activation of bradykinin and VEGF. Finally, Sharma et al. (2011) have shown that amlodipine can prevent binding of native, acylated eNOS complexes to the active domain of Cav-1 in a concentration-dependent fashion, suggesting that amlodipine has an antagonistic effect on the native eNOS/Cav-1 signalling complex.

Amlodipine has also been shown to improve endothelial cell function through anti-oxidant and anti-inflammatory actions [e.g. Toma et al. (2011)]. In human medicine, oxidised low-density lipoprotein and the pro-inflammatory state induced by obesity are important factors contributing to endothelial cell dysfunction, coronary artery disease, hypertension and atherosclerosis development, leading to major cardiovascular morbidity and mortality.

Amlodipine exists as two enantiomers in solution (S and R) with the pharmacodynamic activity residing in the S enantiomer (1000 to 1 potency ratio) in terms of calcium channel blocking activity but in the R enantiomer in terms of nitric oxide releasing activity. There is a vast literature on calcium channel blockers as antihypertensive agents in human medicine where they are used as one of the first-line drugs often as part of combination therapy. In veterinary medicine amlodipine is the drug for which there is most data to support its clinical use, and much of that data is from its use in the management of feline hypertension.

11.2.3.1 Evidence of the Blood Pressure Lowering Effect of Amlodipine in Veterinary Species

Although the dog has been used as a model for development of amlodipine as a drug for use in human medicine, much of the data in the literature relates to its use as a cardioprotective agent in models of ischaemia-induced cardiac injury (e.g. Lucchesi and Tamura 1989). There is little pre-clinical data relating to dog models of hypertension. Although amlodipine is one of the drugs used in the management of canine hypertension in clinical patients (see Chap. 13), there are no randomised controlled clinical studies assessing the response of hypertensive dogs to amlodipine, and most of the evidence for the efficacy of amlodipine as an antihypertensive agent in the dog is in the form of retrospective studies and case reports [e.g. Geigy et al. (2011)].

Thus, this section will focus on the pre-clinical and clinical data providing evidence for the use of amlodipine in the cat. The first study suggesting amlodipine was an effective antihypertensive agent to treat cats appeared in 1997 (Henik et al. 1997). Amlodipine was authorised for use in cats to treat hypertension in Europe in 2014, and the registration dossier used previously published data supplemented by product-specific pharmacokinetic, target animal safety and field safety and efficacy studies.

11.2.3.2 Pharmacokinetics of Amlodipine in Cats

There are no published data in the literature on the pharmacokinetics of amlodipine in cats, and the data that follows is taken (with permission) from the registration studies submitted for product authorisation of amlodipine for use in cats. In a small pharmacokinetic study, four cats (2.9–9 years old, two males 4.8 and 5.4 kg and two females 3.1 and 3.8 kg) were administered a single 1.25 mg dose of a human formulation of amlodipine besylate to target a dose of 0.25 mg/kg. Amlodipine was measurable by a racemic liquid chromatography/mass spectrometry method optimised for cat plasma and still quantifiable in the sample taken 7 days (168 h) after dosing. The time to peak plasma concentration (T_{max}) was between 3 and 6 h after dosing. Dose-normalised C_{max} varied between 18.9 and 35.5 ng/ml. The estimated terminal half-life was between 39.5 and 88.5 h. This exploratory study confirms that amlodipine is slowly cleared from the plasma of the cat, making it similar to the dog and human patient in this respect. In a second study, the same four cats were dosed daily for 14 days with 0.625 mg. Blood samples were collected on day 1, 7 and 14 of dosing at pre-dose, 1, 2, 4, 6, 8, 12 and 24 h after dosing. Drug accumulation occurred with repeated administration as C_{max}, trough plasma concentrations and the AUCs were higher on day 14 compared to day 7. Dose-adjusted steady-state plasma concentrations (trough levels on day 14) varied between 22 and 44 ng/ml. Dose proportionality was tentatively verified when comparing the dose-corrected C_{max} and AUC from both studies. This study also incorporated blood pressure evaluation (by a Doppler method) prior to and during the dosing period and was the source of the blood samples (collected 24 h after the last dose on day 14), but no relationship between blood pressure and plasma concentration could be identified in normotensive cats.

A pivotal bioavailability study was carried out in 12 healthy adult cats (6 M–6 F; 6–11 years old). The target dose given orally was 0.25 mg/kg. Oral absorption was slow leading to a peak plasma concentration at 3–6 h following dosing. A secondary peak was detected in two animals at 8 and 24 h, perhaps representing enterohepatic recirculation of amlodipine has occurred. Examination of the individual curves suggested a definite flattening of the plasma concentration vs. time curve between

8 and 12 h in most cats following oral administration, suggesting this phenomenon might be more common (just not always leading to a secondary peak). The estimated elimination half-life following oral administration was 53 h (range 39.8–81.2 h), which closely matched that calculated following intravenous administration [51.8 h (range 46.1–67.4 h)]. The mean oral bioavailability was calculated to be 74% (range 40–122%). The two methods used to calculate the volume of distribution of amlodipine both gave very high values (V_z—10.3 L/kg and V_{ss}—9.62 L/kg). Such high apparent volumes of distribution for a drug that is heavily plasma protein bound suggest the drug is accumulating or binding within an extravascular site. Slow release from this site or sites might explain the low total body clearance (2.31 mL/min/kg). In conclusion, this pivotal study clearly demonstrates that the final formulated product delivers amlodipine into the plasma in a similar way to the human formulated product. This provides reasonable confidence that the kinetic data on multiple daily dosing obtained with the product formulated for human medicine can probably be extrapolated to the final formulated product.

Amlodipine is a weak base (pKa: 8.6) extensively plasma protein bound (96.8, 96.5 and 96.2% at 100, 400 and 1000 ng/mL, respectively); however, concentrations tested were at least twice the maximum recorded in the pivotal pharmacokinetic study. Hepatic metabolism of amlodipine was compared between species using isolated hepatocytes from rat, human, dog and cat. Amlodipine was incubated at 1 and 10 μM concentrations (10 to 100 times peak plasma concentrations). No species produced a glucuronidated metabolite of amlodipine. Rat hepatocytes were much more efficient at metabolising amlodipine than the other species with 96.8% conversion of amlodipine to metabolites occurring in 180 min for the rat compared to 33.1, 28.1 and 18.8% conversion seen in cat, human and dog hepatocytes, respectively. This suggests that this in vitro system does reflect the in vivo situation to a certain extent. The six metabolites found in cats were common to all species. Using plasma samples banked from the initial PK studies, three metabolites were identified in vivo, two of which were the result of sequential reduction, deamination and sulphonation (both identified from the in vitro study).

The high apparent volume of distribution of amlodipine in cats (10 L/kg) is interesting for a drug that is highly plasma protein bound. This suggests concentration of the drug within an extravascular body compartment. This might be through its association with the voltage-dependent calcium channels which are abundant in many tissues. The high affinity the drug has for the calcium channel would mean the kinetics of dissociation would be slow, and the release of the drug from such binding sites could be one factor limiting its clearance from the body. In addition, the drug is a weak base and so could become ion trapped within intracellular fluid if it penetrates across cell membranes. Additionally, high protein binding reduces hepatic extraction ratio, independently of liver blood flow and intrinsic clearance. Either of these processes could be altered in disease states indicating that the study of the drug's disposition in clinical patients would provide more information on how the drug is handled under the clinical conditions in which it is used.

11.2.3.3 Pharmacodynamic Effects of Amlodipine in Cats: Experimental Studies

Mathur et al. (2002) published a short-term experimental study to assess the efficacy of amlodipine in the management of hypertension associated with partial nephrectomy in healthy young cats (6–12 months of age). The model was described as an 11th/12th nephrectomy model with partial infarction of the left kidney, achieved by tying off the interlobar arteries on day 0 followed 15 days later by total nephrectomy of the right kidney. Blood pressure was measured directly in these cats by radiotelemetry from day 3.

The study was a block randomised controlled study with 10 cats in each group, one group receiving amlodipine at a standard dose of 0.25 mg/kg once a day and the other group receiving placebo. Pairs of cats were produced by ordering all 20 cats on the basis of their mean 8 h blood pressure value on day 3 and pairing the top 2 together and so on. One of each pair was randomly assigned to receive either amlodipine or placebo from day 3 to day 36 at which point the cats were euthanised and a full postmortem examination undertaken. Daily examinations were undertaken by a veterinarian with routine ophthalmic examinations undertaken on day −3 and day 36.

Figure 11.5 graphically illustrates the effect of amlodipine on the systolic blood pressure of the two groups of cats starting 3 days following the initial ligation of

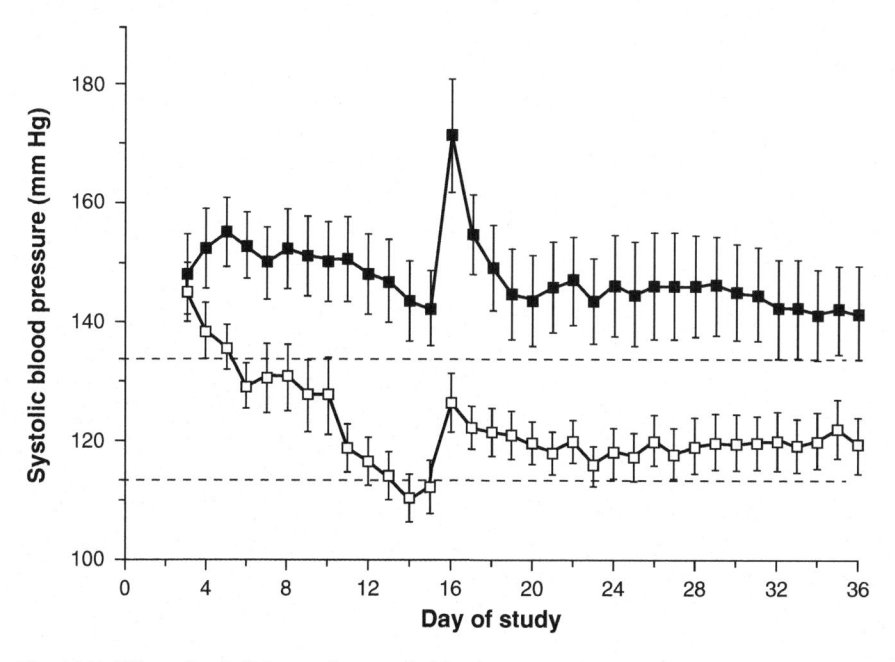

Fig. 11.5 Effect of amlodipine on the systolic blood pressure of cats with experimentally induced CKD. Mean ± SEM values for systolic blood pressure obtained by use of radiotelemetry from 20 cats with surgically induced renal insufficiency. Cats received dextrose-cellulose (control group; solid squares) or 0.25 mg of amlodipine/kg/day (open squares) from days 3 to 36. There was a significant ($P < 0.05$) antihypertensive effect of amlodipine beginning on day 5 and continuing until the end of the study. Dashed lines represent mean ± SD values for similarly obtained data from six clinically normal cats. Reproduced with permission from Mathur et al. (2002)

vessels to the left kidney—the point at which amlodipine dosing commences. Systolic blood pressure decreased in both groups, but the antihypertensive effect of amlodipine was statistically evident by day 5 and remained about 30 mmHg lower in the amlodipine-treated group. A spike in blood pressure occurs on day 16 associated with the removal of the right kidney to complete the 11/12th nephrectomy model. This returns back to the pre-nephrectomy level by day 20. The increase in blood pressure was substantially greater in the placebo-treated cats than in the amlodipine-treated cats. During this period of time, food intake and activity were lower in the amlodipine-treated than the placebo-treated cats, which is interesting in terms of clinical signs of acute hypertension.

Two cats in the placebo group developed severe neurological signs on day 16 of the study associated with severe hypertension and were euthanized. One cat in the amlodipine-treated group developed bloody diarrhoea and was euthanized on day 27. This death was not attributed to hypertension, and the cause of the acute enteritis (confirmed at post-mortem as purulent enteritis) was not determined. Seven of the ten cats in the placebo group had ocular lesions compared with two of the ten cats in the amlodipine-treated group. The difference in prevalence of ocular lesions was statistically significant.

The study showed that amlodipine was able to control blood pressure elevations associated with the renal mass reduction model of chronic kidney disease, leading to approximately 30 mmHg reduction in systolic blood pressure. This represents strong evidence supporting the efficacy of this dose of amlodipine (0.25 mg/kg daily) in hypertension associated with kidney dysfunction. However, one should question whether the model used in this study is representative of hypertension associated with naturally occurring CKD in elderly cats. The model does share one similarity with many cats with naturally occurring hypertension in that it appears to be relatively resistant to the antihypertensive effects of ACE inhibitors such as benazepril – where the blood pressure is only reduced by about 10 mmHg (Brown et al. 2001; see Fig. 11.2).

The model was allowed to run for a relatively short period of time during which ocular lesions appeared in the placebo cats. Amlodipine appeared to protect cats from ocular pathology. The renal pathology found in the remnant kidney at day 36 of the study was no different between the two groups so it is not possible to conclude that amlodipine protected the kidney from hypertensive damage in this model, possibly because insufficient time had been allowed for the damage to become manifest. No difference was found between the groups in terms of proteinuria (UPC) or albuminuria suggesting amlodipine did not manifest an antiproteinuric effect in this model despite the significant reduction in systemic arterial blood pressure. This may be because amlodipine preferentially dilates the afferent arteriole offsetting some of the benefit of the reduction in arterial blood pressure in this model.

In conclusion, this study demonstrates unequivocally that amlodipine at 0.25 mg/kg/day has a marked antihypertensive effect in this model of renal mass reduction. The radiotelemetric method of measuring blood pressure continuously in the freely moving animal rules out any white coat effect or artefact associated with the measurement process. The study also demonstrated a benefit of this degree of blood pressure reduction on ocular pathology associated with hypertension.

11.2.3.4 Clinical Trials to Determine the Antihypertensive Effect of Amlodipine in Feline Patients

The first randomised controlled clinical trial in cats with naturally occurring hypertension was published by Snyder (1998). This was a prospective, randomised, placebo-controlled and blinded trial. Its major limitation is that it only involved nine cats with either renal insufficiency or idiopathic hypertension (six of which were mixed breeds and three of different pedigree breeds). A second limitation is that the initial phase lasted for just 7 days which is not long enough for amlodipine to reach steady state. Cats were all client-owned and were recruited to the study if their systolic blood pressure exceeded 170 mmHg on seven occasions over a 24 h period prior to entry to the study. The technique used to measure blood pressure was standardised to the Doppler method. The cats received either amlodipine (0.625 mg/day orally; $n = 5$) or placebo ($n = 4$) for 7 days and returned to have their blood pressure evaluated on seven occasions over a 24 h period again. If the systolic blood pressure had decreased by 15% or was <170 mmHg on average for the 24 h, the cat remained on the same treatment. If neither of these outcome criteria applied, the cat was switched to the other treatment code at this time point. Those cats that switched treatment codes were evaluated after a further 7 days and if their blood pressure was still not adequately controlled the treatment code was broken and these cats were placed on 1.25 mg amlodipine. Otherwise all cats remained on treatment for 2 weeks at which point they were re-evaluated as after 7 days with seven blood pressure measurements over the course of 24 h to determine the 24 h blood pressure control.

The five cats receiving amlodipine treatment initially received a dose rate of 0.11–0.17 mg/kg. After 7 days, four of these five cats had average daily blood pressures less than 170 mmHg and so remained on amlodipine treatment. One cat was switched to placebo as its blood pressure had decreased by 10.8% and on average over 24 h was still >170 mmHg. By contrast, after 7 days all four cats receiving placebo were switched to amlodipine as they did not meet either efficacy criteria. After 14 days, three of the five cats that were switched still did not meet either efficacy criteria, and the code was broken, and these cats then were treated with 1.25 mg amlodipine. These three cats were significantly heavier than the remaining six cats (6.1 vs. 4.1 kg). Only six of the original nine cats were available for re-evaluation after 16 weeks, three taking 0.625 mg/day and three taking 1.25 mg/day. All six cats had average daily blood pressure below 170 mmHg at the 16 week time point, and there was no apparent difference in blood pressure between the two dose rates although clearly the group sizes were very small. Three of the cats did not complete the study because they were euthanised. One had mammary neoplasia, another had progressive azotaemia and the third had long-standing inappropriate urination. All three of these cats were responders on day 7. This illustrates the major limitation of this study. The small number of cats entered into the study means these three deaths represent one-third of the population. Two of them most likely were not related to the treatment. The deterioration of the azotaemia in one cat might have been related to the treatment if reducing the blood pressure had led to a fall in glomerular filtration rate. However, this would not be expected unless the drug induced systemic hypotension.

The authors report that there were no adverse effects other than the anorexia associated with deteriorating azotaemia in one cat. It is not clear how systematically the owners were questioned to ensure all adverse events would be recorded. Another cat enrolled in the study showed an increase in creatinine from 1.6 mg/dl (which is within the laboratory reference range) to 2.6 mg/dl (which is outside of the reference range). This would represent significant deterioration in kidney function but it is impossible to know whether this was drug related or would have happened in the absence of amlodipine treatment as progression of CKD in the cat occurs at highly variable rates with urinary protein-to-creatinine ratio being one of the factors most predictive of rate of progression. UPC was not measured in this study.

Despite its limitations this study was convincing enough regarding the efficacy (and safety) of amlodipine for many practitioners struggling to manage feline hypertension with ACE inhibitors and beta-blockers at the time to switch to amlodipine. This was despite the fact that the 5 mg human formulation of the drug was difficult to use to dose cats accurately. Subsequently a number of retrospective uncontrolled studies have been published, generally confirming the efficacy of amlodipine using the human formulated drug product (Elliott et al. 2001; Jepson et al. 2007). Jepson et al. (2007) examined the effect of amlodipine treatment on proteinuria in the cat in an attempt to answer the concern that by dilating the afferent arteriole, amlodipine treatment might expose the kidney to high systemic arterial blood pressure. Although this study was uncontrolled, the UPC values could be compared with the pre-treatment value to determine any changes in response to treatment. The paper reports paired UPC data from 105 cats before treatment with amlodipine and when the cats had been stabilised on treatment. In the group as a whole, a significant decrease in UPC was seen with 73 of the 105 cats (69.5%) showing a decrease. The decreases tended to be more marked in cats with the higher UPC at the pre-treatment time point. The data from this paper has been re-analysed in a different way to illustrate this point and is presented in Fig. 11.6, showing the changes in UPC occurring in each of the three IRIS categories of proteinuria and indicating the shift in proteinuria category occurring in response to treatment. These data suggest the majority of cats treated with amlodipine experience a reduction in proteinuria, and Jepson et al. (2007) went on to show that time-averaged UPC on treatment was inversely and independently related to survival of hypertensive cats, indicating the importance of measuring and monitoring UPC in cats that are treated with amlodipine for their hypertension, a practice that appears to be uncommon in primary care practices in the UK at the present time (Conroy et al. 2018).

In 2015, the results of the pivotal field safety and efficacy study to support the registration of a veterinary product containing amlodipine were published (Huhtinen et al. 2015). The study involved 77 cats diagnosed with systemic hypertension in clinical practice (systolic arterial blood pressure persistently $\geq$165 mmHg by high-definition oscillometry) and was randomised, double blinded and placebo controlled. The efficacy phase was for 28 days with an assessment after 14 days, at which point the dose of amlodipine could be doubled from 0.125 to 0.25 mg/kg daily (in the form

11 Pharmacology of Antihypertensive Drugs

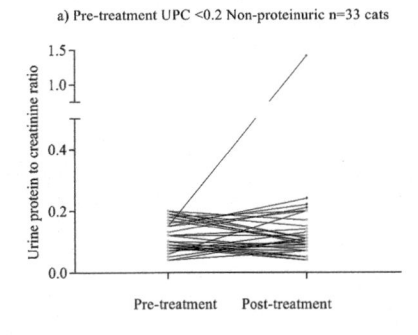

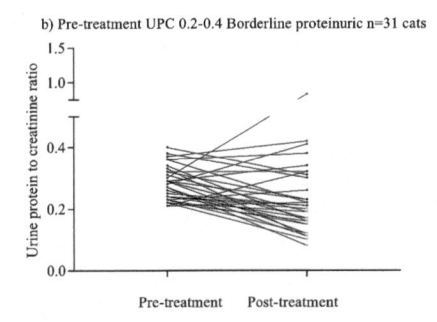

a) Pre-treatment UPC <0.2 Non-proteinuric n=33 cats

b) Pre-treatment UPC 0.2-0.4 Borderline proteinuric n=31 cats

Post-treatment 3/33 (10%) became borderline proteinuric and 1/33 (3%) became proteinuric. 17/29 (58.6%) showed reduction in proteinuria within non-proteinuric substage.

Post-treatment 3/31 (9.7%) became proteinuric. 15/31 (48.4%) had a reduction in substage to non-proteinuric. 6/13 (46.2%) showed a reduction in proteinuria within the borderline substage.

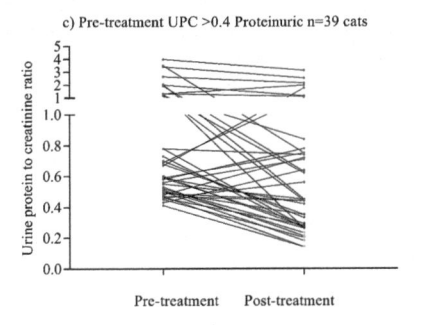

c) Pre-treatment UPC >0.4 Proteinuric n=39 cats

Post-treatment 12/39 (30.8%) had a reduction in substage to borderline proteinuric and 4/39 (10.3%) reduced to the non-proteinuric substage. 16/23 (69.6%) showed a reduction in proteinuria whilst remaining within the proteinuric substage.

Fig. 11.6 Changes in UPC seen in 105 cats in response to amlodipine treatment to control systemic arterial blood pressure

of a chicken-flavoured chewable tablet) if the predefined criteria for treatment success had not been met. These were a 15% reduction in systolic blood pressure relative to the pre-treatment value or systolic blood pressure of 150 mmHg or below. Success of treatment was evaluated at 28 days. At this point those cats randomised to receive placebo were switched to amlodipine (for 90 days with a dose escalation occurring at 14 days if necessary), and those receiving amlodipine continued to receive the drug for a further 60 days as part of the safety evaluation. Cats were not included in the study if they were already taking calcium channel blockers or had started on other antihypertensive drugs or undergone a change in dose with 30 days of the screening visit. Hyperthyroid cats were allowed to be included provided they were stable on the same dose of methimazole for 3 months prior to inclusion (or had been surgically treated and were stable for 3 months). Generally cats that required a change in diet or other medication within 30 days of screening were not included in the study. Cats with blood pressure > 200 mmHg were generally excluded unless the investigator felt they were suitable for inclusion.

Only three cats withdrew during the efficacy phase (1st 28 days) due to adverse events (two from the amlodipine-treated group and one from the placebo group). At day 14, 19 of the 41 cats (46%) receiving amlodipine were classified as responders. The dose rate was doubled for the remaining 22 cats. By contrast in the placebo group 7 of 35 were classified as responders (20%), and the remaining 28 had their 'dose' doubled. By day 28, 25 of the 40 amlodipine-treated cats met the criteria to be classified as responders (63%), whereas the responder rate in the placebo cats was 6 of 34 (18%). Receiving amlodipine increased the chances of the cat being a responder by 7.9 (2.6 to 24) times ($P < 0.001$) and was the only factor that was significantly associated with being a responder. The other covariates tested were whether the cat also received ACE inhibitors, the baseline blood pressure and whether or not the cat had CKD. None of these covariates were significantly associated with whether the cat was classified as a responder or not. Figure 11.7 shows the absolute changes in blood pressure over both phases of this study in the two groups of cats.

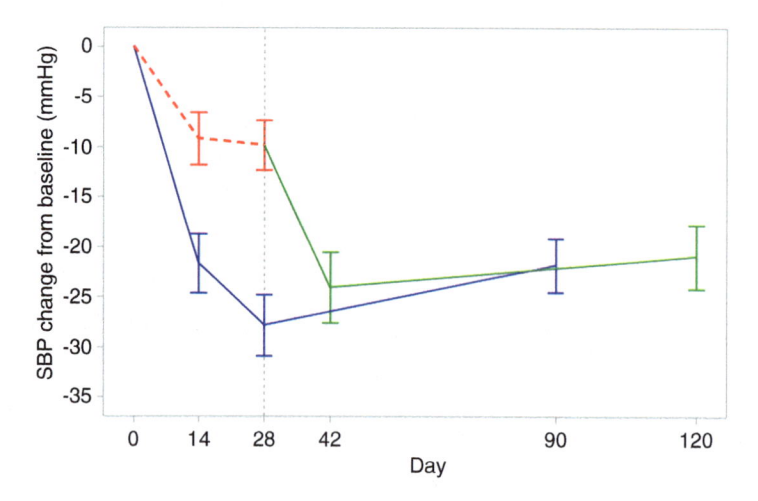

Fig. 11.7 Data from Huhtinen et al. (2015). Absolute changes in blood pressure over time in amlodipine-treated cat (blue line) and placebo-treated cats (dotted red line). At day 14, the dose of amlodipine or placebo for those cats that did not meet the criteria to be classified as a responder was increased. At day 28, 63% of the cats receiving amlodipine were classified as responders, whereas only 18% of the cats receiving placebo were so classified. At day 28, the placebo cats were switched to amlodipine, and dosing continued to day 120, whereas the cats that started on amlodipine continued to receive this for a further 60 days (until day 90)

During the first 28 days of the study, the adverse event rate was identical between the placebo- and the amlodipine-treated groups (28.6%). Nine of the sixty-seven cats followed for 90 days on amlodipine treatment had increases in their plasma creatinine concentrations of greater than 25%. This paper demonstrates the importance of including a placebo group when evaluating the efficacy of antihypertensive agents, and the magnitude of the blood pressure decrease in the cats receiving placebo was remarkably similar to that found in the telmisartan clinical trial (Glaus et al. 2019). The mechanism by which a placebo lowers blood pressure in cats remains to be determined.

Although there were some cats enrolled with systolic blood pressure > 200 mmHg (at the veterinarian's discretion), Huhtinen et al. (2015) did not address the efficacy of amlodipine in managing hypertension in cats with pre-treatment systolic blood pressure that is >200 mmHg. However, the much smaller study published by Snyder (1998) did demonstrate efficacy in cats whose systolic blood pressure was >200 mmHg. A more recent uncontrolled retrospective study suggested that cats with systolic blood pressure above 200 mmHg required a higher dose to control their blood pressure to below 160 mmHg (Bijsmans et al. 2016). In addition, Huhtinen et al. (2015) did not measure urine protein to determine the effect of amlodipine on this important indicator of renal health. As discussed above, a retrospective uncontrolled study has suggested that amlodipine treatment is associated with a reduction in UPC in the majority of cats and that post-treatment time-averaged UPC rather than post-treatment time-averaged blood pressure was associated with survival (Jepson et al. 2007).

The PK/PD relationship of amlodipine in cats has been investigated under clinical conditions. The Huhtinen et al. (2015) study was placebo-controlled with less severely hypertensive cats than the ones from the uncontrolled study from Bijsmans et al. (2016), which followed an optional dose titration design (based on BP response). Amlodipine concentrations were measured at one or two time points for each cat in the Huhtinen study (efficacy +/− safety phase) and only once in the Bijsmans study at the time of blood pressure control. At the time of control (Bijsmans et al. 2016), 61 cats were receiving 0.625 mg (0.17 ± 0.04 mg/kg) (mean ± SD), 42 were receiving 1.25 mg (0.33 ± 0.09 mg/kg) and 1 was receiving 2.5 mg (0.49 mg/kg). Plasma [amlodipine] was measured in samples of 104 cats. Dose and log [Amlodipine] were positively and significantly associated with SBP reduction. Plasma amlodipine concentration at SBP control 0.625 mg: median 33.3 ng/mL [IQR: 25.5, 54.9] 1.25 mg: median 70.5 ng/mL [IQR: 48.9, 98.7]. A hyperbolic relationship between relative blood pressure decrease from baseline when on treatment and plasma [amlodipine] was found. The estimated value of lowest percentage of baseline blood pressure achievable on treatment (E_{max}) was 64.1%, with an EC_{50} value of 15.8 ng/ml (Fig. 11.8). A practical conclusion from this study is that more severely hypertensive cats should be started on the 1.25 mg dose. Two limitations are acknowledged: the data are extremely sparse and the analysis was racemic and not enantiomer selective.

In human medicine patients with hypertension secondary to CKD would normally receive a combination of a calcium channel blocker and RAAS inhibitor, the

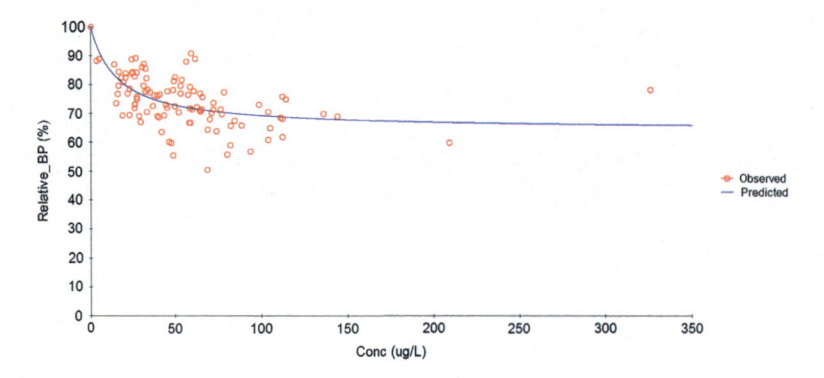

Fig. 11.8 Modelling of the pharmacodynamics of amlodipine in the Bijsmans et al. (2016) study. Blood pressures were normalised to baseline pressure at inclusion. The percentage reduction was modelled with an I_{max} model including the measured plasma concentration of amlodipine at the time of BP control in 61 hypertensive cats. Estimated value of lowest percentage of baseline blood pressure achievable on treatment (E_{max}) was 64.1%, and the concentration that achieved half this reduction (EC$_{50}$) was 15.8 ng/ml

rationale being that the antihypertensive effect of calcium channel blockers is accompanied by activation of RAAS. In cats, amlodipine treatment is accompanied by an increase in plasma renin activity but no change in plasma aldosterone concentration (Jepson et al. 2014). The addition of benazepril to the treatment of cats whose blood pressure has been stabilised on amlodipine has been examined in a randomised controlled crossover study (Elliott et al. 2004). Hypertensive cats were enrolled if they had been stable on amlodipine for at least 12 weeks. They were randomised to receive either placebo or benazepril (0.5–1 mg/kg q24h) for 12 weeks and were then crossed over onto the other arm of the study for a further 12 weeks. Animals were re-examined at 2 to 4, 6 and 12 weeks on each phase when systolic blood pressure (SBP) measurements were made (Doppler method) and a clinical score recorded (0, deteriorated; 1, no change; and 2, improved). Blood and urine samples were taken at 6 and 12 weeks. Urine protein-to-creatinine (UPC) ratios were measured after 12 weeks, and plasma creatinine and potassium concentrations were measured after 6 and 12 weeks and averaged.

Twenty-five cats were enrolled into the study. Five cats died or were euthanised before completing both phases (three during placebo, two during benazepril phase). One further cat was excluded from the analysis due to lack of regular attendance. The average clinical score for each phase tended to be higher when the cats were taking benazepril compared to placebo (1.0 [1.0–1.5] vs. 1.0 [0.5–1.0]; $P = 0.068$, $n = 17$). UPC ratios did not differ significantly between the two phases. Five cats (four receiving placebo and one benazepril) had UPC ratios >0.4, but none had elevated UPCs when on the opposite phase. SBP did not differ significantly between the two phases. However, those cats with an average SBP above 160 mmHg when on placebo ($n = 9$) tended to have lower average blood pressure when taking benazepril (166.4 [162.4–171.5] vs. 158.7 [152.5–165.0]; $P = 0.066$; see Fig. 11.9). Plasma creatinine and potassium concentrations did not differ significantly between the two

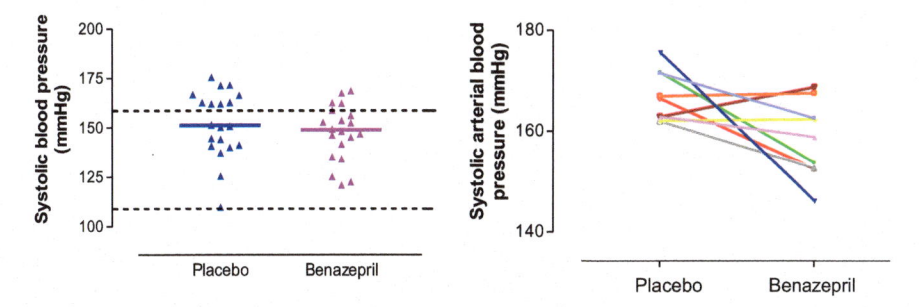

Fig. 11.9 The effect of concomitant administration of amlodipine and benazepril or placebo on the systolic arterial blood pressure of hypertensive cats. Data from Elliott et al. (2004) showing the effect of concomitant administration of amlodipine and benazepril or placebo on the systolic arterial blood pressure of hypertensive cats. The scatter plots show the mean systolic blood pressure from each individual cat (measured at 6 and 12 weeks of each phase) whilst taking placebo and amlodipine (blue triangles) or when taking benazepril and amlodipine (magenta triangles). The solid bar indicates the median value. Note the number of cats with systolic blood pressure above 160 mmHg is nine when taking amlodipine and placebo and only four when taking amlodipine and benazepril. The data presented in the right-hand panel shows the blood pressure readings from these nine cats with a line joining the two mean values recorded for each phase and shows that the blood pressure tended to be lower in seven of the nine cats when receiving benazepril with amlodipine compared to when they received placebo with amlodipine (166.4 [162.4 to 171.5] vs. 158.7 [152.5 to 165.0] mmHg; $P = 0.066$).

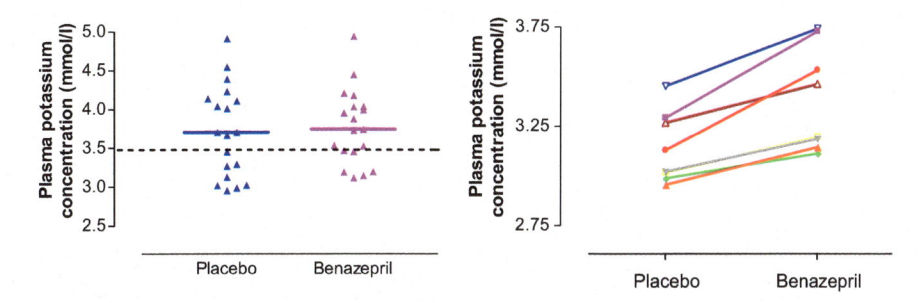

Fig. 11.10 The effect of concomitant administration of amlodipine and benazepril or placebo on the plasma potassium ion concentration of hypertensive cats. Data from Elliott et al. (2004). Plasma potassium ion concentration did not differ significantly between phases; cats that were hypokalaemic (plasma [K] < 3.5 mmol/l) whilst taking placebo had significantly higher plasma [K] when taking benazepril: (3.07 [2.99 to 3.28] vs. 3.32 [3.15 to 3.67] mmol/l; $P = 0.012$)

phases. However, cats which had plasma [K] <3.5 mmol/l ($n = 8$) when taking placebo all had higher plasma [K] when taking benazepril (see Fig. 11.10). These data suggest benazepril can be safely combined with amlodipine in the management of hypertension in cats, and doing this may limit the prevalence of hypokalaemia, improve the control of SBP if amlodipine has not reduced the SBP below 160 mmHg and reduce the prevalence of proteinuria (UPC > 0.4). Larger and longer controlled studies are required to determine the benefit on progressive kidney damage and survival of hypertensive cats.

11.3 Overall Conclusions

The clinical pharmacology of drugs which inhibit RAAS and the calcium channel blocker amlodipine have been well characterised in cats and, to a certain extent, in dogs (RAAS inhibition) to support their therapeutic use in veterinary practice. This chapter has highlighted gaps in our knowledge about these groups of drugs and how they might be combined most effectively to manage hypertension clinically. Furthermore, central inhibitors of the sympathetic nervous system remain to be examined in veterinary patients as a way of managing hypertension associated with CKD in dogs and cats despite the likelihood that overactivity of the sympathetic nervous system play a significant part in the pathophysiology of hypertension in the renal patient (see Chap. 1).

References

Acierno MJ, Brown S, Coleman AE, Jepson RE, Papich M, Stepien RL, Syme HM (2018) ACVIM consensus statement: guidelines for the identification, evaluation, and management of systemic hypertension in dogs and cats. J Vet Intern Med 32(6):1803–1822

Ahmad S, Varagic J, Groban L, Dell'Italia LJ, Nagata S, Kon ND, Ferrario CM (2014) Angiotensin-(1-12): a chymase-mediated cellular angiotensin II substrate. Curr Hypertens Rep 16(5):429

Ames MK, Atkins CE, Lantis AC, zum Brunnen J (2016) Evaluation of subacute change in RAAS activity (as indicated by urinary aldosterone:creatinine, after pharmacologic provocation) and the response to ACE inhibition. J Renin Angiotensin Aldosterone Syst 17(1):1470320316633897

Ames MK, Atkins CE, Eriksson A, Hess AM (2017) Aldosterone breakthrough in dogs with naturally occurring myxomatous mitral valve disease. J Vet Cardiol 19(3):218–227

Batova S, DeWever J, Godfraind T, Balligand JL, Dessy C, Feron O (2006) The calcium channel blocker amlodipine promotes the unclamping of eNOS from caveolin in endothelial cells. Cardiovasc Res 71(3):478–485

Berhane Y, Sent U, Elliott J (2009) Effect of telmisartan on angiotensin II-induced contraction of feline arteries. J Vet Pharmacol Ther 32:232 (abstract)

Bernay F, Bland JM, Häggström J, Baduel L, Combes B, Lopez A, Kaltsatos V (2010) Efficacy of spironolactone on survival in dogs with naturally occurring mitral regurgitation caused by myxomatous mitral valve disease. J Vet Intern Med 24(2):331–341

Bijsmans ES, Jepson RE, Chang YM, Syme HM, Elliott J (2015) Changes in systolic blood pressure over time in healthy cats and cats with chronic kidney disease. J Vet Intern Med 29(3):855–861

Bijsmans ES, Doig M, Jepson RE, Syme HM, Elliott J, Pelligand L (2016) Factors influencing the relationship between the dose of amlodipine required for blood pressure control and change in blood pressure in hypertensive cats. J Vet Intern Med 30(5):1630–1636

Bomback AS, Klemmer PJ (2007) The incidence and implications of aldosterone breakthrough. Nat Clin Pract Nephrol 3(9):486–492

Bomback AS, Rekhtman Y, Klemmer PJ, Canetta PA, Radhakrishnan J, Appel GB (2012) Aldosterone breakthrough during aliskiren, valsartan, and combination (aliskiren + valsartan) therapy. J Am Soc Hypertens 6(5):338–345

Bovee KC, Wong PC, Timmermans PB, Thoolen MJ (1991) Effects of the nonpeptide angiotensin II receptor antagonist DuP 753 on blood pressure and renal functions in spontaneously hypertensive PH dogs. Am J Hypertens 4(4 Pt 2):327S–333S

Brown SA, Langford K, Tarver S (1997) Effects of certain vasoactive agents on the long-term pattern of blood pressure, heart rate, and motor activity in cats. Am J Vet Res 58(6):647–652

Brown SA, Brown CA, Jacobs G, Stiles J, Hendi RS, Wilson S (2001) Effects of the angiotensin converting enzyme inhibitor benazepril in cats with induced renal insufficiency. Am J Vet Res 62(3):375–383

Brown SA, Finco DR, Brown CA, Crowell WA, Alva R, Ericsson GE, Cooper T (2003) Evaluation of the effects of inhibition of angiotensin converting enzyme with enalapril in dogs with induced chronic renal insufficiency. Am J Vet Res 64(3):321–327

Buchwalder-Csajka C, Buclin T, Brunner HR et al (1999) Evaluation of the angiotensin challenge methodology for assessing the pharmacodynamics profile of antihypertensive drugs acting on the renin-angiotensin system. Br J Clin Pharmacol 48:594–604

Chaugai S, Sherpa LY, Sepehry AA, Kerman SRJ, Arima H (2018) Effects of long- and intermediate-acting dihydropyridine calcium channel blockers in hypertension: a systematic review and meta-analysis of 18 prospective, randomized, actively controlled trials. J Cardiovasc Pharmacol Ther 23(5):433–445

Christ DD, Wong PC, Wong YN, Hart SD, Quon CY, Lam GN (1994) The pharmacokinetics and pharmacodynamics of the angiotensin II receptor antagonist losartan potassium (DuP 753/MK 954) in the dog. J Pharmacol Exp Ther 268(3):1199–1205

Coleman AE, Brown SA, Stark M, Bryson L, Zimmerman A, Zimmering T, Traas AM (2018) Evaluation of orally administered telmisartan for the reduction of indirect systolic arterial blood pressure in awake, clinically normal cats. J Feline Med Surg

Coleman AE, Brown SA, Traas AM, Bryson L, Zimmering T, Zimmerman A (2019) Safety and efficacy of orally administered telmisartan for the treatment of systemic hypertension in cats: results of a double-blind, placebo-controlled, randomized clinical trial. J Vet Intern Med 33(2):478–488

Conroy M, Chang YM, Brodbelt D, Elliott J (2018) Survival after diagnosis of hypertension in cats attending primary care practice in the United Kingdom. J Vet Intern Med 32(6):1846–1855

Ebner T, Schänzle G, Weber W, Sent U, Elliott J (2013) In vitro glucuronidation of the angiotensin II receptor antagonist telmisartan in the cat: a comparison with other species. J Vet Pharmacol Ther 36(2):154–160

Edwards LP, Brown-Bryan TA, McLean L, Ernsberger P (2012) Pharmacological properties of the central antihypertensive agent, moxonidine. Cardiovasc Ther 30(4):199–208

Elliott J, Barber PJ, Syme HM, Rawlings JM, Markwell PJ (2001) Feline hypertension: clinical findings and response to antihypertensive treatment in 30 cases. J Small Anim Pract 42(3):122–129

Elliott J, Fletcher MG, Souttar K, Cariese S, Syme HM (2004) Effect of concomitant amlodipine and benazepril therapy in the management of feline hypertension. J Vet Intern Med 18(5):788 (abstract)

EMEA – European Medicines Agency – Veterinary Medicines (2007). https://www.ema.europa.eu/en/documents/scientific-discussion/prilactone-epar-scientific-discussion_en.pdf

Farquharson CA, Struthers AD (2000) Spironolactone increases nitric oxide bioactivity, improves endothelial vasodilator dysfunction, and suppresses vascular angiotensin I/angiotensin II conversion in patients with chronic heart failure. Circulation 101(6):594–597

Ferrario CM, Mullick AE (2017 Nov) Renin angiotensin aldosterone inhibition in the treatment of cardiovascular disease. Pharmacol Res 125(Pt A):57–71

Geigy CA, Schweighauser A, Doherr M, Francey T (2011 Jul) Occurrence of systemic hypertension in dogs with acute kidney injury and treatment with amlodipine besylate. J Small Anim Pract 52(7):340–346

Glaus TM, Elliott J, Herberich E, Zimmering T, Albrecht B (2019) Efficacy of long-term oral telmisartan treatment in cats with hypertension: results of a prospective European clinical trial. J Vet Intern Med 33(2):413–422

Grassi G (2016) Sympathomodulatory effects of antihypertensive drug treatment. Am J Hypertens 29(6):665–675

Grassi G, Mark A, Esler M (2015) The sympathetic nervous system alterations in human hypertension. Circ Res 116(6):976–990

Grassi G, Robles NR, Seravalle G, Fici F (2017) Lercanidipine in the Management of Hypertension: an update. J Pharmacol Pharmacother 8(4):155–165

Grauer GF, Greco DS, Getzy DM, Cowgill LD, Vaden SL, Chew DJ, Polzin DJ, Barsanti JA (2000) Effects of enalapril versus placebo as a treatment for canine idiopathic glomerulonephritis. J Vet Intern Med 14(5):526–533

Guyonnet J, Elliott J, Kaltsatos V (2010) A preclinical pharmacokinetic and pharmacodynamic approach to determine a dose of spironolactone for treatment of congestive heart failure in dog. J Vet Pharmacol Ther 33(3):260–267

Häggström J, Hansson K, Karlberg BE, Kvart C, Madej A, Olsson K (1996) Effects of long-term treatment with enalapril or hydralazine on the renin-angiotensin-aldosterone system and fluid balance in dogs with naturally acquired mitral valve regurgitation. Am J Vet Res 57(11):1645–1652

Henik RA, Snyder PS, Volk LM (1997) Treatment of systemic hypertension in cats with amlodipine besylate. J Am Anim Hosp Assoc 33(3):226–234

Horiuchi M, Mogi M (2011) Role of angiotensin II receptor subtype activation in cognitive function and ischaemic brain damage. Br J Pharmacol 163(6):1122–1130

Huhtinen M, Derré G, Renoldi HJ, Rinkinen M, Adler K, Aspegrén J, Zemirline C, Elliott J (2015) Randomized placebo-controlled clinical trial of a chewable formulation of amlodipine for the treatment of hypertension in client-owned cats. J Vet Intern Med 29(3):786–793

Jaisser F, Farman N (2016) Emerging roles of the mineralocorticoid receptor in pathology: toward new paradigms in clinical pharmacology. Pharmacol Rev 68(1):49–75

James R, Guillot E, Garelli-Paar C, Huxley J, Grassi V, Cobb M (2018) The SEISICAT study: a pilot study assessing efficacy and safety of spironolactone in cats with congestive heart failure secondary to cardiomyopathy. J Vet Cardiol 20(1):1–12

Jenkins TL, Coleman AE, Schmiedt CW, Brown SA (2015) Attenuation of the pressor response to exogenous angiotensin by angiotensin receptor blockers and benazepril hydrochloride in clinically normal cats. Am J Vet Res 76(9):807–813

Jepson RE, Elliott J, Brodbelt D, Syme HM (2007) Effect of control of systolic blood pressure on survival in cats with systemic hypertension. J Vet Intern Med 21(3):402–409

Jepson RE, Syme HM, Elliott J (2014) Plasma renin activity and aldosterone concentrations in hypertensive cats with and without azotemia and in response to treatment with amlodipine besylate. J Vet Intern Med 28(1):144–153

Jeunesse E, Woehrle F, Schneider M, Lefebvre HP (2007) Effect of spironolactone on diuresis and urine sodium and potassium excretion in healthy dogs. J Vet Cardiol 9(2):63–68

King JN, Gunn-Moore DA, Tasker S, Gleadhill A, Strehlau G (2006) Benazepril in renal insufficiency in cats study group. Tolerability and efficacy of benazepril in cats with chronic kidney disease. J Vet Intern Med 20(5):1054–1064

Kittleson MD, Bonagura JD (2010) Re: Efficacy of spironolactone on survival in dogs with naturally occurring mitral regurgitation caused by myxomatous mitral valve disease. J Vet Intern Med 24(6):1245–1246

Lefebvre HP, Toutain PL (2004) Angiotensin-converting enzyme inhibitors in the therapy of renal diseases. J Vet Pharmacol Ther 27(5):265–281

Lefebvre HP, Laroute V, Concordet D, Toutain PL (1999) Effects of renal impairment on the disposition of orally administered enalapril, benazepril, and their active metabolites. J Vet Intern Med 13(1):21–27

Lefebvre HP, Jeunesse E, Laroute V, Toutain PL (2006) Pharmacokinetic and pharmacodynamic parameters of ramipril and ramiprilat in healthy dogs and dogs with reduced glomerular filtration rate. J Vet Intern Med 20(3):499–507

Lefebvre HP, Ollivier E, Atkins CE, Combes B, Concordet D, Kaltsatos V, Baduel L (2013) Safety of spironolactone in dogs with chronic heart failure because of degenerative valvular disease: a population-based, longitudinal study. J Vet Intern Med 27(5):1083–1091

Lenasi H, Kohlstedt K, Fichtlscherer B, Mülsch A, Busse R, Fleming I (2003) Amlodipine activates the endothelial nitric oxide synthase by altering phosphorylation on Ser1177 and Thr495. Cardiovasc Res 59(4):844–853

Lucchesi BR, Tamura Y (1989) Cardioprotective effects of amlodipine in the ischemic-reperfused heart. Am Heart J 118(5 Pt 2):1121–1122

MacDonald KA, Kittleson MD, Kass PH, White SD (2008) Effect of spironolactone on diastolic function and left ventricular mass in Maine Coon cats with familial hypertrophic cardiomyopathy. J Vet Intern Med 22(2):335–341

Mathur S, Syme H, Brown CA, Elliot J, Moore PA, Newell MA, Munday JS, Cartier LM, Sheldon SE, Brown SA (2002) Effects of the calcium channel antagonist amlodipine in cats with surgically induced hypertensive renal insufficiency. Am J Vet Res 63(6):833–839

Mathur S, Brown CA, Dietrich UM, Munday JS, Newell MA, Sheldon SE, Cartier LM, Brown SA (2004) Evaluation of a technique of inducing hypertensive renal insufficiency in cats. Am J Vet Res 65(7):1006–1013

Mishina M, Watanabe T (2008) Development of hypertension and effects of benazepril hydrochloride in a canine remnant kidney model of chronic renal failure. J Vet Med Sci 70(5):455–460

Miura S, Fujino M, Hanzawa H, Kiya Y, Imaizumi S, Matsuo Y, Tomita S, Uehara Y, Karnik SS, Yanagisawa H, Koike H, Komuro I, Saku K (2006) Molecular mechanism underlying inverse agonist of angiotensin II type 1 receptor. J Biol Chem 281(28):19288–19295

Miura S, Karnik SS, Saku K (2011) Review: angiotensin II type 1 receptor blockers: class effects versus molecular effects. J Renin Angiotensin Aldosterone Syst 12(1):1–7

Mochel JP, Danhof M (2015) Chronobiology and pharmacologic modulation of the renin-angiotensin-aldosterone system in dogs: what have we learned? Rev Physiol Biochem Pharmacol 169:43–69

Mochel JP, Fink M, Peyrou M, Desevaux C, Deurinck M, Giraudel JM, Danhof M (2013) Chronobiology of the renin-angiotensin-aldosterone system in dogs: relation to blood pressure and renal physiology. Chronobiol Int 30(9):1144–1159

Mochel JP, Fink M, Bon C, Peyrou M, Bieth B, Desevaux C, Deurinck M, Giraudel JM, Danhof M (2014) Influence of feeding schedules on the chronobiology of renin activity, urinary electrolytes and blood pressure in dogs. Chronobiol Int 31(5):715–730

Mochel JP, Fink M, Peyrou M, Soubret A, Giraudel JM, Danhof M (2015) Pharmacokinetic/Pharmacodynamic modeling of renin-angiotensin aldosterone biomarkers following angiotensin-converting enzyme (ACE) inhibition therapy with benazepril in dogs. Pharm Res 32(6):1931–1946

Narayan H, Webb DJ (2016) New evidence supporting the use of mineralocorticoid receptor blockers in drug-resistant hypertension. Curr Hypertens Rep 18(5):34

Ojima M, Igata H, Tanaka M, Sakamoto H, Kuroita T, Kohara Y, Kubo K, Fuse H, Imura Y, Kusumoto K, Nagaya H (2011) In vitro antagonistic properties of a new angiotensin type 1 receptor blocker, azilsartan, in receptor binding and function studies. J Pharmacol Exp Ther 336(3):801–808

Peterson JC, Alder S, Burkart JM et al (1995) Blood pressure control, proteinuria, and the progression of renal disease. The modification of diet in renal disease study. Ann Intern Med 123:754–762

Pitt B, Zannad F, Remme WJ, Cody R, Castaigne A, Perez A, Palensky J, Wittes J (1999) The effect of spironolactone on morbidity and mortality in patients with severe heart failure. Randomized Aldactone Evaluation Study Investigators. N Engl J Med 341(10):709–717

Putnam K, Shoemaker R, Yiannikouris F, Cassis LA (2012) The renin-angiotensin system: a target of and contributor to dyslipidemias, altered glucose homeostasis, and hypertension of the metabolic syndrome. Am J Physiol Heart Circ Physiol 302(6):H1219–H1230

Reyes S, Varagic J, Ahmad S, VonCannon J, Kon ND, Wang H, Groban L, Cheng CP, Dell'Italia LJ, Ferrario CM (2017) Novel cardiac Intracrine mechanisms based on Ang-(1-12)/Chymase Axis require a revision of therapeutic approaches in human heart disease. Curr Hypertens Rep 19(2):16

Rosen SG, Supiano MA, Perry TJ, Linares OA, Hogikyan RV, Smith MJ, Halter JB (1990) Beta-adrenergic blockade decreases norepinephrine release in humans. Am J Phys 258(6 Pt 1):E999–E1005

Santos RAS, Sampaio WO, Alzamora AC, Motta-Santos D, Alenina N, Bader M, Campagnole-Santos MJ (2018) The ACE2/angiotensin-(1-7)/MAS Axis of the renin-angiotensin system: focus on angiotensin-(1-7). Physiol Rev 98(1):505–553

Sent U, Gössl R, Elliott J, Syme HM, Zimmering T (2015) Comparison of efficacy of long-term oral treatment with telmisartan and benazepril in cats with chronic kidney disease. J Vet Intern Med 29(6):1479–1487

Sharma A, Trane A, Yu C, Jasmin JF, Bernatchez P (2011) Amlodipine increases endothelial nitric oxide release by modulating binding of native eNOS protein complex to caveolin-1. Eur J Pharmacol 659(2–3):206–212

Snyder PS (1998) Amlodipine: a randomized, blinded clinical trial in 9 cats with systemic hypertension. J Vet Intern Med 12(3):157–162

Syme HM, Barber PJ, Markwell PJ, Elliott J (2002) Prevalence of systolic hypertension in cats with chronic renal failure at initial evaluation. J Am Vet Med Assoc 220(12):1799–1804

Takezako T, Unal H, Karnik SS, Node K (2017) Current topics in angiotensin II type 1 receptor research: focus on inverse agonism, receptor dimerization and biased agonism. Pharmacol Res 123:40–50

Toma L, Stancu CS, Sanda GM, Sima AV (2011) Anti-oxidant and anti-inflammatory mechanisms of amlodipine action to improve endothelial cell dysfunction induced by irreversibly glycated LDL. Biochem Biophys Res Commun 411(1):202–207

Toutain PL, Lefebvre HP (2004) Pharmacokinetics and pharmacokinetic/pharmacodynamic relationships for angiotensin-converting enzyme inhibitors. J Vet Pharmacol Ther 27(6):515–525

Toutain PL, Lefebvre HP, King JN (2000) Benazeprilat disposition and effect in dogs revisited with a pharmacokinetic/pharmacodynamic modeling approach. J Pharmacol Exp Ther 292(3):1087–1093

Towfighi A, Ovbiagele B (2008) Partial peroxisome proliferator-activated receptor agonist angiotensin receptor blockers. Potential multipronged strategy in stroke prevention. Cerebrovasc Dis 26(2):106–112

Vongpatanasin W, Kario K, Atlas SA, Victor RG (2011) Central sympatholytic drugs. J Clin Hypertens (Greenwich) 13(9):658–661

Wang W (1994) Chronic administration of aldosterone depresses baroreceptor reflex function in the dog. Hypertension 24(5):571–575

Yao K, Nagashima K, Miki H (2006) Pharmacological, pharmacokinetic, and clinical properties of benidipine hydrochloride, a novel, long-acting calcium channel blocker. J Pharmacol Sci 100(4):243–261

Yasuda N, Miura S, Akazawa H, Tanaka T, Qin Y, Kiya Y, Imaizumi S, Fujino M, Ito K, Zou Y, Fukuhara S, Kunimoto S, Fukuzaki K, Sato T, Ge J, Mochizuki N, Nakaya H, Saku K, Komuro I (2008) Conformational switch of angiotensin II type 1 receptor underlying mechanical stress-induced activation. EMBO Rep 9(2):179–186

Zannad F, Alla F, Dousset B, Perez A, Pitt B (2000) Limitation of excessive extracellular matrix turnover may contribute to survival benefit of spironolactone therapy in patients with congestive heart failure: insights from the randomized aldactone evaluation study (RALES). Rales Investigators. Circulation 102(22):2700–2706

Zannad F, McMurray JJ, Krum H, van Veldhuisen DJ, Swedberg K, Shi H, Vincent J, Pocock SJ, Pitt B, EMPHASIS-HF Study Group (2011) Eplerenone in patients with systolic heart failure and mild symptoms. N Engl J Med 364(1):11–21

Zhang X, Hintze TH (1998) Amlodipine releases nitric oxide from canine coronary microvessels: an unexpected mechanism of action of a calcium channel-blocking agent. Circulation 97 (6):576–580

Zou Y, Akazawa H, Qin Y, Sano M, Takano H, Minamino T, Makita N, Iwanaga K, Zhu W, Kudoh S, Toko H, Tamura K, Kihara M, Nagai T, Fukamizu A, Umemura S, Iiri T, Fujita T, Komuro I (2004) Mechanical stress activates angiotensin II type 1 receptor without the involvement of angiotensin II. Nat Cell Biol 6(6):499–506

Management of Hypertension in Cats

12

Sarah M. A. Caney

12.1 Introduction

Systemic hypertension—a persistent increase in the systemic blood pressure—is now commonly recognised in feline practice. There are several reasons for this including an increased awareness of hypertension as a feline problem, increased access to diagnostic facilities and possibly an increased prevalence of this condition related to the increasing age of the cat population.

An age-related increase in systolic blood pressure has been reported (Bijsmans et al. 2015; Payne et al. 2017) meaning that older cats are at an increased risk of developing systemic hypertension. Systemic hypertension is often associated with other conditions, many of which increase in prevalence with age, which also increases the risk of systemic hypertension in older cats. For example, one study including 58 cats with systemic hypertension reported their mean age to be 13 years (Chetboul et al. 2003). The most commonly diagnosed associated illnesses are chronic kidney disease (CKD) and hyperthyroidism. Prevalence rates quoted have been highly variable—for example, 20–65% of cats with CKD are diagnosed with concurrent systemic hypertension, compared to 9–23% of newly diagnosed hyperthyroid cats (Kobayashi et al. 1990; Littman 1994; Stiles et al. 1994; Syme et al. 2002; Syme and Elliott 2003; Morrow et al. 2009; Williams et al. 2010). Other diseases that have been associated with hypertension in cats include primary hyperaldosteronism (Conn's syndrome) (Ash et al. 2005; Javadi et al. 2005), phaeochromocytoma (Patnaik et al. 1990; Henry et al. 1993; Chun et al. 1997; Calsyn et al. 2010; Wimpole et al. 2010), therapy with erythrocyte-stimulating agents (Cowgill et al. 1998; Chalhoub et al. 2012) and occasionally hyperadrenocorticism (Brown et al. 2012).

S. M. A. Caney (✉)
Vet Professionals Ltd, Midlothian Innovation Centre, Edinburgh, UK
e-mail: sarah@vetprofessionals.com

A link between systemic hypertension and other feline endocrinopathies such as diabetes mellitus and acromegaly has not yet been confirmed.

Idiopathic hypertension (also referred to as primary or essential hypertension) accounts for less than 20% of cases, and most reported cases occur secondary to other medical problems (Maggio et al. 2000; Elliott et al. 2001; Jepson et al. 2007).

12.2 Confirming a Diagnosis of Systemic Hypertension

Before considering management of hypertension, it is vital to ensure that an accurate and reliable diagnosis has been made by ruling out 'false-positive' and 'false-negative' diagnoses of hypertension. The most common cause of 'false-positive' high blood pressure readings is situational (also called 'white-coat' or 'stress-associated') hypertension. Blood pressure readings are extremely labile. On average, the 'white-coat' effect increases SBP by 15–20 mmHg (Belew et al. 1999). However, the effect is highly variable between cats, with SBP decreasing in some and increasing by as much as 75 mmHg in others (Belew et al. 1999, Fig. 12.1). 'False-negative' results may be obtained when assessing blood pressure in a clinically dehydrated patient. The patient should be rehydrated and blood pressure readings repeated to minimise the risk of this possibility.

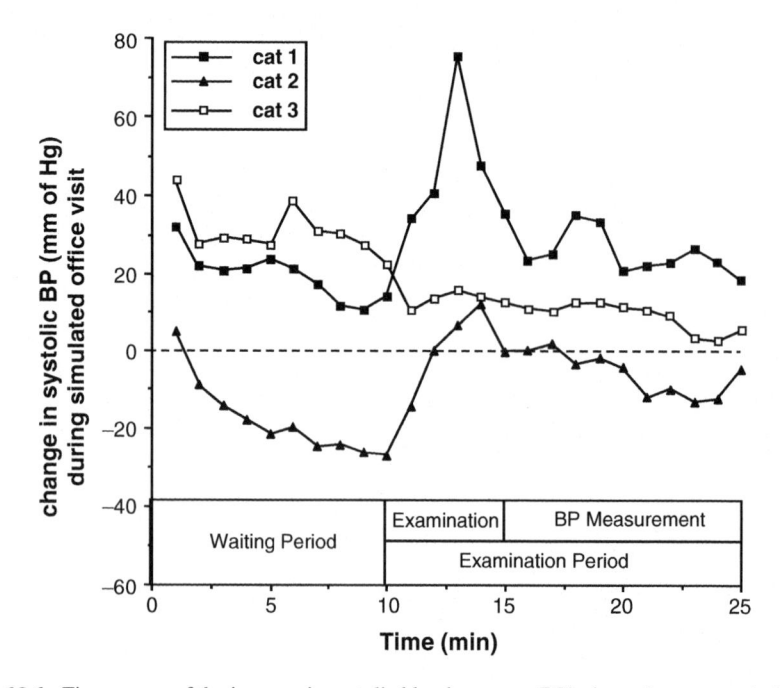

Fig. 12.1 Time course of the increase in systolic blood pressure (BP) above the mean ambulatory systolic BP during the simulated office visit for three cats. Reproduced with permission from Belew et al. (1999)

It is vital to perform blood pressure measurement using appropriate methodology to minimise the possibility of technical errors contributing to 'false' results. Detailed protocols for diagnosis of systemic hypertension are reviewed in greater detail elsewhere (Brown et al. 2007; Stepien 2011; Taylor et al. 2017; Acierno et al. 2018).

Blood pressure assessment should be performed in a quiet room, away from barking dogs and telephones, ideally allowing the cat 10 min to acclimatise to these surroundings before the measurements are taken. This 'acclimatisation' period helps to reduce the incidence of situational hypertension where stress and anxiety stimulate the sympathetic nervous system leading to falsely high blood pressure readings. For some cats, having the owner present limits the effect of stress on blood pressure readings. After the acclimatisation period, the cat is restrained as gently as possible for the procedure to be done—usually all that is required is gentle steadying of the cat whilst the cuff is placed and readings are taken. The forelimb is used most commonly, but care should be taken not to overextend the elbow as this is a common site for osteoarthritis in older cats. Some cats do not like their paws to be held; when this is the case, it can be simplest to use the base of the tail (coccygeal artery) instead. Further information on the measurement of blood pressure can be found in Chap. 2. Other tips for minimising the possibility of situational hypertension and other technical causes of misleading results are presented in Table 12.1.

The presence of target organ damage (TOD) in addition to elevated systolic blood pressure readings assists in confirming the diagnosis. The target organs most vulnerable to hypertensive damage are the brain, heart, kidneys and eyes (Table 12.2). Cats with systemic hypertension may also present with clinical signs referable to their associated systemic disease such as inappetence and weight loss in a patient with CKD.

A detailed ophthalmic examination is extremely helpful in supporting a diagnosis of systemic hypertension. Not all patients with confirmed systemic hypertension will have visible evidence of ocular TOD, but it is currently estimated that up to 80% of patients have ocular manifestations of their high blood pressure (Sansom et al. 2004; Syme et al. 2002; Elliott et al. 2001; Stiles et al. 1994; Carter et al. 2014). Following cats with CKD and measuring their systolic blood pressure every 8 to 16 weeks meant that the prevalence of retinal lesions at diagnosis of hypertension was lower than in a group of cats seen through the same clinics when hypertension was diagnosed at initial presentation (50% vs. 73%; Bijsmans et al. 2015). Typically hypertensive TOD lesions are most often seen in patients with severe hypertension (SBP > 180 mmHg), occur in both eyes and are often pathognomonic when seen (Sansom et al. 1994; Stiles et al. 1994). Unfortunately, some other manifestations of TOD (Table 12.2) such as proteinuria can be seen for reasons other than systemic hypertension so their presence should not be taken as definitive evidence of confirmed hypertension.

The author considers a quick and practical way of performing a fundus examination to be most easily done by using distant indirect ophthalmoscopy. Equipment needed for this includes a handheld lens and a light source. The lens should be around 20–30 dioptre in magnification—inexpensive acrylic options are now

Table 12.1 Tips for minimising situational hypertension and for obtaining reliable readings (applicable to all blood pressure measurement methodologies)

Tip
Use a quiet room for blood pressure measurement: away from barking dogs, telephones and human traffic
Allow the cat 5–10 min to acclimatise to you and the surroundings before starting the procedure
Always measure blood pressure before performing any other assessments in the cat—otherwise add 30 min rest period after procedures before collecting blood pressure readings
It can be helpful having the owner present to gently restrain their cat
Use minimal restraint
Don't rush!
Wear headphones so that the cat is not aware of any noise associated with the procedure
Doppler BP measurements: No need to clip the fur—even in long-haired cats—using clippers may be stressful to the cat. Prepare the area by wiping with surgical alcohol and then applying plenty of ultrasound gel
Doppler BP measurements: Slowly inflate the cuff in a series of gentle puffs—sharp inflations can surprise or stress your cat. Sometimes doing a series of 'practice' cuff inflations and deflations can be helpful in acclimatising your patient to the procedure
Doppler BP measurements: Completely deflate the cuff between readings—otherwise the procedure may start to become uncomfortable and therefore stressful to the cat
Doppler BP measurements: Use a sphygmomanometer that allows slow and gradual cuff deflation
Collect at least five readings: Discard the first reading if very different to the others and average the remainder. If the readings are very variable, continue to collect readings until they 'plateau', averaging the 'plateau' readings. If no plateau is reached, the lowest reading is likely to be most representative of the cat's true systolic BP
Always record the cuff used and site of BP measurement as size of cuff and location affects the readings obtained. The cuff should be at the level of the right atrium: If lower than this, the reading will be higher than the 'true' BP. The ideal cuff width is 30–40% of the limb circumference: If wider than this, the readings obtained will be lower than the 'true' BP
Discard cuffs that inflate unevenly or need securing with tape to stop them from 'popping off'

available in many countries. The ideal light source is a focussed light source of adjustable intensity such as a Finoff transilluminator or the direct ophthalmoscope set to a small circle. Pen torches are often very bright with no ability to reduce the light intensity so can be uncomfortable for patients and therefore disliked.

If possible, the cat should be examined in a dark room. A room with no windows or with blackout blinds covering the windows is most suitable. Light sources including computer screens should be switched off during the eye examination. If a dark room is not available, it may be necessary to dilate the pupils using tropicamide drops. These take up to 15 min to exert their effect which will typically last for several hours.

The cat is restrained on an examination table with the handler gently lifting the cat's chin so that they are not looking down at the floor (Fig. 12.2). The examiner stands at arm's length from the cat and starts by directing the light source towards the cat's eye, adjusting the angle of the light beam until a bright tapetal reflection is seen.

12 Management of Hypertension in Cats

Table 12.2 Potential target organ damage associated with systemic hypertension

Organ affected	Pathology	Clinical findings	Reported approximate prevalence
Brain and spinal cord	Hyperplastic arteriosclerosis of cerebral vessels, oedema of the white matter and microhaemorrhage development resulting in hypertensive encephalopathy +/− stroke	Many changes possible including behavioural changes (e.g. night time vocalisation, signs of confusion or dementia), altered mentation, ataxia, seizures, focal neurological deficits including central blindness, 'headaches', coma	15%
	Ischaemic myelopathy (Theobald et al. 2013; Simpson et al. 2014)	Ambulatory paresis, plegia, +/− spinal pain	
Heart	Left ventricular hypertrophy and thickening of the interventricular septum	New murmur, arrhythmia, and/or gallop rhythm.	50–80%
Kidneys	Glomerular hypertrophy and sclerosis, nephrosclerosis, tubular atrophy and interstitial nephritis resulting in progression of CKD	Reduced urine specific gravity, proteinuria, increasing creatinine and IDEXX SDMA® levels, decreasing GFR	
Eyes	Hypertensive retinopathy/ choroidopathy resulting in many changes including intra-ocular haemorrhage, retinal oedema, retinal detachment, arterial tortuosity, variable diameter of retinal arterioles, papilloedema, glaucoma. Foci of retinal degeneration (hyperreflectivity) may develop where damage has previously occurred	Visual deficits, blindness, hyphaema, mydriasis	60–80%

Once seen, the hand lens is inserted just in front of the eye (Fig. 12.3). An inverted image of the fundus should be visible, and the lens magnification typically ensures that much of the fundus is visible in one view. Thus, this is an ideal technique for visualisation of large portions of the fundus quickly in the clinic. If lesions are evident, then direct ophthalmoscopy can be helpful in examining these at a higher magnification. A number of ocular pathological abnormalities are possible with systemic hypertension (Table 12.2). Annotated images of a variety of abnormalities are available as a Free Download on the author's website (Caney 2018) and elsewhere.

There has been some interest in use of biomarkers of cardiac stress in helping to confirm hypertension in cats (see also Chap. 1). Levels of NT-proBNP have been reported to be increased in cats with systemic hypertension and reduce following

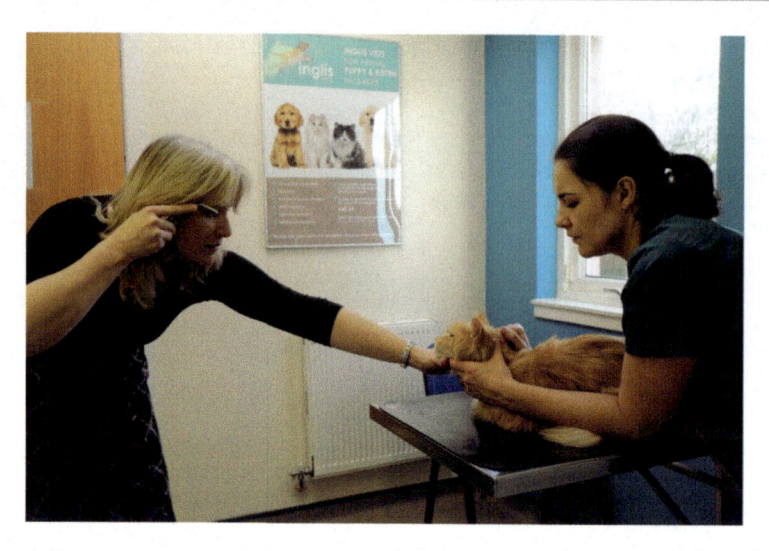

Fig. 12.2 Indirect ophthalmoscopy is a helpful technique for evaluation of the feline fundus. The patient should be examined in a dark room. Standing at arm's length from the patient, direct the light source towards the eye, adjusting the angle of direction until the tapetal reflection is seen

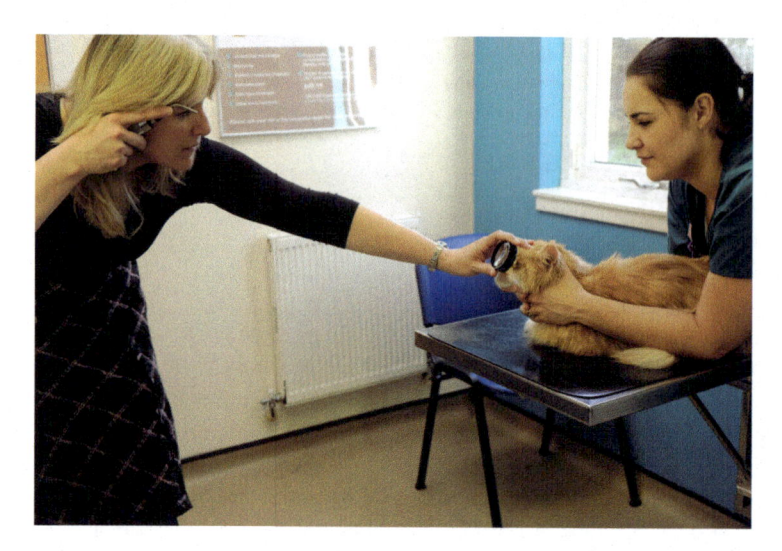

Fig. 12.3 Once a tapetal reflection has been obtained, insert the hand lens just in front of the eye. An inverted view of the fundus should be clearly visible

successful management of hypertension (Lalor et al. 2009). However, a recent paper suggested that none of the biomarkers tested, including NT-proBNP, were suitably reliable to be used in diagnosis and monitoring of patients with systemic hypertension (Bijsmans et al. 2017).

12.3 Interpreting SBP Results

A number of different 'reference ranges' have been published for normal cats citing normal SBP readings from 107 to 181 mmHg in healthy cats (Kobayashi et al. 1990; Sparkes et al. 1999; see Chap. 2). When it is possible to measure it, the DBP of normal cats should be <95 mmHg. As discussed previously, situational hypertension is a significant issue when interpreting blood pressure results in cats. Clinicians should use their judgment when assessing patients taking into consideration the guidelines below and the individual cat perhaps also considering how busy/noisy their clinic is and what impact this might have on blood pressure readings.

The American College of Veterinary Internal Medicine has published a classification system according to risk of TOD (Table 12.3).

12.3.1 SBP > 180 mmHg: Severe Hypertension with Severe Risk of TOD

In general, cats with SBP in excess of 180 mmHg are genuinely hypertensive, and therapy is justified. However, some healthy cats may transiently have SBP above 180 mmHg. Hypertension should therefore *never* be treated solely on the basis of a single abnormal blood pressure reading. If evidence of TOD is present, the diagnosis of hypertension is confirmed, and treatment can be instituted. In the absence of TOD, it is prudent to recheck the SBP on another occasion before pursuing treatment. The author recommends the following steps are taken in cats with SBP readings >180 mmHg. First, ensure that measurements are taken correctly allowing at least 5 (preferably 10) min for acclimatisation before readings are taken. Secondly, assess the patient for evidence of TOD. Ocular TOD is typically pathognomonic; clinical signs consistent with neurological TOD should also be considered. If clear evidence of TOD is present, a diagnosis of systemic hypertension is confirmed. If there is no evidence of TOD or the clinician is uncertain, then repeat assessment of SBP and TOD is indicated on one or two separate occasions within the subsequent 14 days. If, during these repeat assessments, SBP readings remain high, antihypertensive treatment is justified. Further investigations aimed at finding illnesses associated with hypertension should be pursued.

Table 12.3 ACVIM classification of blood pressure in cats (in mmHg) based on the risk of future target organ damage (TOD) (Acierno et al. 2018)

Category	Systolic BP	Risk of future TOD
Normotensive	< 140	Minimal
Prehypertensive	140–159	Mild
Hypertension	160–179	Moderate
Severe hypertension	≥ 180	Severe

12.3.2 SBP 160–179 mmHg: Hypertension with Moderate Risk of TOD

SBP readings that are persistently between 160 and 180 mmHg are believed to pose a moderate risk of TOD. Persistence is defined as being present on several occasions over a 2-month period. If there is evidence of TOD (e.g. hypertensive retinopathy) or if the cat is known to have CKD or any other condition known to be associated with hypertension, then antihypertensive therapy is justified. In the absence of either of these, it might not be possible to rule out situational hypertension, and further monitoring/investigations for potential underlying diseases associated with hypertension might be more appropriate before starting therapy.

12.3.3 SBP 140–159 mmHg: Prehypertension with Mild Risk of TOD

Cats in this group may have mild hypertension, but many normal cats will also give blood pressure readings in this range due to situational hypertension. Treatment is not normally recommended unless there is evidence of TOD or clear evidence supporting an upward trend in SBP readings over time. For those cats with conditions known to predispose to hypertension, 1–3 monthly monitoring of blood pressure and evaluation for evidence of TOD is recommended once readings >140 mmHg are obtained.

12.3.4 SBP Less than 140 mmHg: Normotension

Most normal cats have SBP readings of <140 mmHg. This should be viewed as the ideal 'target range' following treatment for hypertension. Clinical hypotension is uncommon in the author's experience and is typically associated with SBP readings less than 110 mmHg.

12.4 Management of Hypertensive Cats

The main goals of management are to reduce systolic blood pressure readings to an 'ideal' reference range, less than 140 mmHg, and to identify and treat potential underlying/associated medical conditions such as CKD. Treatment of systemic hypertension is given with the aim of preventing or slowing the progression of TOD. Ocular TOD is often slow to resolve, taking several months in some cases. A recent publication reported that 58% of eyes that were blind due to hypertensive chorioretinopathy regained some vision following treatment (Young et al. 2019).

Irrespective of the nature of the underlying or associated disease, specific management with antihypertensive agents is required in cats diagnosed with systemic hypertension (Table 12.4).

Table 12.4 Antihypertensive agents commonly used in cats

Class of agent	Agent/s and oral dosage regime	Effective?	Comments
Ca channel blocker	Amlodipine 0.625–1.25 mg/cat q24h (maximum suggested dose 0.5 mg/kg/day)	Very—typically reduces SBP by 30–70 mmHg	Often effective as sole therapy Consider starting dose of 1.25 mg per cat if SBP $\geq$ 200 mmHg at initial diagnosis
Angiotensin receptor blocker	Telmisartan 1–3 mg/kg/day	Mild to moderate according to dose	When used as a sole agent, a starting dose of 2 mg/kg/day is suggested. If used in combination with amlodipine, start at a dose of 1 mg/kg/day
ACE inhibitor	Benazepril 0.5–1.0 mg/kg q24h; enalapril 0.25–0.5 mg/kg q24h; ramipril 0.125–0.25 mg/kg q24	Mild to moderate: Typically reduce SBP by 10–20 mmHg	Often ineffective as a sole therapy, especially in cases of marked hypertension (SBP > 180 mmHg) Benazepril can be used in combination with amlodipine

Some medications can increase blood pressure. If a patient is receiving any of these medications, then consideration should be given to reducing or stopping therapy, where possible, particularly if it is proving difficult to control blood pressure with standard treatment. Examples include erythrocyte-stimulating agents, mineralocorticoids and glucocorticoids (Cowgill et al. 1998; Brown et al. 2007; Chalhoub et al. 2012; Acierno et al. 2018).

Amlodipine (Amodip®, Ceva Animal Health) is a veterinary authorised treatment for feline hypertension and is the current first-choice antihypertensive medication. Amlodipine is generally extremely effective as a sole treatment. Amlodipine acts on vascular smooth muscle and is a peripheral arterial dilator resulting in a reduction in systemic vascular resistance and blood pressure (see Chap. 11). Vasodilation is induced slowly with a dose-dependent impact on the magnitude and duration of the antihypertensive effect. In general, the SBP falls by 30–70 mmHg following administration of amlodipine, and the majority of patients respond to amlodipine as a once daily oral monotherapy (Henik et al. 1997; Snyder 1998; Elliott et al. 2001; Mathur et al. 2002; Jepson et al. 2007; Huhtinen et al. 2015; Bijsmans et al. 2016).

A typical recommended starting dose is 0.125–0.25 mg/kg (typically 0.625 or 1.25 mg per cat) per day. If SBP remains above 160 mmHg when rechecked 7–14 days later, the dose can be doubled, up to a maximum of 0.5 mg/kg per day (Table 12.4). Cats with severe hypertension tend to require a higher dose of amlodipine than those with mild–moderate hypertension (Bijsmans et al. 2016); therefore, a higher starting dose is justified in patients with higher SBP at initial diagnosis. If the maximum dose of amlodipine is reached, compliance is good and yet the SBP is still above 160 mmHg, then consideration should be given to adding a second medication such as those discussed below.

Amlodipine appears to be well tolerated with very few adverse results reported in a large field study (Huhtinen et al. 2015). Most common of the adverse effects was vomiting (13% of cats); however, concurrent disease including CKD and hyperthyroidism was common in this group of cats which may have also contributed to this clinical sign. TOD such as cardiac hypertrophy and proteinuria should stabilize and may improve following stabilization of SBP using amlodipine (Snyder et al. 2001; Jepson et al. 2007). A reduction in proteinuria is associated with greater survival times in these patients (Jepson et al. 2007).

Whilst effective in reducing blood pressure in hypertensive patients, when administered to normotensive patients at a dose of 0.125 mg/kg, amlodipine has no significant impact on SBP readings (Amodip® Marketing Authorisation file 2015).

Transdermal amlodipine is available in some countries and may be a suitable option for patients that are difficult to medicate. It is probably prudent to stabilize the patient using oral amlodipine before transferring to the transdermal formulation since this route is typically slower to stabilize and may be less effective (Helms 2007). Since transdermal medication has a more variable efficacy and is not veterinary authorised, it is not recommended as a first-line treatment.

Alternative agents with an impact on lowering blood pressure include renin–angiotensin–aldosterone system (RAAS) antagonists such as the angiotensin-converting enzyme (ACE) inhibitor benazepril. Benazepril generally has a low efficacy in comparison to amlodipine, typically lowering the SBP by 10–20 mmHg (Brown et al. 2001; Steele et al. 2002; Watanabe and Mishina 2007; Jenkins et al. 2015), and therefore is not recommended as a first-line therapy for hypertensive cats (Acierno et al. 2018). Benazepril may be sufficient as a sole therapy in mildly affected cats (SBP < 180 mmHg) and may be preferred as initial therapy, for example, in a patient with concurrent CKD, where benazepril is an authorized medication for the management of proteinuria associated with this disease. Benazepril can be safely used in addition to amlodipine, using both agents at their standard doses, and use of two agents is sometimes needed to adequately control blood pressure (Elliott et al. 2004).

Telmisartan, an angiotensin receptor blocker that also suppresses the RAAS, has recently received a veterinary authorisation for treatment of systemic hypertension in cats, and there are a small number of publications reporting its use for this indication (Glaus et al. 2019; Coleman et al. 2019a, b). Efficacy appears to be dose dependent with lower doses of 1 mg/kg/day likely to be effective only in mildly hypertensive cats or in combination with amlodipine. If using telmisartan as the sole agent for management of hypertension, a starting dose of 2 mg/kg/day is suggested. If SBP remains above 160 mmHg when rechecked 7–14 days later, the dose can be increased to 3 mg/kg/day (Table 12.4). With less data and clinical experience available for telmisartan, it is likely to be a second choice for antihypertensive therapy compared to amlodipine other than in patients with evidence of RAAS activation such as proteinuria.

Salt restriction has not been found to be beneficial in feline patients suffering from systemic hypertension, and salt-restricted diets may lead to negative consequences including reduced glomerular filtration rate and activation of the RAAS (Buranakarl et al. 2004). The patient's diet should be selected according to clinical requirements which will differ between patients.

For 'routine' cases, patient monitoring should be undertaken 7–14 days after starting treatment or adjusting the dose and should include a detailed history including discussions of compliance, physical examination, SBP measurement and evaluation for ocular TOD. The dose of medication can be adjusted, to effect, at these reexaminations until the patient is stable. In cats with SBP greater than 200 mmHg, ocular TOD or hypertensive encephalopathy, closer monitoring is justified with repeat SBP measurement and evaluation of TOD 2–7 days after starting treatment and more frequent dose adjustments if needed.

Emergency management of systemic hypertension may be indicated in those patients with severe acute elevations in SBP (SBP > 200 mmHg) or where evidence of severe ongoing TOD (e.g. severe neurological signs) is present. Patients should be hospitalized in a calm and quiet location of the clinic to allow repeated daily SBP assessment. Oral amlodipine is often effective within a few hours and can be administered according to a 'fast-track' protocol. If necessary, additional parenteral medications can be provided (Table 12.5). Once stable, the patient can be managed according to standard recommendations as detailed above.

In successfully treated cases, the blood pressure should decrease to levels between 120 and 140 mmHg. Reducing SBP to below 160 mmHg is an important short-term objective since this will drastically reduce the risk of TOD. In the long term, further reduction of SBP—ideally between 120 and 140 mmHg—is recommended. Posttreatment blood pressure readings between 140 and 160 mmHg

Table 12.5 Emergency treatment options for severely hypertensive patients

Medication (mechanism of action)	Dose and route	Comments
Amlodipine (calcium channel blocker, peripheral arterial dilator)	0.625–1.25 mg per cat, repeated if needed every 4–8 h to a maximum of 2.5 mg per cat q24 h	Fast track protocol: Assess SBP and TOD every 4 h until stable
Hydralazine (direct arterial vasodilator)	0.2–0.5 mg per cat subcutaneously	This dose can be repeated after 15 min if necessary
Acepromazine (phenothiazine, vasodilator)	50–100 mcg per cat intravenously or subcutaneously	
Nitroprusside (peripheral vasodilator)	0.5–1 mcg per kilogram per minute by constant rate infusion	Dose can be increased to 3 mcg/kg/min if needed Constant monitoring of SBP recommended
Esmolol (beta-blocker)	50–100 mg/kg/min constant rate infusion (CRI)	The CRI can be adjusted with dose increases every 10 min according to results from constant monitoring of SBP

Adapted from Taylor et al. (2017)

are acceptable, especially if patients are suspected to be vulnerable to stress-associated increases in SBP, as long as there is no evidence of continued TOD. In some treated cases, SBP readings may decrease to below 120 mmHg. In this situation, it is important to question the carer regarding possible clinical signs of hypotension such as ataxia, weakness or collapse and to examine the patient for tachycardia or other new findings which could be attributable to hypotension. So long as clinical signs of hypotension are not present, a SBP of less than 120 mmHg may be acceptable.

Once blood pressure is within the reference range, patients should be assessed every 4 to 6 weeks, reducing the frequency to a minimum of once every 3 months in very stable patients. Follow-up assessments should include measurement of blood pressure, assessment for evidence of TOD, periodic blood and urinalysis including creatinine levels and assessment of proteinuria. Once stable, these assessments should be done every 6–12 months according to the individual patient's needs.

Where possible, following diagnosis of systemic hypertension, investigations for underlying diseases and those known to have an association with hypertension are indicated. Ruling out CKD and hyperthyroidism is the initial priority so laboratory investigations should include serum thyroxine (T4), blood urea, creatinine, IDEXX SDMA® and urinalysis. Systemic hypertension may result in mild proteinuria so if documented the systemic hypertension should be treated and the patient re-evaluated to determine whether the proteinuria is a persistent abnormality. In cases where the proteinuria persists in spite of normalisation of blood pressure with amlodipine therapy, adding a RAAS suppressant may be indicated. In human medicine, the magnitude of proteinuria is used to determine target SBP for patients with the aim of ensuring renoprotective benefits through lowering blood pressure and proteinuria (Jarraya 2017). At present, no data exists to support this approach in cats.

If initial laboratory work does not reveal an obvious associated illness, additional diagnostic tests which may be considered include more thorough laboratory evaluation. Repeat total T4, free T4 and TSH assessment may be helpful in diagnosing occult hyperthyroidism (Peterson 2013; Peterson et al. 2015). Patients with primary hyperaldosteronism may have hypokalaemia, an adrenal mass/es evident on abdominal ultrasound, often markedly elevated serum aldosterone concentrations with low plasma renin activity (Ash et al. 2005; Bisignano and Bruyette 2012). Clinicians should be aware that many hypertensive cats have mildly elevated aldosterone and low plasma renin activity in the absence of primary adrenal disease (Jepson et al. 2014).

12.5 Management of Patients with Concurrent Disease

In hypertensive patients known to have CKD, it may be beneficial to use a combination of amlodipine with a RAAS suppressant as this will result in vasodilation of both afferent and efferent arterioles within the kidney and hence a reduction in glomerular capillary pressure (see Chap. 11). This approach is also indicated in those patients that remain proteinuric following stabilisation of their SBP using amlodipine.

Amlodipine is cleared by hepatic metabolism so should be used more cautiously, if at all, in patients with liver disease, depending on the severity of this. Although amlodipine metabolites are excreted by the kidneys, there is no contraindication to use of this medication in cats with CKD, and the dose does not need to be reduced (Huhtinen et al. 2015).

There are a number of concurrent medications which may have an impact on blood pressure and therefore increase the risk of hypotension when amlodipine or a RAAS suppressant therapy is prescribed. These include diuretics, beta-blockers, alpha 2 agonists, other calcium channel blockers such as diltiazem and other vasodilators. It is vital to ensure that patients are not dehydrated and that their blood pressure is monitored closely if prescribing an antihypertensive agent to a cat receiving any of these medications concurrently.

It is sensible to be cautious when using amlodipine in cats with cardiac disease, especially those receiving treatment with a negative chronotrope or inotrope such as beta-blockers or diltiazem, a cardioselective calcium channel blocker.

12.6 Prognosis

The long-term prognosis is very dependent on the presence, nature and extent of any associated illnesses. It is usually possible to manage the hypertension and prevent future TOD such as ocular haemorrhage. Diagnosis at an earlier stage helps to reduce the incidence of life-limiting and life-threatening TOD such as progressive ocular pathology and renal injury. Preventative health checks and proactive blood pressure monitoring of 'at-risk' patients are therefore recommended to facilitate an early diagnosis. The biggest challenge remains reliable blood pressure assessment in cats such that this condition is not under- or overdiagnosed. Amlodipine is the current treatment of choice, and many cats can be successfully managed on a once a day dose of this medication. Even in those cats with blindness and other severe hypertensive TOD, a clinical improvement is often reported as an improved quality of life.

References

Acierno MJ, Brown S, Coleman AE, Jepson RE, Papich M, Stepien RL, Syme HM (2018) ACVIM consensus statement: guidelines for the identification, evaluation, and management of systemic hypertension in dogs and cats. J Vet Intern Med 32(6):1803–1822

Amodip® Marketing Authorisation File 2015, on file at Ceva Animal Health

Ash RA, Harvey AM, Tasker S (2005) Primary hyperaldosteronism in the cat: a series of 13 cases. J Feline Med Surg 7:173–182

Belew A, Bartlett T, Brown SA (1999) Evaluation of the white-coat effect in cats. J Vet Intern Med 13:134–142

Bijsmans ES, Jepson RE, Chang YM et al (2015) Changes in systolic blood pressure over time in healthy cats and cats with chronic kidney disease. J Vet Intern Med 29:855–861

Bijsmans ES, Doig M, Jepson RE et al (2016) Factors influencing the relationship between the dose of amlodipine required for blood pressure control and change in blood pressure in hypertensive cats. J Vet Intern Med 30:1630–1636

Bijsmans ES, Jepson RE, Wheeler C et al (2017) Plasma N-terminal pro-brain natriuretic peptide, vascular endothelial growth factor and cardiac troponin 1 as novel biomarkers of hypertensive disease and target organ damage in cats. J Vet Intern Med 31:650–660

Bisignano J, Bruyette DS (2012) Feline hyperaldosteronism: recognition and diagnosis. Vet Med 107:118–125

Brown SA, Brown CA, Jacobs G et al (2001) Effects of the angiotensin converting enzyme inhibitor benazepril in cats with induced renal insufficiency. Am J Vet Res 62:375–383

Brown S, Atkins C, Bagley R et al (2007) Guidelines for the identification, evaluation, and Management of Systemic Hypertension in dogs and cats. J Vet Intern Med 21:542–558

Brown AL, Beatty JA, Lindsay SA et al (2012) Severe systemic hypertension in a cat with pituitary-dependent hyperadrenocorticism. J Small Anim Pract 53:132–135

Buranakarl C, Mathur S, Brown SA et al (2004) Effects of dietary sodium chloride intake on renal function and blood pressure in cats with normal and reduced renal function. Am J Vet Res 65:620–627

Calsyn JD, Green RA, Davis GJ et al (2010) Adrenal pheochromocytoma with contralateral adrenocortical adenoma in a cat. J Am Anim Hosp Assoc 46:36–42

Caney SMA. Ocular manifestations of systemic hypertension (2018). A free to download guide on www.vetprofessionals.com

Carter JM, Irving AC, Bridges JP et al (2014) The prevalence of ocular lesions associated with hypertension in an population of geriatric cats in New Zealand. N Z Vet J 62:21–29

Chalhoub S, Langston CE, Farrelly J (2012) The use of Darbepoetin to stimulate erythropoiesis in anemia of chronic kidney disease in cats: 25 cases. J Vet Intern Med 26:363–369

Chetboul V, Lefebvre HP, Pinhas C et al (2003) Spontaneous feline hypertension: clinical and echocardiographic abnormalities and survival rate. J Vet Intern Med 17:89–95

Chun R, Jakovljevic S, Morrison WB et al (1997) Apocrine gland adenocarcinoma and pheochromocytoma in a cat. J Am Anim Hosp Assoc 33:33–36

Coleman AE, Brown SA, Stark M et al (2019a) Evaluation of orally administered telmisartan for the reduction of indirect systolic arterial blood pressure in awake, clinically normal cats. J Feline Med Surg 21(2):109–114

Coleman AE, Brown SA, Traas AM, Bryson L, Zimmering T, Zimmerman A (2019b) Safety and efficacy of orally administered telmisartan for the treatment of systemic hypertension in cats: results of a double-blind, placebo-controlled, randomized clinical trial. J Vet Intern Med 33 (2):478–488

Cowgill LD, James KM, Levy JK et al (1998) Use of recombinant human erythropoietin for management of anemia in dogs and cats with renal failure. J Am Vet Med Assoc 212:521–528

Elliott J, Barber PJ, Syme HM et al (2001) Feline hypertension: clinical findings and response to antihypertensive treatment in 30 cases. J Small Anim Pract 42:122–129

Elliott J, Fletcher M, Souttar K et al. Effect of concomitant amlodipine and benazepril therapy in the management of feline hypertension, 14th ECVIM CA congress 2004

Glaus TM, Elliott J, Herberich E et al (2019) Efficacy of long-term oral telmisartan treatment in cats with hypertension: results of a prospective European clinical trial. J Vet Intern Med 33 (2):413–422

Helms SR (2007) Treatment of feline hypertension with transdermal amlodipine: a pilot study. J Am Anim Hosp Assoc 43:149–156

Henik RA, Snydrer PS, Volk LM (1997) Treatment of systemic hypertension in cats with amlodipine besylate. J Am Anim Hosp Assoc 33:226–234

Henry CJ, Brewer WG, Montgomery RD et al (1993) Clinical vignette – adrenal phaeochromocytoma in a cat. J Vet Intern Med 7:199–201

Huhtinen M, Derre G, Renoldi HJ et al (2015) Randomised placebo controlled clinical trial of a chewable formulation of amlodipine for the treatment of hypertension in client owned cats. J Vet Intern Med 29:786–793

Jarraya F (2017) Treatment of hypertension: which goal for which patient. Adv Exp Med Biol 956:117–127

Javadi S, Djajadiningrat-Laanen SC, Kooistra HS et al (2005) Primary hyperaldosteronism, a mediator of progressive renal disease in cats. Domest Anim Endocrinol 28:85–104

Jenkins TL, Coleman AE, Schmiedt CW et al (2015) Attenuation of the pressor response to exogenous angiotensin by angiotensin receptor blockers and benazepril hydrochloride in clinically normal cats. Am J Vet Res 76:807–813

Jepson RE, Elliott J, Brodbelt D et al (2007) Effect of control of systolic blood pressure on survival in cats with systemic hypertension. J Vet Intern Med 21:402–409

Jepson RE, Syme HM, Elliott J (2014) Plasma renin activity and aldosterone concentrations in hypertensive cats with and without azotaemia and in response to treatment with amlodipine besylate. J Vet Intern Med 28:144–153

Kobayashi DL, Peterson ME, Graves TK et al (1990) Hypertension in cats with chronic renal failure or hyperthyroidism. J Vet Intern Med 4:58–62

Lalor SM, Connolly DJ, Elliott J et al (2009) Plasma concentrations of natriuretic peptides in normal cats and normotensive and hypertensive cats with chronic kidney disease. J Vet Cardiol 11 (Suppl 1):S71–S79

Littman MP (1994) Spontaneous systemic hypertension in 24 cats. J Vet Intern Med 8:79–86

Maggio F, DeFrancesco TC, Atkins CE et al (2000) Ocular lesions associated with systemic hypertension in cats: 69 cases (1985-1998). J Am Vet Med Assoc 217:695–702

Mathur S, Syme H, Brown CA et al (2002) Effects of the calcium channel antagonist amlodipine in cats with surgically induced hypertensive renal insufficiency. Am J Vet Res 63:833–839

Morrow LD, Adams VJ, Elliott J et al (2009) Hypertension in hyperthyroid cats: prevalence, incidence and predictors of its development (abstract). J Vet Intern Med 23:699

Patnaik AK, Erlandson RA, Lieberman PH et al (1990) Extra-adrenal pheochromocytoma (paraganglioma) in a cat. J Am Vet Med Assoc 197:104–106

Payne JR, Brodbelt DC, Fuentes V (2017) Blood pressure measurements in 780 apparently healthy cats. J Vet Intern Med 31:15–21

Peterson ME (2013) More than just T4: diagnostic testing for hyperthyroidism in cats. J Feline Med Surg 15:767–777

Peterson ME, Guterl JN, Nichols R, Rishniw M (2015) Evaluation of serum thyroid-stimulating hormone concentration as a diagnostic test for hyperthyroidism in cats. J Vet Intern Med 29 (5):1327–1334.

Sansom J, Barnett KC, Dunn KA et al (1994) Ocular disease associated with hypertension in 16 cats. J Small Anim Pract 35:604–611

Sansom J, Rogers K, Wood JLN (2004) Blood pressure assessment in healthy cats and cats with hypertensive retinopathy. Am J Vet Res 65:245–252

Simpson KM, De Risio L, Theobald A et al (2014) Feline ischaemic myelopathy with a predilection for the cranial cervical spinal cord in older cats. J Feline Med Surg 16:1001–1006

Snyder PS (1998) Amlodipine: a randomized, blinded clinical trial in 9 cats with systemic hypertension. J Vet Intern Med 12:157–162

Snyder PS, Sadek D, Jones GL (2001) Effect of amlodipine on echocardiographic variables in cats with systemic hypertension. J Vet Intern Med 15:52–56

Sparkes AH, Caney SMA, King MCA et al (1999) Inter- and intraindividual variation in Doppler ultrasonic indirect blood pressure measurements in healthy cats. J Vet Intern Med 13:314–318

Steele JL, Henik RA, Stepien RL (2002) Effects of angiotensin converting enzyme inhibition on plasma aldosterone concentration, plasma renin activity, and blood pressure in spontaneously hypertensive cats with chronic renal disease. Vet Ther 3(2):157–166

Stepien RL (2011) Feline systemic hypertension: diagnosis and management. J Feline Med Surg 13:35–43

Stiles J, Polzin DJ, Bistner SI (1994) The prevalence of retinopathy in cats with systemic hypertension and chronic renal failure or hyperthyroidism. J Am Vet Med Assoc 30:564–572

Syme HM, Elliott J (2003) The prevalence of hypertension in hyperthyroid cats at diagnosis and following treatment (abstract). J Vet Intern Med 17:754

Syme HM, Barber PJ, Markwell PJ et al (2002) Prevalence of systolic hypertension in cats with chronic renal failure at initial evaluation. J Am Vet Med Assoc 220:1799–1804

Taylor SS, Sparkes AH, Briscoe K et al (2017) ISFM consensus guidelines on the diagnosis and management of hypertension in cats. J Feline Med Surg 19:288–303

Theobald A, Volk HA, Dennis R et al (2013) Clinical outcome in 19 cats with clinical and magnetic resonance imaging diagnosis of ischaemic myelopathy (2000-2011). J Feline Med Surg 15:132–141

Watanabe T, Mishina M (2007) Effects of benazepril hydrochloride in cats with experimentally induced or spontaneously occurring chronic renal failure. J Vet Med Sci 69:1015–1023

Williams TL, Peak KJ, Brodbelt D et al (2010) Survival and the development of azotaemia after treatment of hyperthyroid cats. J Vet Intern Med 24:863–869

Wimpole JA, Adagra CF, Billson MF et al (2010) Plasma free metanephrines in healthy cats, cats with non-adrenal disease and a cat with suspected phaeochromocytoma. J Feline Med Surg 12:435–440

Young WM, Zheng C, Davidson MG et al (2019) Visual outcome in cats with hypertensive chorioretinopathy. Vet Ophthalmol 22:161–167

Management of Hypertension in Dogs

13

Sarah Spencer

13.1 Introduction

Despite the recognised adverse consequences of systemic hypertension (SH) in veterinary patients, there are relatively few published studies on the management of SH in dogs. Recommendations are largely based on experiments investigating antihypertensive agents in normal (nonhypertensive) dogs and those with experimentally induced renal hypertension, human medicine and anecdotal experience. Whilst some general recommendations can be applied to the management of canine SH, it is important to consider therapy and monitoring on an individual patient basis.

13.2 Diagnosis of SH

Various studies have reported arterial blood pressure (BP) values in normal dogs (Höglund et al. 2012; Kallet et al. 1997; Mooney et al. 2017; Stepien and Rapoport 1999). Consensus is that dogs with a systolic blood pressure (SBP) ≥160 mmHg are considered to be hypertensive, based on an increased risk of target organ damage (TOD) (Table 13.1) (Acierno et al. 2018; Brown et al. 2013). In order to correctly diagnose SH, blood pressure (BP) needs to be measured accurately using a standard protocol (see Chap. 2).

Interbreed differences in BP are recognised in dogs, but currently only sighthounds are categorised separately, with a SBP of 10–20 mmHg higher than other breeds (Bodey and Sansom 1998; Cox et al. 1976; Surman et al. 2012). Brachycephalic dogs were reported to have a higher mean BP than meso- or dolichocephalic breeds in one small study (Hoareau et al. 2012). Temperament

S. Spencer (✉)
Department of Comparative Biomedical Sciences, Royal Veterinary College, London, UK
e-mail: sspencer18@rvc.ac.uk

© Springer Nature Switzerland AG 2020
J. Elliott et al. (eds.), *Hypertension in the Dog and Cat*,
https://doi.org/10.1007/978-3-030-33020-0_13

Table 13.1 ACVIM consensus classification of canine hypertension based on the risk of target organ damage

Classification	Systolic blood pressure	Risk of target organ damage
Normotensive	<140 mmHg	Minimal
Prehypertensive	140–159 mmHg	Low
Hypertensive	160–179 mmHg	Moderate
Severely hypertensive	≥180 mmHg	Severe

Table 13.2 Evidence of target organ damage due to systemic hypertension

Tissue	Target organ damage secondary to systemic hypertension	Clinical findings indicative of target organ damage
Cardiovascular system	Left ventricular hypertrophy Less commonly—left-sided congestive heart failure Rarely—aortic aneurysm/dissection	Left ventricular concentric hypertrophy on echocardiography, systolic heart murmur, arrhythmias, signs of left-sided congestive failure (e.g. tachypnoea), haemorrhage (e.g. epistaxis)
Brain	Encephalopathy, stroke	Range of central neurological signs including depression, obtundation, seizures
Eye	Retinopathy/choroidopathy	Acute onset blindness, retinal detachment, retinal haemorrhage/oedema, retinal vessel tortuosity or perivascular oedema, papilloedema, hyphaema, secondary glaucoma, retinal degeneration
Kidney	(Progression of) chronic kidney disease	(Progression of) azotaemia, reduced glomerular filtration rate, proteinuria, microalbuminuria

may also partly account for interbreed variations (Höglund et al. 2012). There is a recognised need for breed-specific ranges (Acierno et al. 2018).

There are three main situations where SH may be diagnosed: firstly, in a 'healthy' dog undergoing routine screening (e.g. geriatric animals); secondly, in a dog with a condition (or receiving drug therapy) known to cause or predispose to increased BP; or thirdly, in a dog presenting with clinical signs of target organ damage (TOD). Potential organs for TOD are the eye, brain, cardiovascular system and kidney (Table 13.2). Once SH has been documented, the clinician should attempt to categorise hypertension as either situational, idiopathic or secondary.

13.2.1 Situational Hypertension

Situational hypertension (sometimes referred to as 'white-coat hypertension') is the term used for transient BP increases in an otherwise normotensive animal, which results from environmental or situational stressors such as anxiety, fear, pain or excitement (Acierno et al. 2018). Hypertension resolves when the physiological stimulus is decreased or eliminated, and therefore, efforts should be made to alter measurement conditions to achieve this. For example, BP measurement may be performed at the dog's home, after provision of pain relief, or following acclimatisation to repeated vet visits (Kallet et al. 1997; Schellenberg et al. 2007). Ambulatory/home BP monitoring is useful in people where situational hypertension is suspected but has not been evaluated in dogs. Whilst there is evidence that people with situational hypertension have an increased risk of developing sustained SH and long-term cardiovascular effects (Martin and Mcgrath 2014), there is no evidence to justify its treatment in canine patients. It is not possible to predict which dogs exhibit marked BP increases due to anxiety, and clinicians should always consider the possibility of excitement- or fear-induced BP increases to avoid the erroneous diagnosis of pathological SH (Kallet et al. 1997).

13.2.2 Idiopathic Hypertension

Idiopathic hypertension is used to describe cases of SH without an identifiable underlying cause. As secondary hypertension comprises the majority of cases of canine SH, 'idiopathic' is preferred over 'primary' or 'essential' hypertension (terms used in human medicine), as many dogs with no discernible underlying cause may have early chronic kidney disease (CKD) or other overlooked primary conditions (Acierno et al. 2018). True essential hypertension, i.e. where no underlying or associated cause can be found, may be difficult to establish but has been described rarely in dogs (Bovee et al. 1989, 1986; Slaughter et al. 1996). Repeated evaluation of SBP and careful examination for evidence of hypertensive TOD may be required to substantiate the diagnosis. Management of dogs with idiopathic hypertension is based on antihypertensive drug therapy but should also involve close monitoring of renal function and for clinical findings which could indicate an underlying condition, especially if empirical treatment is ineffective in controlling BP.

13.2.3 Secondary Hypertension

Secondary hypertension is the persistent and pathological increase in BP concurrent with a disease, therapeutic agent or toxic substance known to cause SH (Acierno et al. 2018). This category is thought to represent over 80% of hypertensive dogs. Diseases associated with canine SH are listed in Box 13.1. Even if dogs with conditions or receiving drug therapy associated with SH are normotensive, wellness and BP evaluation are advised every 3 to 6 months in order to identify SH should it develop.

Several therapeutic agents are known to increase systemic BP (Box 13.2), and such drugs where prescribed may be presumed to be at least partly responsible for increasing BP in a hypertensive dog. The most commonly used agents include steroid hormones, erythropoiesis-stimulating agents, phenylpropanolamine and

Box 13.1 Diseases associated with secondary hypertension in dogs

Chronic kidney disease (CKD) – proteinuric and non-proteinuric
Acute kidney injury (AKI)
Hyperadrenocorticism
Pheochromocytoma
Diabetes mellitus
Neoplasia
Hyperaldosteronism
Brachycephalic obstructive airway syndrome
Hypothyroidism/hyperthyroidism

Box 13.2 Drugs and toxicants associated with increased systemic blood pressure in dogs

Glucocorticoids, e.g. prednisolone, dexamethasone
Mineralocorticoids, e.g. desoxycorticosone pivalate (DOCP)
Phenylpropanolamine
Erythropoietin (EPO)/synthetic EPO analogues, e.g. darbepoetin alfa
Toceranib phosphate (+/− other tyrosine kinase inhibitors, e.g. masitinib)
Topical ocular drugs – prednisolone, phenylephrine
Cocaine
Methamphetamine/amphetamine
5-Hydroxytryptophan

tyrosine kinase inhibitors. Although statistically significant BP increases are reported in dogs receiving glucocorticoid treatment, clinically relevant doses are uncommonly associated with overt hypertension (Nakamoto et al. 1991; Schellenberg et al. 2008). BP-increasing effects are dose-dependent, and therefore, dosages should be reduced where possible if SH is encountered in a dog receiving glucocorticoids. Although desoxycorticosterone pivalate (DOCP) may increase BP experimentally (Ueno et al. 1988), increased BP was not seen at recommended doses for hypoadrenocorticism treatment (Kaplan and Peterson 1995). Phenylpropanolamine, used to treat urinary incontinence, may cause increased BP due to agonistic effects at α-adrenergic receptors and indirect effects on β-adrenergic receptors (Noël et al. 2012). SH is rarely reported at therapeutic doses (Cohn et al. 2002; LeBlanc et al. 2011) but is seen with phenylpropanolamine toxicosis (Peterson et al. 2011). Antihypertensive treatment is appropriate to reduce the risk of TOD in toxicosis cases where BP ≥ 180 mmHg; β-blocker therapy is proposed to be the most suitable agent in this situation (Pentel et al. 1985) but should be avoided in dogs experiencing (reflex) bradycardia. A case report also describes successful management of phenylpropanolamine-induced SH with sodium nitroprusside infusion, alongside short-term phenoxybenzamine, sotalol and esmolol (Ginn et al. 2012). It is not only systemic therapies which may cause increased systemic BP; changes can occur due to topical ocular medications including glucocorticoids and phenylephrine hydrochloride (Herring et al. 2004; Pascoe et al. 1994).

Erythropoietin (EPO) and other erythropoiesis-stimulating agents such as darbepoetin alfa are associated with SH (Cowgill et al. 1998; Fiocchi et al. 2017;

Randolph et al. 2004). Given that no placebo-controlled studies exist, it is difficult to discern a causal effect, however, as these agents are most commonly used in renal disease and observed BP increases may be due to kidney disease progression. In one study, 96% dogs receiving darbepoetin showed a significant BP increase; 32% of dogs required initiation of or an increase in antihypertensive medication, although no cases required darbepoetin discontinuation (Fiocchi et al. 2017). SH was noted in 40% dogs receiving recombinant human EPO (Cowgill et al. 1998). Overall, BP did not significantly increase from baseline in dogs with an initial BP <180 mmHg, although 20% (5/15) of dogs treated with recombinant canine or human EPO developed BP ≥180 mmHg (Randolph et al. 2004). Antihypertensive treatment was only attempted in two dogs: amlodipine was successful in controlling BP in one, while enalapril was unsuccessful in another (Randolph et al. 2004).

SH was reported in 28% normotensive dogs at baseline treated with the tyrosine kinase inhibitor toceranib phosphate for 14 days (Tjostheim et al. 2016); however, an undisclosed number were receiving concurrent glucocorticoid or NSAID therapy. The management of these dogs was not reported. Enalapril monotherapy was unsuccessful in normalising BP in four out of six dogs that were hypertensive at baseline, whereas enalapril and amlodipine co-therapy was successful in the remaining two dogs (Tjostheim et al. 2016). SH has not been reported following masitinib administration in dogs although there are no published reports specifically evaluating BP in treated dogs; as the mechanism of action for toceranib-associated SH is unknown, a potential effect of masitinib on BP is proposed.

Several intoxicants are associated with SH in dogs including cocaine (Thomas et al. 2014), methamphetamine/amphetamine (Bischoff et al. 1998; Pei and Zhang 2014; Tontodonati et al. 2007) and 5-hydroxytryptophan (Jennifer et al. 2017). Specific antihypertensive treatment is not commonly required. When treatment is indicated, i.e. in patients with evidence of TOD, shorter-acting and titratable agents (e.g. fenoldopam, sodium nitroprusside) should be used as effects are expected to be short-lived. When treating SH, care is needed to avoid exacerbating possible detrimental cardiovascular effects associated with some intoxicants (e.g. arrhythmias and tachycardia).

13.3 Decision to Institute Therapy

The absolute BP value at which intervention for SH is required remains the subject of debate. Antihypertensive treatment should be considered in any dog at moderate or severe risk of TOD (BP ≥160 mmHg), and evidence of TOD is almost always a strong indication for antihypertensive treatment. An algorithm describing different scenarios following BP measurement and subsequent management is shown in Fig. 13.1. Following an initial SBP measurement of ≥160 mmHg, treatment should be instigated if TOD (or other more subtle clinical signs of SH such as lethargy and dullness) is present and cannot be ascribed to an alternative condition. If TOD or other clinical signs are absent, persistence of SH must be ascertained, particularly to help rule out situational hypertension as far as possible. The patient should be

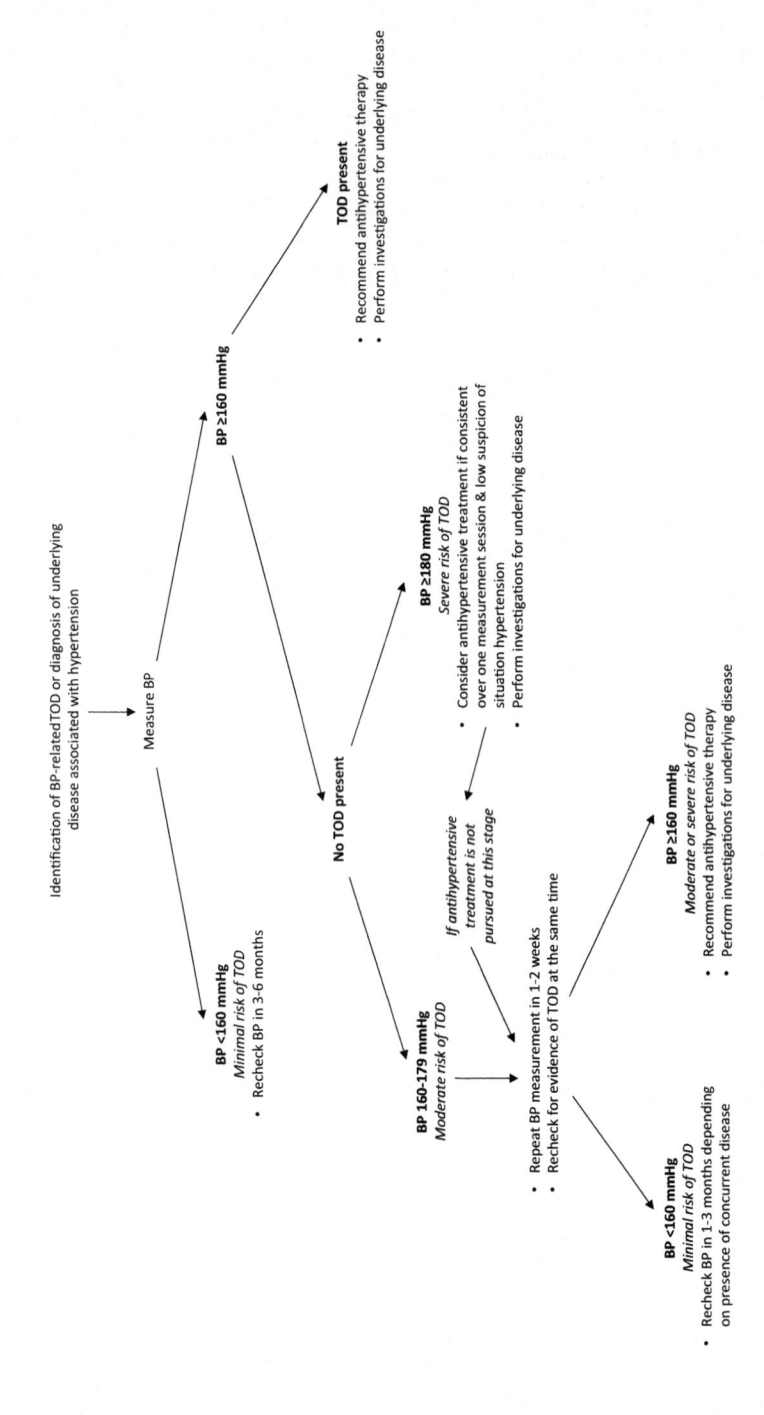

Fig. 13.1 Recommended approach to the evaluation of a potentially hypertensive canine patient

reevaluated and treatment started if BP is persistently ≥160 mmHg and/or TOD develops. The time at which to perform reassessment depends on the initial BP and other patient factors such as presence of concurrent disease, but 1–2 weeks' time is appropriate in most cases. Although documenting persistently elevated BP is desirable, due to the severe risk of TOD, antihypertensive treatment should be considered if SBP ≥180 mmHg for only one measurement session if there is sufficient reason to suspect that this is genuine (Acierno et al. 2018). Dogs may be more resistant to developing ocular lesions indicative of TOD compared to cats (Jacob et al. 2003), and so substantiating a diagnosis of SH may be more difficult in this species. Note that dogs with SBP ≥160 mmHg and CKD or glomerular disease are presumed to have TOD and should receive antihypertensive treatment. Patients considered prehypertensive (SBP ≥140–159 mmHg) do not generally require antihypertensive therapy but should be monitored. Exceptions to the above recommendations are cases that are suspected to be experiencing situational hypertension, where SH is expected to be transient (e.g. toxicosis) or in rare situations when antihypertensive treatment is contraindicated (see below).

These recommendations are based on evidence in hypertensive people where any BP reduction that does not result in overt hypotension lowers the risk of TOD and adverse outcomes (Ettehad et al. 2016; Xie et al. 2016). Although not confirmed in veterinary patients, SH treatment is presumed to be of prognostic benefit and is consistent with studies in dogs with experimentally induced renal disease (Finco 2004; Kang et al. 2018). Prognostic benefit remains challenging to investigate as it is unethical to withhold antihypertensive treatment in affected animals.

SH is a silent disease in most dogs, or low-grade, nonspecific clinical signs such as lethargy, depression and reduced exercise tolerance may be missed by the owner or disregarded as signs of ageing. It can therefore be difficult to persuade owners of hypertensive pets to instigate treatment, as therapy generally aims to avoid adverse consequences rather than overtly improve quality of life. Further challenges regarding compliance may occur as treatment for SH is not curative and is usually lifelong. This should be communicated with owners at diagnosis. The interindividual variability regarding response to medication and the need for frequent monitoring, especially until therapeutic goals are achieved, must also be conveyed.

13.4 Initial Assessment of a Dog with SH

Initial assessment of a hypertensive dog should involve identifying and characterising TOD and determining if there are any underlying causes of SH or concurrent conditions which may complicate antihypertensive therapy (e.g. cardiac disease).

13.4.1 Assessment for TOD

Diagnostic investigations which may be useful to assess for TOD are outlined in Table 13.2. To assess for renal TOD, serum creatinine and urea concentrations, +/− symmetric dimethylarginine (SDMA) should be measured, and urinalysis with quantitative assessment of proteinuria (by urine protein/creatinine ratio [UPC]) and/or albuminuria should be performed. Full ophthalmic evaluation including a funduscopic examination is recommended in all patients to assess for retinopathy or choroidopathy indicative of ocular TOD. Neurological examination should also be performed and magnetic resonance imaging (MRI) considered in patients with central neurological signs (Lowrie et al. 2012; O'Neill et al. 2013b). Careful cardiac auscultation should be performed to assess for abnormal heart rhythms or sounds. Where cardiac TOD is suspected, thoracic radiography, echocardiography (particularly to assess for left ventricular concentric hypertrophy) and electrocardiogram are suitable tests to consider.

13.4.2 Investigations for Causes of Secondary Hypertension

Underlying causes or contributing factors for SH are present in approximately 80% hypertensive dogs. A basic database (haematology, serum biochemistry and urinalysis) should be performed in all cases. Increased BP may cause polyuria ('pressure diuresis'), and reduced urine-specific gravity (<1.030) in a hypertensive patient can be difficult to interpret. Subclinical renal disease can further be assessed for by performing renal ultrasound, SDMA measurement, microalbuminuria (Bacic et al. 2010) and glomerular filtration rate (GFR) quantification. Dogs with persistent or progressive proteinuria (especially when UPC ≥ 2.0) should be suspected of having glomerular disease, and further investigations such as infectious disease testing and renal biopsy may be suitable (Littman et al. 2013). Testing for hyperadrenocorticism (e.g. low-dose dexamethasone suppression test or ACTH-stimulation test) may also be considered on an individual basis if historical, clinical and biochemical findings are suggestive. Serum and urine catecholamine and aldosterone concentrations and adrenal imaging may also be suitable in some patients to investigate for pheochromocytoma and hyperaldosteronism (see Chap. 4). Importantly, hypertensive cats which had investigations for underlying disease had a decreased risk of death (Conroy et al. 2018), and intuitively, the same may be true in dogs. All medications being administered to a hypertensive dog should be reviewed in case they could be contributing to increased BP.

13.5 Goals of Therapy

The ideal target level to which BP should be suppressed in hypertensive dogs is unknown, as no studies have investigated this to date. It is agreed, however, that the primary goal of antihypertensive treatment is to prevent TOD, in order to maximise quality and quantity of life (Acierno et al. 2018). If TOD is already present, the goal should be to slow the progression of further damage and to ameliorate associated clinical signs where possible. Based on TOD risk categorisation (Table 13.1), this means ideally reducing SBP to <140 mmHg regardless of the initial increase. A more realistic aim in challenging cases may be to decrease SBP by at least one risk substage, or to <160 mmHg (Acierno et al. 2018; IRIS 2017). As in people, evidence suggests a continuum between BP and CKD progression in dogs (Finco 2004; Jacob et al. 2003), and therefore, efforts should be made to achieve a BP as low as possible whilst avoiding hypotension (SBP <110 mmHg). It is generally appropriate to reduce BP gradually over a couple of weeks; acute marked decreases in BP should be avoided. The goals of treatment of an emergent hypertensive crisis differ and are discussed below.

An additional goal of antihypertensive treatment is to reduce UPC in proteinuric patients. The coexistence of SH and proteinuria is common in dogs with CKD (Bacic et al. 2010; Jacob et al. 2003, 2005), hyperadrenocorticism (Ortega et al. 1996) and diabetes mellitus (Struble et al. 1998). Cats achieving a UPC <0.2 post-antihypertensive therapy have improved survival (Jepson et al. 2007); although the same benefit has not been shown in dogs, proteinuria and related hypertension are associated with a greater risk of a uraemic crisis, CKD progression and death in dogs with CKD (Jacob et al. 2003, 2005). In proteinuric dogs, the aim is to reduce UPC to <0.5, or by >50% if this is unachievable (Brown et al. 2013).

13.6 General Recommendations for the Management of SH

Pharmacological treatment is the mainstay of canine SH management. There is a paucity of studies investigating antihypertensive agents in dogs, however, especially with regard to efficacy in naturally occurring SH. Many reports of SH associated with underlying disease do not report on specific antihypertensive treatment or their efficacy. Furthermore, pharmaceutical studies and those specifically investigating haemodynamic effects are usually performed in normotensive dogs.

Although general recommendations can be made regarding the management of SH, treatment should be individually tailored and based on comorbidities and suspected underlying mechanisms contributing to SH (which may be multiple). If secondary hypertension is confirmed, treatment of the underlying condition should be implemented immediately, and as patients are at continued risk of TOD whilst the primary condition is being controlled, antihypertensive therapy should not be delayed. Furthermore, specific treatment or cure of the underlying condition is not always feasible (e.g. CKD), and therefore, specific antihypertensive therapy is essential. Antihypertensive agent choice may be dictated by the primary condition; specific management in certain diseases is discussed below. Although successful control of the underlying disease is not expected to resolve SH in most cases, it will help maximise antihypertensive drug efficacy and hopefully allow reduced dosages to be used. Any drugs that may be contributing to or causing SH should be discontinued; if this is not possible, then a dosage decrease should be considered.

Hypertensive patients requiring volume resuscitation should undergo this cautiously before antihypertensive treatment is initiated, unless rapidly progressive TOD and severe SH are present.

A stepwise approach is recommended for nonemergent SH (Fig. 13.2) (Acierno et al. 2018). Generally, antihypertensive drugs should be started at the low end of the dosage range and titrated upwards as required. The stepwise approach is unsuitable for hypertensive emergencies, i.e. severely hypertensive dogs with evidence of severe or progressive ocular or neurological TOD. When the current regimen fails to achieve therapeutic goals, treatment escalation is required. Importantly, SH is often progressive, and therefore, frequent monitoring is needed along with the expectation for treatment intensification over time.

General considerations when deciding on an antihypertensive treatment regimen include dosing requirements, as once daily therapy is more convenient for owners and may result in better compliance, and whether to use one or multiple agents. Fewer medications are generally preferred over polypharmacy; however, utilising drugs from classes with different mechanisms of actions may be desirable in some patients.

13.6.1 Patient Monitoring

Following instigation of antihypertensive treatment or dosage change, dogs should be reevaluated every 7–14 days until BP control is achieved (Fig. 13.1), although timing will depend on SH severity, general patient condition and concurrent disease, with more frequent assessment required in unstable patients or those where concurrent conditions dictate it. The same technique and circumstance should be employed when measuring BP where possible. Reevaluation may also include serum biochemistry and urinalysis, including UPC measurement, depending on the individual. Dogs with IRIS stage 3 or 4 CKD should be reevaluated sooner, within 3–5 days (Brown et al. 2013; IRIS 2017), and serum creatinine and potassium measurement should be performed at a minimum. Dogs with severe or rapidly progressive TOD should be reassessed within 1–3 days (Brown et al. 2007). Hospitalised hypertensive patients should undergo clinical assessment and BP +/− serum creatinine measurement at least daily, especially if they are receiving fluid therapy or pharmacological agents with cardiovascular effects.

Once BP control is achieved, ongoing monitoring is required every 1–4 months to confirm continuing treatment efficacy (especially as signs of SH may be subtle or nonexistent) and to ensure hypotension does not occur. These visits should include repeat physical examination (particularly focusing on the identification of TOD, including fundic examination), BP measurement, serum biochemistry (or at least assessment of renal function) and urinalysis including UPC measurement. Although no studies exist in dogs, epidemiological data suggests that the monitoring of hypertensive cats in primary-care practice is poor but that cats undergoing increased monitoring have increased survival (Conroy et al. 2018).

13 Management of Hypertension in Dogs

Decision to instigate antihypertensive therapy

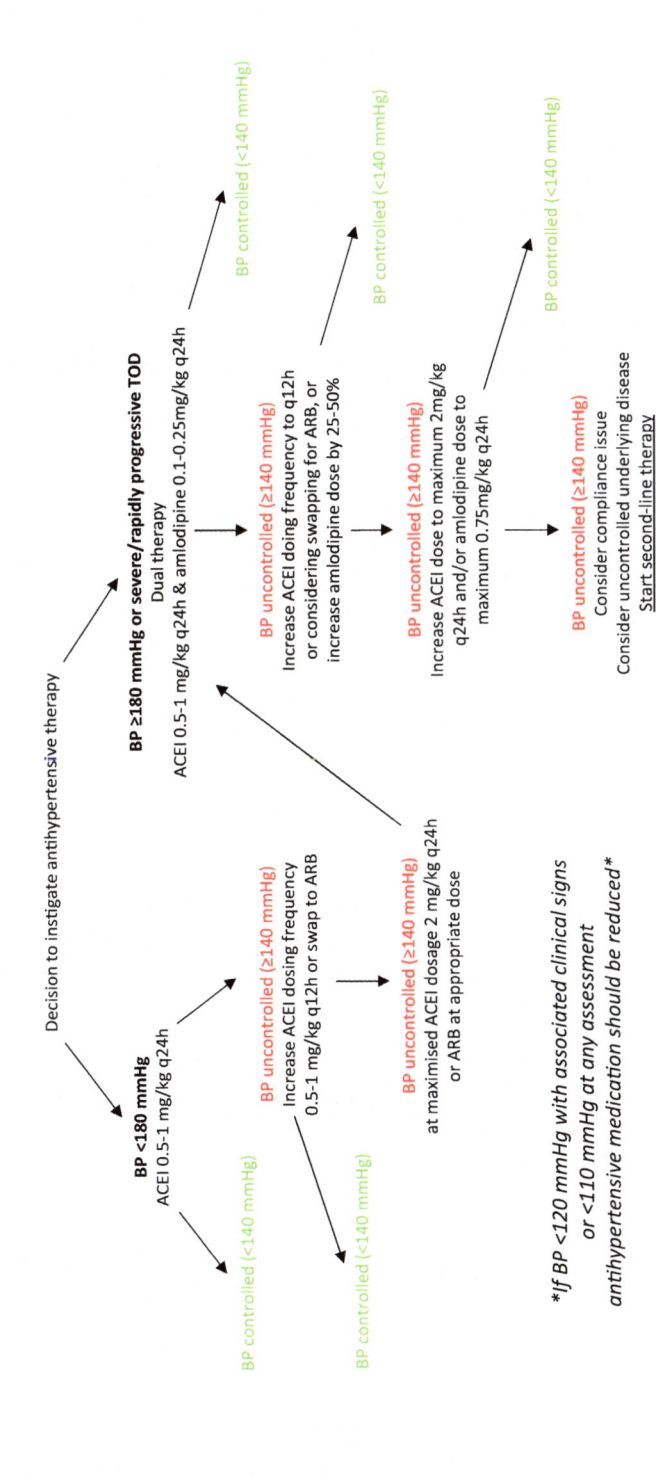

BP <180 mmHg
ACEI 0.5–1 mg/kg q24h

- BP controlled (<140 mmHg)
- BP uncontrolled (≥140 mmHg)
  Increase ACEI dosing frequency 0.5–1 mg/kg q12h or swap to ARB
  - BP controlled (<140 mmHg)
  - BP uncontrolled (≥140 mmHg) at maximised ACEI dosage 2 mg/kg q24h or ARB at appropriate dose

BP ≥180 mmHg or severe/rapidly progressive TOD
Dual therapy
ACEI 0.5–1 mg/kg q24h & amlodipine 0.1–0.25mg/kg q24h

- BP controlled (<140 mmHg)
- BP uncontrolled (≥140 mmHg)
  Increase ACEI dosing frequency to q12h or considering swapping for ARB, or increase amlodipine dose by 25–50%
  - BP controlled (<140 mmHg)
  - BP uncontrolled (≥140 mmHg)
    Increase ACEI dose to maximum 2mg/kg q24h and/or amlodipine dose to maximum 0.75mg/kg q24h
    - BP controlled (<140 mmHg)
    - BP uncontrolled (≥140 mmHg)
      Consider compliance issue
      Consider uncontrolled underlying disease
      Start second-line therapy

If BP <120 mmHg with associated clinical signs or <110 mmHg at any assessment antihypertensive medication should be reduced

Fig. 13.2 Example stepwise approach to initial therapy in canine hypertension

Although antihypertensive treatment is often lifelong, permanent therapy is not required in all cases, and the decision to treat warrants periodic evaluation. If antihypertensive medication is withdrawn, reevaluation (e.g. after 7–14 days) is recommended to monitor for SH recurrence and then at later time points depending on the individual and any ongoing concurrent disease.

13.6.2 Dietary Therapy

Reduction in dietary sodium has been proposed in the management of canine SH (Brown and Henik 1998). This is extrapolated from human medicine, and there is no evidence that reducing sodium intake is beneficial in hypertensive dogs. Sodium restriction may be deleterious by exacerbating RAAS activation and causing kaliuresis and hypokalemia (Krieger et al. 1990; Lovern et al. 2001). This said, it is advisable to avoid high salt intake in hypertensive dogs (Acierno et al. 2018). Although reduced capacity to excrete sodium with renal dysfunction may increase salt sensitivity (Koomans et al. 1982), varying dietary salt intake did not alter BP in dogs with experimentally reduced renal mass (Greco et al. 1994; Hansen et al. 1992). The International Renal Interest Society (IRIS) recommends that dogs with CKD are fed a diet with reduced salt content, however, regardless of IRIS stage (IRIS 2017); if reasons unrelated to SH dictate dietary sodium reduction in a hypertensive dog, it should be performed gradually and alongside antihypertensive therapy. As in people, dogs with nephrotic syndrome especially are considered to be salt-sensitive due to excessive sodium retention (Geers et al. 1984; Klosterman and Pressler 2011).

Several dietary supplements have been evaluated for their possible benefit in hypertensive humans, including fish oil, garlic and vitamin C (Rasmussen et al. 2012); however, their effects and safety are poorly characterised in humans and are wholly unknown in dogs, therefore cannot be recommended.

Obesity is a major risk factor for SH in humans, and reducing body weight has been recommended in obese, hypertensive dogs (Brown and Henik 1998; Montoya et al. 2006). Some studies report an association with obesity (Bodey and Michell 1996), whereas others report no correlation with weight or body condition score (Mooney et al. 2017); it is proposed that any association is related to underlying disease rather than obesity per se (Pérez-Sánchez et al. 2015).

13.6.3 Initial Antihypertensive Drug Therapy

The most widely recommended and utilised antihypertensive agents in dogs are renin-angiotensin-aldosterone system (RAAS) inhibitors and the calcium channel blocker (CCB), amlodipine. RAAS inhibitors, specifically angiotensin-converting enzyme inhibitors (ACEIs), are commonly chosen as the initial agent due to the high frequency of CKD and proteinuria in hypertensive dogs (Acierno et al. 2018). There are no randomised studies comparing RAAS inhibitors and CCBs in dogs with naturally occurring SH, and it is therefore unknown which class is best for the treatment of dogs with non-proteinuric CKD or patients with other underlying diseases. One agent may be initially considered in dogs with modest SH (≥160–179 mmHg), whereas RAAS inhibitor and amlodipine co-therapy should be considered in dogs with more severe SH (≥200 mmHg) (Acierno et al. 2018).

13.6.3.1 RAAS Inhibitors

RAAS inhibitors comprise of ACEIs, angiotensin II receptor blockers (ARBs), mineralocorticoid receptor antagonists (MRAs) and, in human medicine, direct renin inhibitors (see Chap. 11). Clinical experience is greatest with ACEIs, and therefore, an ACEI such as benazepril or enalapril, instigated at or towards the lower end of the dosage range, is a sensible initial drug choice in a hypertensive dog (Table 13.3). ACEIs significantly reduce BP in experimentally induced CKD (Brown et al. 2003, 1993). They are generally expected to reduce SBP by 10–20 mmHg (Grauer et al. 2000; Mishina and Watanabe 2008) and are therefore often insufficient to control severe SH when used alone. Benazepril is the preferred ACEI for dogs with renal dysfunction as it is partially metabolized by the liver, unlike enalapril which relies solely on renal excretion. The standard initial ACEI dosage is 0.5 mg/kg PO once daily (King et al. 1995), but dosage escalation and twice daily therapy are anticipated to be required in most hypertensive dogs, as is expected for their sufficient antiproteinuric effect in glomerular disease (Brown et al. 2013; Grauer et al. 2000). The recommended upper ACEI dosage is 2 mg/kg per day (Brown et al. 2013) although some clinicians may choose to instigate a second agent before approaching this. In healthy dogs with furosemide-induced RAAS activation, high-dose ACEI (1 mg/kg PO q12) leads to comparable ACE activity inhibition to 1 mg/kg q24h (Ames et al. 2016; Lantis et al. 2015); the effect on BP was not specifically examined in these studies.

ARBs, such as telmisartan and losartan, represent a newer class of RAAS inhibitors, with overall actions similar to those of ACEIs. They may be considered as an alternative initial therapy for SH (and proteinuria), although their efficacy and safety in canine patients are currently not well known and no licensed ARBs exist for use in dogs in most countries. Telmisartan, marketed for the management of proteinuria in feline CKD, is an effective antihypertensive treatment in this species (Glaus et al. 2019). Telmisartan is as effective as amlodipine in reducing BP in hypertensive people with CKD and may be preferred due to additional benefit in reducing proteinuria (Nakamura et al. 2018). It is superior to benazepril in inhibiting rises in BP following angiotensin I administration (Jenkins et al. 2015). A dosage of 1 mg/kg PO q24 decreases BP in normal dogs (Konta et al. 2018). A small canine case series describes the successful use of telmisartan in controlling SH in combination with amlodipine, where previous benazepril/amlodipine co-therapy was inadequate (Caro-Vadillo et al. 2018). Telmisartan resolved SH in a dog with CKD where amlodipine had been ineffective (Kwon et al. 2018) and was also used to successfully manage proteinuria refractory to ACEI treatment (Bugbee et al. 2014). There are few reports on the use of other ARBs in dogs; pharmacokinetic studies of

Table 13.3 Oral antihypertensive agents

Drug class	Drug	Recommended dosage	References
Angiotensin-converting enzyme inhibitors (ACEIs)	Benazepril	0.5 mg/kg q12–24 h, maximum 2 mg/kg per day 0.25 mg/kg starting dose in azotaemic dogs	Acierno et al. (2018), Brown et al. (2013)
	Enalapril	0.5 mg/kg q12–24 h, maximum 2 mg/kg per day 0.25 mg/kg starting dose in azotaemic dogs	Acierno et al. (2018), Brown et al. (2013)
Angiotensin receptor blockers (ARB)	Telmisartan	0.4–1 mg/kg q24h, maximum 2 mg/kg described in cats	Acierno et al. (2018), Brown et al. (2013), Bugbee et al. (2014), Caro-Vadillo et al. (2018), Glaus et al. (2019), Kwon et al. (2018)
	Losartan	0.5–1 mg/kg q24h 0.125–0.25 mg/kg q24h in azotaemic dogs	Brown et al. (2013)
Calcium channel blocker (CCB)	Amlodipine	0.1–0.25 mg/kg q24h (maximum 0.75 mg/kg)	Acierno et al. (2018), Brown et al. (2013)
α-Adrenergic antagonist	Prazosin	0.5–2 mg/kg q8–12h	Acierno et al. (2018), Brown et al. (2013)
	Phenoxybenzamine	0.25–1.5 mg/kg q8–12h	Acierno et al. (2018)
	Acepromazine	0.5–2 mg/kg q8h	Acierno et al. (2018)
β-Blocker	Propranolol (non-selective)	0.2–1 mg/kg q8h	Acierno et al. (2018)
	Atenolol (β1-selective)	0.25–1 mg/kg q12h	Acierno et al. (2018)
Direct vasodilator	Hydralazine	0.5–2 mg/kg q12h	Acierno et al. (2018), Pariser and Berdoulay (2011)
Mineralocorticoid antagonist	Spironolactone	1–2 mg/kg q12h	Acierno et al. (2018)
Thiazide diuretic	Hydrochlorothiazide	2–4 mg/kg q12–24 h	Acierno et al. (2018)
Loop diuretic	Furosemide	1–4 mg/kg q8–24h	Acierno et al. (2018)

Generally drugs are instigated at the lower end of the dosage range and titrated upwards to effect.

irbesartan in normal beagles suggest an antihypertensive dose of 5 mg/kg (Carlucci et al. 2013). Losartan is commonly used clinically in dogs (Brown et al. 2013), and its pharmacokinetics and pharmacodynamics have been studied in healthy dogs (Christ et al. 1994).

13.6.3.2 Amlodipine

Amlodipine, a dihydropyridine CCB, reduces BP by causing systemic arteriolar vasodilation (see Chap. 11). It is the recommended first-line treatment for hypertensive cats (Acierno et al. 2018; Huhtinen et al. 2015) and may be a suitable initial alternative to ACEIs or ARBs when RAAS activation is not thought to be a big contributing factor in dogs. Amlodipine reduces BP in hypertensive dogs (Burges et al. 1989; Ishida et al. 2003) although no large clinical trials exist, and its effect is minimal in normal dogs (Atkins et al. 2007). Recommended starting dosage is 0.1–0.25 mg/kg PO q24h, increasing as needed every 1–2 weeks up to 0.75 mg/kg (Acierno et al. 2018; Huhtinen et al. 2015) although the margin of safety of amlodipine in the dog has not been determined through systematic registration studies. The elimination half-life is fairly long in dogs (30 h) (Burges et al. 1989; Stopher et al. 1988); more frequent dosage adjustments should be avoided, and twice daily administration is unnecessary. Cats with higher BP at diagnosis seemingly require a higher amlodipine dosage for adequate BP control (Bijsmans et al. 2016), but it is unknown if the same is true for dogs. Transdermal administration has been investigated in cats but was less effective than oral dosing (Mixon and Helms 2008) and cannot be recommended in dogs.

Although amlodipine can be used as a single agent in dogs, administration of an ACEI alongside amlodipine is often recommended because the former blunts the RAAS activation and potential increase in glomerular capillary pressure seen with CCBs (Atkins et al. 2007; Brown et al. 1993; Gaber et al. 1994; Yue et al. 2001).

13.6.4 Combination Antihypertensive Therapy

If the chosen initial antihypertensive agent is not or only partially effective in achieving therapeutic goals, the next step is to either increase the dosage or add an additional drug. Which of these to perform will depend on the first-line therapy used, current dosage, presence of any adverse effects or likelihood of these occurring with increased dosage or combination therapy, underlying disease and the likely pathophysiological mechanism behind the patient's SH as well as practical aspects such as drug availability and dosing regimen. Controlled canine studies are needed to determine if the antihypertensive effects of RAAS inhibitors and CCBs are optimized by combination therapy or monotherapy with individualised dosage escalation. Generally, if the initial drug has been started at the lower end of the dosage range, the dose may be increased by 25–50%, or in the case of ACEIs, administration may be increased from every 24 h to 12 h.

If the dosage of the initial agent is maximised without achieving therapeutic goals or other reasons prevent a dosage escalation, combination therapy is indicated. Most humans with SH require more than one antihypertensive agent, and the same is considered to be true in dogs (Brown et al. 2007), although there are no studies specifically reporting this. Combination therapy provides the benefit of multiple antihypertensive mechanisms and may mitigate adverse effects of a single antihypertensive agent, as discussed with amlodipine and ACEI co-therapy above. Meta-analyses in human medicine have shown that combination therapy results in a larger drop in BP compared to intensified monotherapy (Law 2003; Wald et al. 2009). The most commonly used combination therapy in dogs is amlodipine alongside a RAAS inhibitor. This approach is also recommended at the outset in cases at severe risk of TOD or those with rapidly progressive TOD. ACEI and amlodipine coadministration has been studied in normotensive dogs and appears to provide a modest additive effect in reducing BP (Atkins et al. 2007). Telmisartan and amlodipine co-therapy is effective and well tolerated in hypertensive people (Littlejohn et al. 2009) and has been described in a small number of cases in dogs (Bugbee et al. 2014; Caro-Vadillo et al. 2018). An alternative option is to coadminister an ACEI and an ARB, as either class alone may provide incomplete RAAS blockade (Ames et al. 2016; Konta et al. 2018; Lantis et al. 2015). ACEI and ARB co-therapy is the standard of care in proteinuric human CKD patients. The safety of dual RAAS blockade has been questioned, however (Chrysant 2010; St. Peter et al. 2013), and the European Medicines Agency endorses restrictions on combination ARB and ACEI treatment due to a higher risk of worsening renal function and hyperkalaemia (European Medicines Agency 2014; McAlister et al. 2011; Yusuf and ONTARGET Investigators 2008). Apart from a case report (Bugbee et al. 2014), there are currently no published reports of the efficacy or safety of this approach in dogs. Because of this lack of information, it is more typical to initially administer a CCB alongside a RAAS inhibitor in cases requiring combination therapy. The exception may be dogs that are severely proteinuric, and renal function should be carefully monitored when this therapeutic regimen is pursued, as there are unpublished data of modest increases in serum creatinine (6/11 dogs) and potassium (4/11 dogs) in dogs receiving ARBs alongside ACEIs and/or CCBs (Brown et al. 2013).

13.6.5 Management of Dogs with Refractory Hypertension

Controlling SH in dogs can be challenging; BP control (<160 mmHg) was achieved in only 1/14 dogs with CKD-associated SH treated with a variety of antihypertensive agents (Jacob et al. 2003), and another study reported no difference in BP between hypertensive dogs that were and were not receiving antihypertensive treatment (LeBlanc et al. 2011). Patients that are not sufficiently controlled with first-line combination therapy (RAAS inhibitor and amlodipine) at optimal dosages may be considered refractory, although this situation is not particularly uncommon. The presence or not of (progressive) TOD may affect the decision to escalate therapy in refractory cases, as the risks of further therapy may be difficult to justify in some individuals.

In a refractory case, or where BP decreases with therapy but does not reach target levels, efforts should be made to ensure compliance. If an underlying disease is present, could it be better controlled or require a different treatment approach? In patients considered to have idiopathic SH at diagnosis, the clinician should revisit the possibility of secondary SH. Once compliance is assured and underlying disease is optimally controlled, an alternative antihypertensive agent should be added to the treatment regime or used in substitution for one of the current agents. Options include α- and β-adrenergic antagonists, mineralocorticoid receptor anatagonists (MRAs), hydralazine and diuretics (Table 13.3). There is no evidence or consensus regarding which agent to add next, and the underlying condition, where known/present, should dictate which class is expected to be optimal. If the clinician's experience of these more uncommonly used agents is limited, patients may be hospitalised initially (e.g. for 24–72 h) for monitoring where appropriate. Referral to an internal medicine specialist may also be considered in dogs with refractory SH, or in patients with concurrent illnesses that are complicating its management.

13.6.5.1 Alpha-Adrenergic Receptor Antagonists

α-Adrenergic receptor blockade in peripheral vascular smooth muscle results in vasodilation and reduced SBP. Prazosin and phenoxybenzamine are the α-blocker agents most used in veterinary patients. Prazosin (an $α_1$-selective blocking agent) acts as a peripheral vasodilator (Kellar et al. 2014); intravenous prazosin, but not phenoxybenzamine, significantly reduced BP in healthy dogs (Fischer et al. 2003) and was also effective in dogs with renal-associated SH (Massingham and Hayden 1975). Recommended dosage is 0.5 mg/kg PO q8-12 (Acierno et al. 2018; Haagsman et al. 2013) although more conservative doses are described (1 mg total q8h in dogs weighing <15 kg; 2 mg total q8h in dogs >15 kg) (Ramsey 2012). Some also advocate administering the first dose in the clinic alongside by BP monitoring, as potent acute vasodilation may occur (Haagsman et al. 2013). It is anecdotally reported that dogs with the *ABCB1* mutation (MDR1) may be overly sensitive to prazosin's effect (Plumb 2015). Phenoxybenzamine is most commonly used in hypertensive dogs with pheochromocytoma and is further discussed below. The phenothiazine acepromazine maleate exerts a BP-lowering effect by antagonising α-adrenergic receptors. Oral treatment has been suggested in hypertensive dogs (Acierno et al. 2018), but there are no reports of its chronic use, and excessive sedation may prove problematic.

13.6.5.2 Beta-Adrenergic Receptor Antagonists

β-Blockers such as atenolol and propranolol can reduce BP due to negative inotropic and chronotropic effects and inhibition of renin release. A β-blocker is a logical addition in refractory cases where tachycardia is present. However, their BP-lowering effect may be fairly limited (Henik et al. 2008), and their use should be reserved for patients with normal cardiac function. As a cardioselective β-blocker, atenolol is preferred over non-cardioselective propranolol, as the latter may increase total peripheral resistance and systemic afterload, negating BP-lowering effects. Drugs which reduce sympathetic nerve activity by acting on the central nervous system are thought to have significant advantages over the peripherally acting sympathoinhibitory drugs, and some β-blockers with central effects are being investigated in human medicine, but their use in veterinary medicine is yet to be reported (see Chap. 11).

13.6.5.3 Mineralocorticoid Receptor Antagonists

Mineralocorticoid receptor activation may contribute to hypertension by sodium (and therefore volume) retention and also direct vascular effects of aldosterone including endothelial dysfunction and vasoconstriction. MRAs are increasingly utilised in hypertensive people (Bazoukis et al. 2018; White et al. 2003) and are effective in reducing BP in CKD patients as well as having additional renoprotective actions (Bianchi et al. 2006; Bolignano et al. 2014). Eplerenone was equivalent to amlodipine in its BP-lowering effect in elderly hypertensive patients (White et al. 2003), and spironolactone was superior to an α_1- or β-blocker in hypertensive people already receiving an ACEI, ARB and CCB (Williams et al. 2015). Spironolactone's BP-lowering effect in dogs is poorly established, and its benefit may only be expected if circulating aldosterone levels are increased, such as in rare cases of primary aldosteronism, or with so-called 'aldosterone breakthrough'. The latter occurs due to incomplete angiotensin II blockade and has been documented in healthy dogs treated with ACEIs or ARBs following RAAS activation induced by furosemide (Ames et al. 2015; Lantis et al. 2015) and amlodipine (Ames et al. 2016; Konta et al. 2018). Spironolactone may therefore be considered in persistently hypertensive dogs already receiving a RAAS inhibitor and amlodipine, especially where increased serum aldosterone concentrations are documented. Hyperkalaemia is a potential adverse effect that should be monitored when RAAS inhibitors are combined with MRAs.

13.6.5.4 Hydralazine

Hydralazine is a potent direct arteriole dilator, often administered in an emergent hypertensive crisis due to its rapid effect (Kittleson and Hamlin 1983). Chronic administration has not been evaluated in hypertensive dogs, although its successful use has been described in a case report (Pariser and Berdoulay 2011) and after feline renal transplantation (Kyles et al. 1999). It should be instigated at the lower end of the dosage range (0.5 mg/kg PO q12) due to the risk of symptomatic hypotension and reflex tachycardia. It is generally used instead of amlodipine rather than alongside it due to the risk of hypotension. Hydralazine is typically given with an ACEI due to its RAAS-activating effect and potential to reduce renal perfusion (Massingham and Hayden 1975; Plumb 2015).

13.6.5.5 Diuretics

Although diuretics are frequently prescribed in hypertensive people, they should be reserved for dogs with SH secondary to overhydration or volume overload only. Diuretic therapy may therefore be useful in dogs with SH and concurrent nephrotic syndrome (Brown et al. 2013) or congestive heart failure (CHF). Diuretics should be avoided in hypertensive CKD patients in particular as dehydration and hypovolaemia are particularly deleterious in this population. Suitable agents include spironolactone, hydrochlorothiazide or furosemide. Diuretics should be combined with an ACEI or ARB to blunt their RAAS-activating and potassium-losing effects.

13.6.6 Adverse Effects of Antihypertensive Medications and Their Management

Hypotension is a possible adverse effect of any antihypertensive medication. Regardless of the patient, clinical signs of weakness or syncope alongside SBP <110 mmHg indicate systemic hypotension, and therapy should be adjusted or discontinued. Anecdotally, hypotension is uncommon if the initial diagnosis of SH was correct, and therefore, the clinician should reevaluate the diagnosis if it occurs. Caution should be exercised when using any antihypertensive agent in patients with CKD (especially those in higher IRIS stages) or those prone to dehydration due to the potential risk of reduced renal perfusion. Cessation of antihypertensive medication, especially RAAS inhibitors, is advised at least temporarily should such a patient present acutely unwell.

Reflex tachycardia is a possible sequel to arterial vasodilation, e.g. caused by hydralazine or amlodipine, although amlodipine's relatively slow onset of action may minimise this effect (Massingham and Hayden 1975). Atenolol administration may be considered in dogs with reflex tachycardia, especially if it is thought to be contributing to ineffective BP control. However, as it may occur as compensatory mechanism for hypotension, the treatment of reflex tachycardia should be carefully considered and not overly aggressive. There is also increased risk of bradycardia, severe hypotension, heart failure and atrioventricular block when CCBs are used with β-blockers (Dodd et al. 1989). Other adverse effects of amlodipine are rare, but include lethargy, inappetence, peripheral oedema (Creevy et al. 2013) and gingival hyperplasia (8.5% prevalence in one case series) (Pariser and Berdoulay 2011; Thomason et al. 2009); these effects resolve with discontinuation. As amlodipine undergoes hepatic metabolism, dosage should be reduced in dogs with significant hepatic dysfunction (Ramsey 2012).

Adverse effects of ACEIs usually centre around lethargy and gastrointestinal signs like anorexia, vomiting and diarrhoea. Due to their potential to reduce GFR, there is concern over RAAS inhibitors inducing or exacerbating azotaemia. Studies in dogs show that GFR is maintained following ACEI administration (Brown et al. 2003, 1993; Tenhündfeld et al. 2009), however, and whilst ACEI therapy commonly causes mild increases in serum creatinine concentrations, acute azotaemia or clinically relevant deterioration in renal function is rare in well-hydrated patients (Atkins et al. 2002, 2007; Brown et al. 2003; Grauer et al. 2000; Mishina and Watanabe 2008; Woodfield 1995). Similarly, hyperkalaemia is a possible adverse effect of ACEIs, but clinically relevant increases are uncommon (Tenhündfeld et al. 2009). When used for the treatment of canine proteinuria due to glomerular disease, ACEI or ARB treatment modification is not indicated when serum creatinine increases by <25% in patients with IRIS stage 1–2 and early stage 3 CKD (Brown et al. 2013), and a similar approach may be adopted in hypertensive dogs. Dosage adjustments are sensible with serum creatinine increases in dogs with later stage 3 and 4 CKD as clinical consequences may be more likely. Adverse effects of ARBs in humans include headache, dizziness and rarely angioedema, deterioration in renal function and hyperkalaemia (Axelsson et al. 2015). No adverse effects are reported with canine use to date (Bugbee et al. 2014; Caro-Vadillo et al. 2018; Konta et al. 2018; Schierok et al. 2001). In cats with CKD, progression of renal dysfunction was reported in only 1.8% cases treated with telmisartan (Sent et al. 2015). Hypotension was documented in 2% hypertensive cats receiving telmisartan (Glaus et al. 2019).

Potential adverse effects of α_1-adrenoreceptor antagonists are many and include hypotension, tachycardia, weakness, vomiting, increased intraocular pressure, anorexia and urinary incontinence. Lethargy was the only reported adverse effect of prazosin in a long-term study in dogs with vesicourethral reflex dyssynergia, although BP was not monitored to assess for hypotension (Haagsman et al. 2013).

13.7 Management of a Hypertensive Crisis

Acute TOD in veterinary patients is most likely to manifest as ocular or neurological signs, and the presence of severe TOD in these body systems indicates a hypertensive crisis, and emergency management is justified. Less common situations requiring aggressive emergency antihypertensive treatment are rapidly progressive acute kidney injury (AKI) or intoxication with a BP-increasing agent. If TOD is not present but the patient is at severe risk of its development (SBP ≥ 180 mmHg), immediate treatment is indicated, but oral therapy instigated as outlined above is suitable. There are no published veterinary studies investigating therapeutic strategies for acute hypertensive crises, and recommendations are largely anecdotal or based on human evidence. As emergent hypertensive cases usually require hospitalisation for treatment, as well as 24 h care and monitoring, referral to a facility where this is possible should be considered early on. Parenteral antihypertensive treatment (Table 13.4) may be needed as it is more rapid in onset and titratable although management with oral agents can be successful . As parenteral agents can be very potent, monitoring of direct (invasive) arterial pressure is desirable. The goal in a hypertensive crisis is to rapidly lower (but not necessarily normalise) BP over several hours to days, without instigating hypotension or other adverse effects, and to relieve acute clinical signs and prevent further TOD (Varon and Marik 2000). BP should be decreased incrementally, for example, by 10% in the first hour and then another 15% in the following 4–6 h (Elliott 2003). Once in the range of 160 to 180 mmHg for 48–72 h, further reduction (usually with oral medication by this stage) may be pursued. Rapid restoration of BP to physiological levels (<140 mmHg) is not desirable because of the risk of inadequate cerebral perfusion in patients that have been chronically hypertensive (Barry 1985; Fujishima et al. 2008).

Hypertensive encephalopathy may be successfully treated with antihypertensive therapy and supportive treatment alone, and although some dogs in one case series also received corticosteroids, there is no evidence to support their use in hypertensive encephalopathy (O'Neill et al., 2013b). Prognosis may depend on underlying disease (Lowrie et al. 2012). In some cases, it may be challenging to differentiate between hypertensive encephalopathy and SH secondary to an acute increase in intracranial pressure, known as the Cushing reflex. This phenomenon occurs due to

Table 13.4 Drugs which may be used for management of a hypertensive crisis

Drug class	Drug	Recommended dosage	References
Dopamine-1 receptor agonist	Fenoldopam	0.1 µg/kg/min CRI with incremental increases of 0.1 µg/kg/min every 10–15 mins (maximum 1.6 µg/kg/min) IV	Acierno et al. (2018), Varon and Marik (2000)
		0.8 µg/kg/min CRI	Bloom et al. (2012), Nielsen et al. (2015)
Direct vasodilator	Sodium nitroprusside	0.5 µg/kg/min CRI titrated upwards every 3–5 mins (maximum 3.5 µg/kg/min) until target BP is achieved	Acierno et al. (2018), Plumb (2015)
		1–2 µg/kg/min CRI then titrate upwards until target BP achieved	Ross (2011)
		0.1–8 µg/kg/min CRI	Kyles et al. (2003)
	Hydralazine	0.5–3 mg/kg q8–12h PO	Acierno et al. (2018)
		0.5–3 mg/kg q12h IV	Plumb (2015), Ross (2011)
		0.1 mg/kg loading dose IV, then 1.5–5 µg/kg/min CRI	Acierno et al. (2018), Ross (2011)
Calcium channel blocker (CCB)	Amlodipine	0.1–0.25 mg/kg q24h PO (maximum 0.75 mg/kg)	Acierno et al. (2018), Brown et al. (2013)
Angiotensin-converting enzyme inhibitor (ACEI)	Enalaprilat	0.2 mg/kg IV q1–2h	Brown et al. (2007)
α-Adrenergic antagonist	Phentolamine (intraoperatively for pheochromocytoma)	0.1 mg/kg loading dose IV, 1–2 µg/kg/min CRI	Kyles et al. (2003)
	Acepromazine	0.05–0.1 mg/kg IV	Pascoe et al. (1994)
β-Blocker	Labetalol	0.25 mg/kg IV over 2 mins repeated up to a maximum 3.75 mg/kg followed by 25 µ/kg/min CRI	Acierno et al. (2018)
	Esmolol	50–75 µg/kg CRI	Acierno et al. (2018)
Diuretics	Hydrochlorothiazide	2–4 mg/kg q12–24 h IV	Acierno et al. (2018)
	Furosemide	1–4 mg/kg q8–12h IV	Acierno et al. (2018)

sympathetic activation secondary to increased intracranial pressure and ischaemia. Bradycardia and obtundation are usual features of the Cushing reflex. As appropriate antihypertensive treatment leads to rapid reversal of CNS signs (within hours to days) in cases of hypertensive encephalopathy (O'Neill et al., 2013b), response to treatment, along with case history and other clinical findings including ocular lesions indicative of hypertensive injury, may help in reaching a diagnosis. Brain MRI may also be useful if the patient is a suitable anaesthetic candidate (Lowrie et al. 2012; O'Neill et al., 2013b).

No specific treatment is indicated for hypertensive ocular lesions unless severe hyphaema is present, in which case topical corticosteroids such as 1% prednisolone acetate or 0.1% dexamethasone are indicated two to four times daily until resolution of hyphaema occurs. However, systemic absorption of ocular corticosteroids may exacerbate SH. The prognosis for return of vision may depend on lesion severity and successful management of SH, although successful BP control does not necessarily resolve ocular abnormalities, especially when injury is chronic (LeBlanc et al. 2011; Maggio et al. 2000).

The first-line treatment recommended for dogs presenting as a hypertensive emergency where oral administration of medication is not possible is fenoldopam, a selective dopamine-1-receptor agonist (Acierno et al. 2018). It causes peripheral artery vasodilation and is used in acutely hypertensive humans (Murphy et al. 2001; Varon and Marik 2000). Its appeal lies in its apparent ability to maintain renal perfusion, unlike many antihypertensive agents; fenoldopam increased renal blood flow, GFR and fractional sodium excretion in healthy dogs but had no effect on BP (Aronson et al. 1990; Bloom et al. 2012; Kelly et al. 2016). Fenoldopam has been retrospectively studied in dogs and cats with AKI but has not been specifically investigated in hypertensive patients (Nielsen et al. 2015). In Nielsen et al.'s (2015) study, hypotension (defined as SBP <90 mmHg) prevalence (7%) was not different between dogs which did and did not receive fenoldopam. It is unknown if any animals in this study were hypertensive prior to treatment. Fenoldopam infusion should be initiated at 0.1 µg/kg/min, increasing by 0.1 µg/kg/min every 15 min (maximum 1.6 µg/kg/min) until the desired BP is achieved (Acierno et al. 2018; Varon and Marik 2000). As the plasma half-life is short, effects diminish within minutes of discontinuation (Bloom et al. 2012).

Sodium nitroprusside is an alternative option if fenoldopam is unavailable or ineffective. As a nonselective arteriolar and venous vasodilator, it is a potent antihypertensive agent which may cause reduced renal blood flow (Aronson et al. 1990) and increased intracranial pressure (Kondo et al. 1984). Infusion should be started at 0.5 µg/kg/min and titrated upwards (maximum 10 µg/kg/min) every 3–5 min until target BP is achieved (Hunter et al. 2003; Kyles et al. 2003). Like fenoldopam, its use requires constant or near-constant BP and heart rate monitoring, and discontinuation results in abrupt cessation of effect (Hunter et al. 2003). The infusion line should be protected from light. Sodium nitroprusside should be administered for no longer than 24 h to avoid potential thiocyanate and cyanide toxicity, which may manifest as methaemoglobinaemia and lactic acidosis. Due to its deleterious effect on renal perfusion, it should be avoided in dogs with renal dysfunction. Considering the risk of toxicity, sodium nitroprusside use is not recommended in human medicine unless there is lack of alternatives (Elliott 2003; Varon and Marik 2000).

If fenoldopam or sodium nitroprusside are unavailable or deemed too labour-intensive or novel, oral or intravenous hydralazine is recommended (Tables 13.3 and 13.4). Subcutaneous hydralazine administration has also been described in cats

(Kyles et al. 1999). Oral hydralazine has a rapid onset of action (within an hour), with maximum effects in 3–5 h (Kittleson and Hamlin 1983). Dosing should start at the lower end of the range as it cannot be abruptly discontinued should hypotension occur. Hydralazine is not recommended in the management of human hypertensive emergencies due to the inability to effectively titrate its effects (Varon and Marik 2000); although prolonged effects are demonstrated in dogs (Kittleson and Hamlin 1983), it is still considered a useful drug in this scenario due to the lack of readily available alternatives. Acepromazine has been proposed as a cheap and widely accessible alternative in an emergent hypertensive crisis. Repeated small intravenous boluses may be given to effect (Pascoe et al. 1994). Its use carries a risk of bradycardia, atrioventricular block and reflex tachycardia (ECG monitoring is advised), and sedative effects may make monitoring of neurological patients challenging. Enalaprilat or esmolol may also be considered in hypertensive crises (Table 13.4). Finally, if the clinician does not have access to any of the aforementioned treatments, amlodipine (at high dosages with caution) may be trialled (Geigy et al. 2011). Rectal amlodipine administration has been described in patients not amenable to oral therapy (Geigy et al. 2011).

Once BP is seen to be successfully decreasing with parenteral treatment, oral medication with amlodipine, RAAS inhibitors or both may be added if the dog is not already receiving them. When BP has reached the target range (160–180 mmHg), the parenteral agent can be tapered and hopefully discontinued. Dogs may be discharged once BP and clinical signs have stabilised and no complications (e.g. reduced renal function) have been observed. Repeat assessment should be performed in 1–3 days, with the aim to reduce BP to <140 mmHg gradually over the subsequent weeks.

13.8 Management of SH in Specific Conditions

13.8.1 Renal Disease

13.8.1.1 Chronic Kidney Disease

CKD is likely the most common cause of canine SH, hence the most widely used antihypertensive treatment in dogs being ACEIs due to the added beneficial antiproteinuric effects of these agents. Prevalence of SH in canine CKD varies depending on the study and frequency of glomerular disease (where it is higher) (Bacic et al. 2010; Buranakarl et al. 2007; Cook and Cowgill 2014; Cortadellas et al. 2006; Guess et al. 2018; Jacob et al. 2003; O'Neill et al. 2013a), but it is estimated that approximately 20% of dogs are hypertensive at CKD diagnosis with an additional 10–20% becoming hypertensive over time (IRIS 2017). Every dog with CKD should be IRIS-staged based on serum creatinine concentrations and substaged depending on the presence of SH and proteinuria (IRIS 2017). Dogs with CKD are more likely to be proteinuric than cats. Circulating creatinine levels are not directly correlated to BP (Buranakarl et al. 2007; Cortadellas et al. 2006), and SH may be present, and should be treated, in all IRIS stages (1–4).

SH is positively associated with severity of histological renal lesions (Finco 2004; Kang et al. 2018) and proteinuria (Bacic et al. 2010; Jacob et al. 2005; Wehner et al. 2008); proteinuria in turn promotes renal injury progression and is associated with adverse outcomes in dogs (Grauer et al. 2000; Grodecki et al. 1997; Jacob et al. 2005; Rudinsky et al. 2018; Wehner et al. 2008). Although there is clearly a relationship between SH, proteinuria and adverse outcomes, there are limited data specifically reporting the benefit of BP control in CKD (Finco 2004; Kang et al. 2018), and the optimal treatment of canine renal hypertension has not been established. Furthermore, as TOD in the kidney manifests as a progressive decline in GFR, it can be challenging to understand if progressive renal dysfunction is due to uncontrolled SH or progression of renal disease itself. SBP was associated with the risk of developing a uremic crisis and death in CKD dogs (Jacob et al. 2003). Although this was probably because the dogs with the highest BP tended to be the most proteinuric (Jacob et al. 2005), it highlights the importance of successful antihypertensive treatment.

As discussed in Chap. 1, the pathophysiology of CKD-induced SH is likely multifactorial, involving RAAS-activation, reduced sodium excretion, increased intravascular volume, sympathetic stimulation, structural arterial changes and endothelial dysfunction. Although RAAS inhibitors have not been systematically evaluated in naturally occurring canine CKD and studies in cats have shown inconsistent results, ACEIs are considered the first-line therapy in hypertensive renal disease patients (Acierno et al. 2018; IRIS 2017). Alongside BP-reducing effects they reduce proteinuria and improve patient outcome in dogs with experimentally induced and naturally occurring CKD (Brown et al. 2003; Tenhündfeld et al. 2009), hereditary nephritis (Grodecki et al. 1997) and glomerulonephritis (Grauer et al. 2000). They may also help to maintain effective renal blood flow (Grodecki et al. 1997). A reduced ACEI starting dose (0.25 mg/kg q24h) is recommended in azotaemic dogs, escalating to the full standard dose in 10–14 days if renal function remains stable. Dogs with glomerular disease will likely already be receiving RAAS inhibitors to reduce proteinuria, but if not and SBP $\geq$160 mmHg, institution of an ACEI or ARB is appropriate (Brown et al. 2013). BP values only tend to be modestly reduced with ACEI therapy in renal-associated SH, however (Brown et al. 2003; 1993; Mishina and Watanabe 2008; Tenhündfeld et al. 2009), and combination therapy may be required. No systematic studies have evaluated ARBs in dogs with SH and/or proteinuric CKD, but there are individual and anecdotal reports of telmisartan's efficacy (Bugbee et al. 2014; Caro-Vadillo et al. 2018). As discussed above, caution is warranted when administering a RAAS inhibitor to a dog in late IRIS CKD stages 3 and 4. Optimisation of hydration status and stabilisation of azotaemia are advised before starting RAAS inhibitors. Dual therapy with an ACEI and ARB may be considered in some cases, especially those with refractory proteinuria. As there is increasing evidence that aldosterone plays an important role in mediating SH and kidney injury, spironolactone may also be considered. When amlodipine is used in patients with CKD, it should be used alongside RAAS inhibition (Atkins et al. 2007; Brown et al. 2013). As it is not uncommon for CKD patients to experience a gradual loss of muscle mass and lean body weight, antihypertensive drug dosages should be decreased accordingly, as long as BP remains controlled.

The therapeutic goals in CKD are the same as those previously discussed. In one study of 45 dogs with naturally occurring CKD, there appeared to be a continuum between BP and median survival time (MST); MST of dogs in the high-BP (>161 mmHg), medium-BP (144–160 mm Hg) and low-BP (<144 mmHg) groups were 154, 348 and 425 days, respectively (Jacob et al. 2003). Survival times were significantly different between the high- and low-BP groups; this reflects human guidelines where it is recommended to achieve a BP as low as possible (whilst avoiding hypotension). Indeed, experimental evidence suggests that BP in the prehypertensive category (BP 140–159 mmHg) may contribute to renal injury in some dogs (Finco 2004). Consequently, antihypertensive therapy may be considered in prehypertensive patients that are proteinuric or have rapidly progressive renal disease. Due to the prognostic significance of proteinuria, monitoring UPC in renal patients is very important, with the goal being to reduce UPC to the non-proteinuric range (<0.2) or by at least 50%.

EPO-stimulating agents are commonly used to treat anaemia in CKD patients. As they are associated with SH (Cowgill et al. 1998; Fiocchi et al. 2017; Randolph et al. 2004), extra vigilance and awareness are needed when using these agents in hypertensive patients or those already at risk of SH due to their renal disease. Subcutaneous fluid therapy (e.g. performed at home by owners) is another adjunctive treatment which has potential to affect BP in CKD patients. Although dogs are more robust to volume overload than cats, it is advisable to reserve subcutaneous fluid therapy for patients that become dehydrated without such treatment (Syme 2011).

13.8.1.2 Acute Kidney Disease

SH is common in acute AKI, affecting around one-third of dogs at admission, increasing to 81% during hospitalisation (Geigy et al. 2011). As with CKD, SH can occur at any AKI grade and should be treated when it is identified due to possible exacerbation of renal injury. The safety concerns related to preservation of renal function during treatment of hypertensive dogs with AKI remain to be studied prospectively or experimentally. Care must be taken to avoid precipitous drops in BP due to the risk of renal hypoperfusion. RAAS inhibitors are generally avoided in critically ill dogs with AKI due to their potential in reducing renal perfusion and GFR. Parenteral antihypertensive therapy is often warranted in AKI patients, as hypertension is often severe (Geigy et al. 2011) and oral therapy may be precluded by vomiting and anorexia. This said, amlodipine is usually the first-line antihypertensive therapy employed in hypertensive AKI patients. Its use has been described in dogs with AKI (0.25 mg/kg q1-3 h until SBP 140–160 mmHg was reached or up to a maximal cumulated dose of 1 mg/kg/day) and may be effective within 24–48 h (Geigy et al. 2011). Although amlodipine successfully corrected severe hypertension in 10/11 dogs, it was associated with reduced survival in this retrospective study, likely because of greater AKI severity in the dogs that received amlodipine (Geigy et al. 2011). Prospective placebo-controlled studies are warranted to evaluate for potential negative effects of amlodipine in this population, however, and close monitoring of renal function is advised. Fenoldopam is the recommended parenteral therapy as it appears to be safe in dogs with AKI and may have additional renoprotective properties (Nielsen et al. 2015). Other agents should be used with great care due to the risk of renal hypoperfusion. Overhydration may exacerbate SH in AKI patients, and care should be taken to avoid discrepancies between intravenous fluid therapy and urine output so that progressive volume loading does not occur. When volume overload is suspected, fluid therapy should be reduced or stopped and diuretic therapy, e.g. furosemide, may be cautiously employed. Dialysis may be indicated in oliguric or anuric dogs, particularly those with volume overload where ultrafiltration for fluid removal may be beneficial in the management of SH.

13.8.1.3 Other Renal Diseases

SH is common (61.5% prevalence in one study) in dogs with glomerular disease associated with leishmaniasis (Cortadellas et al. 2006). SH is also reported in dogs with Lyme nephropathy, but its frequency is uncertain (Littman et al. 2018). Where present, specific treatment of proteinuria and SH is indicated alongside antiprotozoal/antibiotic therapy.

13.8.2 Hyperadrenocorticism

SH is a common feature of dogs with untreated hyperadrenocorticism (Chen et al. 2016; Goy-Thollot et al. 2002; Lien et al. 2010; Ortega et al. 1996), being most prevalent in dogs with unilateral adrenal tumours in this population (Ortega et al. 1996; Reusch et al. 2010). Importantly, SH persists in many patients after hyperadrenocorticism treatment [40% in one study (Ortega et al. 1996)]; other studies have not reported the specific prevalence of persistent SH but report median BP values similar to baseline (Smets et al. 2012a) and significantly higher values in treated dogs compared to healthy controls (Goy-Thollot et al. 2002). The mechanism behind SH in HAC is incompletely understood and is likely multifactorial (see Chap. 4); in humans with hyperadrenocorticism, decreased nitric oxide production and/or increased circulating aldosterone is implicated. As dogs with adrenocortical tumours have significantly higher circulating plasma aldosterone concentrations compared to healthy dogs and those with pituitary-dependent hyperadrenocorticism (Javadi et al. 2005), MRAs have been suggested for SH treatment in this population and in any dog with hyperadrenocorticism that remain hypertensive despite treatment of their endocrinopathy (Ames et al. 2019). However, until further evidence is available, empirical treatment with an ACEI and/or amlodipine alongside specific medical or surgical treatment for hyperadrenocorticism may be trialled. ACEIs or ARBs may be particularly suitable as subclinical renal dysfunction may be present in hyperadrenocorticism (Smets et al. 2012a, b). As proteinuria severity is not correlated with SH in dogs with hyperadrenocorticism (Chen et al. 2016; Lien et al. 2010), UPC may not be a useful surrogate marker of antihypertensive treatment efficacy. SH is also commonly reported in dogs with sudden-acquired retinal degeneration syndrome (SARDs) (Carter et al. 2014). The aetiology of SARDs is unknown but is thought to involve adrenocorticoid hormonal dysfunction.

13.8.3 Pheochromocytoma

Pheochromocytoma results in SH due to excessive catecholamine secretion and α- (and to a lesser extent β-) adrenergic receptor stimulation (see Chap. 4). SH may be sustained or variable/paroxysmal, and around half of dogs with pheochromocytoma experience SH at some stage in their disease (Barthez et al. 1997; Gilson et al. 1994; Herrera et al. 2008). Nevertheless, its prevalence and response to treatment are poorly reported in many studies. The α-adrenoceptor antagonist phenoxybenzamine is the first-line treatment for the clinical consequences of pheochromocytoma, including SH, prior to adrenalectomy or in dogs where surgery is not possible. The optimal dosage and treatment duration prior to surgery is not established; however, the starting dosage of 0.25 mg/kg q12 h increased stepwise until the final dosage of 1 mg/kg is reached (or adverse effects are seen) for 2–3 weeks is commonly employed (Herrera et al. 2008; Kyles et al. 2003). Phenoxybenzamine improved survival despite a similar pattern of peri- and intraoperative BP variability in one retrospective study (Herrera et al. 2008). Prazosin is an alternative selective $α_1$-adrenoceptor antagonist, but its use in pheochromocytoma has not been described. Cases refractory to α-blockade and those with concerning tachyarrhythmias should receive β-blockade (Acierno et al. 2018). β-Blockers should not be used alone as loss of β-adrenergic-mediated vasodilation can result in profound hypertension in the presence of unopposed $α_1$ effects (Ramsey 2012; Varon and Marik 2000).

Potential perioperative complications are numerous in pheochromocytoma, and close patient monitoring is critical for a successful outcome (Herrera et al. 2008). Sodium nitroprusside and the short-acting α-adrenergic blocker phentolamine (loading dose 0.1 mg/kg then 1–2 μg/kg/min IV) have been described in successful management of intraoperative SH during pheochromocytoma resection (Kyles et al. 2003; Lang et al. 2011). Acute hypotension may occur after tumour excision in the continued presence of phenoxybenzamine. As phenoxybenzamine is long-acting (recovery is dependent on synthesis of new receptors as it alkylates the receptors), it should be discontinued postoperatively once BP normalises. In humans, increased preoperative fluid and sodium intake is endorsed to prevent postoperative hypotension (Lenders et al. 2014), but there is no evidence for this approach in dogs.

13.8.4 Diabetes Mellitus

SH is recognised in up to half of canine diabetics (Herring et al. 2014; Struble et al. 1998), although renal function has not been investigated in all studies. Conflicting results exist regarding the effect of disease duration on SH prevalence in diabetic dogs, although glycaemic control does not appear to be associated with SH (Herring et al. 2014; Struble et al. 1998). In most cases, SH tends to be fairly mild (SBP <160 mmHg). In diabetic people, SH is thought to occur mainly due to diabetic nephropathy in type I disease, whereas a range of factors including obesity, insulin resistance and hyperlipidaemia are implicated in type 2 diabetes mellitus. Antihypertensive therapy in diabetic dogs is poorly described but largely involves RAAS inhibition, amlodipine or both. ACEI and CCB co-therapy reduced BP in dogs with experimentally induced diabetes mellitus, and combination therapy had a greater effect on reducing proteinuria than either agent alone (Brown et al. 1993).

13.8.5 Hyperaldosteronism

Idiopathic primary hyperaldosteronism (Breitschwerdt et al. 1985) and aldosterone-secreting tumours (Behrend et al. 2005; Johnson et al. 2006) are very rarely described in dogs (see Chap. 4). Two case reports also describe confirmed or suspected deoxycorticosterone-secreting (aldosterone precursor) adrenal tumours; SH was diagnosed in both dogs and resolved following tumour excision (Gójska-Zygner et al. 2012; Reine et al. 1999). SH associated with feline hyperaldosteronism is also expected to resolve following adrenalectomy (Reusch et al. 2010). Spironolactone is the agent of choice for stabilisation prior to surgery or in cases where surgery is not pursued, although additional antihypertensive treatments are likely to be required to gain sufficient BP control (Ash et al. 2005; Gójska-Zygner et al. 2012).

13.8.6 Thyroid Disease

Although rare, hyperthyroidism does occur in dogs and SH has been reported in this condition. SH secondary to thyroid gland carcinoma resolved following thyroidectomy in one reported case (Simpson and McCown 2009). β-Blockers are ineffective antihypertensive agents when used alone in hyperthyroid cats (Henik et al. 2008), and empirical antihypertensive treatment alongside specific therapy for the thyroid disease is advised. SH has been described in hypothyroid dogs (Gwin et al. 1978; O'Neill et al. 2013b) although a causal relationship is not established and effective treatment in this population is unknown.

13.8.7 Cardiovascular Disease

Unlike in people, cardiovascular disease, including CHF, rarely results in SH in dogs (Petit et al. 2013). When SH does occur in CHF, it is likely to be due to circulatory volume overload, and therefore, treatment for CHF (primarily with diuretics) should be prioritised. Amlodipine therapy has been described in dogs with degenerative mitral valvular disease, following the hypothesis that peripheral arteriolar dilatation sufficient to cause reduced SBP will decrease the volume of mitral regurgitation (Thomason et al. 2009). There is no evidence supporting this theory, and great care is required when using antihypertensive treatment in dogs with cardiac insufficiency as reducing peripheral resistance may lead to a precipitous drop in cardiac output.

13.8.8 Circulatory Volume Overload

SH may occur following excessive intravenous fluid administration. In these cases, fluid therapy should be reduced or stopped, and diuretics may be administered.

13.9 Potential Future Therapeutic Options

There are numerous novel antihypertensive agents used in human medicine which may be potential future options for veterinary patients. Carvedilol is a third-generation non-cardioselective β-blocker used in hypertensive people. It decreases BP by vascular α_1-receptor blockade, but cardiac output and heart rate are maintained (Stafylas and Sarafidis 2008). Its utility in hypertensive dogs has yet to be established; high-dose (0.4 mg/kg) carvedilol (0.4 mg/kg) decreased heart rate and BP but affected renal function in dogs with experimentally induced valvular disease (Uechi et al. 2002). Direct renin inhibitors (e.g. aliskiren) provide another mechanism of RAAS inhibition but have not been used to any great extent in veterinary patients. Newer classes of CCB which target not only L-type but also N- or T-type calcium channels (e.g. efonidipine) are available in human medicine and have superior renoprotective benefits compared to amlodipine, as they more effectively dilate the glomerular efferent arteriole and may not induce RAAS activation (Hayashi et al. 2010; Honda et al. 2001). Whilst these newer agents hold promise, they have not been evaluated in dogs in the clinical setting.

13.10 Summary

Antihypertensive treatment is indicated in dogs with SH, regardless of its aetiology, in order to reduce the risk of TOD. Treatment in most cases is likely to involve RAAS blockade +/− CCBs, although alternative agents will be appropriate in specific disease conditions and may prove useful in refractory cases. In general, treatment is stepwise and aims to reduce BP relatively gradually. Close monitoring of hypertensive dogs is important as achieving therapeutic goals may improve outcome. Substantial gaps in our knowledge regarding optimum therapy and treatment objectives remain. There is a recognised need for prospective controlled clinical studies which examine treatment efficacy and effect on quality of life, as well as potential adverse effects on renal function and overall survival.

References

Acierno MJ, Brown S, Coleman AE et al (2018) ACVIM consensus statement: guidelines for the identification, evaluation, and management of systemic hypertension in dogs and cats. J Vet Intern Med 32:1803–1822

Ames MK, Atkins CE, Lee S et al (2015) Effects of high doses of enalapril and benazepril on the pharmacologically activated renin-angiotensin-aldosterone system in clinically normal dogs. Am J Vet Res 76:1041–1050

Ames MK, Atkins CE, Lantis AC et al (2016) Evaluation of subacute change in RAAS activity (as indicated by urinary aldosterone: creatinine, after pharmacologic provocation) and the response to ACE inhibition. J Renin-Angiotensin-Aldosterone Syst 17:1–12

Ames MK, Atkins CE, Pitt B (2019) The renin-angiotensin-aldosterone system and its suppression. J Vet Intern Med 33:363–382

Aronson S, Goldberg LI, Roth S et al (1990) Preservation of renal blood flow during hypotension induced with fenoldopam in dogs. Can J Anaesth 37:380–384

Ash RA, Harvey AM, Tasker S (2005) Primary hyperaldosteronism in the cat: a series of 13 cases. J Feline Med Surg 7:173–182

Atkins CE, Brown WA, Coats JR et al (2002) Effects of long-term administration of enalapril on clinical indicators of renal function in dogs with compensated mitral regurgitation. J Am Vet Med Assoc 221:654–658

Atkins CE, Rausch WP, Gardner SY et al (2007) The effect of amlodipine and the combination of amlodipine and enalapril on the renin-angiotensin-aldosterone system in the dog. J Vet Pharmacol Ther 30:394–400

Axelsson A, Iversen K, Vejlstrup N et al (2015) Efficacy and safety of the angiotensin II receptor blocker losartan for hypertrophic cardiomyopathy: the INHERIT randomised, double-blind, placebo-controlled trial. Lancet Diabetes Endocrinol 3:123–131

Bacic A, Kogika MM, Barbaro KC et al (2010) Evaluation of albuminuria and its relationship with blood pressure in dogs with chronic kidney disease. Vet Clin Pathol 39:203–209

Barry DI (1985) Cerebral blood flow in hypertension. J Cardiovasc Pharmacol 7:S94–S98

Barthez PY, Marks SL, Woo J et al (1997) Pheochromocytoma in dogs: 61 cases (1984-1995). J Vet Intern Med 11:272–278

Bazoukis G, Thomopoulos C, Tsioufis C (2018) Effect of mineralocorticoid antagonists on blood pressure lowering: overview and meta-analysis of randomized controlled trials in hypertension. J Hypertens 36:987–994

Behrend EN, Weigand CM, Whitley EM et al (2005) Corticosterone – and aldosterone-secreting adrenocortical tumor in a dog. J Am Vet Med Assoc 226:1662–1666

Bianchi S, Bigazzi R, Campese VM (2006) Long-term effects of spironolactone on proteinuria and kidney function in patients with chronic kidney disease. Kidney Int 70:2116–2123

Bijsmans ES, Doig M, Jepson RE et al (2016) Factors influencing the relationship between the dose of amlodipine required for blood pressure control and change in blood pressure in hypertensive cats. J Vet Intern Med 30:1630–1636

Bischoff K, Beier E, Edwards WC (1998) Methamphetamine poisoning in three Oklahoma dogs. Vet Hum Toxicol 40:19–20

Bloom CA, Labato MA, Hazarika S et al (2012) Preliminary pharmacokinetics and cardiovascular effects of fenoldopam continuous rate infusion in six healthy dogs. J Vet Pharmacol Ther 35:224–230

Bodey AR, Michell AR (1996) Epidemiological study of blood pressure in domestic dogs. J Small Anim Pract 37:116–125

Bodey AR, Sansom J (1998) Epidemiological study of blood pressure in domestic cats. J Small Anim Pract 39:567–573

Bolignano D, Palmer SC, Navaneethan SD et al (2014) Aldosterone antagonists for preventing the progression of chronic kidney disease. Cochrane Database Syst Rev 4:CD007004

Bovee KC, Littman MP, Saleh F et al (1986) Essential hereditary hypertension in dogs: a new animal model. J Hypertens Suppl 4:S172

Bovee KC, Littman MP, Crabtree BJ et al (1989) Essential hypertension in a dog. J Am Vet Med Assoc 195:81–86

Breitschwerdt EB, Meuten DJ, Greenfield CL et al (1985) Idiopathic hyperaldosteronism in a dog. J Am Vet Med Assoc 187:841–845

Brown SA, Henik RA (1998) Diagnosis and treatment of systemic hypertension. Vet Clin North Am Small Anim Pract 28:1481–1494

Brown SA, Walton CL, Crawford P et al (1993) Long-term effects of antihypertensive regimens on renal hemodynamics and proteinuria. Kidney Int 43:1210–1218

Brown SA, Finco DR, Brown CA et al (2003) Evaluation of the effects of inhibition of angiotensin converting enzyme with enalapril in dogs with induced chronic renal insufficiency. Am J Vet Res 64:321–327

Brown S, Atkins C, Bagley R et al (2007) Guidelines for the identification, evaluation, and management of systemic hypertension in dogs and cats. J Vet Intern Med 21:542–558

Brown S, Elliott J, Francey T et al (2013) Consensus recommendations for standard therapy of glomerular disease in dogs. J Vet Intern Med 27:S27–S43

Bugbee AC, Coleman AE, Wang A et al (2014) Telmisartan treatment of refractory proteinuria in a dog. J Vet Intern Med 28:1871–1874

Buranakarl C, Ankanaporn K, Thammacharoen S et al (2007) Relationships between degree of azotaemia and blood pressure, urinary protein: creatinine ratio and fractional excretion of electrolytes in dogs with renal azotaemia. Vet Res Commun 31:245–257

Burges RA, Dodd MG, Gardiner DG (1989) Pharmacologic profile of amlodipine. Am J Cardiol 64:10–18

Carlucci L, Song KH, Yun HI et al (2013) Pharmacokinetics and pharmacodynamics (PK/PD) of irbesartan in beagle dogs after oral administration at two dose rates. Pol J Vet Sci 16:555–561

Caro-Vadillo A, Daza-González MA, Gonzalez-Alonso-Alegre E et al (2018) Effect of a combination of telmisartan and amlodipine in hypertensive dogs. Vet Rec Case Rep 6:e000471

Carter RT, Oliver JW, Stepien RL et al (2014) Elevations in sex hormones in dogs with sudden acquired retinal degeneration syndrome (SARDS). J Am Anim Hosp Assoc 45:207–214

Chen H-Y, Lien Y-H, Huang H-P (2016) Association of renal resistive index, renal pulsatility index, systemic hypertension, and albuminuria with survival in dogs with pituitary-dependent hyperadrenocorticism. Int J Endocrinol 2016:3814034

Christ D, Wong P, Wong Y et al (1994) The pharmacokinetics and pharmacodynamics of the angiotensin II receptor antagonist losartan potassium (DuP 753/MK 954) in the dog. J Pharmacol Exp Ther 268:1199–2005

Chrysant SG (2010) Current status of dual renin angiotensin aldosterone system blockade for the treatment of cardiovascular diseases. Am J Cardiol 105:849–852

Cohn LA, Dodam JR, Szladovits B (2002) Effects of selegiline, phenylpropanolamine, or a combination of both on physiologic and behavioral variables in healthy dogs. Am J Vet Res 63:827–832

Conroy M, Chang YM, Brodbelt D et al (2018) Survival after diagnosis of hypertension in cats attending primary care practice in the United Kingdom. J Vet Intern Med 32:1846–1855

Cook A, Cowgill L (2014) Clinical and pathological features of protein-losing glomerular disease in the dog: a review of 137 cases (1985-1992). J Am Anim Hosp Assoc 32:313–322

Cortadellas O, Del Palacio MJF, Bayón A et al (2006) Systemic hypertension in dogs with leishmaniasis: prevalence and clinical consequences. J Vet Intern Med 20:941–947

Cowgill LD, James KM, Levy JK et al (1998) Use of recombinant human erythropoietin for management of anemia in dogs and cats with renal failure. J Am Vet Med Assoc 212:521–528

Cox R, Peterson L, Detweiler D (1976) Comparison of arterial hemodynamics in the mongrel dog and the racing greyhound. Am J Physiol 230:211–218

Creevy KE, Scuderi MA, Ellis AE (2013) Generalised peripheral oedema associated with amlodipine therapy in two dogs. J Small Anim Pract 54:601–604

Dodd MG, Gardiner DG, Carter AJ et al (1989) The hemodynamic properties of amlodipine in anesthetised and conscious dogs: comparison with nitrendipine and influence of beta-adrenergic blockade. Cardiovasc Drugs Ther 3:545–545

Elliott WJ (2003) Management of hypertension emergencies. Curr Hypertens Rep 5:486–492

Ettehad D, Emdin CA, Kiran A et al (2016) Blood pressure lowering for prevention of cardiovascular disease and death: a systematic review and meta-analysis. Lancet 387:957–967

European Medicines Agency (EMA) (2014) Combined use of medicines affecting the renin-angiotensin system (RAS) to be restricted – CHMP endorses PRAC recommendation

Finco DR (2004) Association of systemic hypertension with renal injury in dogs with induced renal failure. J Vet Intern Med 18:289–294

Fiocchi EH, Cowgill LD, Brown DC et al (2017) The use of darbepoetin to stimulate erythropoiesis in the treatment of anemia of chronic kidney disease in dogs. J Vet Intern Med 31:476–485

Fischer JR, Lane IF, Cribb AE (2003) Urethral pressure profile and hemodynamic effects of phenoxybenzamine and prazosin in non-sedated male beagle dogs. Can J Vet Res 67:30–38

Fujishima M, Ibayashi S, Fujii K et al (2008) Cerebral blood flow and brain function in hypertension. Hypertens Res 18:111–117

Gaber L, Walton C, Brown S et al (1994) Effects of different antihypertensive treatments on morphologic progression of diabetic nephropathy in uninephrectomized dogs. Kidney Int 46:161–169

Geers AB, Koomans HA, Boer P et al (1984) Plasma and blood volumes in patients with the nephrotic syndrome. Nephron 38:170–173

Geigy CA, Schweighauser A, Doherr M et al (2011) Occurrence of systemic hypertension in dogs with acute kidney injury and treatment with amlodipine besylate. J Small Anim Pract 52:340–346

Gilson SD, Withrow SJ, Wheeler SL et al (1994) Pheochromocytoma in 50 dogs. J Vet Intern Med 8:228–232

Ginn JA, Bentley E, Stepien RL (2012) Systemic hypertension and hypertensive retinopathy following PPA overdose in a dog. J Am Anim Hosp Assoc 49:46–53

Glaus TM, Elliott J, Herberich E et al (2019) Efficacy of long-term oral telmisartan treatment in cats with hypertension: results of a prospective European clinical trial. J Vet Intern Med 33:413–422

Gójska-Zygner O, Lechowski R, Zygner W (2012) Functioning unilateral adrenocortical carcinoma in a dog. Can Vet J 53:623–625

Goy-Thollot I, Péchereau D, Kéroack S et al (2002) Investigation of the role of aldosterone in hypertension associated with spontaneous pituitary-dependent hyperadrenocorticism in dogs. J Small Anim Pract 43:489–492

Grauer GF, Greco DS, Getzy DM et al (2000) Effects of enalapril versus placebo as a treatment for canine idiopathic glomerulonephritis. J Vet Intern Med 14:526–533

Greco DS, Lees GE, Dzendzel G et al (1994) Effects of dietary sodium intake on blood pressure measurements in partially nephrectomized dogs. Am J Vet Res 55:160–165

Grodecki KM, Gains MJ, Baumal R et al (1997) Treatment of X-linked hereditary nephritis in Samoyed dogs with angiotensin converting enzyme (ACE) inhibitor. J Comp Pathol 117:209–225

Guess SC, Yerramilli M, Obare EF et al (2018) Longitudinal evaluation of serum symmetric dimethylarginine (SDMA) and serum creatinine in dogs developing chronic kidney disease. Int J Appl Res Vet Med 16:122–130

Gwin RM, Gelatt KN, Terrell TG (1978) Hypertensive retinopathy associated with hypothyroidism, hypercholesterolemia and renal failure in a dog. J Am Anim Hosp Assoc 26:897–894

Haagsman AN, Kummeling A, Moes ME et al (2013) Comparison of terazosin and prazosin for treatment of vesico-urethral reflex dyssynergia in dogs. Vet Rec 173:41

Hansen B, DiBartola SP, Chew DJ et al (1992) Clinical and metabolic findings in dogs with chronic renal failure fed two diets. Am J Vet Res 53:326–334

Hayashi K, Homma K, Wakino S et al (2010) T-type Ca channel blockade as a determinant of kidney protection. Keio J Med 59:84–95

Henik RA, Stepien RL, Wenholz LJ et al (2008) Efficacy of atenolol as a single antihypertensive agent in hyperthyroid cats. J Feline Med Surg 10:577–582

Herrera MA, Mehl ML, Kass PH et al (2008) Predictive factors and the effect of phenoxybenzamine on outcome in dogs undergoing adrenalectomy for pheochromocytoma. J Vet Intern Med 22:1333–1339

Herring IP, Jacobson JD, Pickett JP (2004) Cardiovascular effects of topical ophthalmic 10% phenylephrine in dogs. Vet Ophthalmol 7:41–46

Herring IP, Panciera DL, Werre SR (2014) Longitudinal prevalence of hypertension, proteinuria, and retinopathy in dogs with spontaneous diabetes mellitus. J Vet Intern Med 28:488–495

Hoareau GL, Jourdan G, Mellema M et al (2012) Evaluation of arterial blood gases and arterial blood pressures in brachycephalic dogs. J Vet Intern Med 26(4):897–904

Höglund K, Hanås S, Carnabuci C et al (2012) Blood pressure, heart rate, and urinary catecholamines in healthy dogs subjected to different clinical settings. J Vet Intern Med 26:1300–1308

Honda M, Hayashi K, Matsuda H et al (2001) Divergent renal vasodilator action of L- and T-type calcium antagonists in vivo. J Hypertens 19:2031–2037

Huhtinen M, Derré G, Renoldi HJ et al (2015) Randomized placebo-controlled clinical trial of a chewable formulation of amlodipine for the treatment of hypertension in client-owned cats. J Vet Intern Med 29:786–793

Hunter SL, Culp LB, Muir WW et al (2003) Sodium nitroprusside-induced deliberate hypotension to facilitate patent ductus arteriosus ligation in dogs. Vet Surg 32:336–340

International Renal Interest Society (2017) Treatment recommendations for CKD in dogs. Accessed online at http://www.iris-kidney.com/pdf

Ishida Y, Tomori K, Nakamoto H et al (2003) Effects of antihypertensive drugs on peritoneal vessels in hypertensive dogs with mild renal insufficiency. Adv Perit Dial 19:10–14

Jacob F, Polzin DJ, Osborne CA et al (2005) Evaluation of the association between initial proteinuria and morbidity rate or death in dogs with naturally occurring chronic renal failure. J Am Vet Med Assoc 226:393–400

Jacob F, Polzin DJ, Osborne CA et al (2003) Association between initial systolic blood pressure and risk of developing a uremic crisis or of dying in dogs with chronic renal failure. J Am Vet Med Assoc 222:322–329

Javadi S, Djajadiningrat-Laanen SC, Kooistraa HS et al (2005) Primary hyperaldosteronism, a mediator of progressive renal disease in cats. Domest Anim Endocrinol 28:85–104

Jenkins T, Coleman A, Schmiedt C (2015) Attenuation of the pressor response to exogenous angiotensin by angiotensin receptor blockers and benazepril hydrochloride in clinically normal cats. Am J Vet Res 76:807–813

Jennifer H, Mariana P, Karyn B (2017) Serotonin syndrome from 5-hydroxytryptophan supplement ingestion in a 9-month-old labrador retriever. J Med Toxicol 13:183–186

Jepson RE, Elliott J, Brodbelt D et al (2007) Effect of control of systolic blood pressure on survival in cats with systemic hypertension. J Vet Intern Med 21:402–409

Johnson KD, Henry CJ, McCaw DL et al (2006) Primary hyperaldosteronism in a dog with concurrent lymphoma. J Vet Med A Physiol Pathol Clin Med 53:467–470

Kallet AJ, Cowgill LD, Kass PH (1997) Comparison of blood pressure measurements obtained in dogs by use of indirect oscillometry in a veterinary clinic versus at home. J Am Vet Med Assoc 210:651–654

Kang JG, Yu MY, Lee H et al (2018) Blood pressure management and progression of chronic kidney disease in a canine remnant kidney model. Gen Physiol Biophys 37:243–252

Kaplan AJ, Peterson ME (1995) Effects of desoxycorticosterone pivalate administration on blood pressure in dogs with primary hypoadrenocorticism. J Am Vet Med Assoc 206:327–331

Kellar KJ, Quest JA, Spera AC et al (2014) Comparative effects of urapidil, prazosin, and clonidine on ligand binding to central nervous system receptors, arterial pressure, and heart rate in experimental animals. Am J Med 77:87–95

Kelly KL, Drobatz KJ, Foster JD (2016) Effect of fenoldopam continuous infusion on glomerular filtration rate and fractional excretion of sodium in healthy dogs. J Vet Intern Med 30:1655–1660

King JN, Mauron C, Kaiser G (1995) Pharmacokinetics of the active metabolite of benazepril, benazeprilat, and inhibition of plasma angiotensin-converting enzyme activity after single and repeated administrations to dogs. Am J Vet Res 56:1620–1628

Kittleson MD, Hamlin RL (1983) Hydralazine pharmacodynamics in the dog. Am J Vet Res 44:1501–1505

Klosterman ES, Pressler BM (2011) Nephrotic syndrome in dogs: clinical features and evidence-based treatment considerations. Top Companion Anim Med 26:135–142

Kondo T, Brock M, Bach H (1984) Effect of intra-arterial sodium Nitroprusside on intracranial pressure and cerebral autoregulation. Jpn Heart J 25:231–237

Konta M, Nagakawa M, Sakatani A et al (2018) Evaluation of the inhibitory effects of telmisartan on drug-induced renin-angiotensin-aldosterone system activation in normal dogs. J Vet Cardiol 20:376–383

Koomans HA, Roos JC, Boer P et al (1982) Salt sensitivity of blood pressure in chronic renal failure: evidence for renal control of body fluid distribution in man. Hypertension 4:190–197

Krieger JE, Liard JF, Cowley AW (1990) Hemodynamics, fluid volume, and hormonal responses to chronic high-salt intake in dogs. Am J Physiol 259:1629–1636

Kwon YJ, Suh GH, Kang SS et al (2018) Successful management of proteinuria and systemic hypertension in a dog with renal cell carcinoma with surgery, telmisartan, and amlodipine. Can Vet J 59:759–762

Kyles AE, Gregory CR, Wooldridge JD et al (1999) Management of hypertension controls postoperative neurologic disorders after renal transplantation in cats. Vet Surg 28:436–441

Kyles AE, Feldman EC, De Cock HEV et al (2003) Surgical management of adrenal gland tumors with and without associated tumor thrombi in dogs: 40 cases (1994-2001). J Am Vet Med Assoc 223:654–662

Lang JM, Schertel E, Kennedy S et al (2011) Elective and emergency surgical management of adrenal gland tumors: 60 cases (1999–2006). J Am Anim Hosp Assoc 47:428–435

Lantis AC, Ames MK, Atkins CE et al (2015) Aldosterone breakthrough with benazepril in furosemide-activated renin-angiotensin-aldosterone system in normal dogs. J Vet Pharmacol Ther 38:65–73

Law MR (2003) Value of low dose combination treatment with blood pressure lowering drugs: analysis of 354 randomised trials. Br Med J 326:1427

LeBlanc NL, Stepien RL, Bentley E (2011) Ocular lesions associated with systemic hypertension in dogs: 65 cases (2005–2007). J Am Vet Med Assoc 238:915–921

Lenders JWM, Duh QY, Eisenhofer G et al (2014) Pheochromocytoma and paraganglioma: an endocrine society clinical practice guideline. J Clin Endocrinol Metab 99:1915–1942

Lien YH, Hsiang TY, Huang HP (2010) Associations among systemic blood pressure, microalbuminuria and albuminuria in dogs affected with pituitary – and adrenal-dependent hyperadrenocorticism. Acta Vet Scand 12:61

Littlejohn TW, Majul CR, Olvera R et al (2009) Results of treatment with telmisartan-amlodipine in hypertensive patients. J Clin Hypertens 11:207–213

Littman MP, Daminet S, Grauer GF et al (2013) Consensus recommendations for the diagnostic investigation of dogs with suspected glomerular disease. J Vet Intern Med 27:S19–S26

Littman MP, Gerber B, Goldstein RE et al (2018) ACVIM consensus: update on Lyme borreliosis in dogs and cats. J Vet Intern Med 32:887–903

Lovern CS, Swecker WS, Lee JC et al (2001) Additive effects of a sodium chloride restricted diet and furosemide administration in healthy dogs. Am J Vet Res 62:1793–1796

Lowrie M, De Risio L, Dennis R et al (2012) Concurrent medical conditions and long-term outcome in dogs with nontraumatic intracranial hemorrhage. Vet Radiol Ultrasound 53:381–388

Maggio F, DeFrancesco TC, Atkins CE et al (2000) Ocular lesions associated with systemic hypertension in cats: 69 cases (1985–1998). J Am Vet Med Assoc 217:695–702

Martin CA, Mcgrath BP (2014) White-coat hypertension. Clin Exp Pharmacol Physiol 41:22–29

Massingham R, Hayden ML (1975) A comparison of the effects of prazosin and hydrallazine on blood pressure, heart rate and plasma renin activity in conscious renal hypertensive dogs. Eur J Pharmacol 30:121–124

McAlister FA, Zhang J, Tonelli M et al (2011) The safety of combining angiotensin-converting-enzyme inhibitors with angiotensin-receptor blockers in elderly patients: a population-based longitudinal analysis. Can Med Assoc J 183:655–663

Mishina M, Watanabe T (2008) Development of hypertension and effects of benazepril hydrochloride in a canine remnant kidney model of chronic renal failure. J Vet Med Sci 70:455–460

Mixon W, Helms SR (2008) Treatment of feline hypertension with transdermal amlodipine: a pilot study. J Am Anim Hosp Assoc 43:149–156

Montoya JA, Morris PJ, Bautista I et al (2006) Hypertension: a risk factor associated with weight status in dogs. J Nutr 136:S2011–S2013

Mooney AP, Mawby DI, Price JM et al (2017) Effects of various factors on Doppler flow ultrasonic radial and coccygeal artery systolic blood pressure measurements in privately-owned, conscious dogs. Peer J 22:e3101

Murphy MB, Murray C, Shorten GD (2001) Fenoldopam — a selective peripheral dopamine-receptor agonist for the treatment of severe hypertension. N Engl J Med 345:1548–1557

Nakamoto H, Suzuki H, Kageyama Y et al (1991) Characterization of alterations of hemodynamics and neuroendocrine hormones in dexamethasone induced hypertension in dogs. Clin Exp Hypertens 13:587–606

Nakamura R, Yabuki A, Ichii O et al (2018) Changes in renal peritubular capillaries in canine and feline chronic kidney disease. J Comp Pathol 160:79–83

Nielsen LK, Bracker K, Price LL (2015) Administration of fenoldopam in critically ill small animal patients with acute kidney injury: 28 dogs and 34 cats (2008-2012). J Vet Emerg Crit Care 25:396–404

Noël S, Massart L, Hamaide A (2012) Urodynamic and haemodynamic effects of a single oral administration of ephedrine or phenylpropanolamine in continent female dogs. Vet J 192:89–95

O'Neill DG, Elliott J, Church DB et al (2013a) Chronic kidney disease in dogs in UK veterinary practices: prevalence, risk factors, and survival. J Vet Intern Med 27:814–821

O'Neill J, Kent M, Glass EN et al (2013b) Clinicopathologic and MRI characteristics of presumptive hypertensive encephalopathy in two cats and two dogs. J Am Anim Hosp Assoc 49:412–420

Ortega TM, Feldman EC, Nelson RW et al (1996) Systemic arterial blood pressure and urine protein/creatinine ratio in dogs with hyperadrenocorticism. J Am Vet Med Assoc 209:1724–1729

Pariser MS, Berdoulay P (2011) Amlodipine-induced gingival hyperplasia in a great Dane. J Am Anim Hosp Assoc 47:375–376

Pascoe PJ, Ilkiw JE, Stiles J et al (1994) Arterial hypertension associated with topical ocular use of phenylephrine in dogs. J Am Vet Med Assoc 205:1562–1564

Pei Z, Zhang X (2014) Methamphetamine intoxication in a dog: case report. BMC Vet Res 24:139

Pentel PR, Asinger RW, Benowitz NL (1985) Propranolol antagonism of phenylpropanolamine-induced hypertension. Clin Pharmacol Ther 37:488–494

Pérez-Sánchez AP, Del-Angel-Caraza J, Quijano-Hernández IA et al (2015) Obesity-hypertension and its relation to other diseases in dogs. Vet Res Commun 39:45–51

Peterson KL, Lee JA, Hovda LR (2011) Phenylpropanolamine toxicosis in dogs: 170 cases (2004–2009). J Am Vet Med Assoc 239:1463–1469

Petit AM, Gouni V, Tissier R et al (2013) Systolic arterial blood pressure in small-breed dogs with degenerative mitral valve disease: a prospective study of 103 cases (2007-2012). Vet J 197:830–835

Plumb DC (2015) Plumb's veterinary drug handbook, 8th edn. Wiley-Blackwell

Ramsey I (2012) BSAVA small animal formulary, 7th edn. British Small Animal Veterinary Association, Gloucester

Randolph JF, Scarlett J, Stokol T et al (2004) Clinical efficacy and safety of recombinant canine erythropoietin in dogs with anemia of chronic renal failure and dogs with recombinant human erythropoietin-induced red cell aplasia. J Vet Intern Med 18:81–91

Rasmussen CB, Glisson JK, Minor DS (2012) Dietary supplements and hypertension: potential benefits and precautions. J Clin Hypertens 14:467–471

Reine NJ, Hohenhaus AE, Peterson ME et al (1999) Deoxycorticosterone-secreting adrenocortical carcinoma in a dog. J Vet Intern Med 13:386–390

Reusch CE, Schellenberg S, Wenger M (2010) Endocrine hypertension in small animals. Vet Clin N Am 40:335–352

Ross L (2011) Acute kidney injury in dogs and cats. Vet Clin N Am Small Anim Pract 41(1):1–14

Rudinsky AJ, Harjes LM, Byron J et al (2018) Factors associated with survival in dogs with chronic kidney disease. J Vet Intern Med 32:1977–1982

Schellenberg S, Glaus TM, Reusch CE (2007) Effect of long-term adaptation on indirect measurements of systolic blood pressure in conscious untrained beagles. Vet Rec 161:418–421

Schellenberg S, Mettler M, Gentilini F et al (2008) The effects of hydrocortisone on systemic arterial blood pressure and urinary protein excretion in dogs. J Vet Intern Med 22:273–281

Schierok H, Pairet M, Hauel N et al (2001) Effects of telmisartan on renal excretory function in conscious dogs. J Int Med Res 29:131–139

Sent U, Gössl R, Elliott J et al (2015) Comparison of efficacy of long-term oral treatment with telmisartan and benazepril in cats with chronic kidney disease. J Vet Intern Med 29:1479–1486

Simpson AC, McCown JL (2009) Systemic hypertension in a dog with a functional thyroid gland adenocarcinoma. J Am Vet Med Assoc 235:1474–1479

Slaughter JB, Padgett GA, Blanchard G et al (1996) Canine essential hypertension: probable mode of inheritance. J Hypertens 4:S170–S171

Smets PMY, Lefebvre HP, Meij BP et al (2012a) Long-term follow-up of renal function in dogs after treatment for ACTH-dependent hyperadrenocorticism. J Vet Intern Med 26:565–574

Smets PMY, Lefebvre HP, Kooistra HS et al (2012b) Hypercortisolism affects glomerular and tubular function in dogs. Vet J 192:532–534

St. Peter WL, Odum LE, Whaley-Connell AT (2013) To RAS or not to RAS? The evidence for and cautions with renin-angiotensin system inhibition in patients with diabetic kidney disease. Pharmacotherapy 33:496–514

Stafylas PC, Sarafidis PA (2008) Carvedilol in hypertension treatment. Vasc Health Risk Manag 4:23–30

Stepien RL, Rapoport GS (1999) Clinical comparison of three methods to measure blood pressure in non-sedated dogs. J Am Vet Med Assoc 215:1623–1628

Stopher DA, Beresford AP, Macrae PV et al (1988) The metabolism and pharmacokinetics of amlodipine in humans and animals. J Cardiovasc Pharmacol 12:55–59

Struble AL, Feldman EC, Nelson RW et al (1998) Systemic hypertension and proteinuria in dogs with diabetes mellitus. J Am Vet Med Assoc 213:822–825

Surman S, Couto CG, Dibartola SP et al (2012) Arterial blood pressure, proteinuria, and renal histopathology in clinically healthy retired racing greyhounds. J Vet Intern Med 26:1320–1329

Syme H (2011) Hypertension in small animal kidney disease. Vet Clin N Am 41:63–89

Tenhündfeld J, Wefstaedt P, Nolte IJA (2009) A randomized controlled clinical trial of the use of benazepril and heparin for the treatment of chronic kidney disease in dogs. J Am Vet Med Assoc 234:1031–1037

Thomas EK, Drobatz KJ, Mandell DC (2014) Presumptive cocaine toxicosis in 19 dogs: 2004-2012. J Vet Emerg Crit Care 24:201–207

Thomason JD, Fallaw TL, Carmichael KP et al (2009) Gingival hyperplasia associated with the administration of amlodipine to dogs with degenerative valvular disease (2004-2008). J Vet Intern Med 23:39–42

Tjostheim SS, Stepien RL, Markovic LE et al (2016) Effects of toceranib phosphate on systolic blood pressure and proteinuria in dogs. J Vet Intern Med 30:951–957

Tontodonati M, Fasdelli N, Moscardo E et al (2007) A canine model used to simultaneously assess potential neurobehavioural and cardiovascular effects of candidate drugs. J Pharmacol Toxicol Methods 56:265–275

Uechi M, Sasaki T, Ueno K et al (2002) Cardiovascular and renal effects of carvedilol in dogs with heart failure. J Vet Med Sci 64:469–475

Ueno Y, Mohara O, Brosnihan KB et al (1988) Characteristics of hormonal and neurogenic mechanisms of deoxycorticosterone-induced hypertension. Hypertension 11:S172–S177

Varon J, Marik PE (2000) The diagnosis and management of hypertensive crises. Chest 118:214–227

Wald DS, Law M, Morris JK et al (2009) Combination therapy versus monotherapy in reducing blood pressure: meta-analysis on 11,000 participants from 42 trials. Am J Med 122:290–300

Wehner A, Hartmann K, Hirschberger J (2008) Associations between proteinuria, systemic hypertension and glomerular filtration rate in dogs with renal and non-renal diseases. Vet Rec 162:141–147

White WB, Duprez D, St Hillaire R et al (2003) Effects of the selective aldosterone blocker eplerenone versus the calcium antagonist amlodipine in systolic hypertension. Hypertension 41:1021–1026

Williams B, Macdonald TM, Morant S et al (2015) Spironolactone versus placebo, bisoprolol, and doxazosin to determine the optimal treatment for drug-resistant hypertension (PATHWAY-2): a randomised, double-blind, crossover trial. Lancet 386:2059–2068

Woodfield JA (1995) Controlled clinical evaluation of enalapril in dogs with heart failure: results of the cooperative veterinary Enalapril (COVE) study group. J Vet Intern Med 9:243–252

Xie X, Atkins E, Lv J et al (2016) Effects of intensive blood pressure lowering on cardiovascular and renal outcomes: updated systematic review and meta-analysis. Lancet 387:435–443

Yue W, Kimura S, Fujisawa Y et al (2001) Benidipine dilates both pre- and post-glomerular arteriole in the canine kidney. Hypertens Res 24:429–436

Yusuf S, ONTARGET Investigators (2008) Telmisartan, ramipril, or both in patients at high risk for vascular events. N Engl J Med 358:1547–1559

Part IV

Future Perspectives

Future Perspectives

14

Harriet M. Syme, Rosanne E. Jepson, and Jonathan Elliott

Over the last 30 years, significant progress has been made in diagnosing and managing hypertension in cats and dogs, but there are still many gaps in our knowledge. The application of non-invasive blood pressure measuring devices, although not perfect for companion animals, has enabled clinical researchers to measure blood pressure in populations of cats and dogs with different diseases and document risk factors for hypertension in these species. The variability of blood pressure from one minute to the next is a physiological fact. This, together with the published information on the different devices which shows the level of agreement with the gold standard of direct measurement by telemetry is not as good as it might be means some veterinarians feel measuring blood pressure using practical indirect devices does not give them meaningful data. However, the published data reviewed in this book suggests indirect devices can be used to identify veterinary patients who are at risk of target organ damage because of their high blood pressure. One unmet need, however, is the availability of circulating biomarkers which can be used to differentiate transient situational hypertension from hypertension that is damaging the vasculature and so leading to target organ damage. The advent of well-validated biomarkers of hypertensive target organ damage (or risk thereof) would increase the confidence of veterinary practitioners in recognising and treating hypertensive patients. Preliminary data have been published on such biomarkers in cats (Bijsmans et al. 2017), but further work is needed before this approach can be used to help veterinarians identify patients that they need to monitor closely and treat to prevent target organ damage.

H. M. Syme · R. E. Jepson (✉)
Department of Clinical Sciences and Services, Royal Veterinary College, University of London, Hertfordshire, UK
e-mail: rjepson@rvc.ac.uk

J. Elliott
Department of Comparative Biomedical Sciences, Royal Veterinary College, University of London, London, UK

The more epidemiological data we gather from our pet animals relating to blood pressure, the more parallels it seems we are able to draw with the published data on people. Hypertension is a problem that is seen in human populations following industrialisation, leading to more sedentary lives and excessive consumption of calories and sodium. Domestication of dogs and cats has led to their lifestyles becoming like ours, and the data gathered from cats where large enough populations are studied show quite clearly that blood pressure increases with age, just as it does in people in developed countries. The same is probably true in dogs where the relationship between obesity and increased blood pressure has been demonstrated experimentally. Given our pets age faster than we do, there is a real opportunity to take a One Health approach and identify novel ways that could be more effective and acceptable socially to reduce the rise in blood pressure with age. By studying these interventions in dogs and cats over a 5 year period, this could provide proof-of-concept evidence for their application as public health measures to human populations. Professor Bob Michell was a strong advocate of this One Health approach and extolled the benefits of using pet dogs in this way in a review article published in 2000 (Michell 2000).

Study of the pathophysiology of hypertension in the CKD patient is in its infancy in cats and has not been studied in dogs with naturally occurring CKD despite the fact that dogs have been used as experimental models for hypertension over the last century. In cats, much focus has been on the RAAS system, but the role of sympathetic nervous system dysfunction has been largely ignored. Dysfunction of renal afferent nerves and the resulting overactivity of the sympathetic nervous system are important in human medicine (see Chap. 1) and open alternative approaches to the management of hypertension, including non-pharmacological methods of treatment. Centrally acting drugs which lower sympathetic nerve activity (see Chap. 11) have not been used in veterinary medicine to control blood pressure and should be investigated to determine whether they add to the quality of treatment that can be achieved. This is particularly pertinent to cats where amlodipine, a peripherally acting vasodilator drug, will lead to further increases in sympathetic nerve activity as a result of its antihypertensive effect.

A number of approaches targeting the overactivity of the sympathetic nervous system driven by renal afferent nerve and baroreceptor dysfunction have been developed for use in human patients and are showing promise (Lobo et al. 2017). Methods to ablate renal sympathetic nerves (afferents and efferents), electrical field stimulation of the carotid sinus to reduce sympathetic outflow and increase vagal parasympathetic tone, and direct vagal nerve stimulation are some of the approaches that have had some success in pharmacoresistant hypertensive patients. As we learn more about sympathetic nerve activity in hypertensive dogs and cats and determine what situations overactivity is relevant to the pathophysiology of hypertension, these methods, which are new to human medicine, may well help us to achieve better control of hypertension.

There has been a huge investment in sequencing the human genome, and much effort and research funding have gone into unravelling the complex interaction between the genome and the environment in determining prevailing blood pressure and predisposition to hypertension. To make progress in understanding genomics of human hypertension has required studies on a scale which are inconceivable for veterinary medicine. To date, no attempt has been made to undertake comparative genomic studies using the pets we expose to the very same lifestyle factors that have contributed to the epidemic of human hypertension in developed countries. As the cost of genome sequencing comes down and more and more dog and cat genomes are being sequenced and made publically available, comparative genomics of hypertension will become much more possible to study, to the benefit of both human and veterinary medicine. Within the next 30–40 years, it is likely that whole-genome sequencing will be undertaken on veterinary patients in clinic and will be used to determine what gene mutations are contributing to hypertension in an individual patient and to predict how that patient will respond to particular antihypertensive agents.

The history of the management of hypertension in human patients has been one of decreasing the target blood pressure that is viewed as optimal to achieve with antihypertensive treatment. It seems highly likely that the current recommendations for veterinary species will continue to be lowered over time as veterinarians become more confident in the diagnosis and more studies become available providing evidence of the benefits of effective treatment. Two relatively large multicentre randomised controlled clinical trials (RCCT) for different antihypertensive medications for use in cats have been published in the last 5 years. Neither of these RCCTs combined blood pressure reduction and control of proteinuria as end points, despite the evidence from human medicine that the more proteinuric the patient the lower the blood pressure needs to be reduced to maximise the renoprotective benefit of the treatment. This is an issue that needs to be addressed in future RCCTs, where combinations of drugs might prove necessary to achieve the more stringent future blood pressure targets particularly in the more proteinuric patients. It is probably even more important for dogs because hypertension appears to be relatively common in dogs with proteinuric CKD. Anecdotally, blood pressure seems to be harder to control in dogs than in cats, and so we definitely need multicentre RCCT in dogs to generate evidence on which we can base therapeutic decision-making in the future.

In conclusion, hypertension is an active field of clinical research for canine and feline medicine. There is still much to learn about our species and much that we can share with our medical colleagues, now and in the future, to the benefit of both disciplines.

References

Bijsmans ES, Jepson RE, Wheeler C, Syme HM, Elliott J (2017) Plasma N-terminal probrain natriuretic peptide, vascular endothelial growth factor, and cardiac troponin i as novel biomarkers of hypertensive disease and target organ damage in cats. J Vet Intern Med 31 (3):650–660

Lobo MD, Sobotka PA, Pathak A (2017) Interventional procedures and future drug therapy for hypertension. Eur Heart J 38(15):1101–1111

Michell AR (2000) Hypertension in dogs: the value of comparative medicine. J R Soc Med 93 (9):451–452

Printed by Printforce, the Netherlands